NEUROMETHODS

Series Editor
Wolfgang Walz
University of Saskatchewan
Saskatoon, SK, Canada

For further volumes:
http://www.springer.com/series/7657

Neuromethods publishes cutting-edge methods and protocols in all areas of neuroscience as well as translational neurological and mental research. Each volume in the series offers tested laboratory protocols, step-by-step methods for reproducible lab experiments and addresses methodological controversies and pitfalls in order to aid neuroscientists in experimentation. *Neuromethods* focuses on traditional and emerging topics with wide-ranging implications to brain function, such as electrophysiology, neuroimaging, behavioral analysis, genomics, neurodegeneration, translational research and clinical trials. *Neuromethods* provides investigators and trainees with highly useful compendiums of key strategies and approaches for successful research in animal and human brain function including translational "bench to bedside" approaches to mental and neurological diseases.

Contemporary Approaches to the Study of Pain

From Molecules to Neural Networks

Edited by

Rebecca P. Seal

Department of Neurobiology, Pittsburgh Center for Pain Research, University of Pittsburgh, Pittsburgh, PA, USA

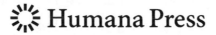 Humana Press

Editor
Rebecca P. Seal
Department of Neurobiology
Pittsburgh Center for Pain Research
University of Pittsburgh
Pittsburgh, PA, USA

ISSN 0893-2336   ISSN 1940-6045   (electronic)
Neuromethods
ISBN 978-1-0716-2041-0   ISBN 978-1-0716-2039-7   (eBook)
https://doi.org/10.1007/978-1-0716-2039-7

Cover Caption: Confocal image of a multi-color in situ hybridization performed on the dorsal root ganglion of adult Rhesus macaque. Messenger RNA for mu opioid receptor is visualized in green, cholecystokinin B receptor in red and cholecystokinin A receptor in blue. Adapted from Chapter 3.

This Humana imprint is published by the registered company Springer Science+Business Media, LLC part of Springer Nature.
The registered company address is: 1 New York Plaza, New York, NY 10004, U.S.A.

Preface to the Series

Experimental life sciences have two basic foundations: concepts and tools. The *Neuromethods* series focuses on the tools and techniques unique to the investigation of the nervous system and excitable cells. It will not, however, shortchange the concept side of things as care has been taken to integrate these tools within the context of the concepts and questions under investigation. In this way, the series is unique in that it not only collects protocols but also includes theoretical background information and critiques which led to the methods and their development. Thus it gives the reader a better understanding of the origin of the techniques and their potential future development. The *Neuromethods* publishing program strikes a balance between recent and exciting developments like those concerning new animal models of disease, imaging, *in vivo* methods, and more established techniques, including, for example, immunocytochemistry and electrophysiological technologies. New trainees in neurosciences still need a sound footing in these older methods in order to apply a critical approach to their results.

Under the guidance of its founders, Alan Boulton and Glen Baker, the *Neuromethods* series has been a success since its first volume published through Humana Press in 1985. The series continues to flourish through many changes over the years. It is now published under the umbrella of Springer Protocols. While methods involving brain research have changed a lot since the series started, the publishing environment and technology have changed even more radically. Neuromethods has the distinct layout and style of the Springer Protocols program, designed specifically for readability and ease of reference in a laboratory setting.

The careful application of methods is potentially the most important step in the process of scientific inquiry. In the past, new methodologies led the way in developing new disciplines in the biological and medical sciences. For example, Physiology emerged out of Anatomy in the nineteenth century by harnessing new methods based on the newly discovered phenomenon of electricity. Nowadays, the relationships between disciplines and methods are more complex. Methods are now widely shared between disciplines and research areas. New developments in electronic publishing make it possible for scientists that encounter new methods to quickly find sources of information electronically. The design of individual volumes and chapters in this series takes this new access technology into account. Springer Protocols makes it possible to download single protocols separately. In addition, Springer makes its print-on-demand technology available globally. A print copy can therefore be acquired quickly and for a competitive price anywhere in the world.

Saskatoon, SK, Canada   *Wolfgang Walz*

Preface

Pain remains a major clinical problem due to its high prevalence and a lack of adequate treatment options. Opioids and nonsteroidal anti-inflammatory agents are the most pre-scribed therapies, but their use remains limited by a lack of efficacy or serious side effects including the nearly intractable problem of addiction. The dearth of more effective, nonad-dictive treatment options is due in large part to the heterogeneous nature of pain, which presents in many forms, with different mechanisms and underlying neural circuitry. Efforts to unravel the molecular, cellular, circuit, and network complexities which involve not only neurons but also nonneuronal cells, such as glia and immune cells, will require ever more sophisticated approaches that permit interrogations with unprecedented levels of temporal and spatial resolution and control. Current research efforts are centered on understanding the bases for the many different physiological responses that contribute to the formation of the normal healthy pain percept and critically, how perturbations at any of these levels-molecular, cellular, circuit, and network, result in the maladaptive signaling that drives chronic pain.

In this volume, we present detailed methods and cutting-edge approaches that address these efforts and are revolutionizing our understanding of the neurobiology of pain. Several chapters describe methods to identify the gene expression profiles of individual cells or of defined groups of cells as transcriptomes or translatomes, as well as provide information about the bioinformatic tools that are available to organize and compare gene expression profiles. Since the use of human tissue is critical to translational efforts, chapters also include protocols for the isolation of single cell/nuclei from human and nonhuman primate tissue. Similarly, the preparation and use of progenitor cells and human induced pluripotent stem cells are also described. These undifferentiated cell types can be used to rewire neural circuits in vivo and to model pain pathology in vitro for the purpose of uncovering basic mechanisms and screening drug candidates. A few of the chapters describe harvesting human dorsal root ganglion neurons for electrophysiological recordings, as well as preparing *ex vivo* skin-nerve or nerve-muscle preparations for electrophysiological recordings in rodents. To gain access to specific cell populations in rodent models for the purpose of manipulating or imaging neuronal activity, researchers are taking advantage of adeno-associated virus (AAV), which is nontoxic and does not elicit an immune response in rodents. While intracranial injection of viruses is common, intraspinal injection is more specialized. Thus, protocols for intraspinal delivery of AAV are also included. This volume also includes several chapters on how to measure neural activity and synaptic connectivity in the brain, spinal cord, and primary sensory neurons *in vivo* and *ex vivo* using calcium imaging, optogenetics, or chemogenetics. Lastly, both pain researchers and clinicians are in need of better methods to measure pain reliably and objectively. The final chapter of this volume describes a method to measure reflexive somatosensory behaviors in rodents with greater precision by taking advantage of high-speed videography and machine learning. All of the chapters provide step-by-step instructions with helpful notes and figures as well as background information on how these highlighted approaches are transforming our understanding of pain biology.

Pittsburgh, PA, USA   *Rebecca P. Seal*

Contents

Contributors

ISHMAIL ABDUS-SABOOR • *Department of Biology, Zuckerman Mind Brain Behavior Institute, Columbia University, NY, USA*

BIAFRA AHANONU • *Department of Anatomy, University of California, San Francisco, San Francisco, CA, USA*

KATHRYN M. ALBERS • *Department of Neurobiology, University of Pittsburgh School of Medicine, Pittsburgh, PA, USA; Pittsburgh Center for Pain Research, University of Pittsburgh, Pittsburgh, PA, USA*

CYNTHIA M. AROKIARAJ • *Department of Neurobiology, University of Pittsburgh School of Medicine, Pittsburgh, PA, USA; Pittsburgh Center for Pain Research, University of Pittsburgh School of Medicine, Pittsburgh, PA, USA*

ALLAN I. BASBAUM • *Department of Anatomy, University of California, San Francisco, San Francisco, CA, USA*

TEMUGIN BERTA • *Department of Anesthesiology, Pain Research Center, University of Cincinnati Medical Center, Cincinnati, OH, USA*

JOÃO M. BRAZ • *Department of Anatomy, University of California, San Francisco, San Francisco, CA, USA*

ALEXANDER CHAMESSIAN • *Department of Anesthesiology and Neurology, Washington University in Saint Louis School of Medicine, Saint Louis, MO, USA*

BIN CHEN • *Department of Neurobiology, Duke University, Durham, NC, USA; Department of Brain and Cognitive Sciences, McGovern Institute for Brain Research, Massachusetts Institute of Technology, Cambridge, MA, USA*

KIM I. CHISHOLM • *Neurorestoration Group, Wolfson Centre for Age-Related Diseases, King's College London, London, UK*

JONATHAN A. COHEN • *Pittsburgh Center for Pain Research, Pittsburgh Healthcare System, Pittsburgh, PA, USA; Department of Dermatology, University of Pittsburgh School of Medicine, Pittsburgh, PA, USA*

GREGORY CORDER • *Departments of Psychiatry and Neuroscience, University of Pennsylvania, Philadelphia, PA, USA*

BRIAN M. DAVIS • *Department of Neurobiology, University of Pittsburgh School of Medicine, Pittsburgh, PA, USA; Pittsburgh Center for Pain Research, University of Pittsburgh, Pittsburgh, PA, USA*

XINZHONG DONG • *The Solomon H. Snyder Department of Neuroscience and the Center for Sensory Biology, Johns Hopkins University School of Medicine, Baltimore, MD, USA; Howard Hughes Medical Institute, Johns Hopkins University School of Medicine, Baltimore, MD, USA*

ARIEL EPOUHE • *Department of Neurobiology, University of Pittsburgh School of Medicine, Pittsburgh, PA, USA; Pittsburgh Center for Pain Research, Pittsburgh Healthcare System, Pittsburgh, PA, USA*

DAVID W. FERREIRA • *Department of Neurobiology, University of Pittsburgh School of Medicine, Pittsburgh, PA, USA; Pittsburgh Center for Pain Research, University of Pittsburgh School of Medicine, Pittsburgh, PA, USA*

MICHAEL S. GOLD • *Department of Neurobiology, University of Pittsburgh School of Medicine, Pittsburgh, PA, USA*

BRETT A. GRAHAM • *Brain Neuromodulation Program, Hunter Medial Research Institute and School of Biomedical Sciences and Pharmacy, University of Newcastle, Newcastle, NSW, Australia*

KAREN HAENRAETS • *Institute for Pharmacology and Toxicology, University of Zürich, Zürich, Switzerland*

MARTIN HÄRING • *Division of Molecular Neurobiology, Department of Medical Biochemistry and Biophysics, Karolinska Institutet, Stockholm, Sweden; Division of Molecular Neurosciences, Center for Brain Research, Vienna, Austria*

DAVID I. HUGHES • *Spinal Cord Group, Institute of Neuroscience and Psychology, University of Glasgow, Glasgow, UK*

MICHAEL P. JANKOWSKI • *Division of Pain Management, Department of Anesthesia, Cincinnati Children's Medical Center, Cincinnati, OH, USA; Department of Pediatrics, College of Medicine, University of Cincinnati, Cincinnati, OH, USA*

MARSHA RITTER JONES • *Pittsburgh Center for Pain Research, Pittsburgh Healthcare System, Pittsburgh, PA, USA; Department of Anesthesiology and Veterans Administration, Pittsburgh Healthcare System, Pittsburgh, PA, USA*

DANIEL H. KAPLAN • *Pittsburgh Center for Pain Research, Pittsburgh Healthcare System, Pittsburgh, PA, USA; Department of Dermatology, University of Pittsburgh School of Medicine, Pittsburgh, PA, USA*

H. RICHARD KOERBER • *Department of Neurobiology, University of Pittsburgh School of Medicine, Pittsburgh, PA, USA; Pittsburgh Center for Pain Research, Pittsburgh Healthcare System, Pittsburgh, PA, USA*

JUSSI KUPARI • *Division of Molecular Neurobiology, Department of Medical Biochemistry and Biophysics, Karolinska Institutet, Stockholm, Sweden*

MARK LAY • *The Solomon H. Snyder Department of Neuroscience and the Center for Sensory Biology, Johns Hopkins University School of Medicine, Baltimore, MD, USA*

EMANUEL LOEZA-ALCOCER • *Department of Neurobiology, University of Pittsburgh School of Medicine, Pittsburgh, PA, USA; Pittsburgh Center for Pain Research, University of Pittsburgh, Pittsburgh, PA, USA*

WENQIN LUO • *Department of Neuroscience, University of Pennsylvania, Philadelphia, PA, USA*

Z. DAVID LUO • *Department of Pharmacology, University of California Irvine, School of Medicine, Irvine, CA, USA; Department of Anesthesiology and Perioperative Care, University of California Irvine, School of Medicine, Irvine, CA, USA; Reeve-Irvine Research Center, Chao Family Comprehensive Cancer Center, University of California Irvine, Irvine, CA, USA*

STEPHEN B. MCMAHON • *Neurorestoration Group, Wolfson Centre for Age-Related Diseases, King's College London, London, UK*

SALIM MEGAT • *School of Behavioral and Brain Sciences and Center for Advanced Pain Studies, University of Texas at Dallas, Richardson, TX, USA*

JAMIE K. MOY • *Department of Neurobiology, University of Pittsburgh School of Medicine, Pittsburgh, PA, USA*

SARAH A. NAJJAR • *Department of Neurobiology, University of Pittsburgh School of Medicine, Pittsburgh, PA, USA; Pittsburgh Center for Pain Research, University of Pittsburgh, Pittsburgh, PA, USA*

MYUNG-CHUL NOH • *Department of Neurobiology, Pittsburgh Center for Pain Research, University of Pittsburgh School of Medicine, Pittsburgh, PA, USA*

JEAN-CHARLES PATERNA • *Viral Vector Facility, University of Zurich and Swiss Federal Institute (ETH) Zurich, Zurich, Switzerland*

TERESA N. PATITUCCI • *Department of Cell Biology, Neurobiology and Anatomy, Medical College of Wisconsin, Milwaukee, WI, USA*

THEODORE J. PRICE • *School of Behavioral and Brain Sciences and Center for Advanced Pain Studies, University of Texas at Dallas, Richardson, TX, USA*

LUIS F. QUEME • *Division of Pain Management, Department of Anesthesia, Cincinnati Children's Medical Center, Cincinnati, OH, USA*

KATELYN E. SADLER • *Department of Cell Biology, Neurobiology and Anatomy, Medical College of Wisconsin, Milwaukee, WI, USA*

KATRIN SCHRENK-SIEMENS • *Department of Pharmacology, University of Heidelberg, Heidelberg, Germany*

REBECCA P. SEAL • *Department of Neurobiology, Pittsburgh Center for Pain Research, University of Pittsburgh, Pittsburgh, PA, USA*

KELLY M. SMITH • *Pittsburgh Center for Pain Research and Department of Neurobiology, University of Pittsburgh, Pittsburgh, PA, USA*

KRISTEN M. SMITH-EDWARDS • *Department of Neurobiology, University of Pittsburgh School of Medicine, Pittsburgh, PA, USA; Pittsburgh Center for Pain Research, University of Pittsburgh, Pittsburgh, PA, USA; Mayo Guggenheim Bldg, Rochester, MN, USA*

CHERYL L. STUCKY • *Department of Cell Biology, Neurobiology and Anatomy, Medical College of Wisconsin, Milwaukee, WI, USA*

JUN TAKATOH • *Department of Neurobiology, Duke University, Durham, NC, USA; Department of Brain and Cognitive Sciences, McGovern Institute for Brain Research, Massachusetts Institute of Technology, Cambridge, MA, USA*

DIANA TAVARES-FERREIRA • *School of Behavioral and Brain Sciences and Center for Advanced Pain Studies, University of Texas at Dallas, Richardson, TX, USA*

ANDREW J. TODD • *Spinal Cord Group, Institute of Neuroscience and Psychology, University of Glasgow, Glasgow, UK*

FAN WANG • *Department of Neurobiology, Duke University, Durham, NC, USA; Department of Brain and Cognitive Sciences, McGovern Institute for Brain Research, Massachusetts Institute of Technology, Cambridge, MA, USA*

HENDRIK WILDNER • *Institute for Pharmacology and Toxicology, University of Zürich, Zürich, Switzerland*

YANHUI PETER YU • *Department of Pharmacology, University of California Irvine, School of Medicine, Irvine, CA, USA*

Chapter 1

Single-Cell RNA Sequencing of Somatosensory Neurons

Martin Häring and Jussi Kupari

Abstract

Chronic pain is one of the major health issues of today, yet our understanding of the mechanisms behind pain perception and sensation in general remains limited. However, single-cell transcriptomics has provided a completely new insight into the cellular complexity of sensory tissue. Thus, recent publications describe the transcriptome of spinal cord and dorsal root ganglia including the expression profile of cell types. Essential for successful and high-quality sequencing is the preparation of healthy single-cell suspensions that are free of debris. However, depending on whether the tissue of interest consists of sensitive cells, like the spinal cord, obtaining viable cells might prove difficult. In this chapter, we will describe in detail how such a cell suspension can be obtained from tissue involved in somatosensation such as dorsal spinal cord and dorsal root ganglia (DRG). Additionally, we will highlight potential pitfalls and strategies when working with this sensory tissue, in particular spinal cord with its highly sensitive neurons. In summary, we present a comprehensive description on how to perform single-cell RNA-sequencing including suggestions on how to choose and design the most fitting strategy as well as point out major limitations of tissue dissociation.

Key words Single-cell transcriptomics, Spinal cord, DRG, Ganglia, Dissociation

1 Introduction

One of the most relevant public health issues today is chronic pain as it negatively impacts the quality of life of million individuals and involves an enormous socioeconomic cost. However, knowledge of how somatosensory information, such as pain, is processed in sensory tissue is still very limited [1]. In somatosensation, specialized terminals of peripheral sensory neurons detect the respective stimuli (e.g., touch, temperature, pain) and forward it to the spinal cord where the information is processed and relayed to the midbrain, thalamus, and subsequently to the somatosensory cortex. How the different sensory modalities are encoded remains unclear. There are two recognized theories: "labeled lines" and "population coding" [2].

Labeled lines suggests there are specialized neuron types along these sensory routes, which together convey one type of sensation. Population coding does not deny the existence of specialized

Rebecca P. Seal (ed.), *Contemporary Approaches to the Study of Pain: From Molecules to Neural Networks*, Neuromethods, vol. 178, https://doi.org/10.1007/978-1-0716-2039-7_1, © Springer Science+Business Media, LLC, part of Springer Nature 2022

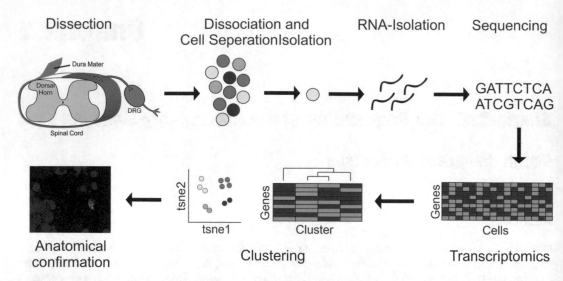

Fig. 1 Schematic Illustration of a single cell sequencing workflow. Following the dissection and dissociation of the sensory tissue, the cells are separated and lysed to obtain the RNA molecules. The transcriptome is sequenced by the platform of choice and subsequently analyzed using available bioinformatics tools. After cell types (clusters) are defined the results are confirmed typically by in multiplex fluorescent in situ hybridization (*see* Chapter 4)

pathways however it argues that the crosstalk among these pathways encodes somatosensory perception. The dorsal horn of the spinal cord especially plays a critical role for processing distinct modalities of noxious and innocuous sensation, but little is known about the neuronal subtypes involved, hampering efforts to explain principles governing somatic sensation. Thus, one major task to understand the mechanism underlying sensory coding is to understand the cellular complexity.

Single-cell transcriptomics has emerged in recent years, providing the scientific community with an increasing amount of information on cellular identities and expression profiles [3, 4] (*see* Fig. 1).

Currently, the number of cells that can be processed within a day has risen to several thousands, thereby revealing unknown cell diversities. In fact, recent publications describe the transcriptome of the entire central nervous system including both neuronal and non-neuronal cell types [5]. Other studies have focused on more in depth analysis of sensory tissues, namely dorsal spinal cord [6, 7] and DRG [8, 9].

A major obstacle to successfully performing single cell transcriptome analysis of a sensitive tissue such as the spinal cord is obtaining a clean and vital single-cell suspension. In this chapter, we describe a gentle method that has been shown to produce high-quality cell suspensions for spinal cord tissue as well as ganglionic tissue [5, 7, 10, 11]. In the case where obtaining a suspension of single cells is not possible, the isolation of single-cell nuclei is

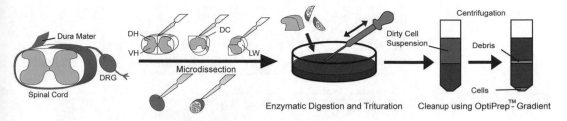

Fig. 2 Schematic Illustration of sensory tissue dissociation workflow

performed, which also has its advantages and limitations as discussed below and in Chapter 2. Our protocol for isolating single cells is based on a simple Optiprep® gradient diluted in modified artificial cerebral spinal fluid and neuronal culture medium combined with low speed centrifugation (*see* Fig. 2).

To obtain single cells, the primary sensory ganglia or spinal cord tissue has to be dissected and cut into appropriate pieces that allow a dissociation using enzymatic digestion and physical force by trituration. An important first step in the dissection is the removal or disruption of the dura mater as it causes problems with microdissection and enzymatic treatment of the target tissue. For spinal cord tissue, we recommend to carefully peel off the dura mater, so that the spinal cord structure stays intact. Ganglionic neurons are very robust, requiring more physical force for a successful dissociation. Thus, we recommend cutting or ripping the ganglia open with small scissors or fine forceps. The method of dissection and the composition and length of the enzymatic treatment depend on the type of tissue, which has to be stronger for ganglionic compared to spinal cord tissue.

2   Preparations

2.1   Materials

2.1.1   Consumables

Bovine serum albumin (BSA)	Sigma	Cat# A3059
TrypLE™ express	Life technologies	Cat#12605-010
Papain	Worthington	Cat#LK003172
Collagenase/Dispase	Roche	Cat#11097113001
Neurobasal-A	Gibco	Cat#10888
L-glutamine	Gibco	Cat#25030-123
B27	Gibco	Cat#17504-044
Penicillin/streptomycin	Sigma	Cat#P4458
Optiprep density gradient M.	Sigma	Cat#D1556
DNase I	Worthington	Cat#LK003172
Conical tube 15 ml	Sarstedt	Cat#NC9531248

(continued)

30-μm cell strainer	Falcon	Cat#352340
35-mm cell culture dish	Corning	Cat#353801
Glass Pasteur pipettes	Sigma-Aldrich	Cat#CLS7095B5X

2.1.2 Surgery Tools

Scissors, ToughCut Straight sharp-blunt, 17.5 cm	AgnTho's	Cat# 14130-17
Narrow pattern forceps curved 2 mm tip, 12 cm	AgnTho's	Cat# 11003-12
Standard pattern forceps Curved, 2.5 mm tip, 12 cm	AgnTho's	Cat# 11001-12
Vannas spring scissors Sharp straight, 2 mm cutting, 8 cm	AgnTho's	Cat# 15000-03
Dissecting knife, straight 40 mm cutting, 12.5 cm	AgnTho's	Cat# 10055-12
Bonn-strabismus scissors Tough-cut, straight, 9 cm	AgnTho's	Cat# 14084-09
Fine scissors Tough-cut, straight, 9 cm	AgnTho's	Cat# 14058-09

2.1.3 Chemicals

NaCl	Sigma	Cat# S3014
KCl	Sigma	Cat# P9541
NaH_2PO_4	Sigma	Cat# S0751
$NaHCO_3$	Sigma	Cat# S5761
Sucrose	Sigma	Cat# S0389
Glucose	Sigma	Cat # G8270
$MgSO_4$	Sigma	Cat# M7506
$CaCl_2$	Sigma	Cat# C1016

2.1.4 Equipment

Zeiss Stemi 2000 (stereo microscope)	Zeiss
CL 6000 LED (light source)	Zeiss
Heraeus Megafuge 1.0 (centrifuge)	Heraeus
Scanlaf Mars (biosafety cabinet)	Scanlaf

2.2 Solutions

2.2.1 Stock Solution: 4× Concentrated Pre-cutting Solution (Pre-CS) in 1 l

20.34 g NaCl	(348 mM)
0.75 g KCl	(10 mM)
0.6 g NaH_2PO_4	(5 mM)
8,74 g $NaHCO_3$	(104 mM)
102,7 g sucrose	(300 mM)
7,21 g glucose	(300 mM)

Pre-CS can be prepared up to 4 weeks in advance and kept at 4 °C.

2.2.2 Cutting Solution (CS) 100 ml (Prepare Shortly Before Use; see Chapter 5, **Note 1**)

25ml of 4× Pre-CS

750ml of autoclaved water

400 μl of 1 M $MgSO_4$ (4 mM)

50 μl of 1 M $CaCl_2$ (0.5 mM)
    (*Oxygenate for 30 min*)

2.2.3 Enzymes (Prepare Shortly Before Use to the Following Concentrations; see Chapter 5, **Note 2**)

Papain + DNase	(25 U/ml)
Dnase	(1 mM)
Collagenase/Dispase	(20 mg/ml)
TrypLE	(ready to use)

2.3 Tissue Isolation

2.3.1 Animals

Mice up to 7 weeks (Spinal Cord) or 12 weeks (DRG) of age.

2.3.2 Surgery/Dissection Tools

Prepare surgery tools by sterilizing with 70% ethanol.

2.3.3 Polish Pasteur Pipettes

Prepare 4–6 Paster pipettes with decreasing (70–15%) outlet diameter (polished with flame). Incubate the inside and tip of the pipettes with 0.5% BSA in PBS.

3 Method

3.1 Dissection

Prior to surgery, a ketamine/xylazine solution (20 mg/kg and 2 mg/kg body weight for ketamine and xylazine, respectively) is administered via intraperitoneal injection. Once the animal is fully anaesthetized (failure to respond to a toe-or tail pinch), the animal is placed on a cooled (4 °C) flat surface and perfused with cutting solution (4 °C). The perfusion surgery has been described for rodents in great detail [12]. After the perfusion, animals are decapitated, and the spinal cord and DRG removed as described below. It is essential that the spinal cord stays intact and that the meninges (mainly dura mater) are removed before starting the microdissection. This allows the gray matter microdissection of the dorsal horn (*see* Fig. 2).

3.1.1 Spinal Cord Removal

- Decapitate the animal and place the animal belly up on the dissection area.
- Remove fur and ribcage, pin down the body and remove/detach all organs so that ventral side of the spinal cord is visible.
- Start removing the ventral spine starting from the cervical opening (cut with a fine but stable scissor horizontally and alternating without damaging the spinal cord).
- When the desired length of the ventral spine is removed, carefully remove the spinal cord by cutting the dura mater and dorsal roots.
- Remove the *dura mater* that is enveloping the spinal cord.

3.1.2 DRG Removal

- Decapitate the animal and place the animal belly up on the dissection area.
- Remove fur and ribcage, pin down the body, and remove/ detach all organs, so that the ventral side of the spinal cord is visible.
- Start removing the ventral spine starting from the cervical opening (cut with a fine but stable scissor horizontally and alternating without damaging the spinal cord).
- Use spinal cord/dura mater to pull out single DRGs and cut first the peripheral nerves and subsequently the dorsal root, so that DRGs can be collected.

3.1.3 Dorsal Horn Microdissection

- Make 2–3 three coronal cuts to obtain 3–4 spinal cord pieces.
- Cut each piece along the midline.
- Place each hemisphere on the lateral side and start peeling away the dorsal column.
- Turn around the hemisphere and remove the lateral white matter by carefully inserting a micro scalpel between white and gray matter and cutting toward the central canal.
- Transfer the dorsal column into the dissociation medium and cut it into ~2-mm pieces.

3.1.4 Ganglia Microdissection

- Remove nerve fibers as cleanly as possible from the ganglia.
- Only if readily accessible, remove pieces of *dura mater* from the ganglia and transfer them intact to the dissection media.

3.2 Dissociation

The dissociation of tissue can be done using mechanical force, enzymatic treatment, or as described here as a combination of both. The length of the incubation times and concentration of enzymes in the dissociation solution has to be adjusted depending on the type of tissue (*see* Chapter 5, **Note 3**).

3.2.1 Dorsal Horn Dissociation

- Prepare dissociation solution and incubate at 37 °C (enough for both dorsal columns from cervical to lumbar).
    - 300 μl Tryp-LE.
    - 2000 μl Papain solution.
    - 100 μl DNAse I.
    - 100 μl Cutting Solution.
- Incubate for 10 min at 37 °C.
- Carefully pipette 10 times up and down with first Pasteur pipette.
- Incubate for 10–15 min at 37 °C.
- Carefully pipette 10 times up and down with second Pasteur pipette.
- Incubate for 10–15 min at 37 °C.
- Carefully pipette 10 times up and down with third Pasteur pipette.
- Incubate for 10–15 min at 37 °C.
- Carefully pipette 10 times up and down with fourth Pasteur pipette.

3.2.2 Ganglia Dissociation

- Prepare dissociation solution and incubate at 37 °C (enough for both dorsal columns from cervical to lumbar).
    - 400 μl Tryp-LE.
    - 100 μl Collagenase/Dispase.
    - 1900 μl Papain solution.
    - 100 μl DNAse I (Stock:1 mM).
- Incubate for 20 min at 37 °C.
- Carefully pipette 10 times up and down with first Pasteur pipette; Use forceps to carefully rip open the ganglia.
- Incubate for 20 min at 37 °C.
- Carefully pipette 10 times up and down with second Pasteur pipette.
- Incubate for 20 min at 37 °C.
- Carefully pipette 10 times up and down with third Pasteur pipette; Should bigger ganglia pieces still remain use again forceps to carefully rip the piece open.

- Incubate for 20–30 min at 37 °C.
- Carefully pipette 10 times up and down with fourth Pasteur pipette.

For both dorsal horn and ganglia dissociation, it is essential that no large pieces of tissue remain. If there are large pieces, incubate longer, add TrypLE, and repeat last step. Smaller cell aggregates might be ignored if incubation time reaches 60 min for spinal cord and 90 min for ganglia, as cell quality will be reduced with longer incubation.

3.3 Removal of Cell Debris

Essential for a useful single-cell suspension is its "cleanness," or in other words, how well cell debris like axon fibers or myelin has been removed from the suspension. The first step, however, is to remove the enzyme solution and cell clusters that did not dissociate. This can be achieved using cell strainers (FALCON; 30 μm for dorsal horn and 40 μm for ganglia) in order to filter the cell suspension into a 15-ml Falcon tube in which 2.5 ml of cutting solution is added to dilute the enzyme solution. Mix by inverting 3×. Subsequently, the cell suspension is centrifuged at $150 \times g$ at 8 °C for 4 min. The supernatant is removed without disturbing the pellet and a mix (1:1; 1 ml) of cutting solution and Neurobasal-A media is added to resuspend the cell pellet. During enzyme removal, prepare OptiPrep™ gradient in 15-ml Falcon tube and keep on ice. Depending on the amount of cell debris and the weight of your cells, different final concentrations of OptiPrep™ can be used (*see* Chapter 5, **Note 4–6**).

3.3.1 Dorsal Horn Gradient

160 μl Optiprep™ + 920 μl Neurobasal-A + 920 μl cutting solution (mix well).

3.3.2 Ganglia Gradient

100 μl Optiprep™ + 450 μl Neurobasal-A + 450 μl cutting solution (mix well).

Centrifuge at $100 \times g$ for 10 min in cooled (4–10 °C) centrifuge. Cells can be found in the pellet or last 100–200 μl. A ring of debris should be visible at the border between the phases. Evaluate the viability of the cell suspension and dilute or concentrate to obtain the desired cell density.

4 Experimental Approach

4.1 Before the Start

Before setting up an scRNA-seq experiment, it is critical to consider some important points regarding the study to optimize the use of resources and quality of data. Choosing the most suitable experimental strategy to answer a specific scientific question is essential. There are several major questions one should ask oneself in advance.

4.1.1 What Is the Biological Question I Want to Answer?

This could be for example determining (1) the cell type composition in a specific part of the nervous system, (2) the role of a perturbation (e.g., a disease) on the gene expression profiles of a single- or multiple-cell types, (3) the gene regulatory networks in several interconnected cell types, (4) the developmental trajectory of a specific cell type (5). Am I interested in only defining or identifying cell types or am I interested in extracting biological meaning from the expression of lowly expressed genes (e.g., some transcription factors, long noncoding RNAs)? (6) Am I interested in identifying rare populations? (7) Am I interested in gene isoforms, single-nucleotide polymorphisms (SNP), mutations or just molecular counts of genes?

4.1.2 How Many Cells Can I Collect from a Sample?

Do I have a lot of cells to work with from each sample? How ubiquitous or rare are my cells of interest in the tissue? How many biological samples do I have to work with?

4.1.3 How Sensitive Are the Cells or Tissues with Respect to the Dissociation?

Are my cells known to be vulnerable when dissociated (e.g., spinal motoneurons)? Am I interested in the expression of genes that are affected by the dissociation procedure (e.g., immediate early genes for neuronal activation studies)?

4.1.4 What Is my Budget?

Will I be working with only naïve samples for example to describe cell types (fewer samples needed, relatively inexpensive) or with possibly several levels of treatment (more samples needed, expenses build up fast)?

The answers to these questions will determine the dissociation and sequencing strategy. Considering the significant costs associated with single-cell sequencing, it is important not to underestimate this step.

4.2 scRNA-Seq Platforms

Over the last decade, single-cell sequencing has developed dramatically from covering only a few dozen cells to literally thousands of cells at one time. In the last few years, many different sc-RNAseq platforms and cell capturing technologies have been developed [13, 14]. Below we will briefly address some of the technologies based on the approach to quantification and cell capture (Fig. 3).

4.2.1 Quantification

Full-Length Transcript Methods

scRNA-seq methods such as Smart-seq2 [15], SMARTer [16], SUPeR-seq [17], and MATQ-seq [18] enable the capture of full-length transcripts and therefore give access to a wide array of research questions related to gene isoforms, alternative isoform usage, editing, mutation studies, SNPs/eQTLS, etc. The most common method, Smart-seq2, uses the Moloney murine leukemia virus reverse transcriptase (MMLV) for reverse transcription (RT) and template switching (TS). In TS, a few untemplated nucleotides are added at the 3′ end of the new cDNA strand that

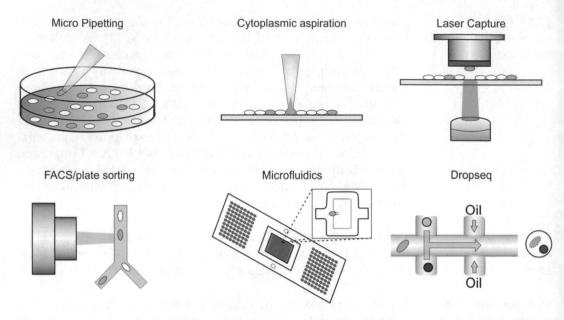

Fig. 3 Schematic Illustration of different low (top row) and high throughput (lower row) sequencing platforms

act as a docking site for a template-switching oligonucleotide (switching from RNA to DNA) that enables the addition of a 3′ end for the cDNA molecule that is identical to the 5′ end of the oligo-DT primer used for RT. This way, the full transcriptome can then be amplified in a single PCR, tagmented (tagged and fragmented), and sequenced [15].

Tag-Based Methods

Several protocols such as CEL-seq2 [19], MARS-seq [20], STRT-2i-seq [21], Dropseq [22], and 10xGenomics Chromium-3′ [23] use an approach where only the 3′ or 5′ end of an RNA molecule is captured and sequenced. Since MARS-seq, these methods have used a unique molecular identifier (UMI) barcode to tag unique transcripts for identification. The upsides of most tag-based methods over full-length transcripts are higher throughput, easier demultiplexing of samples, and lower per-cell cost; however, as the information concerning the rest of the transcript is lost, these methods are more limited in the research questions that can be applied. High-throughput tag-based methods also have a lower sequencing depth and gene detection rate compared to the full-transcript methods.

4.2.2   *Cell Capturing*

Several low- and high-throughput techniques and approaches are available to fit any given task. Below we will address the most common approaches, but for additional reading, see following reviews [3, 24, 25].

Low Throughput

Low-throughput approaches allow collecting up to several dozen cells per day and transferring them into microplates in which they can be individually processed. While very slow to obtain reasonable numbers, these techniques allow the scientist to be highly selective. Under optimal conditions, the cells of interest are labeled with visible markers such as fluorescent dyes; hence, transgenic mouse lines expressing enhanced green fluorescent protein (eGFP) or tdTomato are highly suited for these approaches. The low-throughput approaches are suitable to be used with the full-transcript methods such as Smart-seq2.

(a) Micropipetting

This method describes the manual or semi-automated collection of single cells by micropipettes. This method has been first successfully applied for single-cell sequencing on dorsal root ganglia neurons [8].

(b) Cytoplasmic aspiration

This technique is often incorporated and combined with an electrophysiological reading (Patch-seq). Instead of collecting the whole cell by suction, intracellular material is collected by puncturing the cell with a micropipette and aspirating only the cytoplasm for transcriptome analysis [26, 27].

(c) Laser capture microscopy

This approach employs a low-power infrared laser beam heating a thermolabile polymer to capture single cells from a tissue section. One advantage of this approach is that the exact cell location within the tissue can be visualized after the dissection [28, 29].

High Throughput

High-throughput platforms enable scientists to analyze the cellular composition of the entire tissues as they allow for the collection of hundreds to thousands of cells at one time.

(a) Microfluidic

In this approach, specific capturing chips separate the cells by microfluidics, arresting them at a specific capturing site. The technique itself is nonselective; however, preselection of cell populations can be done using FACS sorting [7, 10].

(b) Automated FACS or limiting dilution dispensing in microwell plates

Here single cells are directly transferred in single wells of capture plates following FACS sorting or limiting dilution dispensing. Most commonly, cells are collected in 384 cell plates; however, the Wafergen/Takara ICell8 system uses plates with >5000 nanowells. If fully automated, a high number of single cells can be captured and separately processed [30–32).

(c) Microdroplets

This approach is the most novel technical advancement in the field of single-cell sequencing as it allows the capture of thousands of single cells at one time. This approach on its own is also nonselective; however, again a preselection can be done using FACS sorting [5, 11].

4.3 Single Cell Vs. Single Nucleus Sequencing

Many studies have successfully employed a single-cell sequencing strategy. However, in cases where cells are less able to remain viable through the isolation procedure, single-nucleus sequencing has emerged as an alternative strategy as nuclei are more resistant to mechanical force. With respect to sensory tissue, the spinal cord is a prime example of this vulnerability. With increased age, cellular integrity decreases dramatically. In our experience, survival of neurons is minimal for animals over 8 weeks of age. Even at 4 weeks of age, basically no projection neurons or motor neurons were found after sequencing the spinal cord [5, 7]. DRG neurons, on the other hand, seem to be very robust, and cell suspensions can be prepared from mice up to several months of age [5].

The advantage of sequencing from single cells rather than single nuclei is the better representation of RNA molecules. However, the lengthy process of protease treatment and mechanical force can cause aberrant gene activation. This can be particularly problematic for studies of neuronal activity (i.e., c-Fos activation). Due to the relatively rapid isolation protocol for cell nuclei, which does not require prolonged protease treatment above 4 °C, studies on neuronal activation seem to benefit from this approach [33]. Another advantage of isolating nuclei is that frozen tissue or less readily available tissue (e.g., human tissue) for which cell dissociation is tremendously difficult can be utilized [34, 35]. Single-nuclear sequencing has been used successfully on adult spinal cord [6]. While sequencing depth was lower compared to the single-cell dataset, this study managed to detect motor and other ventral horn populations. A detailed comparison between the published datasets will be needed as future publications rely on a scientific consent.

4.4 scRNA-Seq Analysis

4.4.1 Data Processing

Quality Control

After sequencing, the analysis of scRNA-seq data begins with the read quality control and demultiplexing (assigning reads to correct cells and molecules) followed by genome alignment and read or count matrix generation. These steps are generally dealt with automatized pipeline tools (such as CellRanger or zUMIs [36]). Using the read/count matrix, the analysis goes through the following steps: quality control, normalization, removal of unwanted effects (technical and biological), feature selection (highly variable genes), dimension reduction, visualization, clustering, and further downstream analyses. The full analysis workflow is beyond the scope of this protocol and has been recently described in great detail [37].

Software

scRNA-seq analyses rely mostly on software packages written in the R- or Python-programming languages. The choice of language depends on personal preference, as all the necessary tools are available for each language and the possibility to seamlessly mix both languages is entirely possible. For both R and Python, excellent software packages are available that enable an end-to-end analysis of a standard scRNA-seq project (see Seurat/R and Scanpy/Python) [38, 39]. The authors and maintainers of these packages offer excellent in-depth tutorials on the software websites that will set even a novice analyst on a good path on their journey (https://satijalab.org/seurat/articles/get_started.html or https://scanpy.readthedocs.io/en/stable/tutorials.html).

Hardware

Finally, for a smooth analysis experience, the level of computational hardware performance needs to be considered. Smaller scRNA-seq datasets (perhaps some thousands of cells) can usually be run on a personal computer with 16GB RAM without problems; however, as the dataset size increases to tens- or hundreds-of-thousands of cells, the requirement for more computing power becomes essential. In this case, the simplest solution is to get access to a High-Performance Computing Cluster associated with the working institute or set up an account in a cloud-based computing service.

4.4.2 In Vivo Confirmation

Independent of the analysis and sequencing strategy used, the most essential subsequent task for such a study is the confirmation of the identified cell types in vivo. What the researcher needs to keep in mind is that a given transcriptomic analysis is dependent on sample quality or parameter settings used in the analysis; the clustering algorithm can only provide the results based on the data given to it. Besides technical limitations, similar but separated cell types might simply represent the same cell type but in different physiological states.

In such cases, follow up studies could include utilizing transgenic mouse models to specifically target cell types of interest, evaluating electrophysiological properties and in situ confirmation. In vast sequencing projects, this might be problematic and impractical especially if physiological characteristics should be addressed using transgenic animals. The most straightforward approach is RNA in situ hybridization to confirm identified cell clusters in the tissue of interest. Thanks to recent advances in fluorescent in situ hybridization techniques, however, commercially available RNA detection kits have become available. Novel sophisticated approaches allow the consecutive staining of dozens of transcripts on a single-tissue section. This technique not only allows the confirmation of the sequencing results but also provides a detailed anatomical map of the gene transcripts representing the clusters [40–42].

5 Notes

1. Depending on neuronal tissue or age CS composition might be adjusted.

2. Depending on neuronal tissue or age enzymatic mix might be adjusted (e.g., using more TrypLE with increased age).

3. The time intervals presented here are optimized for sensory tissue from animals 2–7 weeks of age and might require adjustment for younger or older animals. Spinal cord tissue from animals older than 9 weeks of age will probably yield insufficient cell quality.

4. Before attempting sequencing, practice dissection and dissociation until the workflow is performed in an undisrupted way yielding a viable single cell suspension.

5. In order to obtain best ration between debris removal and cell enrichment, the OptiPrep™ concentration can be adjusted.

6. Volume of Neurobasal-A Medium and Supplements used in the cleanup process can be adjusted if necessary and potentially exchanged with other buffers supporting cell survival. For the author, a sole use of CS solution in the enzyme removal step and gradient prevented the formation of a complete cell pellet.

References

1. Peirs C, Seal RP (2016) Neural circuits for pain: recent advances and current views. Science 354:578–584

2. Ma Q (2012) Population coding of somatic sensations. Neurosci Bull 28:91–99

3. Poulin J-F, Tasic B, Hjerling-Leffler J, Trimarchi JM, Awatramani R (2016) Disentangling neural cell diversity using single-cell transcriptomics. Nat Neurosci 19:1131–1141

4. Hwang B, Lee JH, Bang D (2018) Single-cell RNA sequencing technologies and bioinformatics pipelines. Exp Mol Med 50:96

5. Zeisel A et al (2018) Molecular architecture of the mouse nervous system. Cell 174:999–1014.e22

6. Sathyamurthy A et al (2018) Massively parallel single nucleus transcriptional profiling defines spinal cord neurons and their activity during behavior. Cell Rep 22:2216–2225

7. Häring M et al (2018) Neuronal atlas of the dorsal horn defines its architecture and links sensory input to transcriptional cell types. Nat Neurosci 21:869–880

8. Usoskin D et al (2015) Unbiased classification of sensory neuron types by large-scale single-cell RNA sequencing. Nat Neurosci 18:145–153

9. Li C et al (2016) Erratum: somatosensory neuron types identified by high-coverage single-cell RNA-sequencing and functional heterogeneity. Cell Res 26:967–967

10. Furlan A et al (2016) Visceral motor neuron diversity delineates a cellular basis for nipple- and pilo-erection muscle control. Nat Neurosci 19:1331–1340

11. Kupari J, Häring M, Agirre E, Castelo-Branco G, Ernfors P (2019) An atlas of vagal sensory neurons and their molecular specialization. Cell Rep 27:2508–2523.e4

12. Gage GJ, Kipke DR, Shain W (2012) Whole animal perfusion fixation for rodents. J Vis Exp (65):3564

13. Ziegenhain C et al (2017) Comparative analysis of single-cell RNA sequencing methods. Mol Cell 65:631–643.e4

14. Svensson V et al (2017) Power analysis of single-cell rnA-sequencing experiments. Nat Methods 14:381–387

15. Picelli S et al (2013) Smart-seq2 for sensitive full-length transcriptome profiling in single cells. Nat Methods 10:1096–1100

16. Verboom K et al (2019) SMARTer single cell total RNA sequencing. Nucleic Acids Res 47: e93

17. Fan X et al (2015) Single-cell RNA-seq transcriptome analysis of linear and circular RNAs in mouse preimplantation embryos. Genome Biol 16:148

18. Sheng K, Cao W, Niu Y, Deng Q, Zong C (2017) Effective detection of variation in single-cell transcriptomes using MATQ-seq. Nat Methods 14:267–270

19. Hashimshony T et al (2016) CEL-Seq2: sensitive highly-multiplexed single-cell RNA-Seq. Genome Biol 17:77

20. Jaitin DA et al (2014) Massively parallel single-cell RNA-seq for marker-free decomposition of tissues into cell types. Science 343:776–779

21. Hochgerner H et al (2017) STRT-seq-2i: dual-index 5′ single cell and nucleus RNA-seq on an addressable microwell array. Sci Rep 7:16327

22. Macosko EZ et al (2015) Highly parallel genome-wide expression profiling of individual cells using nanoliter droplets. Cell 161: 1202–1214

23. Zheng GXY et al (2017) Massively parallel digital transcriptional profiling of single cells. Nat Commun 8:1–12

24. Choi JR, Yong KW, Choi JY, Cowie AC (2020) Single-cell RNA sequencing and its combination with protein and DNA analyses. Cell 9: 1130

25. Alessio E, Bonadio RS, Buson L, Chemello F, Cagnin S (2020) A single cell but many different transcripts: a journey into the world of long non-coding RNAs. Int J Mol Sci 21:302

26. Fuzik J et al (2016) Integration of electrophysiological recordings with single-cell RNA-seq data identifies neuronal subtypes. Nat Biotechnol 34:175–183

27. Cadwell CR et al (2016) Electrophysiological, transcriptomic and morphologic profiling of single neurons using patch-seq. Nat Biotechnol 34:199–203

28. Cummings M, Mappa G, Orsi NM (2018) Laser capture microdissection and isolation of high-quality RNA from frozen endometrial tissue. Methods Mol Biol 1723:155–166

29. Lovatt D, Bell T, Eberwine J (2015) Single-neuron isolation for RNA analysis using pipette capture and laser capture microdissection. Cold Spring Harb Protoc 2015:60–68

30. Handley A, Schauer T, Ladurner AG, Margulies CE (2015) Designing cell-type-specific genome-wide experiments. Mol Cell 58: 621–631

31. Okaty BW, Sugino K, Nelson SB (2011) Cell type-specific transcriptomics in the brain. J Neurosci 31:6939–6943

32. Kupari et al (2021) Single cell transcriptomics of primate sensory neurons identifies cell types associated with chronic pain. Nat Commun 2021 12:1510

33. Lacar B et al (2016) Nuclear RNA-seq of single neurons reveals molecular signatures of activation. Nat Commun 7:11022

34. Lake BB et al (2016) Neuronal subtypes and diversity revealed by single-nucleus RNA sequencing of the human brain. Science 352: 1586–1590

35. Habib N et al (2017) Massively parallel single-nucleus RNA-seq with DroNc-seq. Nat Methods 14:955–958

36. Parekh S, Ziegenhain C, Vieth B, Enard W, Hellmann I (2018) zUMIs - a fast and flexible pipeline to process RNA sequencing data with UMIs. Gigascience 7:giy059

37. Luecken MD, Theis FJ (2019) Current best practices in single-cell RNA-seq analysis: a tutorial. Mol Syst Biol 15:e8746

38. Wolf FA, Angerer P, Theis FJ (2018) SCANPY: Large-scale single-cell gene expression data analysis. Genome Biol 19:15

39. Stuart T et al (2019) Comprehensive integration of single-cell data. Cell 177: 1888–1902.e21

40. Moffitt JR, Zhuang X (2016) RNA imaging with multiplexed error-robust fluorescence in situ hybridization (MERFISH). Methods Enzymol 572:1–49

41. Lignell A, Kerosuo L, Streichan SJ, Cai L, Bronner ME (2017) Identification of a neural crest stem cell niche by spatial genomic analysis. Nat Commun 8:1830

42. Codeluppi S et al (2018) Spatial organization of the somatosensory cortex revealed by osmFISH. Nat Methods 15:932–935

Preparation of Human and Rodent Spinal Cord Nuclei for Single-Nucleus Transcriptomic Analysis

Alexander Chamessian and Temugin Berta

Abstract

The spinal cord is a complex and heterogeneous tissue that is composed of numerous neuronal and non-neuronal cell types. Single-cell RNA-seq has emerged as a powerful method to study heterogeneous tissues by allowing for the capture and analysis of individual cells. A key step in the analysis of solid tissues is the dissociation of the tissue into single cells. While many tissues can be readily dissociated by enzymatic or mechanical methods, highly interconnected tissues such as the spinal cord are especially difficult to dissociate, leading to reductions in cellular viability or large biases in the representation of cell types. Moreover, it has been shown that dissociation of live cells can induce gene expression changes that influence downstream analysis. Single-nucleus RNA-seq (snRNA-seq) offers an alternative method of studying the transcriptomes of individual cells that circumvents many of these issues. An additional benefit of snRNA-seq is that it can be performed on frozen tissues, thus opening the door to the study of biobanked pathological human tissues. We present here a straightforward protocol to isolate both murine and human spinal cord nuclei for transcriptomic analysis on multiple platforms, including 10× Genomics Chromium.

Key words Fluorescence-activated nuclear sorting (FANS), Transcriptional analysis, Spinal cord

1 Introduction

The nervous system is an enormously complex network comprising hundreds of billions of neuronal and non-neuronal cells. In particular, it is well-documented that neuronal and non-neuronal cells are key participants in the perception of pain and are a central aim of preclinical pain research [1–3]. Although RNA-sequencing (RNA-seq) is a powerful tool to uncover the transcriptional profile and function of these cells, most studies have used RNA isolated from bulk spinal tissue that represents a mixture of genes expressed by multiple neuronal and non-neuronal cells. Recent studies employing high-throughput RNA-seq (e.g., drop- and well-based RNA-seq) from single spinal cells have dramatically expanded our understanding of the contribution of various cells to pain [4–7]. Dissociation of specific cell populations from neuronal tissues

Rebecca P. Seal (ed.), *Contemporary Approaches to the Study of Pain: From Molecules to Neural Networks*, Neuromethods, vol. 178, https://doi.org/10.1007/978-1-0716-2039-7_2, © Springer Science+Business Media, LLC, part of Springer Nature 2022

coupled with fluorescence-activated cell sorting has been used to profile neurons and glial cells [5, 8], but dissociation of live cells from spinal cord requires protease treatments at warm temperatures, which may induce artificial alterations in gene expression due to processing [9, 10]. To circumvent these challenges, we and others have used dissociated nuclei rather than whole cells [11, 12]. Nuclei, like cells, can be profiled using single-cell genomics methods and platforms such as the $10\times$ Chromium, Drop-seq [13], or Smart-seq3 [14]. Nuclei are especially useful for studying human tissues since nuclei can be readily isolated from frozen and biobanked tissues [15, 16], enabling the study of diseases affecting the spinal cord such as multiple sclerosis, amyotrophic lateral sclerosis, spinal cord injury, and chronic pain conditions.

Several protocols for the isolation of nuclei from neural tissues have been published [15, 17–20] and have been shown to be effective in various genomics assays. Neural tissues with abundant axonal tracts such as spinal cord generate a large amount of myelin and debris when making nuclei suspensions. The myelin and debris must be separated from the nuclei in order to produce pure inputs for genomics assays. Many protocols address this issue by using density gradients made from either sucrose [6, 19] or iodixanol [17, 21]. While these gradient separation methods do indeed remove contaminants and produce purified nuclei, they add additional time, labor, cost, and manipulations to the nuclei and often require access to an ultracentrifuge. Moreover, in our experience, even with optimized protocols, we have encountered significant nuclei clumping. An alternative to density gradient purification is fluorescence-activated nuclei sorting (FANS). FANS has several advantages over density gradients. Sorter instruments can easily separate intact, single nuclei from debris and broken nuclei. The use of DNA-binding dyes such as dihydrochloride (DAPI), Hoescht, or DRAQ5 makes this task especially straightforward and assures the isolation of single, intact nuclei. Because the sorting does the work of purification, upstream processing can be minimal, reducing the amount of manipulation that the nuclei undergo. In addition, FANS enables one to stain for nuclear-localized cell markers such as NeuN (neurons) [22], SOX9 (astrocytes) [23], or OLIG2 (oligodendrocytes) [24], allowing for the enrichment of specific or rarer cell types. Moreover, because sorter instruments allow for the isolation of single nuclei into 96- or 384-well plates for plate-based single-cell methods [14, 25]. The major disadvantage of FANS is the requirement for a cell sorter, which is an expensive and specialized instrument that most laboratories do not have in house. Thus, for most labs, using FANS requires working with an institutional core facility, which constrains one to work within the availability and scheduling of the shared instrument. Nevertheless, the benefits outweigh the disadvantages, in our opinion.

Here, we describe an optimized protocol to isolate nuclei from both murine and human spinal cord that we have successfully applied for both single-nucleus and population-level RNA-seq [11, 12]. While the protocol has been validated in RNA-seq, it is also likely to be suitable for other genomics applications such as whole-genome sequencing, Assay for Transposase-Accessible Chromatin using sequencing (ATAC-seq), Chromatin Immunoprecipitation sequencing (ChIP-seq), and the like.

2 Materials

- Phosphate-Buffered Saline (10×), pH 7.4 (Thermo Fisher Scientific, AM9625).
- Tris (1 M), pH 8.0 (Thermo Fisher Scientific, AM9855G).
- KCl (2 M) (Thermo Fisher Scientific, AM9640G).
- $MgCl_2$ (1 M) (Thermo Fisher Scientific, AM9530G).
- UltraPure™ DNase/Distilled Water (Thermo Fisher Scientific, 10977015) or equivalent molecular biology grade water (*see* **Note 1**).
- RNasin Plus (Promega, N2615).
- Superasin (Thermo Fisher Scientific, AM2696).
- DL-Dithiothreitol, BioUltra (Sigma-Aldrich, 43816-50ML).
- cOmplete™, Mini, EDTA-free Protease Inhibitor Cocktail (Roche, 11836170001).
- OmniPur® BSA, Fraction V, Cold Alcohol Isolation (Sigma-Aldrich, 2905-5GM).
- Triton™ X-100 solution ~10% in H_2O (Sigma-Aldrich, 93443-100ML).
- RNase AWAY; Suace Decontaminant (Thermo Fisher Scientific, 21-402-178).
- Molecular Probes DAPI (4′,6-Diamidino-2-Phenylindole, Dihydrochloride, Thermo Fisher Scientific, D1306).
- Dimethyl sulfoxide (DMSO) (Sigma-Aldrich D2650-5X5ML).
- Diethyl pyrocarbonate, ≥97% (Sigma-Aldrich, D5758-25ML).
- Sucrose, Molecular Biology Grade (Sigma-Aldrich, 84097-1KG).
- 2 mL Dounce tissue grinder set (Sigma-Aldrich, D8938-1SET) or 7 mL for large samples (*see* **Note 2**).
- C Chip Hemacytometer Disposable (INCYTO, 22-600-100) or equivalent hemocytometer.
- 15-mL Centrifuge tubes.

- 50-mL Centrifuge tube.
- 1.5 mL DNA LoBind Tubes (Eppendo 022431021).
- 2.0 mL DNA LoBind Tubes (Eppendo, 022431048).
- P1000 barrier tips.
- P200 barrier tips.
- P10 barrier tips.
- P2 barrier tips.
- P2, P20, P200, and P1000 Size Pipettor (dedicated set for RNA work).
- Serological pipette control (e.g., Drummond Pipet-Aid).
- 50-mL serological pipets.
- 25-mL serological pipets.
- 10-mL serological pipets.
- 5-mL serological pipets.
- Vacuum Filter System (0.22 μm filter, Bottle, e.g., Genesee Scientific # 25-225).
- Falcon® Round-Bottom Tubes, 5 mL, sterile, individually packed (Corning, 352003).
- CellTrics® filters - 20 μm (Sysmex, 04-004-2325).
- CellTrics® filters - 50 μm (Sysmex, 04-004-2327).
- DEPC (Sigma D5758).
- Disposable Lab Coat (Kimberly-Clark Professional™ Kimtech Pure™ A7 Cleanroom Lab Coat 47653 or equivalent).
- Disposable procedure mask.
- Nitrile gloves.
- Goggles and/or face shield.
- Disposable, single-use sterile forceps (Evergreen 222–1130-B1I or equivalent).
- 1 L glass bottles with caps (Nalgene).
- Waste containers.
- 10× Chromium controller (10× Genomics).
- Single-Cell Gene Expression Solution (v3.1, 10× Genomics).
- RNAqueous™-Micro Total RNA Isolation Kit (Thermo Fisher, AM1931).

2.1 Buffer Preparation

2.1.1 General Comments

1. Use RNase-free plastics and reagents for all steps. Where indicated, use glassware that has been decontaminated with RNase AWAY and rinsed with DEPC-treated and/or UltraPure water.

2. When preparing reagents, always wear the appropriate PPE (described below), to avoid contamination of reagents and to protect yourself.

3. Work in a low traffic area that at minimum has been decontaminated as described below. An ideal workspace is one that is dedicated to RNA work or even a laminar flow hood.

4. Use dedicated pipettors if available.

5. The amount of buffer and the size of the Dounce homogenizer will depend on the amount of tissue used. Small-scale isolation is for samples of <150 mg and use the 2 mL Dounce Homogenizer set. For samples >150 mg, the 7 mL Dounce Homogenizer set is used (see **Note 3**).

2.2 Workspace Decontamination

1. Spray down all work surfaces and equipment liberally with RNase AWAY including benchtop, pipettors, ice buckets, centrifuges. Wipe residual RNase AWAY with fresh lab tissues (Kimwipe).

2.3 Personal Protective Equipment (PPE)

1. Wear a fresh disposable lab (e.g., coat and face mask) (see **Note 4**).

2. When working with human or primate tissue, use a face shield or goggles.

3. Always wear fresh gloves and change them often. Spray with RNase AWAY before starting to work with a fresh pair.

1. Determine the amount of each reagent in Table 1 that you will need for a given volume of NIM (see **Note 5**).

2. In a pre-cleaned, dry, volumetric flask, add UltraPure water to 50% of the final volume (e.g., 50 mL if the end volume is 100 mL) using a serological pipette.

3. Weigh out the desired amount of sucrose in a fresh 50-mL conical tube.

4. Add 10% of the final volume of water to the sucrose in the 50-mL conical tube to dissolve the sucrose. Shake well until no sucrose crystals are visible.

5. Pour the sucrose solution carefully into the flask.

Table 1
Nuclei isolation medium (NIM)—Adapted from [17]

Component	Stock concentration/amount	Final concentration/amount
Tris–HCl (pH 8.0)	1000 mM	10 mM
Sucrose	N/A	250 mM
KCl	2000 mM	25 mM
$MgCl_2$	1000 mM	5 mM
UltraPure water		To final desired volume

6. Add the other components to the flask (Tris–HCl, KCl, $MgCl_2$) one by one. For volumes <1 mL, use a p1000 pipette. For volumes >1 mL, use a fresh serological pipette with the nearest capacity to the volume of liquid to be transferred (e.g., 2 mL pipette for 1.5 mL of liquid). Add the components into the center of the bottleneck to minimize adherence to the side wall.

7. Insert the clean bottle stopper on the flask opening and swirl the flask to fully dissolve the sucrose. Monitor the sucrose granules and swirl until there are no longer any visible granules.

8. Using a serological pipette (25 mL or 50 mL), add water to bring the solution to the final volume.

9. Pour out the solution into a fresh filter flask system (0.22 μm). Apply vacuum suction to filter the solution. Remove the vacuum bottle top and place a clean cap. Label the date of preparation on the side of the bottle.

10. Store the NIM at 4 °C for up to 30 days.

1. On the day of experiment, just before sample processing, prepare Homogenization Buffer (HB) according to Table 2. Determine the volume of each component you need depending on the number of samples and scale.

Table 2
Homogenization buffer (HB)—Adapted from [22]

Component	Stock concentration	Final concentration
NIM	1×	1×
RNasin plus RNase inhibitor (Because the RNase inhibitors are expensive, the volume you make should be only as much as you need for the samples you have. In this section, I provide an example (for 10 mL) but one should use the concentrations to make only the amount one needs)	40 U/μL	0.4 U/μL
Triton X-100	10%	0.1%
Protease inhibitor (To make protease inhibitor cocktail stock, add 1 cOmplete-mini tablet to 1 mL of UltraPure water. This is 10×. Let it fully dissolve. This can be stored at −20 °C until use. Alternatively, it can be prepared fresh on the day of experiment)	10×	1×
DTT	100 mM	0.1 mM

2. Add the above components using an appropriately sized tube. Maintain on ice at all times.

3. For small-scale experiments, each sample will require 2 mL total of HB. For large-scale experiments, 4 mL total of HB is required.

2.4 DEPC-Treated Water

1. Transfer 1000 mL of Milli-Q water to a clean and dry 1 L glass bottle (e.g., Nalgene). Add 1 mL of DEPC to the water, for a final concentration of 0.1% DEPC. Cap the bottle.

2. Shake vigorously and let the bottle sit at room temperature overnight (8–12 h).

3. Autoclave the solution for 15–45 min at 15 psi (to degrade DEPC).

4. Allow the solution to cool at room temperature.

5. DEPC-treated water can be stored at room temperature until further use (see **Note 6**).

2.5 DAPI Solution

2.5.1 Stock Solution

1. Dissolve 10 mg DAPI (in original vial) in 2 mL of UltraPure water, to a final concentration of 5 mg/mL (14.3 mM). Mix well by vortexing.

2. Aliquot the solution into 25 µL volumes in 1.5-mL tubes.

3. Store at −20 °C until use.

2.5.2 DAPI Work Solution

1. On the day of experiment, thaw a tube of DAPI stock.

2. In a new tube, add 21 µL of the 14.3 mM DAPI stock solution to 979 µL water to make a 300 µM DAPI solution. Mix well by vortexing. Protect from light.

2.6 10 mM DTT Buffer

1. Dilute 1 M DTT (Sigma) 1:100 in UltraPure water in a 50-mL conical tube or larger volume plastic bottle depending on the needed volume.

2.7 10% BSA

1. On the day of experiment, weigh out 1 g BSA into a 50-mL conical tube.

2. Add 10 mL water using a serological pipette.

3. Let the BSA dissolve at room temperature over 10–20 min. Vortex to completely dissolve.

4. Place on ice until use (see **Note 7**).

1. Prepare buffer NRB according to Table 3 immediately before performing the nuclei isolation on the day of experiment.

Table 3
Nuclei wash/resuspension buffer (NRB)—Adapted from [18]

Component	Stock concentration	Final concentration
Phosphate-buffered saline (PBS)	10×	1×
RNasin plus RNase inhibitor	40 U/µL	0.4 U/µL
BSA	10%	1%
Water (UltraPure)		To final volume

2. Add the RNase inhibitor as the last component and aim to use the buffer within 1 h.

3. Maintain on ice at all times.

2.8 NRB with DAPI

1. Add 10 µL of 300 µM DAPI working solution to 990 mL of NRB for a total of 1 mL. Prepare enough NRB with DAPI for the number of samples you will have and the scale of your preparation (e.g., 0.5 mL × 10 small-scale preparations = 5 mL total).

2.9 Equipment Preparation

2.9.1 Surface Decontamination

1. Spray all surfaces and equipment (i.e., pipettors, tube racks, chillers, centrifuge) with 70% ethanol. Wipe away with fresh Kimwipes. Then spray RNase AWAY liberally on the surfaces. Wipe with a fresh Kimwipe.

2.10 Dounce Homogenizer Cleaning and Preparation

2.10.1 Decontamination

1. Place the Dounce homogenizers in a plastic sealable bag (e.g., Ziplock) and spray liberally with RNase AWAY, making sure to expose both the inside and outside of the tube to the reagent. Let incubate for at least 1 h; can be left overnight if desired.

2.10.2 Washing

1. Before washing, gather and prepare the reagents you will need. You will need DEPC water, 25-mL serological pipettes, and pipette controller (e.g., Pipet-Aid). Take the cap off the DEPC water and place the cap face down on a clean Kimwipe. Unwrap a serological pipette at the opening, just enough to load into the pipette controller. Leave the rest of the serological pipette wrapped until you are ready to use it. Position a waste collection container (i.e., a 1 L beaker) close to where you will work. Set up fresh 15-mL and 50-mL conical tubes in a decontaminated tube rack or holder. These will be used to hold the glassware after cleaning.

2. Remove the wrapping of the serological pipette tip. Aspirate 30 mL of DEPC water (2-mL tube) or 50 mL (7-mL tube). Keep this pipette controller in your dominant hand.

3. With your free gloved hand, remove a single tube from the bag containing the tubes. Invert the tube to remove the majority of RNase AWAY in a waste container.

4. Wash the inside and outside of the tube with DEPC water. Fill the tube fully with water and then invert to decant the waste water. Alternate with washing the outside of the vessel. Do this until no water remains in the pipette.

5. Shake out the remaining water from inside the tube in one of the clean 15-mL conical tubes (2-mL tube) or 50-mL conical tube (7-mL tube).

6. Repeat **steps 3–7** on the remaining tubes.

7. After all tubes have been washed with DEPC water, perform a single wash of the inside of each tube with UltraPure water. Shake out any residual water and place each tube in the conical tubes to dry.

8. Now wash a pair of pestles. With gloves hands, identify and remove a pair of pestles from the bag. Each pair is indicated by a loose ("A") and tight ("B"). Place on a clean Kimwipe.

9. Aspirate 50 mL of DEPC water into a clean serological pipette. Grab the pestles near the handles. Dispense the water onto the pestles, letting the waste water drip off into the waste container.

10. Shake off pestles and place in a 50-mL conical tube.

11. Repeat steps 9–11 for all pestle pairs.

12. Give a final wash to all pestles with UltraPure water. Shake off well. Place in the 50-mL conical tubes and let dry (*see* **Note 8**).

2.11 Centrifuge

1. On the day of the experiment, set the centrifuge temperature to 4 °C and 500 × g and give it sufficient time to cool down to temperature before using.

3 Methods

3.1 Tissue Collection and Sample Factors

The common starting point for nuclei isolation is a frozen sample of spinal cord tissue from any species. Detailed instructions on the extraction of spinal cord tissue is beyond the scope of this protocol. Several detailed protocols describing the collection of spinal cord tissue from rodents have been published [26, 27].

Many labs will already have established protocols for collecting spinal cord tissue from mice and rats. These will all likely suffice so long as they can achieve the rapid removal of tissue, with a euthanasia to freezing time of less than 10 min. Samples should be snap frozen in dry ice or liquid nitrogen in clean tubes such as a 1.5-mL microcentrifuge tube. In general, we do not use saline or any other buffers to perfuse the rodents first, although this may be performed if desired, especially if the experimental interpretation relies on removing circulating immune cells from the vasculature of the spinal cord. Tools are cleaned well with 70% ethanol, and the animal's exterior is also sprayed with ethanol. Gloves are changed between each animal, and the dissection area and tools are cleaned with 70% ethanol between collections.

Human spinal cord may come from dedicated procurement services, human tissue banks, or local procurement. Thus, it will come frozen. The critical factors for successful single-nucleus RNA-seq from postmortem human spinal cord have not been systematically investigated, so it is difficult to predict with likelihood, which samples will be suitable for this protocol. RNA integrity value (RIN) is a commonly used metric to judge the adequacy of purified RNA for RNA-seq, with a threshold value of >6–7 typically used. A recent study of brain snRNA-seq used a RIN of 6.5 as a cutoff [16]. In our hands, RIN value is helpful in indicating which samples will most likely not produce high quality data; in other words, they set a floor value. Samples with a RIN value < ~4 will likely produce poor or no data by snRNA-seq. But on the contrary, samples with good RIN values (>6) do not always produce high-quality data, indicating that other factors are at play. We recommend performing a RIN analysis on bulk RNA isolated from a subsample of the tissue using standard RNA isolation methods (e.g., Trizol, Guanidinium Chloride) followed by Tapestation or Bioanalyzer (Agilent). A cutoff of RIN >5 is recommended, although it is up to the discretion of the investigator to increase stringency (e.g., RIN >7 or 8). We note, however, that useful data can still be gained even from samples with suboptimal RIN values, and whether to use such samples is dependent on available resources, the rarity of the sample and the study goals.

Another commonly used criterion is the postmortem interval (PMI), which is the time since a patient has died to the time that organs are collected. In general, we aim to procure tissue with the lowest PMI possible, preferably in the 6–12 h window, although longer PMIs may be tried if RIN is adequate. Some patient factors may not necessarily lead to RNA degradation but can alter the composition of the nuclei that ultimately are detected by sequencing. While this question has not been systematically addressed, our experience suggests that long periods of hypoxia/anoxia before death may differentially affect some cell types, especially in the lumber spinal cord [28]. Thus, we recommend that investigators obtain as many of these patient metadata as possible and integrate these factors into their analyses.

3.2 Nuclei Isolation   Note: The following instructions are written for the small-scale preparation. For the large-scale preparation, use the indicated volumes in Table 4.

Table 4
Small- and large-scale volumes

	Small scale (<150 mg)	Large scale (>150–1000 mg)
Dounce homogenizer	2 mL (sigma D8938)	7 mL (sigma D9063)
Homogenization	2 mL	5 mL
Washes	1 mL	3 mL
Resuspension	0.5 mL	1–2 mL

3.3 Tissue Preparation (Human)

If using rodent spinal cord, no preparation is necessary. Tissue can be advanced directly from the sample tube to homogenization. For human spinal cord, the sample will likely be much larger than is appropriate for homogenization, and thus will need preprocessing. If, however, the spinal cord tissue has already been cut to the desired size at the time of collection, these steps can be skipped and you can move on to *Homogenization*.

1. Fill a clean petri dish about 75% full with dissection buffer. Cover the dish and place in the freezer (either −20 °C or −80 °C) to freeze the buffer. The frozen buffer provides a solid, cold base on which to section spinal cord tissue. When ready, remove the plates from the freezer and bring to the benchtop (*see* **Note 9**).

2. Remove the tissue piece from its original container using clean forceps.

3. Hold the section steady with the forceps, positioning the long axis parallel to the plate (Fig. 1). Allow the sample to thaw slightly for 30 s to 1 min, so that cutting is possible.

4. Using a clean razor blade, make cuts perpendicular to the longitudinal axis with about 1 mm thickness (*see* Fig. 1). Discard the initial end pieces. Take one or more clean sections from the mid-section of the tissue. If you do not plan to use all of the tissue, quickly transfer the remaining spinal cord tissue into a chilled tube and place on dry ice. By working quickly, the tissue should not have thawed much.

5. Using the forceps, transfer the desired section to a clean, pre-chilled (on dry ice) 5-mL tube that has already been tared on a laboratory scale. Weigh and record the mass of the spinal cord section.

6. Using the forceps again, return the section to the cold petri dish. The base should still be frozen.

7. Sections can be further trimmed based on the needs of the experiment. For example, white matter can be trimmed if gray matter cells are the desired targets. Sections may be bisected into dorsal and ventral sections as well. Any dural pieces should

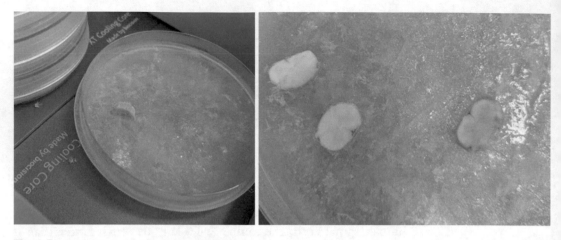

Fig. 1 Frozen human spinal cord sections

be removed gently if possible, as these are fibrous and tough and do not homogenize well.

8. Once the desired portion of a section has been obtained, use the razor blade to cut the section into smaller pieces.

9. Use the forceps to transfer the pieces into the prechilled Dounce tube.

3.4 Homogenization

1. Place the cleaned Dounce tubes in ice to cool them, being careful not to get any ice in the vessel.

2. With a p1000 pipette, transfer 1 mL of HB into the Dounce tubes. Keep on ice.

3. Using clean disposable forceps, transfer the tissue into the tube. If the tissue is very small, aliquot a small amount (~100 μL) of HB and transfer to the tube containing the tissue. Use the pipette to aspirate the tissue piece and transfer to the homogenizer.

4. Let the tissue thaw for 1 min in the HB before starting to homogenize (*see* **Note 10**).

5. Insert the loose pestle ("A") into the homogenizer and perform 20 strokes. Strokes should be smooth and gentle, with a 180° turn of the handle of the pestle as it descends to the bottom of the tube. On the upstroke, turn the handle back 180° to the starting position. Do not use excessive force and do not bring the pestle above the liquid level, as this will create bubbles and frothing. Keep the tube on ice while grinding. After 20 strokes with the loose pestle, perform an additional 20 strokes with the tight pestle ("B"). You will note more resistance with this pestle, but it will decrease as the stroke number increases (*see* **Note 11**).

6. Prepare a 2-mL LoBind tube by placing a 50-μm Sysmex filter into the opening and placing the tube on ice. Make sure it is well-supported.

7. Use a p1000 pipet to aspirate the homogenate from the homogenizer. Dispense the homogenate directly onto the center of the filter, pressing the tip against the filter mesh. If there is abundant debris, as there often is for human spinal cord, filtration can be aided by taking your thumb and covering the top of the filter opening. This will create positive pressure that pushes the filtrate through the filter.

8. Aspirate 1 mL of fresh HB and transfer into the homogenizer. Aspirate up and down five times along the sidewall to capture any residual nuclei. Aspirate the buffer and then apply it to the filter to wash any remaining nuclei. Perform the positive-pressure aid again if needed.

9. Remove the filter and discard. Cap the tube containing the filtered nuclei and keep on ice.

10. Repeat **steps 3–10** for all other samples.

11. Spin down the nuclei suspensions at $500 \times g$ for 5 min at 4 °C.

12. Gently remove the tubes from the centrifuge and place on ice. Aspirate the supernatant, being careful to not disturb the pellet. If the pellet is very small, you can leave ~50 μL of supernatant so as to not risk aspirating the pellet.

13. Dispense 1 mL of NRB into each tube and resuspend the pellet by pipetting gently five times. Do this for all the samples.

14. Spin down the nuclei suspensions at $500 \times g$ for 5 min at 4 °C.

15. Repeat **steps 13–15** twice more, for a total of three washes.

16. Prepare NRB with DAPI by adding.

17. Remove the supernatant as in **step 13**. Resuspend the pellets in each tube in 500 μL of NRB with DAPI. Pipette gently 10 times on ice (*see* **Note 12**).

18. Place a 20-μm Sysmex filter on the opening of a 5-mL polypropylene tube on ice. Aspirate the nuclei suspension from a sample tube and dispense onto the center of the filter. Use positive pressure from your thumb to push all the suspension through the filter. Cap the tube and place on ice. Repeat for all samples. Leave all samples on ice.

19. Remove an aliquot (5 μL) of nuclei from one or a few samples and dispense onto a glass slide. Inspect on an epifluorescence microscope with a DAPI filter to confirm that there are nuclei and to observe their morphology. Nuclei may be round or oblong, depending on the cell of origin. Membranes should be smooth. Broken nuclei or nuclei with blebbing indicates nuclear damage. Debris will also be abundant, especially in larger samples and those with high amounts of white matter (Fig. 2).

20. Proceed to FANS.

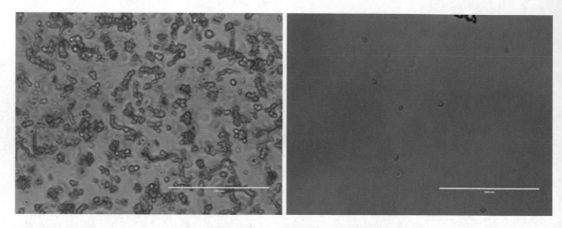

Fig. 2 Left: Human nuclei suspension before FANS. Right: Human nuclei after sorting. Nuclei are stained blue with DAPI

3.5 Fluorescence Activated Nuclear Sorting (FANS)

3.5.1 Preparation of Collection Tubes and Media for Downstream Applications

The nuclei prepared above are a common input for multiple downstream applications. The nuclei may be used for various transcriptomic assays, including:

1. Single-nucleus RNA-seq on the 10× Chromium platform.
2. Single-nucleus RNA-seq in plate-based assays (e.g., Smart-seq3) [29].
3. Bulk RNA-seq.
4. Single-nucleus DNA-seq.

The downstream assay will determine what kind of medium you sort the nuclei into and the type of container (i.e., plate vs. tube). We will focus on the 10× 3′ single-cell RNA-seq assay and bulk RNA-seq here.

For the 10× Chromium, we have had the most success when there is minimal manipulation of the nuclei post-sort. In particular, we aim to avoid any centrifugation or extensive pipetting in order to avoid clumping of nuclei and leakage of nucleic acids into the surrounding buffer.

For the Chromium, there are two options:

1. Sort into the 10× RT Buffer (excluding RT enzyme), as described in the "Frankenstein" protocol by Martelotto [18].
2. Sort into a small volume NRB and then process the sorted nuclei suspension according to the standard 10× Chromium workflow.

Directly sorting into the RT buffer has the advantage of minimizing post-sort manipulations. It also decreases handling time from sort to running on the 10× Chromium, which is important for the integrity of the nuclei. The less time between sorting and running the 10× assay, the less RNA will leak out into the milieu, thus reducing background RNA signal in the final data.

However, if it is not feasible to sort directly into 10× reagents, then sorting into NRB is the next best option. Indeed, we have noticed no difference in data quality between the two methods. The major difference is that sorting into NRB will dilute the sorted nuclei. The 10× Chromium 3′ RNA-seq assay has a fixed maximum sample input volume of 43.2 μL (v3.1). Sorting adds volume from the sheath fluid (see below). So if you are trying to achieve the maximum number of nuclei in the 10× assay, sorting directly into the RT buffer is the best route, since it adds no additional buffer volume besides the sheath fluid.

One could sort into a completely dry tube and achieve the same, but there is the concern that sorting into a tube without a "cushion" of buffer will lead to damage and/or breakage of nuclei or drying on the side wall. Therefore, for the second option, we recommend a cushion of 10 μL NRB to sort into.

For bulk RNA-seq, instead of sorting into NRB, nuclei are sorted into RNA lysis buffer. Our preferred RNA isolation kit is the RNAqueous™-Micro Total RNA Isolation Kit from Thermo Fisher, which excels in isolating the small amount of RNA that one gets from even thousands of nuclei. Other low-volume RNA isolation kits can be substituted at the experimenter's discretion.

Specifically, for the three options above, the following volumes in Table 5 are used:

To prepare the 10× RT Buffer Mix, combine the following for each reaction. If you will run multiple 10× reactions, scale up the following by the number of reactions plus an additional 10% to account for error. Note, the volumes here are for the Chromium v3.1 assay only. If you are using a different version, you must change the volumes accordingly based on the recommended volumes for the RT "Master Mix" in the official 10× documentation.

Component	Volume (μL)
RT Reagent B	18.8
Template switch oligo	2.4
Reducing agent B	2.0

3.6 Preparation of Collection Tubes for Bulk RNA-Seq

1. Using clean scissors or a razor blade, cut the cap off at the hinge. Reserve the cap and keep clean until use (*see* **Note 13**).

2. Add 100 μL of RNA-aqueous Lysis buffer to a 1.5 mL LoBind tube.

3. Place the cap back on the tube. Label the tube.

4. Prepare enough tubes for all samples and keep on ice until use.

Table 5
Sort collection volumes

	Buffer	Volume	Tube
10× chromium 3' RNA-seq (v3)	NRB	10 μL	1.5 mL LoBind 96-well plate
	10× RT buffer mix (without RT enzyme)	23.1 μL	1.5 mL LoBind 96-well plate
Bulk RNA-seq	RNaqeous micro lysis buffer (Ambion)	100 μL	1.5 mL LoBind

3.7 Preparation of Collection Tubes for 10× Chromium 3' RNA-Seq v3

1. Using clean scissors or a razor blade, cut the cap off at the hinge. Reserve the cap and keep it clean until use.

2. Fill the tube with either NRB or RT Buffer Mix (as above).

3. Place cap on the tube and label.

4. Spin down briefly to bring the liquid to the bottom of the tube.

5. Carefully remove the tubes from the centrifuge, making sure not to disturb the buffer at the bottom of the tube.

6. Keep tubes on ice until use.

3.8 FANS Instrument Setup

For FANS, one may either use the services of a trained operator, usually employed by a core facility, or operate a sorter instrument independently. We recommend working with your institution's core facility and having a trained operator perform the sort, as this option has several advantages. First, while you are completing the steps above, the operator will prepare the instrument in advance of your run. Second, it is technically challenging to operate a sorter instrument, and usually it requires prolonged training. For most labs, having an operator perform the sort is the most efficient and straightforward option. However, if you are trained to operate a FACS instrument and you have access to an instrument, independent sorting is an option. The major benefit is the ability to perform experiments at any time and not have to be restricted by scheduling. Indeed, scheduling can be one of the most challenging steps for labs using this protocol, and trying to coordinate two core facilities can be especially challenging. If you do choose to run the instrument yourself, we still recommend having an assistant help you on the day of experiment, since you will be engaged in the nuclei isolation steps above.

There are several instruments available for nuclei sorting. We have used the BD FACSAria II with excellent results. Other instruments are also suitable for this protocol, but it will be important to determine drop volume on each instrument (as described below).

Before performing this full protocol for an actual experiment, we strongly recommend that you perform a practice run on the sorter instrument that you intend to use. It is important to confirm that your specific instrument and operator can isolate nuclei using your gating strategy. It is also important to determine the specific droplet volume (discussed below) in advance.

The major factors that you will manipulate are:

1. Nozzle diameter.
2. Sheath pressure.
3. Gating strategy.

Manipulating these factors affects the droplet volume, event rate, and integrity of the nuclei.

Overall, the goal is to find the combination of settings that does the least damage to the nuclei in the fastest time and with a droplet volume that is consistent with your experimental goals.

Nozzle Diameter and Sheath Pressure: A smaller diameter nozzle produces smaller droplet volumes, and conversely, a larger nozzle produces larger volume droplets. Nuclei are generally quite small, in the 5–10 μm range. Thus, a 70-μm nozzle provides adequate space for nuclei to pass through without causing much damage. We have had success using a 70-μm nozzle with 40 PSI sheath pressure, as well as 85 μm and 45 PSI. The 70 μm/40 PSI configuration produces small droplets, approximately 1 nl per droplet on a BD FACSAria II.

Sheath pressure also affects droplet volume and affects the integrity of sorted nuclei. Higher pressure deforms the nuclei more and may lead to diminished integrity of the membrane, resulting in leakage of RNA and DNA.

Droplet volume is especially important to consider for this protocol because there is a maximum limit on the volume of sorted nuclei that can be loaded into the 10× instrument. For example, if one uses 10 μL of NRB in a collection tube, one can add only 33.2 μL to reach the maximal volume that can be loaded directly into the Chromium assay (43.2 μL). Thus, if one droplet is ~1 nL, a maximum of 33,200 droplets can be sorted directly into the collection tube. One could sort more than this, but this would only serve to dilute the collection suspension, since only 43.2 μL can be loaded into the Chromium assay. Moreover, our experience has been that FANS tends to overestimate the number of captured nuclei by about 20–40%, so this should also figure into the nominal target number of sorted nuclei that one aims for.

In practice, we find that the most accurate way to determine droplet volume is to sort some fluorescent QC beads into a dry tube and to measure the sorted volume. The volumes will only be accurate for a given nozzle and pressure combination.

3.9 Gating Strategy   Standard doublet discrimination should be performed using forward scatter (FSC) and side scatter (SSC) to identify singlets. Excitation with a 407 nm laser can be used to identify DAPI+ events. This is especially important to separate the abundant debris present in spinal cord nuclei suspensions from intact nuclei. A typical gating scheme for human spinal cord is shown in Fig. 1. We recommend working with your local instrument operator to devise a gating scheme prior to executing a complete experiment. If additional dyes or antibodies are used to label nuclei [17, 30], you will also need to incorporate these into the gating strategy.

1. Place the collection tube for either single-nucleus or bulk RNA-seq in a chilled holder on the sorter.

2. Load the nuclei suspension onto the sorter.

3. Find intact nuclei using the predetermined gating scheme (Fig. 3).

4. Collect the desired number of nuclei into the collection tube (*see* **Note 14**).

5. After collecting the desired number of nuclei, remove the collection tube and place on ice. If you plan to sort additional samples, move quickly on to the next sample (*see* **Note 15**).

3.10 10× Chromium 3′ RNA-Seq Assay

After all samples are collected, move directly to the 10× Chromium assay without delay. A maximum of 43.2 µL of the sorted nuclei can be run per reaction. Less volume (i.e., few nuclei) can be loaded, but will need to be adjusted according to the 10× documentation. Carry out the entire protocol according to the 10× documentation. Completed 10× libraries should be run on an Illumina sequencer according to the 10× documentation. We typically aim for 50,000–100,000 reads/nucleus.

3.11 Bulk RNA-Seq

In contrast to the 10× assay, with bulk RNA-seq, the nuclei are sorted directly into lysis buffer. Therefore, the sorted nuclei samples are stable. The samples can be either processed immediately or stored at −80 °C for at least 6 months before processing to completion according to the manufacturer's documentation. RNA-seq libraries must be prepared from the isolated RNA. Many commercial kits are available. We have had success with the NuGEN Ovation SoLo RNA-Seq kit [11] as well as Smart-seq2 [31]. Sequencing should be carried out according to the documentation of the library preparation method.

4 Notes

1. The 10-pack is much cheaper per bottle than buying individually.

2. The number of sets will depend on the size of your study. For each sample in a batch that you process, you should have a clean Dounce grinder set. Ideally you would have as many Dounce tubes as you have samples. The size of the Dounce grinder set

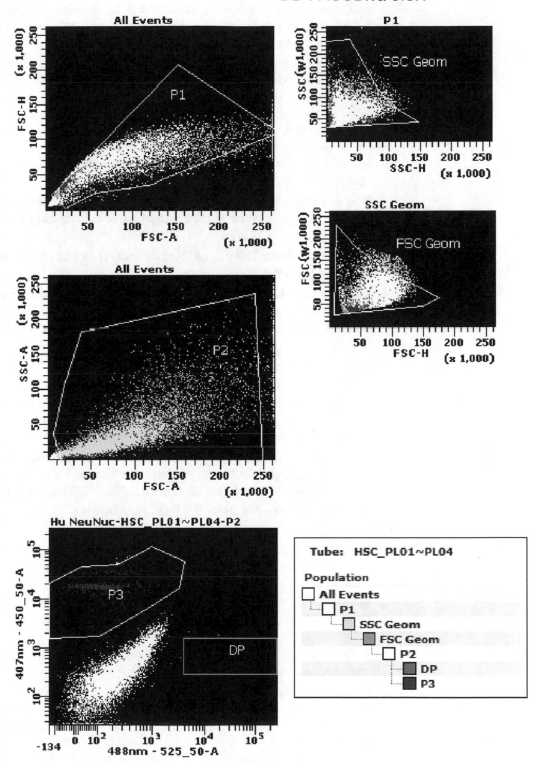

Fig. 3 Gating Strategy for Human Nuclei. P3 (bottom, red) contains the DAPI+ nuclei that are collected into the collection tube. Note the abundance of debris

will depend on the sample size you use. For most studies, we recommend the 2 mL size, which can accommodate samples ≤150 mg. For samples 150–1000 mg, the 7 mL Dounce grinder set is appropriate.

3. For the vast majority of mouse spinal cord experiments, small scale will be adequate. Also, for most human and nonhuman primates, small scale will also be preferred. A 1–2 mm transverse section of human lumbar spinal cord will generate several million nuclei, which will be more than sufficient for most applications. Large-scale preparation would be appropriate when an entire murine spinal cord or large piece of human spinal cord is processed.

4. The lab coat should ideally have thumb holders, so that the glove covers your wrist skin to avoid hair or skin getting into the sample.

5. We typically make 250 mL of NIM at once. It can be stored at 4 °C for 1 month.

6. DEPC water should be prepared in advance of the experiment in order to allow time for incubation with DEPC and sufficient cooling for use. We recommend making large batches in 1 L bottles.

7. Leftover BSA can be aliquoted and stored at −20 °C for future use.

8. If performing in advance (e.g., day before), pestles and tubes should be left in a low traffic area, or ideally a laminar flow hood. UV decontamination can also be applied.

9. We recommend one plate per sample. We typically place the plates on chilled cold blocks, which keep the ice base frozen while cutting. If ice blocks are not available. Working quickly on the benchtop, one plate at a time, is adequate.

10. If using fresh tissue, no waiting is necessary.

11. If there are residual pieces of dense, fibrous tissue, namely the dura, these may block the passage of the pestle. Dural pieces are generally resistant to homogenization with the Dounce grinder. Therefore, they can be removed before advancing to the tight pestle using a p1000 pipette or even the loose pestle itself.

12. Pellets can be resuspended in larger volumes if more dilute nuclei are needed.

13. Cutting the cap is advised to facilitate proper placement of the tube in the tube chiller on the FACS instrument. This step can be omitted if the FACS instrument can accommodate a capped tube.

14. The number of nuclei that you will collect depends on the experimental goals and the volume of the droplets from the sorter. For 10×, refer to the documentation to determine the target number, taking into account that the nominal number sorted will be about 20–40% less what actually ends up in the sorting tube. So, for example, if the sorter says 10,000 nuclei were sorted, the actual number in the tube is likely between 6000–8000. For bulk RNA-seq using the RNA-aqueous micro kit, we have sorted a range of nuclei, from 500 to 50,000 with excellent results.

15. Minimizing the time from sorting to loading on the 10× is critical. We recommend aiming to load sorted samples onto the 10× within 30 min of sorting. Longer post-sort durations may result in higher amounts of ambient RNA captured by the 10× assay or few nuclei captured because of breakdown of fragile, sorted nuclei.

References

1. Ji R, Berta T, Nedergaard M (2013) Glia and pain: is chronic pain a gliopathy? Pain 154:S10–S28

2. Ji R-R, Chamessian A, Zhang Y Q (2016) Pain regulation by non-neuronal cells and inflammation. Science 354:572–577

3. Peirs C, Seal RP (2016) Neural circuits for pain: Recent advances and current views. Science 354:578–584

4. Zeisel A, Hochgerner H, Lönnerberg P et al (2018) Molecular architecture of the mouse nervous system. Cell 174:999–1014.e22

5. Häring M, Zeisel A, Hochgerner H et al (2018) Neuronal atlas of the dorsal horn defines its architecture and links sensory input to transcriptional cell types. Nat Neurosci 61:1

6. Sathyamurthy A, Johnson KR, Matson KJE et al (2018) Massively parallel single nucleus transcriptional profiling defines spinal cord neurons and their activity during behavior 22:2216–2225

7. Rosenberg AB, Roco CM, Muscat RA et al (2018) Single-cell profiling of the developing mouse brain and spinal cord with split-pool barcoding. Science 12:eaam8999

8. Denk F, Crow M, Didangelos A et al (2016) Persistent alterations in microglial enhancers in a model of chronic pain. Cell Rep 15:1771–1781

9. Machado L, Geara P, Camps J et al (2021) Tissue damage induces a conserved stress response that initiates quiescent muscle stem cell activation. Cell Stem Cell 28(6):1125–1135.e7

10. Lacar B, Linker SB, Jaeger BN et al (2016) Nuclear RNA-seq of single neurons reveals molecular signatures of activation. Nat Commun 7:11022

11. Chamessian A, Young M, Qadri Y et al (2018) Transcriptional profiling of somatostatin interneurons in the spinal dorsal horn. Sci Rep 8:6809

12. Serafin EK, Chamessian A, Li J et al (2019) Transcriptional profile of spinal dynorphin-lineage interneurons in the developing mouse. Pain 160:2380–2397

13. Macosko EZ, Basu A, Satija R et al (2015) Highly parallel genome-wide expression profiling of individual cells using nanoliter droplets. Cell 161:1202–1214

14. Hagemann-Jensen M, Ziegenhain C, Chen P et al (2020) Single-cell RNA counting at allele and isoform resolution using smart-seq3. Nat Biotechnol 38(6):708–714

15. Slyper M, Porter CBM, Ashenberg O et al (2020) A single-cell and single-nucleus RNA-Seq toolbox for fresh and frozen human tumors. Nat Med 26:792–802

16. Schirmer L, Velmeshev D, Holmqvist S et al (2019) Neuronal vulnerability and multilineage diversity in multiple sclerosis. Nature 388:1

17. Krishnaswami SR, Grindberg RV, Novotny M et al (2016) Using single nuclei for RNA-seq to capture the transcriptome of postmortem neurons 11:499–524

18. Martelotto L 'Frankenstein' protocol for nuclei isolation from fresh and frozen tissue for snRNAseq. https://www.protocols.io/view/

frankenstein-protocol-for-nuclei-isolation-from-f-3fkgjkw

19. Matson KJE, Sathyamurthy A, Johnson KR et al (2018) Isolation of adult spinal cord nuclei for massively parallel single-nucleus RNA sequencing. J Vis Exp (140):e58413

20. Nott A, Schlachetzki JCM, Fixsen BR et al (2021) Nuclei isolation of multiple brain cell types for omics interrogation. Nat Protoc 16(3):1629–1646

21. Grindberg RV, Yee-Greenbaum JL, McConnell MJ et al (2013) RNA-sequencing from single nuclei. Proc Natl Acad Sci U S A 110: 19802–19807

22. Hodge RD, Bakken TE, Miller JA et al (2019) Conserved cell types with divergent features in human versus mouse cortex. Nature 573: 61–68

23. Sun W, Cornwell A, Li J et al (2017) SOX9 is an astrocyte-specific nuclear marker in the adult brain outside the neurogenic regions. J Neurosci 37:4493–4507

24. Okada S, Saiwai H, Kumamaru H et al (2011) Flow cytometric sorting of neuronal and glial nuclei from central nervous system tissue. J Cell Physiol 226:552–558

25. Sasagawa Y, Danno H, Takada H et al (2018) Quartz-Seq2: a high-throughput single-cell RNA-sequencing method that effectively uses limited sequence reads. Genome Biol 19: 14049

26. Malin SA, Davis BM, Molliver DC (2007) Production of dissociated sensory neuron cultures and considerations for their use in studying neuronal function and plasticity. Nat Protoc 2:152–160

27. Richner M, Jager SB, Siupka P et al (2017) Hydraulic extrusion of the spinal cord and isolation of dorsal root ganglia in rodents. J Vis Exp (119):55226

28. Duggal N, Lach B (2002) Selective vulnerability of the lumbosacral spinal cord after cardiac arrest and hypotension. Stroke 33:116–121

29. Hagemann-Jensen M, Ziegenhain C, Chen P et al (2019) Single-cell RNA counting at allele- and isoform-resolution using smart-seq3. Biorxiv 817924

30. Hodge RD, Bakken TE, Miller JA, Smith KA, Barkan ER, Graybuck LT, Close JL, Long B, Johansen N, Penn O et al (2019) Conserved cell types with divergent features in human versus mouse cortex. Nature 536:1–8

31. Picelli S, Björklund ÅK, Faridani OR et al (2013) Smart-seq2 for sensitive full-length transcriptome profiling in single cells. Nat Methods 10:1096–1098

Chapter 3

Multiplex In Situ Hybridization of the Primate and Rodent DRG and Spinal Cord

David W. Ferreira, Cynthia M. Arokiaraj, and Rebecca P. Seal

Abstract

Fluorescence in situ hybridization (FISH) has become an important tool in laboratory experimentation by providing a qualitative or semi-quantitative technique to detect nucleic acids across different sample types and species. It also serves as a promising platform for the discovery of novel RNA biomarkers and the development of molecular diagnostic assays. While technologies to detect hundreds or thousands of gene transcripts in situ with single-cell resolution are rapidly coming online, smaller scale FISH analysis continues to be highly useful in neuroscience research. In this chapter, we describe a robust, relatively fast and low cost, turnkey in situ hybridization technology (ISH) to identify one or more RNA targets together with immunohistochemical analyses. Specifically, we present a customized version of the protocol that works particularly well for spinal cord and primary sensory ganglia tissues.

Key words Fluorescence in situ hybridization, FISH, RNAscope, Spinal cord, DRG, Rodent, Primate

1   Introduction

Fluorescence in situ hybridization (FISH) is a technique that allows spatial detection, localization, and potential quantification of nucleic acids inside cells or tissues for the analysis and study of genomic sequences and transcriptomic expression profiles. This technique relies on the hybridization of sequence-specific probes to their complementary RNA or DNA target, and has often been used for medical applications such as the detection and diagnosis of cytological, and genetic abnormalities [1]. It has also contributed to basic research, most commonly for gene mapping and expression [2].

Historically, the first in situ hybridization (ISH) was performed in 1969 using radioactively labeled probes to study the binding of ribosomal RNA (rRNA) to the DNA which codes for it (rDNA) in a

Supplementary Information The online version of this chapter (https://doi.org/10.1007/978-1-0716-2039-7_3) contains supplementary material, which is available to authorized users.

Rebecca P. Seal (ed.), *Contemporary Approaches to the Study of Pain: From Molecules to Neural Networks*, Neuromethods, vol. 178, https://doi.org/10.1007/978-1-0716-2039-7_3, © Springer Science+Business Media, LLC, part of Springer Nature 2022

cytological preparation of oocytes [3]. Although radioisotope-labeled nucleotides were used as the gold standard for ISH; low sensitivity and resolution, high background, and long turn-around times were some of the disadvantages that led to the further development of non-radioisotopic labeling approaches including the use of hapten-labeled probes and enzymatically or fluorescently conjugated antibodies [4–6]. This was followed by direct fluorescent detection through the use of fluorescence-labeled probes. The initial technique of fluorescence in situ hybridization (FISH) continued to be further improved by technical advances that increased cellular sensitivity and speed of processing [7]. These improvements were mainly in probe design and labeling strategies, such as the synthetic oligonucleotide-based FISH probes, which can be specifically designed to detect splice variants, single-nucleotide polymorphisms, and for the simultaneous detection of multiple transcripts with high sensitivity and specificity in multiple types of biological samples [8–12].

Until recently, the implementation of FISH required procedures that needed to be optimized empirically for each experiment (probes, samples, etc.). However, many new technologies and protocols have emerged in the past few years that have reduced the hands-on time and the need for extensive training, while at the same time have increased the consistency and reproducibility across different sample types. In this chapter, we will focus on RNAscope® [10], a cutting-edge ISH technology with unique probe design and hybridization-based signal amplification strategies that result in amplify signals and suppressed background [10]. It has become one of the most highly utilized techniques by the neuroscience field for the detection of target RNAs within tissues. The field continues to rapidly evolve, including new methods to detect hundreds or thousands of expressed genes in situ with single-cell resolution [13]. This is a fast, reliable method for the common experiment, where a simple FISH with or without immunohistochemistry (IHC) is needed.

1.1 Technology

The RNAscope® technology has advantages over traditional ISH techniques by offering shorter protocol times, high sensitivity for the detection of even lowly-level expressed genes, and the ability to probe for multiple RNA targets simultaneously. Moreover, unlike past ISH methods, RNAscope® works well for tissue perfused with fixative, as well as fresh frozen tissue. This technology utilizes a Z-shaped probe, the bottom part is designed to be complementary and specific to the target RNA and is ~18–25 bases. The top and bottom parts of the Z probe are separated by 14 base tail spacer sequence. There are 20 pairs of Z probes per target mRNA, but it is possible to achieve a minimum hybridization with six pairs of Z probes. Thus, mRNA targets of >300 bases can be detected, resulting in greater specificity. The signal amplification event is initiated by pre-amplifiers that detect and bind to the 28 base sequence formed by two adjacent Z probes. This is followed by a series of amplification steps in which pre-amplifiers are bound by amplifiers that contain binding sites for fluorophores or for chromogenic enzymes [10].

2 Materials

The following items may be purchased directly from ACDBio:

- RNAscope® Multiplex v2 Fluorescent kit.
- HybEZ™ Hybridization System.
- Catalog Probes or Made-to-Order Probes. ACDBio can design custom probes provided you give them the exact sequence or nucleotide accession number from GenBank. For optimal results, the target sequence should be >300 bases in length. If you plan to detect more than one gene at a time in the same slide, remember to assign different channels to each of the probes.
- Probe diluent.
- Three-plex control probes.

    Items from other manufacturers:

- ImmEdge Hydrophobic Barrier PAP Pen (Vector Laboratories).
- TSA® Reagents or Opal dyes (Akoya Biosciences).
- Superfrost Plus microscope slides (Fisherbrand).
- ProLong™ Gold Antifade Mountant with DAPI (Invitrogen). *Note*: You can also purchase the mountant without DAPI if you want to use the DAPI that comes along with the Multiplex kit from ACDBio—recommended by the manufacturer.
- Coverslips.
- Fluorescent microscope that can visualize the fluorescein, Cy3, and Cy5 fluorophores.
- Water bath that can be set at 40 °C.
- 20× Saline–sodium citrate buffer (Sigma-Aldrich).
- O.C.T compound (Tissue-Tek).
- EasyDip™ Slide Staining System.
- 1× PBS.
- Milli-Q water.
- 100% Ethanol.

3   Method

Below is our modified protocol for RNAscope ISH that works consistently across different tissue samples (Fig. 1). Please also visit the ACDBio website for the manufacturer's instructions.

3.1   Tissue Preparation

The manufacturer's protocol provides three different methods of tissue preparation. Here, we focus on mice perfused with a fixative.

1. Perfuse the animal with nuclease-free $1\times$ phosphate-buffered saline ($1\times$ PBS, pH 7.4), followed by freshly prepared cold 4% paraformaldehyde (4% PFA, pH 7.4) in $1\times$ PBS. Working with fixed tissue is advantageous as the RNAscope® ISH can then be combined with immunohistochemistry (IHC) using the Multiplex v2 kit. Tissue expressing a viral reporter or tissue harvested from transgenic reporter mice can also be probed with this protocol.

2. Dissect out the tissue of interest and transfer it to 4% PFA. Post-fix for 2 h at 4 °C.

3. Then transfer the tissue to a tube of freshly prepared 10% sucrose in $1\times$ PBS (*see* **Note 1**).

4. After the tissue sinks to the bottom of the tube, transfer it to 20% sucrose in $1\times$ PBS. Repeat this step with 30% sucrose in $1\times$ PBS.

5. Let the tissue remain in this solution for 48 h before embedding it in OCT. Freeze the tissue in the OCT mold using dry ice and then store in an airtight container at -80 °C until you are ready to section it.

6. Prior to sectioning, remove the tissue from the -80 °C and place it in the cryostat at -20 °C for 30–45 min.

7. Cut the tissues onto Superfrost Plus slides leaving room on the slide to draw a hydrophobic barrier with the Pap Pen. As per the manufacturer's protocol, it is not ideal to cut tissues at a thickness greater than 20 µm.

8. After sectioning, let the slides sit in -20 °C for 30 min before storing them at -80 °C or proceed directly to **step 2** under Pretreatment.

3.2   Before Starting the Assay

- Reconstitute the fluorophore and store the stock solutions as per the manufacturer's instructions (Akoya Biosciences). Make note of which fluorophore you would like to assign to each of the probes. Do not assign the same fluorophore to probes of two different channels.

- Prepare fresh $1\times$ Wash Buffer as per the manufacturer's instructions.

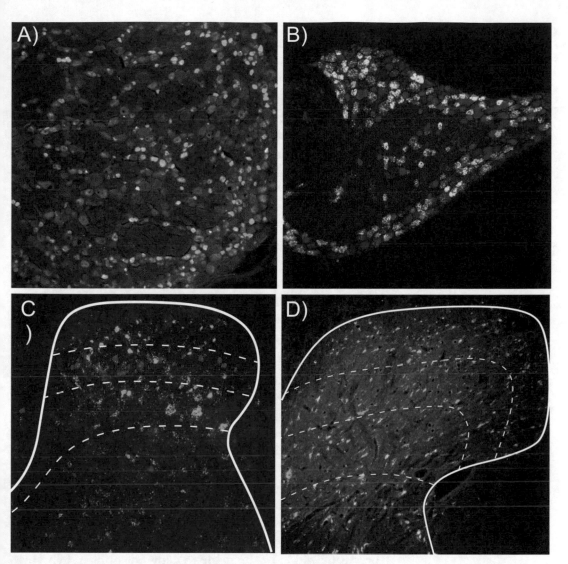

Fig. 1 Fluorescence in situ hybridization (FISH) in macaque and mouse L4/L5 DRG and spinal cord: (**a**) Representative microscopic images (10×) depicting three mRNA targets in fresh frozen monkey DRG: Opioid receptor mu 1 probe (Mmu-OPRM1, Cat No. 518941-C3) is visualized in green, cholecystokinin B receptor probe (Mmu-CCKBR, Cat No. 838061-C1) is visualized in red, and cholecystokinin A receptor probe (Mmu-CCKAR, Cat No. 838051-C2) is visualized in blue. (**b**) Representative microscopic images (20×) depicting 1 mRNA target in perfused injured mouse DRG: cholecystokinin B receptor probe (Mm-Cckbr, Cat No. 439121-C1) is visualized in green, tdTomato autofluorescence is visualized in red, and Dapi is visualized in blue. (**c**) Representative microscopic image (10×) of 2 mRNA targets in perfused macaque spinal cord tissue: Prodynorphin probe (Mmu-Pdyn, Cat No. 838041-C1) is visualized in red and Paired Box 2 probe (Mmu-Pax2, Cat No. 837981-C2) is visualized in green. (**d**) Representative microscopic image (10×) of two mRNA targets in fresh frozen mouse spinal cord tissue: Copine4 probe (Mm-Cpne4, Cat No. 47421-C1) and thyrotropin-releasing hormone probe (Mm-Trh, Cat No. 436811-C3)

3.3 Pretreatment   Prior to hybridization, there are a few pretreatment steps to help unmask the target nucleic acid and allow probe penetration. We use the EasyDip™ Slide Staining System for easy transfer of slides from one solution to another.

1. Before starting, turn on the HybEZ™ oven and make sure the temperature is at 40 °C. Ensure that the humidity control tray with the wet filter paper is placed inside the oven (*see* **Note 2**).

2. When you are ready to begin the RNAscope® in situ hybridization, take the slides out of the −80 °C and keep it outside for 10–15 min.

3. Wash the slides with 1× PBS for 2 min with gentle agitation in order to remove the excess OCT. Repeat this wash with fresh 1× PBS for 2 min.

4. Immerse the slides in freshly prepared 50% EtOH using the EasyDip™ Slide Staining System for 5 min at room temperature (RT).

5. Immerse the slides in 70% EtOH for 5 min at RT.

6. Immerse the slides in 100% EtOH for 5 min at RT. Repeat this step with fresh 100% EtOH.

7. Allow the slides to dry at RT for 5 min. Using the special PAP pen, draw a barrier around the sections.

8. To block endogenous peroxidase activity, add hydrogen peroxide from the kit to the tissue sections and incubate for 10 min at RT. Keep the slides in a hydrated chamber to prevent the tissues from drying up.

9. Wash the slides with 1× PBS for 2 min with gentle agitation. Repeat this wash with fresh 1× PBS for 2 min.

10. Add Protease IV from the kit to the tissue sections and incubate for 30 min at RT in a hydrated chamber. This step causes permeabilization of the tissue to allow for optimal hybridization (*see* **Note 3**).

11. Wash the slides with 1× PBS for 2 min with gentle agitation. Repeat again with fresh 1× PBS. During this time, take out your probes from the 4 °C and incubate them at 40 °C in a water bath or incubator for at least 10 min.

12. Let the probes cool down to RT. C2 and C3 probes must be spun down. As per the manufacturer's protocol, mix 1 volume of C2 and 1 volume of C3 to 50 volumes of C1 probe or the probe diluent in an Eppendorf tube. This is ~4 μL of C2/C3 probes to ~7 drops of C1 probe or ~7 drops of the probe diluent (*see* **Note 4**). Invert to mix.

13. Ensure that the slides do not have any excess liquid before adding the mixed probe solution to completely cover the sections.

14. Place the slides in the EZ-Batch™ Slide rack and insert the rack into the oven.

15. Let the probes hybridize for 2 h at 40 °C.

16. After the hybridization, remove the slide rack from the oven (*see* **Note 5**).

17. Wash the slides with 1× Wash Buffer for 2 min at RT with gentle agitation. Repeat this step with fresh 1× Wash Buffer.

*At this point, you have the option to stop and store the slides in freshly prepared 5× SSC buffer overnight (O/N) at RT or else proceed directly to **step 4** under Amplification.*

3.4 Amplification

For the subsequent amplification steps, remember to equilibrate the reagents (Amp1, Amp2, Amp3, HRPs, and HRP blocker) at RT prior to starting. Do not let the slides dry out at any point during the assay.

1. Turn on the HybEZ™ oven if continuing with the protocol the next day and make sure that the temperature is at 40 °C. Place the humidity control tray with the wet filter paper inside the HybEZ™ oven.

2. After the temperature of the oven is stabilized at 40 °C for at least 30 min, proceed to the next steps.

3. Remove the slides from the 5× SSC Buffer and wash with 1× Wash Buffer in the EasyDip™ Slide Staining System. Repeat this step with fresh 1× Wash Buffer.

4. Pour off any excess liquid from the slides and place the slides in the EZ-Batch™ slide rack. Add enough drops of Amp1 to properly cover the tissue sections on the slide.

5. Carefully insert the EZ-Batch™ slide rack into the oven. Incubate for 30 min at 40 °C in the oven.

6. Remove the slide rack and pour off any excess liquid from the slides.

7. Wash the slides with 1× Wash Buffer for 2 min at RT with gentle agitation. Repeat the wash with fresh 1× Wash Buffer.

8. Pour off any excess liquid from the slides and place the slides in the EZ-Batch™ slide rack. Add enough drops of Amp2 to properly cover the tissue sections on the slide.

9. Carefully insert the EZ-Batch™ slide rack into the oven. Incubate for 30 min at 40 °C in the oven.

10. Remove the slide rack and pour off any excess liquid from the slides. Wash the slides with 1× Wash Buffer for 2 min at RT with gentle agitation. Repeat the wash with fresh 1× Wash Buffer.

11. Pour off any excess liquid from the slides and place the slides in the EZ-Batch™ slide rack. Add enough drops of Amp3 to properly cover the tissue sections on the slide.

12. Carefully insert the EZ-Batch™ slide rack into the oven. Incubate for 15 min at 40 °C in the oven.

13. Remove the slide rack and pour off any excess liquid from the slides. Wash the slides with 1× Wash Buffer for 2 min at RT with gentle agitation. Repeat the wash with fresh 1× Wash Buffer.

The following HRP signal development steps will depend on the channels assigned to the probes. Develop the HRP signal one by one for each probe used in the experimental run. For example, develop the HRP-C1 signal for the C1-assigned probe followed by the HRP-C2 for the C2-assigned probe.

3.5 Preparation of the TSA® Fluorophores

Use the TSA buffer provided by the RNAscope® kit to dilute the fluorophores. About 150–200 μL of the TSA buffer is sufficient to cover the slide. Follow the recommended dilution range (1:750 to 1:3000) of the TSA® dyes fluorophore as mentioned in the manufacturer's protocol.

1. Pour off any excess liquid from the slides and place the slides in the EZ-Batch™ slide rack. Add enough drops of the HRP-C1 to properly cover the tissue sections on the slide and incubate in the oven for 15 min at 40 °C.

2. After incubation, remove the EZ-Batch™ slide rack from the oven and wash the slides in 1× Wash Buffer for 2 min at RT. Repeat the wash with fresh 1× Wash Buffer for 2 min at RT.

3. Pour off any excess liquid from the slide and add the prepared TSA® Plus fluorophore to the slides placed in the EZ-Batch™ slide rack. Incubate in the oven for 30 min at 40 °C. This fluorophore is assigned to the C1 probe.

4. After incubation, remove the EZ-Batch™ slide rack from the oven and wash the slides in 1× Wash Buffer for 2 min at RT. Repeat the wash with fresh 1× Wash Buffer.

5. Add enough drops of HRP blocker (provided in the kit) to properly cover the tissue sections on the slide. Incubate in the oven for 15 min at 40 °C.

6. After incubation, remove the EZ-Batch™ slide rack from the oven and wash the slides with 1× Wash Buffer for 2 min at RT. Repeat the wash with fresh 1× Wash Buffer.

Similarly, develop the signal for HRP-C2 and HRP-C3 if you have probes assigned to those channels. If you plan to do IHC following ISH, proceed to the next step, otherwise go to DAPI Counterstain and Slide Mounting.

4   Immunohistochemistry

It is important to test your antibody by staining with the buffers that the manufacturer recommends prior to running the immuno-histochemistry in combination with ISH.

4.1   Materials

- $1\times$ TBS (pH 7.4).
- TBS-T Wash Buffer: 500 µL of 10% Tween®20 is added to 1 L of $1\times$ TBS Buffer.
- TBS-1% BSA: Add 0.5 g to 50 mL of $1\times$ TBS. Aliquot and store in −20 °C.
- Normal Donkey Serum (if secondary antibody is raised in donkey).

4.2   Method

- If needed, redraw the hydrophobic barrier with the PAP pen. Do not let the slides dry out.
- Wash the slides in TBS-T Wash Buffer for 5 min at RT with gentle agitation. Repeat with fresh TBS-T Wash Buffer.
- Prepare blocking solution by adding 10% Normal Donkey Serum (NDS) to TBS-1% BSA. About 150–200 µL of solution covers the slide, so prepare it accordingly. If a particular primary antibody requires further permeabilization, Triton™ X-100 can be added to the blocking solution at a recommended maximum concentration of 0.1%.
- Add the blocking solution to the slides for 1 h at RT. Make sure the slides are kept in a humidified chamber.
- Prepare the primary antibody solution by adding the primary antibody to TBS-1% BSA at a concentration that has already been optimized.
- Incubate the slides O/N at 4 °C (*see* **Note 6**).
- Wash the slides in TBS-T Wash Buffer for 5 min at RT with gentle agitation. Repeat with fresh TBS-T Wash Buffer.
- Add the respective Alexa Fluor-conjugated secondary antibody in TBS-1% BSA. If using HRP-conjugated secondary antibodies, refer to the manufacturer's protocol for dilution and incubation time.
- Incubate the slides in the secondary antibody solution for 2 h at RT. Remember to keep the slides in a humidified chamber.
- Wash the slides with TBS-T Wash Buffer for 2 min at RT with gentle agitation. Repeat with fresh TBS-T Wash Buffer.
- Remove the excess liquid from the slides. Add two or three drops of Prolong Gold antifade mounting medium with DAPI on the slides and coverslip.

- Dry the slides in the dark for 30 min.
- Store the slides at 4 °C or continue to image after the slides have dried.

5 Quantification

RNAscope® does not provide an actual quantitative measure of how many transcripts of a gene are present in the tissue. It can be used in a semi-quantitative manner by providing scores depending on the number of puncta present per cell and the number of dots lying within clusters. Refer to the ACD Scoring system for this metric. If the goal is to determine the approximate number of positive cells, a threshold of at least 3–5 puncta or more within a cell is considered positive. For human or nonhuman primate tissue, we usually recommend that the green channel remains blank due to the presence of lipofuscin (fatty deposits), which may lead to false-positive signal. The background in the green channel can thus be subtracted from the other channels. Alternatively, if all channels are used to detect real signals, lipofuscin can be imaged specifically to create a lipofusin mask by imaging with custom settings that span the spectrum and thus will not be detected by the individual fluorophores (i.e., a filter set to excite at 405 nm and emit at 647 nm) [14, 15]. This lipofuscin mask is then subtracted from each individual channel to remove the lipofusin signal. Newer software platforms like dotdotdot offer features to mask the lipo-fuscin autofluorescence as well as quantify and analyze the mRNA [16]. Software such as ImageJ, Halo™, and FISH Finder [17] can be used to delimit areas of interest and count the cells in a more automated fashion. ImageJ also has a plugin called Cell Counter for the manual counting of cells.

6 Notes

1. After tissue fixation, the use of nuclease-free 1× PBS solution is optional.
2. Make sure the oven is at 40 °C for at least 30 min prior to the probe hybridization (step 15).
3. If you see over-digestion of the tissue, use Protease III from the kit instead of Protease IV.
4. One drop of the C1 probe or probe diluent is ~30 μL.
5. We save the probes in an Eppendorf tube at 4 °C. The probes can be reused a second time in a future experiment. Remember to place them at 40 °C for at least 10 min after removing it from

the 4 °C. However, reusing the probes is not recommended by the manufacturer, so do a trial run before attempting this for an actual experiment.

6. Incubation times may vary according to the manufacturer's recommendation time for the primary antibody.

References

1. Huber D, Voith von Voithenberg L, Kaigala GV (2018) Fluorescence in situ hybridization (FISH): history, limitations and what to expect from micro-scale FISH? Micro Nano Eng 1:15–24. https://doi.org/10.1016/j.mne.2018.10.006

2. Jensen E (2014) Technical review: in situ hybridization. Anat Rec (Hoboken) 297(8):1349–1353. https://doi.org/10.1002/ar.22944

3. Pardue ML, Gall JG (1969) Molecular hybridization of radioactive DNA to the DNA of cytological preparations. Proc Natl Acad Sci U S A 64(2):600–604. https://doi.org/10.1073/pnas.64.2.600

4. Rudkin GT, Stollar BD (1977) High resolution detection of DNA-RNA hybrids in situ by indirect immunofluorescence. Nature 265(5593):472–473. https://doi.org/10.1038/265472a0

5. Manning JE, Hershey ND, Broker TR, Pellegrini M, Mitchell HK, Davidson N (1975) A new method of in situ hybridization. Chromosoma 53(2):107–117. https://doi.org/10.1007/bf00333039

6. Langer-Safer PR, Levine M, Ward DC (1982) Immunological method for mapping genes on Drosophila polytene chromosomes. Proc Natl Acad Sci U S A 79(14):4381–4385. https://doi.org/10.1073/pnas.79.14.4381

7. Levsky JM, Singer RH (2003) Fluorescence in situ hybridization: past, present and future. J Cell Sci 116(Pt 14):2833–2838. https://doi.org/10.1242/jcs.00633

8. Raj A, van den Bogaard P, Rifkin SA, van Oudenaarden A, Tyagi S (2008) Imaging individual mRNA molecules using multiple singly labeled probes. Nat Methods 5(10):877–879. https://doi.org/10.1038/nmeth.1253

9. Lubeck E, Coskun AF, Zhiyentayev T, Ahmad M, Cai L (2014) Single-cell in situ RNA profiling by sequential hybridization. Nat Methods 11(4):360–361. https://doi.org/10.1038/nmeth.2892

10. Wang F, Flanagan J, Su N, Wang LC, Bui S, Nielson A, Wu X, Vo HT, Ma XJ, Luo Y (2012) RNAscope: a novel in situ RNA analysis platform for formalin-fixed, paraffin-embedded tissues. J Mol Diagn 14(1):22–29. https://doi.org/10.1016/j.jmoldx.2011.08.002

11. Beliveau BJ, Boettiger AN, Avendaño MS, Jungmann R, McCole RB, Joyce EF, Kim-Kiselak C, Bantignies F, Fonseka CY, Erceg J, Hannan MA, Hoang HG, Colognori D, Lee JT, Shih WM, Yin P, Zhuang X, Wu CT (2015) Single-molecule super-resolution imaging of chromosomes and in situ haplotype visualization using Oligopaint FISH probes. Nat Commun 6:7147. https://doi.org/10.1038/ncomms8147

12. Kwon S (2013) Single-molecule fluorescence in situ hybridization: quantitative imaging of single RNA molecules. BMB Rep 46(2):65–72. https://doi.org/10.5483/bmbrep.2013.46.2.016

13. Waylen LN, Nim HT, Martelotto LG, Ramialison M (2020) From whole-mount to single-cell spatial assessment of gene expression in 3D. Commun Biol 3(1):602. https://doi.org/10.1038/s42003-020-01341-1

14. Curley AA, Arion D, Volk DW, Asafu-Adjei JK, Sampson AR, Fish KN, Lewis DA (2011) Cortical deficits of glutamic acid decarboxylase 67 expression in schizophrenia: clinical, protein, and cell type-specific features. Am J Psychiatry 168(9):921–929. https://doi.org/10.1176/appi.ajp.2011.11010052

15. Fish KN, Rocco BR, Lewis DA (2018) Laminar distribution of subsets of GABAergic axon terminals in human prefrontal cortex. Front Neuroanat 12:9. https://doi.org/10.3389/fnana.2018.00009

16. Maynard KR, Tippani M, Takahashi Y, Phan BN, Hyde TM, Jaffe AE, Martinowich K (2020) Dotdotdot: an automated approach to quantify multiplex single molecule fluorescent in situ hybridization (smFISH) images in complex tissues. Nucleic Acids Res 48(11):e66. https://doi.org/10.1093/nar/gkaa312

17. Shirley JW, Ty S, Takebayashi S, Liu X, Gilbert DM (2011) FISH finder: a high-throughput tool for analyzing FISH images. Bioinformatics 27(7):933–938. https://doi.org/10.1093/bioinformatics/btr053

Chapter 4

Using Translating Ribosome Affinity Purification (TRAP) to Understand Cell-Specific Translatomes in Pain States

Diana Tavares-Ferreira, Salim Megat, and Theodore J. Price

Abstract

Translating ribosome affinity purification (TRAP) is a method that allows the study of translational changes through identification of mRNAs associated with tagged ribosomes in specific cell types, in vivo. We have used Nav1.8-Cre mice crossed with mice that have a floxed L10a-GFP fusion. Because L10a protein is associated with translating ribosomes, this crossing generates mice with tagged ribosomes exclusively expressed in Nav1.8-positive neurons, most of which are nociceptors. These mice give new insight into the functional properties of nociceptors and how the translatome of these cells changes following injury. Herein, we describe a detailed protocol that can be used to identify differentially translated mRNAs and study translational control mechanisms in nociceptors in vivo. This protocol can be used to investigate nociceptor translatomes in other chronic pain contexts, and it can be applied to other cell types within sensory ganglia using a variety of genetic lines.

Key words Pain, Ribosome, Translation, Translatome, TRAP, Nociceptors

1 Introduction: Why Use TRAP in Pain Research?

1.1 Transcriptomes Do Not Equal Translatomes: mRNA Levels Do Not Always Equate to Protein Levels

Chronic pain is a complex disease characterized by persistent plasticity in sensory neurons [1–3]. Great progress has been made in characterizing the transcriptome of sensory ganglia and how this changes in chronic pain states [4–11]; however, mRNA abundance is often weakly correlated with protein abundance and this discrepancy can be attributed to translational control mechanisms [12]. Therefore, other techniques are needed to identify mRNAs that are being translated in cells in vivo. Translation control mechanisms are critical for nociceptor plasticity [13], but relatively little is known about the downstream mRNAs associated with ribosomes (the active translatome) that are translated to produce the new proteins that mediate this plasticity [13].

The ribosome (a large ribonucleoprotein complex that in eukaryotes consists of a large (60S) and small (40S) ribosomal subunits) mediates the translation of mRNAs. The mRNA is

Rebecca P. Seal (ed.), *Contemporary Approaches to the Study of Pain: From Molecules to Neural Networks*, Neuromethods, vol. 178, https://doi.org/10.1007/978-1-0716-2039-7_4, © Springer Science+Business Media, LLC, part of Springer Nature 2022

initially recruited by initiation factors to the 40S subunit, which scans the mRNA from 5' to 3' until reaching the initiation codon (initiation step). Then, the 60S subunit joins the complex and the assembled 80S ribosome together with other elongation factors run through the coding region of the mRNA and produce the encoded polypeptide (elongation step). Ribosomal protein L10a is part of the large 60S subunit and is necessary for assembling of the 40S and 60S subunits into a functional 80S ribosome [14, 15]. When reaching the termination codon, peptide chain-release factors cause the release of the nascent polypeptide and separate ribosomes from the mRNA (termination step) [16].

Using the TRAP approach, investigators can stall protein synthesis, immunoprecipitate ribosomes and isolate the associated mRNAs. These mRNAs can then be identified through a number of high-throughput methods, most commonly RNA sequencing (RNA-seq). This methodology permits the study of the translatome in a cell-type specific fashion. We have used the method to more directly investigate the molecules that are potentially driving changes in nociceptor physiology in pain states [17, 18].

1.2 History of Development of the TRAP Method, Advantages, and Limitations

The TRAP approach was first reported by Heiman and colleagues in 2008 to study the complexity and heterogeneity of cell types in the central nervous system (CNS) [19]. It was developed in response to the need for a methodology that can identify genes enriched in specific cells. A transgenic mouse line that expresses enhanced green fluorescent protein (EGFP) fused with the N terminus of RPL10a was created [19]. Among several ribosomal proteins, RPL10a was chosen because it was demonstrated that is present in polysomes using immunoelectron microscopy [19], and it has the advantage of being associated with functional ribosomes [14, 15]. These mice can be crossed with any Cre-line, and by using antibodies that specifically target the EGFP tag, mRNAs associated with ribosomes are isolated from a specific cell type of interest. The mammalian brain, in particular, is composed of diverse cell types and multiple neuronal populations. Using the TRAP methodology Doyle and colleagues have characterized the translatome of 24 CNS cell types [20] demonstrating the potential of TRAP to reveal novel insight into the diversity of neuronal translatomes.

In 2009, Sanz and colleagues reported a similar approach in which a transgenic mouse, called RiboTag, that has a hemagglutinin (HA) tag associated with the ribosomal protein Rpl22 [21]. When crossed with a cre-driver in a particular mouse Cre-line, expression of HA in ribosomes is activated. Immunoprecipitation of HA-tagged ribosomes, and isolation of the associated mRNA, in specific cell types can be obtained by using a monoclonal antibody against HA, in similar way as EGFP tag. Theoretically, either approach can be used to characterize the translatome of nociceptors in vivo. We chose to use the L10a-EGFP approach, and the method described here is specialized for that mouse line.

The biggest advantage of TRAP is isolation of mRNA from translating ribosomes in specific cell types. However, unlike other techniques such as laser capture microdissection (LCM) or fluorescence-activated cell sorting (FACS), it does not damage cell integrity and, therefore, better preserves RNA quality [22]. TRAP is a simple and rapid strategy to purify translating ribosomes and the associated mRNA without any specialized equipment. In addition, it can be applied to any Cre-line to study any cell population. One limitation is the availability of a Cre-line for the study of the cell of interest, but there are increasingly more lines available [23, 24]. Another limitation of this methodology is that it cannot be applied to the study of human tissues, as it requires a genetic component to generate cell specificity [23].

1.3 Transgenic Mice Versus Viral Vector Approaches

The transgenic mouse approach requires the crossing of two different mouse lines (TRAP and a specific Cre-line) to breed mice that express EGFP only in the cell type of interest. Currently, three TRAP or related lines are available: (1) Rosa26^{fsTRAP} mice wherein the Rosa26 locus is targeted with a construct containing a strong CAG promoter followed by a floxed neomycin resistance cassette, a transcriptional stop signal, and EGFP-L10a cDNA (CAG:fs:EGFP-L10a) [24]; (2) BacTRAP. Bacterial artificial chromosome (BAC, an F-plasmid-based tool that allows the combination of inserts into the host genome). BacTRAP is prepared by inserting an EGFP-L10a transgene, which will replace the original gene [19]; (3) Ribo-Tag has an Rpl22 allele with a floxed wild-type C-terminal exon followed by an identical C-terminal exon that has three copies of the hemagglutinin (HA) epitope inserted before the stop codon [21].

A Cre-dependent adeno-associated virus expressing an EGFP-tagged ribosomal protein (AAV-FLEX-EGFPL10a) is available and can be used to isolate mRNAs associated with ribosomes using TRAP methodology [22]. The use of viral vectors has the advantage of bypassing the need to cross different mouse strains which can be time consuming and costly. However, it is not known if this technique leads to equally robust expression of tagged ribosomes in cell types of interest. Again, we have focused our efforts on using a purely transgenic approach, and this will be the focus of the methods described here.

1.4 TRAP Versus Whole Tissue Translation Profiling

The transcriptomic or translatomic analysis of the CNS or PNS presents a challenge due to the heterogeneity of cells present. RNA-seq is now widely used to profile nervous system tissue transcriptomes but whole tissue RNA-seq has the shortfall of lacking cell-type specificity. Single-cell RNA-seq largely overcomes this issue and is now widely applied in the study of nervous system form and function [25, 26]. However, single-cell RNA-seq still does not give insight into the translational profile of these cells, and this is an important issue because mRNA levels do not necessarily correspond to protein levels [13]. Another recent advance in translational profiling of tissues that does not require genetic tagging is called ribosome profiling (or ribosome footprinting). This technique is a modification of the RNase protection assay that is scaled for compatibility with RNA-seq. The technique allows for the isolation of ribosome-protected mRNAs from whole tissues which can then be profiled with RNA-seq [27, 28]. This approach has recently been applied to studying translational changes that occur in the DRG following nerve injury. Multiple mRNAs involved in the ERK pathway were found to be differentially translated in response to sciatic nerve injury (SNI) in DRG, providing an overview of translational changes in whole DRG [29]. A shortcoming of the technique is that it does not allow for cell specificity and may mask significative changes that arise from small-cell populations within a tissue like the DRG [30]. The TRAP methodology, as described above, overcomes this challenge because it provides precise control of cellular expression of the tagged ribosomes.

1.5 TRAP Allows for the Study of Translation Regulation in Nociceptors in Vivo

At least two groups have now applied TRAP to the study of peripheral nervous system (PNS) [17, 18, 31]. As already noted above, this methodology can be used with any Cre-line to target the cell type of interest. The DRG (where the cell bodies of nociceptors are located) consists of several types of neurons, satellite glial cells, macrophages, Schwann cells, and fibroblasts, all of which could be translationally profiled using this technology with appropriate Cre-lines. Several Cre-lines specific to neuronal cell types in the DRG have been created (e.g., Nav1.8-Cre, TRPV1-cre, PIRT-Cre), and others are available that label particular cells in the DRG but lack DRG-specific expression profiles. We have focused our efforts on Nav1.8 because it is expressed in most nociceptors, and therefore, Nav1.8-cre is an ideal line to cross with TRAP mice and study translation regulation in nociceptors. We have thoroughly characterized EGFP expression in the DRG of Nav1.8-TRAP mice demonstrating that EGFP is expressed only in neuronal cells, most of which are small to medium diameter ([17, 18] Fig. 1).

The following section gives a detailed protocol for using TRAP on DRG with the Nav1.8-based approach. This protocol varies slightly from the protocol we have described previously [17, 18] because we have further optimized the technique to allow for isolation of tagged ribosomes and RNA-seq from single animals rather than from pools of animals. This advancement requires fewer animals, reduces costs and opens up the possibility of using this technology in chronic pain models where only a select number of DRGs are injured (e.g., the spared nerve injury model).

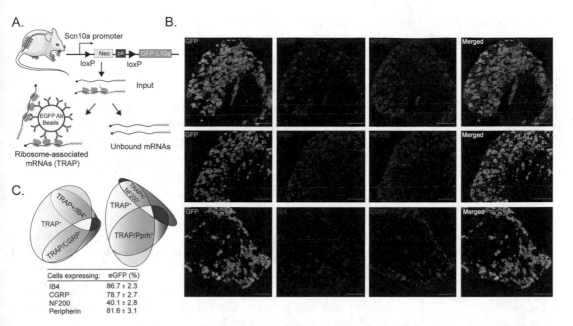

Fig. 1 Expression of the fused EGFP-L10a protein in Nav1.8-positive nociceptors. (**a**) Schematic representation of TRAP shows isolation of translating ribosomes (IP) from input using anti-eGFP-coated beads. (**b, c**) Immunostaining for CGRP, IB4, NF200, and TRPV1 on L4-DRG sections from Nav1.8cre/TRAPfloxed mice. A total of 9528, 8258, and 8526 neurons were counted for CGRP, NF200, and Prph staining, respectively; $n = 3$ mice/group. Scale bars, 100 μm. (Figure and legend taken from Megat et al., 2019 [17])

2 Materials

2.1 Reagents

All reagents should be acquired nuclease-free certified or autoclaved (or filter-sterilized) when applicable before using. All instruments and lab space should be cleaned with 70% ethanol and RNaseZap. Only certified nuclease-free pipette tips and microcentrifuge tubes should be used. It is, however, possible to use autoclaved ones. A key to this procedure is ensuring strict RNase-free conditions throughout the protocol.

2.1.1 Magnetic Beads and Specific Antibodies

- Dynabeads™ Protein G for Immunoprecipitation (Invitrogen, 10004D).

- GFP antibodies (Memorial Sloan-Kettering Monoclonal Antibody Facility; clone names: Htz-GFP-19F7 and Htz-GFP-19C8, bioreactor supernatant purity). *It is recommended to order a lot big enough to cover all planned purifications within the same TRAP experiment and to save the antibodies in single-experiment aliquots [23].*

2.1.2 *Ribosome Stalling*
Reagents/Protein Synthesis
Inhibitors

- Cycloheximide, CHX (Sigma-Aldrich, C7698-5G): protein synthesis inhibitor in eukaryotes (inhibits translation elongation). CHX is toxic and hazardous for the environment [23]. Preparation: dissolve 100 mg of CHX in 1 mL of methanol. CHX should be prepared fresh every time.

- Emetine dihydrochloride hydrate (MedChem Express, HY-B1479B): inhibits protein synthesis by irreversibly blocking ribosome movement along the mRNA. Emetine powder should be kept at -20 °C. Preparation: dissolve emetine in DMSO to a concentration of 20 mg/mL. Emetine in DMSO is stable for 1 month at -20 °C or 6 months at -80 °C.

2.1.3 *Protein Solubilizing*
and Stabilizing Reagents

- Nonylphenyl polyethylene glycol 10%, NP-40 (Abcam, ab142227): detergent; used in cell lysis and to extract and solubilize proteins.

- DL-Dithiothreitol, DTT (Sigma-Aldrich, D9779): reducing agent; reduces protein disulfide bonds (more stable bonds). DTT is a harmful irritant [23].

- 1,2-Diheptanoyl-sn-glycero-3-phosphocholine, DHPC (Avanti Polar Lipids, 850306P). DHPC preserves the native conformation of solubilized proteins. The powder should be stored at -20 °C and thawed to room temperature before opening to reconstitute. After reconstitution, it is good for 7 days at 4 °C [23].

- Protease Inhibitor Cocktail (Sigma-Aldrich, P8340).

2.1.4 *RNA-Related*
Reagents

- Recombinant RNasin Ribonuclease Inhibitor 40 U/µL (Promega, N2515): inhibits most eukaryotic RNases.

- Trizol reagent (Invitrogen, 15596026).

- Direct-zol RNA Microprep (R2060, Zymo Research).

- Turbo DNase 2 U/µL (Invitrogen, AM2238).

- RNaseZap™ RNase Decontamination Solution (Invitrogen, AM9782).

2.1.5 *Stock Solutions*

Stock solutions can be bought nuclease-free certified, if commercially available. In our lab, we prepare the stock solutions using nuclease-free water and autoclave or filter-sterilize them, as appropriate, before use.

- UltraPure™ DNase/RNase-Free Distilled Water (Invitrogen, 10977015). *Water used throughout the whole experiment.*

- 1 M HEPES (powder, Sigma, H4034).

- 1 M $MgCl_2$ (powder, Sigma, M8266) (*see* **Note 1**).

- 2 M KCl (powder, Fisher Chemical, P217).
- 1 M Glucose (powder, Sigma-Aldrich, G7528).
- 1× HBSS (Gibco, 14170112).

2.2 Buffers

Buffers can be kept at 4 °C for several months [23]. Some chemicals should, however, be added immediately before experiment as described below.

- Lysis buffer: mix 20 mM HEPES-NaOH [pH 7.4], 12 mM $MgCl_2$, 150 mM KCl, in RNase free water. Immediately before use add: 0.5 mM DTT, Protease Inhibitors Cocktail (10 μL), 100 μg/mL CHX, 20 μg/mL Emetine, 2 U/mL Turbo DNAse, and 80 U/mL Promega Rnasin (*see* **Note 2**). *Do not vortex after adding Rnasin.*

- Dissection Buffer: mix 1× HBSS, 2.5 mM HEPES–NaOH (pH 7.4), 35 mM Glucose and 5 mM $MgCl_2$. Immediately before use add 100 μg/mL CHX, and 0.2 mg/mL emetine.

- 0.15 M KCl IP Wash Buffer: mix 20 mM HEPES–NaOH (pH 7.4), 12 mM $MgCl_2$, 150 mM KCl, and 1% NP-40 in RNAse-free water. Immediately before use add 0.5 mM DTT and 100 μg/mL CHX.

- 0.35 M KCl IP Wash Buffer: mix 20 mM HEPES–NaOH (pH 7.4), 12 mM $MgCl_2$, 350 mM KCl, 1% NP-40 in RNAse-free water. Immediately before use, add 0.5 mM DTT and 100 μg/mL CHX.

2.3 Equipment

- DFP Flashlight with royal blue and green LEDs (Electron Microscopy Sciences, Cat. # DFP-1).

- Precellys® Minilys Tissue Homogenizer (Bertin-instruments, P000673-MLYS0-A).

- Precellys Tissue homogenizing CKMix—2 Ml (Bertin Instruments, Cat. No.: P000918-LYSK0-A) (*see* **Note 3**).

- Dumont #5 Forceps—Biologie/Dumoxel (Fine Science Tools, 11252-30). *For DRG dissection, it is important to use sharp forceps, as they make it easier to pull out the DRGs*.

- Small scissors Extra Fine Bonn Scissors—Straight/8.5 cm (Fine Science Tools, 14084-08). *Important for an easier and cleaner hemisection of the spinal column*.

- Dumont SS Fine Forceps (Fine Science Tools, 11200-33).

- Large scissors.

- DynaMag-2 magnet (Invitrogen, 12321D).

- NanoDrop 2000C spectrophotometer (Thermo Scientific, ND-2000C).

- AATI Fragment Analyzer.

2.4 Transgenic Animals: Nav1.8(het)-TRAP(EGFP-Homo) Mice

Rosa26^{fsTRAP} mice can be purchased from The Jackson Laboratory (stock #022367). Transgenic mice expressing Cre-recombinase under the control of the Scn10a (Nav1.8) promoter can be obtained from Infrafrontier (EMMA ID: 04582). Characterization of these mice confirmed that the introduction of the Cre-recombinase in heterozygous animals does not affect pain behavior, and their DRG neurons have normal electrophysiological properties [17, 32]. Rosa26^{fsTRAP} mice were crossed with Nav1.8-cre to generate the Nav1.8-TRAP mice that express a fused EGFP-L10a protein in Nav1.8-positive neurons [17, 18]. Mice that are going to be used for TRAP experiments should be genotyped before any procedure in order to make sure the right genotyping was obtained. In addition, it is recommended to use the DFP Flashlight, when dissecting, to confirm the genotyping.

3   Methods

3.1 Protocol Optimization

We have optimized the TRAP protocol for use in DRG tissue by changing the homogenization method. The new homogenization method (described below) was developed in order to increase the RNA yield and not have to use multiple animals per replicate. This also gives for more options in downstream applications because of the higher RNA yield obtained.

3.2 Preparation of Beads and Antibodies (Affinity Matrix)

- Resuspend Dynal Protein G magnetic beads thoroughly.
- Transfer 200 μL beads into an autoclaved microcentrifuge tube at room temperature.
- Place tube in Dynal MPC-S magnet for 1 min to separate the beads from the solution.
- Pipet off supernatant.
- Remove tube from magnet and resuspend beads in 1000 μL 0.15 M KCl IP wash buffer; repeat washing procedure three times.
- Add 100 μg total of anti-GFP (50 μg 19C8 and 50 μg 19F7) antibody to the tube of beads in 200 μL of 0.15 M KCl buffer.
- Incubate with slow end-over-end mixing for 2 h at RT.
- At the end of 2 h, place tube on the magnet and remove supernatant.
- Rinse with 1000 μL of wash buffer.
- Resuspend beads in 200 μL 0.15 M KCl IP wash buffer and store at 4 °C (see **Notes 4** and **5**).

3.3 Tissue (DRG) Dissection

- Clean bench and tools with 70% ethanol and RnaseZap.
- Anesthetize mice with isoflurane and decapitate the mouse. Using the DFP Flashlight, confirm EGFP is expressed and mouse has the correct genotyping (for instance, by flashing the light directly on a tissue that is expected to express EGFP).
- Perform hand dissection of DRGs and place tissue in dissection buffer on ice (15-mL tube) (*see* **Note 6**). *Limit dissection time to maximum of 10 min. Time taken to complete DRG dissection is crucial for this technique to work. If DRGs need to be pooled from multiple animals, one person per mouse is required for dissection. It is important to avoid blood contamination as much as possible since blood contains RNAses.*.
- Centrifuge the 15-mL tube with the collected DRGs in dissection buffer for ~1 min and transfer DRGs into the Precellys homogenizing CKMix tube, in the cold room, using the long Dumont SS Fine forceps.

3.4 Tissue Homogenization

This step (in addition to the time taken to perform DRG dissection) is crucial. It is important to choose a method that is able to break the cell membranes and release the mRNAs and ribosomes (but at the same time keep the integrity and quality of both RNA and ribosomes).

- Homogenize at 10-s intervals (medium speed) for 10 s, in the cold room, in Precellys homogenizing CKMix tube (repeat seven more times).
- Prepare a post-nuclear supernatant (S2) by centrifugation at $2000 \times g$ in the cold room (4 °C) for 5 min (rpm 4).
- Transfer supernatant into a new microcentrifuge tube on ice.
- Add 1/9 sample volume 10% NP-40 to S2 (final concentration = 1%); mix gently by inversion of tube.
- Add 1/9 sample volume 300 mM DHPC (final concentration = 30 mM); mix gently by inversion and incubate on ice for 5 min.
- Prepare post-mitochondrial supernatant (S20) by centrifugation at 4 °C, 10 min, $15,000 \times g$ (rpm 14, max speed).
- Take supernatant (S20) into a new microcentrifuge tube on ice and proceed with immunoprecipitation (IP).
- Save aliquot of 200 μL for input and add 1 μL of DNase into the input.

3.5 Incubation with Affinity Matrix

- Add IP fraction to affinity matrix (after removing supernatant buffer).

- Pipet up and down slowly to resuspend beads in lysate/IP fraction; *AVOID BUBBLES*.
- Incubate at 4 °C for 3 h with end-over-end mixing in the cold room (*see* **Note 7**).
- At the end of 3 h incubation, collect affinity matrix with magnet on an ice bucket (keep affinity matrix cold) for 1 min.
- Remove/save supernatant (unbound fraction) (*see* **Note 8**).
- Resuspend affinity matrix in 1000 μL of 0.35 M KCl IP wash buffer and collect with magnet as above.
- Repeat wash three more times.

3.6 RNA Extraction

- After the last wash, remove all wash buffer from affinity matrix on magnet at room temperature.
- Resuspend affinity matrix in Trizol (1:3, affinity matrix:Trizol) at room temperature. We add 300 μL of Trizol to the affinity matrix, which we estimate to be 100 μL.
- Add Trizol (1:3, Input:Trizol) to the input fraction. We add 600 μL of Trizol to the input fraction which is 200 μL saved from Subheading 3.4 last step.
- Let it sit for 10 min at room temperature.
- At the end of 10 min, vortex affinity matrix for 30 s and quickly spin down before next step.
- Place affinity matrix on magnet for 1 min.
- Collect the supernatant that has the RNA eluted.
- Proceed to RNA extraction and follow RNA extraction kit protocol (*see* **Note 9**).

3.7 Direct-Zol RNA Extraction Kit Protocol

- Add equal volume of ethanol (95–100%) to IP/TRAP and INPUT fractions and mix thoroughly.
- Transfer the mixture into a Zymo-Column in a collection tube and centrifuge (10–16,000 × *g* for 30 s).
- Transfer the column into a new collection tube and discard the flow through.
- Add 400 μL Direct-zol RNA PreWash to the column and centrifuge.
- Discard the flow through and repeat this step.
- Add 700 μL RNA Wash Buffer to the column and centrifuge for 2 min.
- Centrifuge again for an additional 30 s to ensure complete removal of the wash buffer.

- Transfer the column carefully into an autoclaved RNase-free tube.

- To elute RNA, add 15 μL of UltraPure DNase/RNase-free water directly to the column matrix and centrifuge for 2 min at maximum speed.

- Measure RNA quality and quantity with NanoDrop and Fragment Analyzer (or qPCR for quality control).

- Store RNA at −80 °C.

3.8 Downstream Processing

RNA isolated using TRAP methodology can have several applications (e.g., qPCR, microarray, RNA sequencing). The choice is mostly dependent on the RNA yield, which is, furthermore, based on the abundance of the cell of interest [23] and homogenization success. Using the method described here, we usually get above 30 ng/μL for the IP/TRAP fraction from one mouse (collecting all cervical, thoracic, and lumbar DRGs). When proceeding with RNA-sequencing in particular, RNA samples need to be checked on the Fragment Analyzer (or Bioanalyzer). We usually get an RQN of 7 or above for The IP/TRAP samples. This demonstrates that the TRAP approach and homogenization method here described produce RNA of good quality (*see* Fig. 2). Performing TRAP in Nav1.8-TRAP mice should show an enrichment of neuronal markers and a de-enrichment of non-neuronal markers in the TRAP fraction (*see* Fig. 3).

Previously we did QuantSeq 3′ mRNA sequencing library preparation as it was required for the lower amount of RNA that was isolated using the previous method [17, 18]; but with the method described here, there are more options for library preparation because the RNA yield is greater. We are now able to do, for instance, standard Illumina total RNA sequencing as we usually get a yield higher than 100 ng (which is the minimum required). Our samples obtained using this optimized protocol are handed to

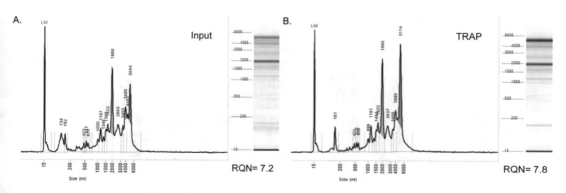

Fig. 2 Fragment analyzer. Fragment analyzer reports show the RNA quality for the Input (**a**) and TRAP/IP (**b**) fractions

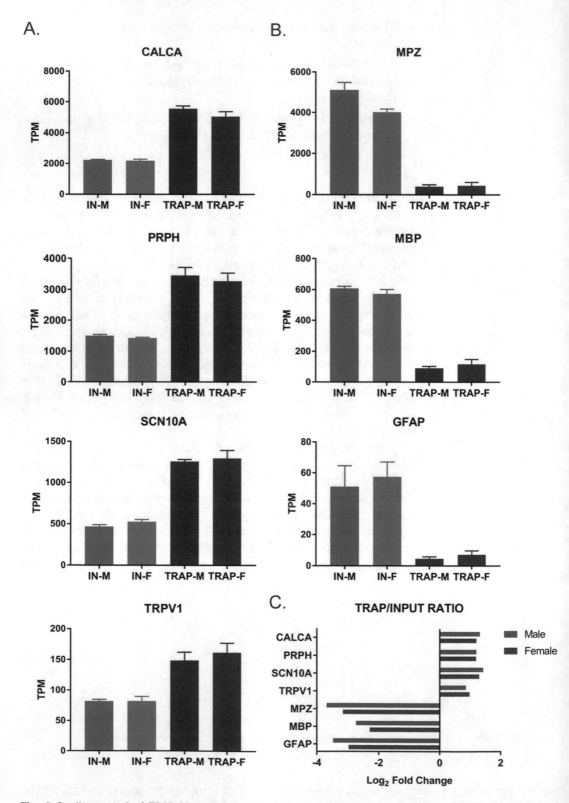

Fig. 3 Quality control of TRAP Methodology. (**a**, **c**) Enrichment of neuronal markers CALCA, SCN10A, and TRPV1 in TRAP fractions. (**b**, **c**) De-enrichment of non-neuronal markers in TRAP fractions

our Core facility (Genome center at University of Texas at Dallas) for cDNA library preparation (total RNA, including ribosomal RNA depletion) and sequencing (high-output, single end with 75 bp of read length). When submitting samples for sequencing, investigators should consider batch effects. Batch effects are a primary source of variation that may affect the RNA sequencing data [33, 34], so the investigator should plan in advance in order to avoid identification of potential false biological changes in gene expression. Ultimately, best practices for sequencing and analysis of RNA samples should be used [35]. For instance, if many samples need to be sequenced and several batches are required, it is important to include controls and randomizing the samples [35, 36].

4   Notes

1. Magnesium is essential for the stabilization of ribosomes [23]. Use only EDTA-free chemicals and solutions. EDTA will chelate magnesium and prevent a good immunoprecipitation [23].

2. Different RNase inhibitor concentrations can be tested. When there are extreme cases of RNase activity, heparin (0.75 mg/mL) can be added to the lysis buffer [23]. However, it is required to do lithium chloride precipitation as heparin inhibits downstream enzymes [23, 37, 38].

3. There are different homogenization materials available that can be adapted to each tissue. Investigators should use the most adequate homogenization technique to the tissue they are studying. The one described in this chapter was the one that gave better outcomes when homogenizing DRG and allowed the use of one mouse per replicate.

4. The affinity matrix can be used immediately or it can be kept for up to 2 weeks at 4 °C by adding 0.02% sodium azide [23]. In that case, the pre-made affinity matrix should be rapidly washed three times in the 0.15 M KCl buffer before use [23]. It may be more challenging to resuspend the pre-made affinity matrix and it is, therefore, recommended to keep it rotating overnight on a tube rotator [23]. Do not vortex the affinity matrix after incubating with the antibody [23].

5. For specific tissues, the investigator can also prevent nonspecific binding by blocking with 5% BSA (nuclease-free). For instance, tissues rich in myelin produce higher nonspecific RNA background and make it more challenging to perform TRAP [23].

6. Quick tissue dissection is a key factor for the success of this technique (even more so when studying PNS). It is important

to dissect the tissue as fast as possible and keeping it in dissection buffer until ready to homogenize. Recommended dissection method for fast DRG extraction:

- Anesthetize the mouse with isoflurane.
- Lay the mouse on its belly on a blue pad and check for a paw reflex.
- Proceed with decapitation by cutting promptly at the edge of the mouse neck with big scissors. Let the mouse bleed.
- Spray the fur of the back with 70% ethanol and cut off the skin.
- At the bottom of the spinal vertebrae, cut near the ilium bone.
- Cut along both sides of the spinal column and remove anything that may be adhering to the column.
- After the spinal column has been removed, cut any extra tissue on the dorsal side so that the spinal column is exposed. *This is important for a good hemisection of the spinal column.*.
- Hold the spinal column, so that the dorsal side faces up, and place one tip of the small scissors in the opening containing the spinal cord. Make a cut along the dorsal side of the spinal column; try to cut as straight as possible and try to go along the spinal cord without taking out the scissors. Then, cut along the spinal cord on the ventral side to complete the hemisection. Place one of the two sections in a tube with dissection buffer. This tube will be used to temporarily store sections of the spinal column.
- With the other section, carefully remove the spinal cord and using sharp forceps (Dumont #5) to pull out the DRGs and place them in another tube with the dissection buffer. This tube (15 mL) will be used as collection buffer. *Do not let the DRGs dry out.*
- As DRGs can be difficult to spot, it is helpful to anticipate the location of the DRGs. *A good spinal column hemisection will make this easier.* Magnifying glasses or microscope are also recommended.*

7. Different incubation times of affinity matrix and lysate can be tested and compared for best outcome.

8. The unbound fraction can be saved for comparison purposes, when optimizing the protocol.

9. Different RNA extraction protocols can be used, we use the direct-zol method described in Subheading 3.7.

Final considerations that are most critical for a successful TRAP:

Best laboratory practices should be in place to avoid sample contamination (e.g., wear and change gloves frequently). It is crucial to use RNase-free equipment and reagents or autoclaved when applicable. Laboratory space area should be cleaned with RNaseZap. Beads with antibody (affinity matrix) should be resuspended by pipetting up and down slowly with the tube on ice. Samples should be kept ice-cold, and care must be taken to avoid bubble formation. This technique can be performed in the cold room (4 °C) or, at least, by keeping the samples ice-cold at all times (until ready for RNA extraction). Only when ready to add Trizol, should samples be handled at room temperature. A quick tissue dissection and keeping tissue in dissection buffer until ready for homogenization are key factors. In addition, choosing the appropriate homogenization method and proper lysis buffer (with magnesium, RNase inhibitor and protein synthesis inhibitors) are fundamental to obtain good results.

5 What Have We Learned from the TRAP Methodology?

TRAP gives the advantage of cell-type specificity and captures translation efficiency, making it a very useful approach to study cellular populations in different chronic pain states. To our knowledge, the Fainzilber [31] and our laboratory [17, 18] are thus far the only groups that have used this technology to better understand the physiology of nociceptors and/or to understand how these cells change their phenotype as chronic pain develops. One goal of this chapter is to make this procedure as clear as possible, so that other investigators can utilize it to gain further insight into their own research directions.

Our laboratory's first foray into use of the TRAP technique was to apply this technology to the study of neuropathic pain [17]. Previous work from our group provided strong evidence that translation regulation signaling is a key contributor to persistent pain, including neuropathic pain [13, 39, 40], but our work had not yet identified specific mRNAs that displayed increase translation in a chronic pain state. We used paclitaxel treatment to produce a model of chemotherapy-induced peripheral neuropathy (CIPN) and extracted DRGs from neuropathic and vehicle treated mice to characterize the nociceptor translatome in CIPN. One of our first observations was that this model produced very few changes in the overall DRG transcriptome, but there were quite substantial changes in the nociceptor translatome [17]. Many of the mRNAs with enhanced ribosome association, almost certainly reflecting enhanced mRNA translation, were previously described genes in the CIPN literature. These included caspase 6 [41] and TRPV1

[42], among others. The fact that so many previously identified genes from the CIPN literature were so clearly identified as increased using the TRAP technique demonstrates the power of the technique for target discovery and also suggests that translation regulation plays a key role in many of the pathways and targets that have previously been identified in CIPN.

In our work using nociceptor TRAP in the CIPN model, we also discovered that eukaryotic initiation factor (eIF) 4E phosphorylation (an effect mediated by mitogen-activated protein kinase interacting kinase or MNK) regulates *Rraga* mRNA translation, which in turn controls mechanistic target of rapamycin complex 1 (mTORC1) activation [17]. This is important because *Rraga* encodes a protein, called RagA that is required for mTORC1 activation. The protein operates as a GTPase that bridges lysosomal amino acid efflux to mTORC1 activation and null mutation of RagA leads to a complete loss of mTORC1 activity in cells [43, 44]. The involvement of mTORC1 in neuropathic pain has been known for some time [45, 46], and we had recently reported on a role for MNK-eIF4E signaling in several types of chronic pain [40, 47, 48]; however, the link between these four proteins (MNK-eIF4E, RagA, and mTORC1) in CIPN was a direct and unexpected result of using the TRAP technique. Correspondingly, we found that inhibition of MNK activity with the small molecule kinase inhibitor eFT508 reduced spontaneous activity in DRG nociceptors collected from mice treated with paclitaxel and reversed hyperalgesic behavior and spontaneous pain. Collectively, this work demonstrated that MNK-eIF4E signaling is a potential target for drug development to treat neuropathic pain [17]. eFT508 (Tomivosertib) is currently entering phase 3 clinical trials for solid tumors and lymphoma [49]. Successes in these trials could lead to approval of a new cancer therapy that can also be used for CIPN. The mechanistic underpinnings of this insight were discovered using the TRAP methodology.

Interestingly, a recent study from the Fainzilber laboratory used RiboTag mice crossed with different Cre-lines (sensory neuron specific: advilin-cre and Isl1-cre; proprioceptive neuron specific: Runx3-cre; and satellite glial cell specific: Dhh-cre) to characterize changes in the translatome of DRG cells after peripheral nerve injury. Their key finding was that nerve injury induced changes in translation for mRNAs enriched with the eIF4E-responsive cytosine-enriched regulator of translation (CERT) motif [31]. This finding highlights the key role that eIF4E plays in regulating translation in the setting of cellular plasticity in the DRG. It is possible that eIF4E, independently of phosphorylation, controls the injury and regeneration response whereas eIF4E phosphorylation plays a critical role in neuronal hyperexcitability driving persistent pain.

We have also used the TRAP technique to compare the translatomes of trigeminal ganglia (TG) and DRG nociceptors. These nociceptors have high functional similarity, but there are important differences [50, 51]. Unexpectedly, we found that mRNAs involved in the regulation of the mTORC1 pathway had higher translation efficiency in the TG compared to DRG [18]. These mRNAs included the mRNA encoding the key regulator of mTORC1 function, RagA. Based on these findings, and the long-standing evidence that increased mTORC1 activity increases nociceptor excitability, we sought to find functional evidence for increased nociceptor excitability in the TG. We used the capsaicin test to assess this possibility. Capsaicin injection into the cheek or hindpaw produced far higher facial grimace scores and nocifensive behaviors in the innervation area of the TG. This effect was reversed by an mTORC1 inhibitor, suggesting a direct link between increased mTORC1 activity and greater nociceptor excitability in the mouse TG [18].

Using the TRAP technique to compare TG and DRG nociceptor translatomes, we also were able to provide independent confirmation of previous findings. A recent study used fluorescence activated cell sorting (FACS) to investigate potential differences between TG and DRG transcriptomes [7]. They found that more than 99% of mRNAs had consistent expression between TG and DRG neurons [7], identifying only 24 mRNAs with differential expression between the two tissues [7, 18]. Differentially expressed genes included Hox genes, which were also found using the TRAP technique [7, 18]. Importantly, again, the TRAP technique revealed differences that were not evident when studying the whole transcriptome of these cells.

We have optimized the TRAP methodology for sensory ganglion (DRG and TG [17, 18]), which we hope will make this technique accessible to additional investigators who want to use this powerful technology to study the physiology and remarkable plasticity of these cells. The technique can also be applied to other areas of pain research to reveal novel insights. One area where we anticipate that TRAP could be advantageously used is for profiling of microglial cells in the spinal cord after nerve injury [52–54]. Profound sex differences have been found in the potential contribution of these cells to neuropathic pain, but the underpinning mechanisms, especially in females, are obscure. Application of TRAP to this pressing question could reveal new insight needed to advance the field.

Acknowledgments

We thank the Genome Center at The University of Texas at Dallas for the services to support our research.

Funding Sources: This work was supported by NIH grants R01NS065926 (TJP), R01NS102161 (TJP).

References

1. Woolf CJ, Salter MW (2000) Neuronal plasticity: increasing the gain in pain. Science 288(5472):1765–1768

2. Price TJ, Gold MS (2018) From mechanism to cure: renewing the goal to eliminate the disease of pain. Pain Med 19(8):1525–1549; Epub 2017/10/28

3. Gold MS, Gebhart GF (2010) Nociceptor sensitization in pain pathogenesis. Nat Med 16(11):1248–1257

4. Kogelman LJA, Christensen RE, Pedersen SH, Bertalan M, Hansen TF, Jansen-Olesen I et al (2017) Whole transcriptome expression of trigeminal ganglia compared to dorsal root ganglia in Rattus Norvegicus. Neuroscience 350: 169–179; Epub 2017/04/01

5. LaPaglia DM, Sapio MR, Burbelo PD, Thierry-Mieg J, Thierry-Mieg D, Raithel SJ et al (2018) RNA-Seq investigations of human post-mortem trigeminal ganglia. Cephalalgia 38(5):912–932; Epub 2017/07/13

6. Li CL, Li KC, Wu D, Chen Y, Luo H, Zhao JR et al (2016) Somatosensory neuron types identified by high-coverage single-cell RNA-sequencing and functional heterogeneity. Cell Res 26(1):83–102; Epub 2015/12/23

7. Lopes DM, Denk F, McMahon SB (2017) The molecular fingerprint of dorsal root and trigeminal ganglion neurons. Front Mol Neurosci 10:304; Epub 2017/10/12

8. Manteniotis S, Lehmann R, Flegel C, Vogel F, Hofreuter A, Schreiner BS et al (2013) Comprehensive RNA-Seq expression analysis of sensory ganglia with a focus on ion channels and GPCRs in trigeminal ganglia. PLoS One 8(11): e79523

9. North RY, Li Y, Ray P, Rhines LD, Tatsui CE, Rao G et al (2019) Electrophysiological and transcriptomic correlates of neuropathic pain in human dorsal root ganglion neurons. Brain 142(5):1215–1226; Epub 2019/03/20

10. Thakur M, Crow M, Richards N, Davey GI, Levine E, Kelleher JH et al (2014) Defining the nociceptor transcriptome. Front Mol Neurosci 7:87; Epub 2014/11/27

11. Ray P, Torck A, Quigley L, Wangzhou A, Neiman M, Rao C et al (2018) Comparative transcriptome profiling of the human and mouse dorsal root ganglia: an RNA-seq-based resource for pain and sensory neuroscience research. Pain 159(7):1325–1345; Epub 2018/03/22

12. Tian Q, Stepaniants SB, Mao M, Weng L, Feetham MC, Doyle MJ et al (2004) Integrated genomic and proteomic analyses of gene expression in mammalian cells. Mol Cell Proteomics 3(10):960–969

13. Khoutorsky A, Price TJ (2018) Translational control mechanisms in persistent pain. Trends Neurosci 41(2):100–114; Epub 2017/12/19

14. del Campo EM, Casano LM, Barreno E (2013) Evolutionary implications of intron–exon distribution and the properties and sequences of the RPL10A gene in eukaryotes. Mol Phylogenet Evol 66(3):857–867

15. Eisinger DP, Dick FA, Trumpower BL (1997) Qsr1p, a 60S ribosomal subunit protein, is required for joining of 40S and 60S subunits. Mol Cell Biol 17(9):5136–5145

16. Jackson RJ, Hellen CUT, Pestova TV (2010) The mechanism of eukaryotic translation initiation and principles of its regulation. Nat Rev Mol Cell Biol 11:113

17. Megat S, Ray PR, Moy JK, Lou TF, Barragan-Iglesias P, Li Y et al (2019) Nociceptor translational profiling reveals the ragulator-rag GTPase complex as a critical generator of neuropathic pain. J Neurosci 39(3):393–411; Epub 2018/11/22

18. Megat S, Ray PR, Tavares-Ferreira D, Moy JK, Sankaranarayanan I, Wanghzou A et al (2019) Differences between dorsal root and trigeminal ganglion nociceptors in mice revealed by translational profiling. J Neurosci 39(35): 6829–6847; Epub 2019/06/30

19. Heiman M, Schaefer A, Gong S, Peterson JD, Day M, Ramsey KE et al (2008) A translational

profiling approach for the molecular characterization of CNS cell types. Cell 135(4):738–748

20. Doyle JP, Dougherty JD, Heiman M, Schmidt EF, Stevens TR, Ma G et al (2008) Application of a translational profiling approach for the comparative analysis of CNS cell types. Cell 135(4):749–762

21. Sanz E, Yang L, Su T, Morris DR, McKnight GS, Amieux PS (2009) Cell-type-specific isolation of ribosome-associated mRNA from complex tissues. Proc Natl Acad Sci 106(33): 13939–13944

22. Nectow AR, Moya MV, Ekstrand MI, Mousa A, McGuire KL, Sferrazza CE et al (2017) Rapid molecular profiling of defined cell types using viral TRAP. Cell Rep 19(3): 655–667

23. Heiman M, Kulicke R, Fenster RJ, Greengard P, Heintz N (2014) Cell type–specific mRNA purification by translating ribosome affinity purification (TRAP). Nat Protoc 9(6):1282

24. Zhou P, Zhang Y, Ma Q, Gu F, Day DS, He A et al (2013) Interrogating translational efficiency and lineage-specific transcriptomes using ribosome affinity purification. Proc Natl Acad Sci 110(38):15395–15400

25. Usoskin D, Furlan A, Islam S, Abdo H, Lonnerberg P, Lou D et al (2015) Unbiased classification of sensory neuron types by large-scale single-cell RNA sequencing. Nat Neurosci 18(1):145–153; Epub 2014/11/25

26. Zeisel A, Hochgerner H, Lonnerberg P, Johnsson A, Memic F, van der Zwan J et al (2018) Molecular architecture of the mouse nervous system. Cell 174(4):999–1014.e22; Epub 2018/08/11

27. Ingolia NT, Brar GA, Rouskin S, McGeachy AM, Weissman JS (2012) The ribosome profiling strategy for monitoring translation in vivo by deep sequencing of ribosome-protected mRNA fragments. Nat Protoc 7(8): 1534

28. Ingolia NT (2016) Ribosome footprint profiling of translation throughout the genome. Cell 165(1):22–33

29. Uttam S, Wong C, Amorim IS, Jafarnejad SM, Tansley SN, Yang J et al (2018) Translational profiling of dorsal root ganglia and spinal cord in a mouse model of neuropathic pain. Neurobiol Pain 4:35–44

30. Kulkarni A, Anderson AG, Merullo DP, Konopka G (2019) Beyond bulk: a review of single cell transcriptomics methodologies and applications. Curr Opin Biotechnol 58: 129–136

31. Rozenbaum M, Rajman M, Rishal I, Koppel I, Koley S, Medzihradszky KF et al (2018) Translatome regulation in neuronal injury and axon regrowth. eNeuro 5(2). https://doi.org/10.1523/ENEURO.0276-17.2018

32. Stirling LC, Forlani G, Baker MD, Wood JN, Matthews EA, Dickenson AH et al (2005) Nociceptor-specific gene deletion using heterozygous NaV1. 8-Cre recombinase mice. Pain 113(1–2):27–36

33. Leigh D, Lischer H, Grossen C, Keller L (2018) Batch effects in a multiyear sequencing study: false biological trends due to changes in read lengths. Mol Ecol Resour 18(4):778–788

34. Auer PL, Doerge R (2010) Statistical design and analysis of RNA sequencing data. Genetics 185(2):405–416

35. Conesa A, Madrigal P, Tarazona S, Gomez-Cabrero D, Cervera A, McPherson A et al (2016) A survey of best practices for RNA-seq data analysis. Genome Biol 17(1):13

36. Crow M, Denk F (2019) RNA-seq data in pain research-an illustrated guide. Pain 160(7): 1502–1504; Epub 2019/06/21

37. Jung R, Lübcke C, Wagener C, Neumaier M (1997) Reversal of RT-PCR inhibition observed in heparinized clinical specimens. BioTechniques 23(1):24–28

38. Gauthier D, Murthy MV (1987) Efficacy of RNase inhibitors during brain polysome isolation. Neurochem Res 12(4):335–339

39. Melemedjian OKAM, Tillu DV, Sanoja R, Yan J, Lark A, Khoutorsky A, Johnson J, Peebles KA, Lepow T, Sonenberg N, Dussor G, Price TJ (2011) Targeting adenosine monophosphate-activated protein kinase (AMPK) in preclinical models reveals a potential mechanism for the treatment of neuropathic pain. Mol Pain 7:70

40. Moy JK, Khoutorsky A, Asiedu MN, Black BJ, Kuhn JL, Barragan-Iglesias P et al (2017) The MNK-eIF4E signaling axis contributes to injury-induced nociceptive plasticity and the development of chronic pain. J Neurosci 37(31):7481–7499

41. Berta T, Perrin FE, Pertin M, Tonello R, Liu YC, Chamessian A et al (2017) Gene expression profiling of cutaneous injured and non-injured nociceptors in SNI animal model of neuropathic pain. Sci Rep 7(1):9367; Epub 2017/08/26

42. Li Y, Adamek P, Zhang H, Tatsui CE, Rhines LD, Mrozkova P et al (2015) The cancer chemotherapeutic paclitaxel increases human and rodent sensory neuron responses to TRPV1 by

activation of TLR4. J Neurosci 35(39): 13487–13500; Epub 2015/10/02

43. Efeyan A, Zoncu R, Chang S, Gumper I, Snitkin H, Wolfson RL et al (2013) Regulation of mTORC1 by the rag GTPases is necessary for neonatal autophagy and survival. Nature 493(7434):679–683; Epub 2012/12/25

44. Efeyan A, Schweitzer LD, Bilate AM, Chang S, Kirak O, Lamming DW et al (2014) RagA, but not RagB, is essential for embryonic development and adult mice. Dev Cell 29(3):321–329; Epub 2014/04/29

45. Terenzio M, Koley S, Samra N, Rishal I, Zhao Q, Sahoo PK et al (2018) Locally translated mTOR controls axonal local translation in nerve injury. Science 359(6382):1416–1421

46. Melemedjian OK, Asiedu MN, Tillu DV, Sanoja R, Yan J, Lark A et al (2011) Targeting adenosine monophosphate-activated protein kinase (AMPK) in preclinical models reveals a potential mechanism for the treatment of neuropathic pain. Mol Pain 7(1):70

47. Moy JK, Khoutorsky A, Asiedu MN, Dussor G, Price TJ (2018) eIF4E phosphorylation influences Bdnf mRNA translation in mouse dorsal root ganglion neurons. Front Cell Neurosci 12:29; Epub 2018/02/23

48. Moy JK, Kuhn J, Szabo-Pardi TA, Pradhan G, Price TJ (2018) eIF4E phosphorylation regulates ongoing pain, independently of inflammation, and hyperalgesic priming in the mouse CFA model. Neurobiol Pain 4:45–50

49. Reich SH, Sprengeler PA, Chiang GG, Appleman JR, Chen J, Clarine J et al (2018) Structure-based design of pyridone–aminal eFT508 targeting dysregulated translation by selective mitogen-activated protein kinase interacting kinases 1 and 2 (MNK1/2) inhibition. J Med Chem 61(8):3516–3540

50. Price TJ, Flores CM (2007) Critical evaluation of the colocalization between calcitonin gene-related peptide, substance P, transient receptor potential vanilloid subfamily type 1 immunoreactivities, and isolectin B4 binding in primary afferent neurons of the rat and mouse. J Pain 8(3):263–272

51. Rodriguez E, Sakurai K, Xu J, Chen Y, Toda K, Zhao S et al (2017) A craniofacial-specific monosynaptic circuit enables heightened affective pain. Nat Neurosci 20(12):1734–1743; Epub 2017/12/01

52. Sorge RE, Mapplebeck JC, Rosen S, Beggs S, Taves S, Alexander JK et al (2015) Different immune cells mediate mechanical pain hypersensitivity in male and female mice. Nat Neurosci 18(8):1081–1083

53. Mapplebeck JCS, Dalgarno R, Tu Y, Moriarty O, Beggs S, Kwok CHT et al (2018) Microglial P2X4R-evoked pain hypersensitivity is sexually dimorphic in rats. Pain 159(9): 1752–1763; Epub 2018/06/22

54. Moriarty O, YuShan T, Sengar A, Salter MW, Beggs S, Walker S (2019) Priming of adult incision response by early life injury: neonatal microglial inhibition has persistent but sexually dimorphic effects in adult rats. J Neurosci 39(16):3081–3093

Chapter 5

Ex Vivo Skin-Teased Fiber Recordings from Tibial Nerve

Katelyn E. Sadler, Teresa N. Patitucci, and Cheryl L. Stucky

Abstract

Ex vivo skin-nerve recordings are an electrophysiological technique used to characterize and measure primary sensory afferent activity. In this procedure, glabrous or hairy rodent skin is dissected along with the innervating peripheral nerve. After transferring the "skin-nerve" preparation to a two-chambered recording bath, the nerve is teased into small bundles in order to isolate single units from which spontaneous and stimulus-evoked activity are recorded using an extracellular microelectrode. Since its original description in the late twentieth century, this technique has been employed in rodents to characterize non-neuronal-to-neuronal cell communication within the epidermis, verify novel ion channel involvement in somatosensory signaling, and measure acute and chronic pain-related changes in primary afferent activity. In this chapter, we discuss the history and common applications of this technique, compare and contrast this technique to similar approaches, and provide a detailed description of the glabrous hindpaw skin and tibial nerve dissection and recording procedure, so that other labs can incorporate this approach into their studies of peripheral somatosensation.

Key words Tibial nerve, Extracellular recording, Teased fiber, Mechanosensation, Pain, Glabrous skin, Primary sensory afferent, Aβ fiber, Aδ fiber, C fiber

---

1 Introduction

Pain is defined as an unpleasant sensory and emotional experience associated with actual or potential tissue damage or described in terms of such damage. As the initial transducers of somatosensory information, primary afferents detect changes in chemical gradients, temperature, and mechanical force associated with tissue damage and subsequently signal these changes to the central nervous system through action potential firing. A promising pain therapeutic strategy involves selective dampening of spontaneous or stimulus-evoked activity in primary afferents without altering processes in the central nervous system, in order to avoid unwanted drug side effects like addiction. Many technical approaches, some of which are discussed at length in the accompanying chapters of this book, are used to study activity in individual primary afferent neurons. In rudimentary in vitro studies, neuronal or

Rebecca P. Seal (ed.), *Contemporary Approaches to the Study of Pain: From Molecules to Neural Networks*, Neuromethods, vol. 178, https://doi.org/10.1007/978-1-0716-2039-7_5, © Springer Science+Business Media, LLC, part of Springer Nature 2022

single-channel activity is measured through calcium imaging or patch-clamp electrophysiology after the cell bodies of afferent neurons are isolated and cultured. Caveats of this approach include direct exposure to exogenous chemicals during the culturing process and the removal of peripheral and central processes ("axotomy") which induces an injury-like transcriptomic phenotype in cell bodies [1]. Ex vivo (also referred to as in vitro) teased fiber recordings which are the focus of this chapter, maintain afferent cytoarchitecture and connectivity in peripheral tissue, but omit the signal processing that occurs at the T-junction [2] (i.e., axon bifurcation at the cell body). Compared to single-cell in vitro approaches, ex vivo recordings also require significantly more technical training and practice to achieve proficiency. More complex ex vivo somatosensory preparations consisting of skin, nerve, dorsal root ganglia (DRG), and spinal cord tissues [3], and in vivo teased fiber recordings [4] maintain the peripheral and central connectivity of individual afferents but require an even higher degree of technical finesse for dissecting out multiple tissues or the added complexity of maintaining the depth of anesthesia for extended periods of time. We encourage readers to consider the caveats of each approach when designing an experiment, and oftentimes, we find that the complementary use of two or more of these techniques provides the most comprehensive answer to an experimental question.

In this chapter, we provide a detailed method for performing teased-fiber tibial nerve recordings in mice (often referred to as "skin-nerve" recordings). The skin is our largest sensory organ and primary detector of environmental stimuli. As such, it is relatively easy to describe skin or somatic sensation with discrete terminology. Frequently used chronic pain descriptors like "allodynia" and "hyperalgesia" also refer to changes in somatic sensation; the former is used when a normally innocuous stimulus is perceived as noxious, and the latter is used when a normally noxious stimulus is perceived as even more painful. Teased-fiber preparations allow for the measurement and correlation of primary afferent activity with changes in sensitivity that can be inferred from behavioral testing. Complementary ex vivo preparations also exist for measuring primary afferent activity during deep tissue stimulation; examples include the bladder-pelvic nerve [5, 6], colon-pelvic, -splanchnic, or -vagus nerve [7, 8], jejunum-vagus nerve [9], plantar muscle-tibial nerve [10], and forepaw muscle-medial or -ulnar nerve preparations (described in next chapter). Collectively, data obtained with these techniques inform our understanding of peripheral afferent activity during tissue stimulation, following tissue damage, or in disease settings.

The first teased-fiber recordings were described in non-mammalian species in the early 1950s; skin-nerve recordings in frogs [11, 12] were followed by recordings in toads [13], salamanders [14], and leeches [15]. Professor Peter Reeh was the first to develop this technique for use in mammals; he published the first teased saphenous nerve recordings from rat and guinea pig in 1986 [16]. The team of Martin Koltzenburg, Cheryl Stucky, and Gary Lewin then adapted this preparation for mouse in 1997 [17]. The saphenous nerve recording variation, described in explicit detail by Zimmerman, Reeh, and colleagues [18], is the most frequently used preparation for measuring hairy skin-associated nerve activity. As illustrated in Fig. 1, the saphenous nerve is a terminal sensory branch of the femoral nerve and contains the axons of neurons whose cell bodies are found in the dorsal root ganglia (DRG) at levels L2 and L3 in mouse. Saphenous fibers innervate ~80% of the dorsal hairy skin surface of the hindpaw while the other 20% is innervated by the common peroneal and sural nerves [19]. Use of this preparation and the recently described median/ulnar preparation has allowed for extensive characterization of afferents that terminate in specialized hairy skin end organs including Merkel cell touch domes [20, 21] and various hair follicles [19]. Nerve activity in these preparations, however, is difficult to correlate with rodent pain-like behaviors, the majority of which are reflexive responses triggered by the application of a stimulus to the glabrous surface of the hindpaw (e.g., Hargreaves radiant heat assay [22], acetone [23] or plantar dry ice [24] cold assays, von Frey, pinprick, and light touch assays [25]). For this reason, additional skin-nerve variations were developed to record activity in branches of the sciatic nerve that innervate glabrous hindpaw skin. The sciatic nerve is a mixed sensory/motor nerve whose cell bodies are predominantly found in L4 and L5 in rats and in L3 and L4 in mice [26]. Sciatic fibers intermingle with other spinal nerves within the lumbosacral plexus, then emerge, traveling along the lateral aspect of the thigh. The nerve trifurcates into the sural, tibial, and common peroneal branches at the level of the popliteal fossa. The common peroneal nerve (also known as fibular nerve or external popliteal nerve) is a mixed sensory/motor branch that innervates dorsal hindpaw skin [27]. Since the saphenous nerve, which is strictly sensory, is easier to dissect than the common peroneal and has a broader terminal innervation pattern in hairy skin, common peroneal preparations are not often used. The sural nerve is a mixed sensory/motor branch that innervates the lateral edge of the paw and transduces information from both hairy and glabrous regions of the skin. The third branch of the sciatic nerve is the tibial nerve, a mixed sensory/motor nerve that innervates the entirety of the hindpaw plantar surface [19, 28] and, therefore, is most relevant for behavioral test correlations. Preparation selection should consider the skin/cell type of interest and, if using an inducible pain model, the site of injury. For example, when examining afferent activity following spared nerve injury (i.e., spared sural, ligated common peroneal, and tibial branches), experimenters used a

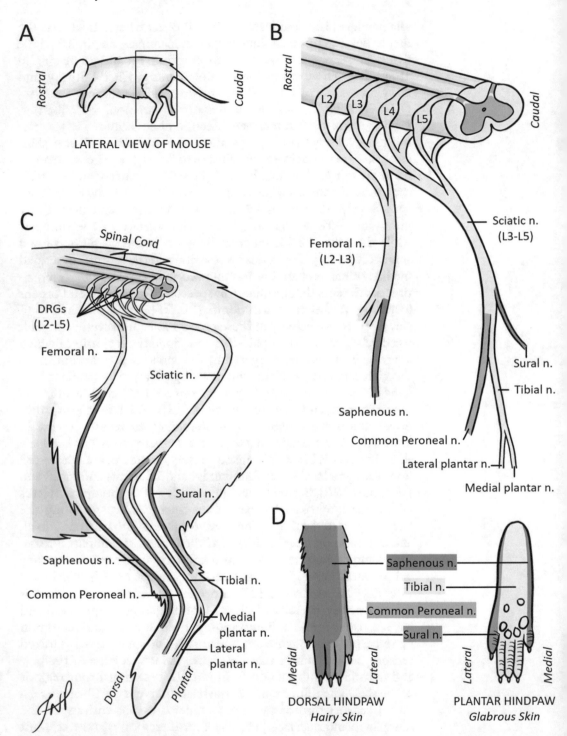

Fig. 1 Hindpaw innervation patterns. (**a**, **b**) Rodent hindpaw skin is innervated by the saphenous, common peroneal, tibial, and sural nerves. (**c**) The saphenous nerve is the terminal sensory branch of the femoral nerve, containing primary sensory afferents that primarily synapse at the L2-L3 level of the spinal cord. Peripheral endings of saphenous fibers detect and transmit sensory information from the medial aspect of the dorsal hairy surface of the hindpaw. The sciatic nerve contains sensory and motor fibers that synapse at the L3-L5 level of the spinal cord in mouse (L4-L5 in rat). At the popliteal fossa, the sciatic nerve branches into the

sural nerve preparation to record activity from fibers innervating hypersensitive regions of the paw [29]. Similar considerations were made for recordings following postoperative incision injury [30]; experimenters used the tibial nerve preparation since the glabrous skin is the primary site of hypersensitivity [31]. Multiple preparations can be utilized to measure activity changes in widespread tissue injuries like Complete Freund's Adjuvant injections [32], diabetic- [33], or chemotherapy-induced [34] neuropathies.

All skin-nerve variations contain heterogeneous populations of primary sensory afferents. As described in previous chapters, primary sensory afferents can be categorized by molecular profiles and anatomical connectivity; physiological properties assessed in teased-fiber recordings are another common method of classification (Table 1). The most basic criteria for sensory afferent classification is conduction velocity, a parameter dictated by axon diameter and extent of axonal myelination; larger, thickly myelinated $A\beta$ axons transmit signals more rapidly than medium diameter lightly myelinated $A\delta$ fibers or unmyelinated C fibers. In addition to signal transmission, different classes of fibers are intrinsically responsive to select stimulus modalities as a result of ion channel expression patterns. In this chapter, we will describe how to characterize the mechanical responsiveness of a fiber by stimulating its cutaneous terminals. For thermal stimulation protocols, readers are referred to detailed protocols from the Reeh [18] and Lewin [19] labs. Chemical responsiveness can be measured by directly adding a compound to the recording bath, thereby stimulating all receptive fields of the skin simultaneously, or by delivering compound to a moat that has been sealed around the receptive field of the fiber of interest [45, 46]. Responses to mechanical stimuli are further classified by the pattern of stimulus-evoked action potential firing; afferent fibers that respond only at the on- and off-set of a stimulus are called rapidly adapting (RA), and fibers that maintain action potential firing throughout the duration of a mechanical stimulus are called slowly adapting (SA). Combined with molecular profiling and anatomical mapping, these physiological parameters allow for the precise identification of sensory afferent populations that are involved in the onset or maintenance of acute, persistent, or chronic pain and, subsequently, provide novel opportunities for therapeutic development.

Fig. 1 (continued) common peroneal, sural, and tibial branches. (**d**) The saphenous nerve receives sensory information from the dorsal aspect of the hindpaw, while the common peroneal nerve relays sensory information from the dorsolateral aspect, the sural relays information from lateral hairy and glabrous skin, and the tibial relays sensory information from the central and medial aspects of the glabrous skin of the hindpaw

Table 1
Physiological classification of sensory afferents

Class	Conduction velocity (m/s)[a]	Modality specificity	Adaptation profile	Skin type	End organ/type
C	<1.2	Polymodal Thermal[b] Mechanical (high threshold[c])	SA	Glabrous, hairy	Free nerve ending Nociceptive Schwann cell?[d]
		Mechanical (low threshold[c])		Hairy	Lanceolate ending [35] Free nerve ending [36]
Aδ	1.2–10	Polymodal	SA	Glabrous, hairy	Free nerve ending Nociceptive Schwann cell?
		Thermal[e]		Glabrous, hairy	Free nerve ending Nociceptive Schwann cell?
		Mechanical		Glabrous	Free nerve ending Nociceptive Schwann cell?
				Hairy	Circumferential ending [37]
		Mechanical (low threshold)	RA	Hairy	Down hair lanceolate ending
Aβ	>10	Mechanical (high threshold)[f]	SA	Glabrous, hairy	Unknown
		Mechanical (low threshold)	SA (type I[d])	Glabrous	Merkel cell
				Hairy	Merkel cell touch dome
			SA (type II)	Glabrous	Ruffini ending [e]
				Hairy	Unknown
			RA (type I)	Glabrous	Meissner corpuscle
				Hairy	Lanceolate ending
			RA (type II)	Glabrous	Pacinian corpuscle

[a]Conduction velocities defined in mouse; additional references for rat [38] and human [39]

[b]Select C fibers only respond to cold, heat, or warming [18]

[c]In mouse, high-threshold fibers are those that respond to von Frey forces >5.7mN; low-threshold fibers are those that respond to von Frey forces <5.7 mN; if mechanical threshold is not specified, fiber can respond to low and high forces

[d]Type I fibers exhibit irregular stimulus-evoked firing patterns; Type II fibers exhibit regular firing upon receptive field stimulation

[e]Ruffini structure has not been identified in rodent skin

[f]Recently identified nociceptive Schwann cells [40] encapsulate the free nerve endings of generic C and A fibers; the prevalence of these cells in hairy and glabrous skin and the modality-specificity of these cells need to be systematically characterized with quantitative imaging and ex vivo or in vivo recordings, respectively

[g]Mechanically insensitive afferents (MIA) often become mechanically responsive following injury or ex vivo preparation dissection [41, 42]; alternatively, naïve mechanical thresholds may be too high to detect with traditional force applications

[h]Aβ nociceptors are often overlooked despite electrophysiological identification in several species. See reviews on this topic by Lawson et al. [43, 44]

2 Materials

1. Extracellular recording setup:
   (a) Equipment (Fig. 2; **Note 1**):
       - Extracellular recording electrodes (Fig. 3b; **Note 2**).
       - Oscilloscope.
       - Extracellular AC Recording System (e.g., Digitimer NeuroLog System):
         - Differential preamplifier headstage.
         - AC amplifier.
         - Band pass and notch filters.
         - Data acquisition device (e.g., AD Instruments PowerLab).
         - Data acquisition analysis software (e.g., AD Instruments LabChart).
         - Computer.

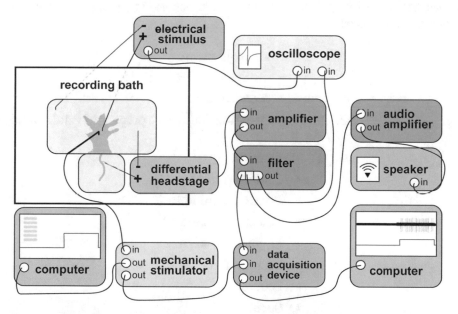

Fig. 2 Electrical components of skin-nerve rig. Fiber bundle extracellular potentials (μV range) are detected by a single microelectrode and compared to recording bath reference activity within the differential headstage. Signal is typically amplified by 5-100 K before being bandpass filtered (100-3000 Hz). Filtered signal is (1) visually observed on an oscilloscope, (2) amplified through a loudspeaker to assist in fiber identification, and (3) transformed and recorded by a data acquisition device and associated computer program. To measure conduction velocity, an electrical stimulator generates current that is applied to the skin through a second electrode; stimulus artifacts are visually observed on an oscilloscope. An externally controlled mechanical stimulator is used to deliver ramp or discrete forces to the skin; stimulator feedback data are transformed and recorded by the same data acquisition device and program used for fiber firing analysis

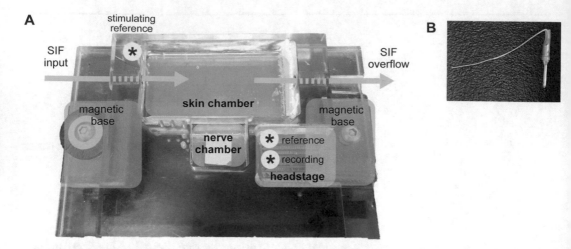

Fig. 3 Recording bath and electrode schematics. (**a**) The recording bath consists of two chambers: a larger chamber for skin superfusion and a smaller chamber for nerve teasing and recording. SIF is pumped into one side of the larger chamber and flows out of the opposite side. SIF flows between the skin and nerve chambers by passing underneath an acrylic bridge that separates the edges of these chambers. A well for the headstage is adjacent to the nerve chamber, and a connector pin for the stimulating electrode reference is located in the rear corner of the skin chamber. The bath is secured to a stable platform by magnetic bases. (**b**) Extracellular recording electrodes consist of gold, silver, or titanium wire soldered to a gold pin. Heat wrap structurally reinforces the pin–wire junction

- Audio equipment:
    - Audio amplifier.
    - Loudspeaker.
2. Synthetic interstitial fluid (SIF) preparation and flow regulation (Fig. 4):
    (a) Reagents:
        - Sodium chloride.
        - Potassium chloride.
        - Calcium chloride.
        - Magnesium sulfate.
        - Monosodium phosphate.
        - Glucose.
        - Sucrose.
        - Sodium gluconate.
        - HEPES.
        - Sodium hydroxide.
    (b) Equipment:
        - Two ≥3 L graduated plastic buckets (one for SIF reservoir, one for SIF flow-through).

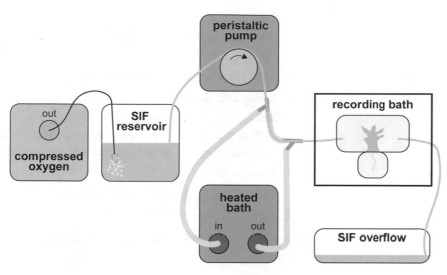

Fig. 4 Synthetic interstitial fluid (SIF) flow pattern. The fresh SIF reservoir is continuously oxygenated throughout the experiment via an air stone cylinder. Oxygenated SIF is pumped into the left side of the recording bath at a rate of $\geq$3.5 mL/min. Between the peristaltic pump and the recording bath, the SIF tubing is surrounded by a thermal water jacket for heat transfer. SIF temperature is monitored in the skin chamber throughout the experiment. Excess SIF flows out of the recording bath and into an overflow bucket located under the rig via gravity

- pH meter.
- Compressed oxygen cylinder and regulator.
- Air stone cylinder (commonly used to oxygenate aquariums; purchase at pet stores).
- Heated circulating bath (e.g., Julabo heating circulator).
- Peristaltic pump.
- Plastic tubing and connectors of various sizes.

3. Surgical dissection of glabrous hindpaw skin and tibial nerve:
   (a) Reagents:
      - Isoflurane.
   (b) Equipment:
      - Anesthesia induction chamber.
      - Stereomicroscope.
      - LED ring light for stereomicroscope or comparable light source.
      - Dissecting board; we cover Styrofoam box lid with fresh aluminum foil for each dissection.
      - Electric hair trimmer.
      - Vacuum for hair removal.

- 10-cc syringe.
- Dissecting needles; bend 30-G needle so that hub is at 90° angle to barrel.
- Two pairs of fine-tipped forceps.
- Dissecting scissors.
- Two or more pairs of microspring scissors (e.g., Westcott rounded blade scissors; one pair reserved for cutting skin, one pair reserved for cutting muscle).
- Cotton thread.
- Delicate task wipes (e.g., Kimwipes).

(c) Adult mouse (*see* **Note 3**).

4. Cleaning dissection in preparation for recording:

(a) Reagents:

- Vacuum grease.
- Mineral oil.

(b) Equipment:

- Custom built acrylic recording bath (Fig. 3a) consisting of a larger chamber for skin superfusion and a smaller adjacent chamber for nerve teasing/recording. A 1 mm thick by 1 cm tall piece of acrylic spans the top edges between the two chambers, providing ~7 mm of space at the bottom of the chambers for SIF flow. Two ports in the lateral edges of the larger chamber allow for SIF superfusion and overflow; the floor of the larger chamber is coated in ~5 mm of silicon elastomer (e.g., Sylgard) for pinning down the skin preparation. Immediately adjacent to the nerve chamber is a headstage well. This positioning allows for the least possible space between the branching point of the teased fiber and the recording electrode. A female pin connector is positioned along the rear left edge of the bath for the stimulating electrode reference.
- Mirror for teasing. A small mirror is required for teasing in the nerve chamber; ours is 2 × 2 cm and secured to a brass block in order to raise the nerve off of the chamber floor.
- Stainless steel insect pins for securing edges of dissection to Sylgard floor.
- Floating benchtop isolation table.
- Light source; we use lamp with arm illuminators for easy light focusing.
- Binocular arm stereomicroscope.

- Battery operated continuous temperature recording device (e.g., Physitemp).
- Two pairs of fine-tipped forceps reserved for placing/manipulating pins and electrodes.
- Microspring scissors.
- 3/10-cc syringe.
- Liquid vacuum with collection bottle for cleanup.

5. Fiber teasing, receptive field identification, and calculating conduction velocity:

   (a) Equipment:

   - Two pairs of fine-tipped forceps reserved for teasing. Forceps should be sharpened with sharpening stone prior to use; tips must meet in order to delicately manipulate epineurium.
   - Pulled glass rods for identifying mechanically active receptive fields; 0.5-cm diameter rods are heated in an open flame and pulled to create a rounded focal point at the tip.
   - Electrical stimulus generator (e.g., NeuroLog pulse generator #NL301, pulse buffer #NL510, and stimulus isolator #NL800).
   - Stimulating tungsten electrode attached to three-way micromanipulator with magnetized base.
   - Digital calipers.

6. Measuring mechanical responsiveness of fiber:

   (a) Equipment:

   - Calibrated von Frey filaments.
   - Automated mechanical stimulator (e.g., 300C-I mechanical stimulator from Aurora Scientific).

3   Methods

1. Extracellular recording setup:

   (a) Assemble recording equipment according to Fig. 2. Key considerations:

   - The differential headstage should have low noise and high impedance to allow for transmission of extracellular potentials (μV range). The reference signal is used as the ground in this setup; to omit a third ground electrode, reference electrode signal is fed into the "- IN" and "GROUND" inputs using a U-shaped connector.

- Extracellular potentials need to be amplified ~100,000-fold for data acquisition software detection and analysis. Amplification can occur at the level of the headstage and the amplifier; we typically set our amplifier gain between 5 K (5000) and 10 K and adjust as needed for each fiber.

- A band pass filter is used to isolate 100–3000 Hz frequency components of electrical signal; a notch filter is used to remove 50 or 60 Hz mains noise (e.g., band-reject filter, line filter; removes "electrical noise" coming from lights, etc.).

- To assist in the initial steps of receptive field identification, filtered signal is emitted through an audio speaker, so that the experimenter can listen to fiber firing while visually identifying the receptive field location on skin. A given fiber or "unit" will often have a characteristic sound. Audio amplifier should have manual controls for adjusting signal threshold and volume.

- An electrical impulse generator is required for conduction velocity measurements and, if warranted by an experiment, can be used to identify receptive fields via electrical stimulation (*see* **Note 21**).

2. Synthetic interstitial fluid (SIF) preparation and flow regulation:

   (a) Construct tubing system for SIF flow similar to schematic in Fig. 4. Key considerations:

   - New SIF needs to be continuously supplied to the recording bath. To regulate flow, we use a peristaltic pump (*see* **Notes 4** and **5**).

   - SIF needs to be oxygenated for the duration of the experiment. To do this, we deliver compressed oxygen (100%) to the SIF reservoir through an air stone cylinder.

   - SIF temperature needs to be 32 ± 0.5 °C in the recording bath. To do this, we ensheath the SIF tubing in a water jacket after exiting the peristaltic pump. Water jacket solution is warmed by a heated circulating bath (*see* **Note 6**).

   - SIF exits the recording bath through an overflow port. Overflow is collected via gravity flow in a collection bucket located underneath the workstation.

   (b) Prepare SIF by dissolving the following in ultrapure distilled water: sodium chloride (123 mM), potassium chloride (3.5 mM), calcium chloride (2 mM), magnesium sulfate (0.7 mM), monosodium phosphate (1.7 mM), glucose (5.5 mM), sucrose (7.5 mM), sodium gluconate (9.5 mM), and HEPES (10 mM). pH to 7.45 ± 0.5 using sodium hydroxide. We prepare 3 L of SIF for ~8 h of recording (*see* **Note 7**).

(c) Start SIF flow prior to beginning dissection, so that recording bath is filled with warm, oxygenated SIF when dissection is complete.

3. Surgical dissection of glabrous hindpaw skin and tibial nerve (~45 min).

(a) Anesthetize the animal with >2% isoflurane. Once the righting reflex is gone, perform cervical dislocation.

(b) Remove fur on either hindleg with electric hair trimmers and vacuum; remove hair from the ankle to inguinal crease (medial) and to the midline of the back (lateral) (*see* **Note 8**).

(c) With animal in supine position, pin fifth digit of hindpaw (i.e., "pinky toe") to dissection board; insert pin through skin adjacent to nail so as to limit damage to proximal phalanx skin covering. Insert additional stabilizing pins through lateral abdominal skin to minimize specimen movement.

(d) Starting with first digit (i.e., hallux, "thumb"), insert tip of microspring scissor blade into nail cuticle. Cut a circle around the base of the nail to completely excise the claw from the surrounding skin.

(e) While holding nail or distal edge of the hairy skin, cut a straight line through the hairy skin down the length of the digit from the cuticle to center of the hindpaw (i.e., metatarsal region). While cutting, use microspring scissors and forceps to separate skin hypodermis from underlying bone/muscle fascia, taking care to not cut any blood vessels. Repeat this and the previous step on the remaining digits (*see* **Note 9**).

(f) Remove hairy skin from dorsal surface of hindpaw to result in pattern observed in Fig. 5.

(g) Use forceps to grab the proximal edge of each interdigital skin flap; pull the flap toward the distal end of the digit and use microspring scissors to separate the hypodermis from the fascia in the interdigital folds (Fig. 5).

(h) Cut a straight line through the hairy skin from the proximal edge of removed dorsal skin patch to the ankle. Separate hypodermis from fascia along metatarsal and tarsal bones. At this point, all of the dorsal hairy skin should be completely flayed from the underlying tissue (Fig. 5).

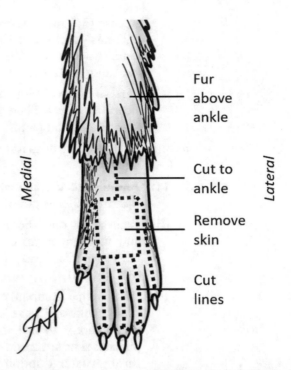

Fig. 5 Cutting pattern on dorsal surface of hindpaw. To maintain the integrity of the glabrous skin for tibial nerve recordings, all dissection cuts are made on the dorsal hairy skin. To start the dissection, circular cuts are made around each of the nails. Linear cuts are then made from the distal tips of each digit to the center of the paw. The hairy skin overlying the metatarsals is excised. An additional cut is made from the proximal edge of the excised skin to the distal edge of the hindlimb fur. These cuts allow the dorsal skin to be flayed from the underlying bone and muscle

(i) Turn attention to the upper portion of the ipsilateral hindleg and use dissecting scissors to cut through the skin from the inguinal crease to the lateral edge of the hip.

(j) Remove stabilizing pins, place the animal into prone position, then re-insert the stabilizing pins into the same areas of tissue. Complete the circular incision from the hip to the inguinal crease, so that the hindleg skin is no longer connected to the abdominal/hip skin.

(k) Use microspring scissors to make a superficial cut through the skin along the posterior edge of the calcaneus bone. Complete a circular cut around the ankle, connecting with the incision made on the anterior side of the ankle in step h. At this point, the hindpaw skin should no longer be connected to the hindleg skin.

(l) Return to calcaneus bone. Cut through two of the tendons connected to the bone: calcaneal (i.e., "Achilles tendon") and superficial flexor (medial to calcaneal).

When cutting through the superficial flexor, do not cut too deep as tibial nerve lies just beneath tendon (*see* **Note 10**).

(m) Grab the newly freed ends of both tendons with forceps then pull the tendons toward the popliteal fossa, separating the medial and lateral gastrocnemius muscles from the underlying tissues, exposing an intact tibial nerve in the process. Use the 10-cc syringe to wet the nerve with SIF then return the gastrocnemius to cover the nerve. There is no need to remove hindleg skin before this step; it will roll up the leg to the popliteal fossa/knee as the gastrocnemius muscles are raised.

(n) Starting with the fifth digit and with the animal sill in the prone position, completely remove the hindpaw skin from the underlying toe bones without touching/grabbing the skin. Use forceps to grab the deep digital flexor tendon in each digit. Cut through the distal tip of the tendon where it attaches to the distal phalanx, then pull on the tendon to separate the overlying tissue from the bone; use microspring scissors to cut through additional tendon connections at the interphalangeal joints and the metatarsophalangeal joints. Continue this process through the metatarsal and tarsal regions of the paw until the hindpaw skin is completely freed from underlying bone and muscle; the tibial nerve should be attached to the proximal end of the hindpaw skin (*see* **Notes 11** and **12**).

(o) Remove the ring of hairy skin that should be at the popliteal fossa region of the hindleg following step m. Using dissecting scissors, cut through the skin overlying the femur, following the route of the sciatic nerve (Fig. 1).

(p) Use forceps to pull the hind leg muscles apart, exposing the entire length of the tibial nerve from where it innervates the hindpaw skin to where it branches from the sciatic nerve. Continue this process to expose the length of the sciatic nerve up to the lumbosacral plexus.

(q) Use cotton thread to tie a knot around the proximal end of the sciatic nerve. Cut the proximal end of the nerve with microspring scissors. Grab onto the knot with forceps and proceed to separate the nerve from muscle fascia, using microspring scissors to cut away the sural and common peroneal branches. After completing this step, the hindpaw skin and tibial/sciatic nerve should be completely free (*see* **Note 13**).

(r) Immediately transfer the skin-nerve preparation to the recording bath which, by this point, is filled with warm, oxygenated SIF.

4. Cleaning dissection in preparation for recording.

(a) Apply a layer of vacuum grease to skin chamber floor immediately adjacent to nerve chamber; grease will assist in stabilizing skin position.

(b) Grabbing only the knotted thread, loop the nerve underneath the chamber divider and into the smaller recording chamber. The proximal end of the skin should nearly about the chamber divider, leaving room for SIF buffer to blow under the divider.

(c) Press the skin, epidermis side down, into the vacuum grease (see **Note 14**). Stabilize the skin position by pulling the skin taught, and pinning the hairy edges in place; we typically insert two pins into the lateral, distal edges of each toe, and several more pins along the lateral edges of the ankle skin. At this point, the preparation should resemble Fig. 6.

(d) Grab the proximal end of the deep digital flexor tendon and tear the tendon away from the corium (i.e., dermis), removing the tendon branches from the individual digits in the process. Re-pin the skin if necessary. Use micro-spring scissors to trim muscle, ligaments, and additional connective tissue from corium as these can block nerve conduction if left intact.

(e) Place the raised mirror very close to the divider in the nerve chamber. Manipulating only the knotted thread, gently bring the nerve to rest on the top of the mirror. Position the recording electrode over the mirror, perpendicular to the nerve. Be sure that the electrode does not directly contact the mirror or the side of the nerve chamber. Position the reference electrode underneath the SIF surface in the skin chamber. Position the stimulating electrode reference underneath the SIF surface in the rear corner of the skin chamber.

(f) Slowly add mineral oil to the nerve chamber until the nerve and entire surface of the mirror are completely submerged in oil (see **Note 15**).

(g) Remove the knotted thread from the end of nerve. Using teasing forceps, gently slide the epineurium up the nerve trunk, taking great care to not stretch or tear the nerve (see **Note 16**).

(h) Allow the tissue to rest for >15 min. Turn on all recording equipment.

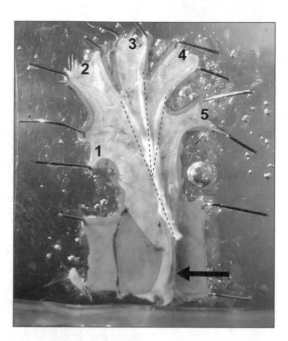

Fig. 6 Uncleaned hindpaw skin in recording bath. The plantar skin and innervating tibial nerve were removed from the left hindpaw of a mouse and placed corium side up in the recording bath. Each of the digits (1–5; 1: hallux, "thumb") is stabilized by pins in the lateral distal tips. Additional pins are used to secure the lateral edges of the ankle skin. Hairy skin of the ankle and along the dorsal edges of the digits is shaded brown to illustrate the regions in which no tibial activity should be detected. The tibial neve (arrow) is threaded underneath the skin/nerve chamber border at the bottom of the image. Before recording, the deep digital flexor tendon (dotted lines; white tissue) and associated muscle tissue should be removed from the preparation

5. Fiber teasing, receptive field identification, and calculating conduction velocity:

(a) Using sharpened, fine-tipped forceps, tease the nerve into approximately eight bundles (*see* **Notes 17** and **18**).

(b) Lay one of the bundles onto the recording electrode. Switch the amplifier into differential recording mode and visualize bundle activity on the oscilloscope (*see* **Notes 19** and **20**).

(c) Use a pulled glass rod to stimulate the corium of the glabrous skin. A quick up/down poking motion will enable identification of rapidly and slowly adapting fibers. At this stage of teasing, it is highly unlikely that mechanical stimulation will result in activation of only one unit, but this step will allow for the assessment of overall bundle activity. Mechanically induced activity should be >3 times larger than background noise levels (*see* **Notes 21** and **22**).

(d) If necessary, remove the fiber bundle from the recording electrode, tease into smaller bundles, and repeat mechanical stimulation until single unit activity is observed. Save the single-unit waveform on the oscilloscope as it will be used for comparison during conduction velocity measurements.

(e) Use a blunt tipped von Frey filament (≤4 mN) to find the most sensitive spot of the receptive field. Position the stimulating electrode directly above this spot; lower the electrode and verify correct positioning via mechanically induced unit firing. Raise the electrode slightly, so that it is still making contact with the corium but not causing the unit to fire.

(f) Use the pulse generator to stimulate the corium with electrical current. Pulses will appear as artifacts on the oscilloscope; adjust the oscilloscope horizontal axis to allow for C fiber identification (typically 20–200 ms, but depends on the position of the receptive field). Increase pulse amplitude until a waveform identical to that saved in *step d* appears in the same x-axis position on the oscilloscope after three repeated pulse artifacts. Stop data acquisition and record the latency between the stimulus artifact and the beginning of the waveform upswing (*see* **Notes 23** and **24**).

(g) Use digital calipers to measure the distance (in mm) between the stimulating electrode and the closest point of the recording electrode that is in contact with the fiber. Divide this distance by the spike latency to calculate the fiber conduction velocity. We classify mouse fibers using the following conduction velocity ranges (m/s): Aβ >10, Aδ 1.2–10, C < 1.2.

(h) Remove the stimulating electrode from the skin.

6. Measuring mechanical responsiveness of fiber:

(a) Determine the fiber's mechanical threshold using calibrated von Frey filaments. Starting with the smallest filament, apply the tip of the filament to the most sensitive spot of the receptive field for 3 s. Record the smallest force that elicits ≥3 spikes during stimulation as the mechanical threshold.

- Similar force applications can be performed with an automated mechanical stimulator. Using an automated force ramp will allow for threshold detection along a continuous range of forces as opposed to the discrete force thresholds that result from von Frey filament stimulation.

(b) Using data acquisition software, record spontaneous fiber activity for 3 min prior to recording mechanical responses.

(c) Position the automated mechanical stimulator probe over the most sensitive spot of the receptive field. Record the mechanically induced activity over a range of forces, starting with the smallest and increasing force with subsequent applications. Allow for ≥ 1 min to pass between force applications (*see* **Notes 25–27**).

7. Data analysis and representative results:

(a) Once recording is complete, use spike discriminator function in data acquisition software (e.g., Spike Histogram in LabChart) to identify and count single-unit waveforms during each force application (*see* **Notes 28** and **29**).

4 Notes

1. If a differential preamplifier headstage and mains notch filter are used, a Faraday cage and/or additional 50/60 Hz noise remover (e.g., HumBug) are not required in the setup. All electrical devices around the setup and the experimenter should be properly grounded.

2. Extracellular electrodes are typically made from silver (does not need to be chlorinated), gold, or platinum wire (0.15 mm diameter). Wire is soldered into a pin contact solder cup; the solder cup/wire junction is heat-wrapped for stabilization and to reduce the surface area exposed to extraneous electrical contact. After soldering, cut the recording and reference electrodes to the same length. Measure resistance with an ohmmeter (should be $<1\ \Omega$).

3. We use mice >8 weeks of age, however, this protocol can be adapted for use in juvenile mice, rats, or other rodent species. Ex vivo rodent teased fiber recordings were originally described in rat [16], and the first murine saphenous recordings were completed in juvenile animals [17].

4. In our original description of this method [17], SIF was superfused over the preparation at a rate of 15 mL/min; our lab currently superfuses SIF at a rate of 3.5 mL/min and is equally successful in keeping preparations alive and recording single unit activity.

5. The peristaltic pump used for SIF flow regulation can be a source of electrical noise. To combat this, we ground the pump chassis and the buffer after it has exited the pump. Other groups combat this issue by omitting a peristaltic pump and using gravity-based flow systems.

6. The bath is held at this temperature to mimic in vivo conditions. However, a recent publication by Paricio-Montesinos et al. [47] reports that mouse forepaw temperatures are ~27 °C in vivo; when the skin-nerve bath is held at 27 °C, spontaneous activity is noted in a population of TRPM8-positive fibers that are required for warming sensation. Therefore, if your experimental question is related to thermosensation, we encourage you to maintain the bath at 27 °C to more accurately model in vivo conditions and prevent biased sampling.

7. SIF needs to be made fresh at the beginning of each recording day; using day-old SIF can lead to spontaneous activity in the preparation, even if the SIF was stored at 4 °C overnight.

8. If readers are unfamiliar with gross hindpaw anatomy, we encourage them to reference *The Anatomy of the Laboratory Mouse* or the microscopic descriptions reported by Wong et al. [48].

9. For all fascia/hypodermis separation steps, we encourage a gentle pulling or spreading strategy rather than blind cutting. Cutting increases the likelihood of snipping nerve fascicles as they split from the main fiber bundle to innervate skin. Cutting also risks snipping blood vessels; pooled blood can obscure dissection. We use forceps to tightly grasp one edge of the cut hairy skin (dorsal hairy skin is not innervated by tibial nerve so this should not damage any receptive fields of interest) and to pull skin away from bone/muscle. While pulling, carefully use scissors to cut through fascia and not through hypodermis.

10. Discriminating tendons and nerves can be challenging at first. Tendons are more opaque, shiny, white in color and flatter in shape; nerves are more translucent in color and rounder in shape.

11. This step can also be performed prior to steps k–m (i.e., prior to manipulating gastrocnemius and visually identifying tibial nerve); just be sure to keep the nerve moistened and limit air exposure.

12. In our opinion, the most difficult part of the dissection is clearing the posterior aspect of the calcaneus bone. The tibial nerve abuts the medial aspect of the bone with essentially no intermediary connective tissue. As a result, there is no tendon or other anchor to pull on to assist in tissue separation; the dissector is blindly cutting into the area. We find that if we make very small cuts that essentially outline the shape of the bone, the tibial nerve is not damaged.

13. Handling of the nerve at any location other than the knot can result in irreversible nerve damage; electrical signals cannot be properly conducted through pinched nerve segments. When

first learning this preparation, many dissectors do not remove enough of the proximal sciatic nerve. It is critical to remove as much sciatic nerve length as possible as this will assist in teasing/reaching the recording electrode in later steps.

14. This orientation (corium/dermis facing up) is referred to as "inside-out". If using the "outside-out" (i.e. epidermis facing up) orientation, be sure to regularly wash the corium with SIF as described by Walcher et al. [19].

15. Mineral oil is less dense/viscous than SIF, therefore, when it is added to the nerve chamber, SIF will pass under the divider and into the skin chamber. If the nerve and mirror are not submerged in oil, it will be nearly impossible to tease individual fibers and electrical signals will be shunted. Occasionally, SIF will puddle on the mirror. If SIF wicks up to the nerve on the electrode, this can cause dampening or shunting of the nerve signal. We use a bevel-needled 3/10 cc syringe to gently suck SIF off of the mirror before teasing. To assess mineral oil depth, place forceps into the nerve chamber and visualize the oil:SIF interface against opaque background. Once the nerve is submerged in oil, it should remain in this solution for the duration of the recording (i.e., not raised into the air during electrode placement).

16. Additional tips for removing epineurium:

    (a) If available, grab subcutaneous fat, connective tissue, or stem of cut nerve branches that are attached to the epineurium as opposed to the epineurium itself. The epineurium is quite thin and difficult to separate from the rest of the nerve trunk at the cut end, so these accessory tissues can be helpful for epineural manipulation.

    (b) Stabilize the cut end of the nerve with one set of forceps; the hand holding those forceps should be stabilized on the edge of the recording bath. Use your second hand/forceps to roll epineurium away from cut end (similar to pushing up sleeves on a long-sleeved shirt).

    (c) Ensure that the common peroneal and sural branches are no longer attached to the sciatic nerve; if these branches are left intact, one could mistakenly tease the proximal end of fibers contained within, assuming they are fibers from the tibial branch.

17. Main strategy for fiber teasing: Grab half of the proximal nerve bundle (where the knot was tied) with each pair of teasing forceps. Stabilize hands on the edge of the recording bath and gently pull the halves away from one another; this pulling motion should not stretch the fibers, but it should raise the fibers off of the mirror. Continue to tease in this fashion in the oil; thin, long fibers are ideal for recording. Keep in mind that

repeated raising/lowering or gross repositioning of fiber bundles can rub the nerve trunk against the edge of the mirror which will shear fibers in the outer edges of the nerve bundle.

18. The tibial nerve consists of two major fiber bundles of unequal size; identifying these bundles will make the first teasing step significantly easier than randomly teasing the nerve into approximately equally sized bundles.

19. Position the recording electrode so that it is as close to the nerve trunk as possible. Typically, the greater the contact distance between the nerve fiber and the recording electrode, the larger the signal amplitude. Ensure that there is no SIF bubble/pool or air bubbles along the length of the recording electrode as these can interfere with/dampen signal transduction. Additionally, be sure to remove all fibers from the electrode once a recording is complete; small fibers tend to stick to the electrode and rip off of the main nerve trunk. If fibers consistently slide off of the electrode, use fine grit sandpaper to slightly roughen the surface of the electrode.

20. Most fibers from uninjured animals do not exhibit spontaneous activity unless they are damaged. If unavoidable, spontaneous activity can be subtracted from stimulus-induced activity during analysis.

21. By using a mechanical search stimulus, we are automatically biasing our recorded fiber populations. To circumvent biased recordings, an electrical search stimulus can be used to identify single-unit receptive fields. In this procedure, an electrical pulse is delivered to the corium through a low impedance microelectrode; if a fiber's receptive field is stimulated, spikes will follow the electrical artifact on the oscilloscope. Readers interested in this approach should read the detailed protocol from Zimmerman et al. [18].

22. To increase the signal-to-noise ratio, adjust amplifier gain or tease the fiber bundle apart further. Be cautious as fiber of interest can be damaged in teasing process. Note that multiple C fibers are ensheathed by a non-myelinating Schwann cell in a given Remak bundle; Remak bundles cannot be teased into individual C fibers. Readers are referred to Zimmerman et al. [18] for detailed protocol on "marking" C fibers within the same bundle.

23. We start stimulating with 100 μA pulses, then gradually increase current until an identical waveform is obtained (typically <10 mA current is required). Repeated stimulation will send concentric circles of electrical current through the skin, thus stimulating all receptive fields near the electrode, some of which may be innervated by fibers on the recording electrode. This is why it is critical that the waveform for the fiber of

interest is saved to the oscilloscope and be readily distinguishable from other waveforms. It is also critical to ensure that the mechanically or thermally evoked spike is the same as the electrically evoked and CV-characterized unit.

24. When measuring conduction velocity, Aβ fibers can have such a short latency to the onset of the action potential that the spike blends into the stimulus artifact. If this happens, lower the stimulus amplitude and attempt to isolate the fiber waveform from the artifact.

25. Our lab uses a custom-built mechanical stimulator as opposed to the 300C stimulator from Aurora Scientific. Critical considerations when designing or choosing a stimulator include the following:

    (a) Ability to apply gradual force ramp for threshold detection.

    (b) Reliable, reproducible application of low (≤ 2 mN) to high (≥ 400 mN) forces; even higher forces may be required for recordings of rat tissue.

    (c) Stimulator probe is made of material that will not slip on corium.

26. A piezo actuator can be modified to deliver a vibrating mechanical stimulus if desired [19].

27. Multiple units can be recorded from the same fiber bundle if limited teasing was performed; however, we encourage readers to tease fibers into the very thinnest bundles possible. In order to include >1 unit per bundle, units should have unique waveforms (amplitude and/or shape), different conduction velocities, and receptive fields that are >3 mm apart if both units are C fibers. Only one RA-Aβ fiber can be recorded from each bundle since the receptive fields of these units are so large.

28. We analyze action potential firing rates independently for the "dynamic" and "stable" periods of mechanical stimulation. We define the dynamic phase as the very short time frame during which the mechanical stimulator is ramping up to the maximum force; the stable period is the time during which the force application remains consistent.

29. Detailed descriptions of functional fiber classification are provided in Table 1 and reviewed by Zimmerman et al. [18] and Walcher et al. [19].

References

1. Wangzhou A, McIlvried LA, Paige C et al (2020) Pharmacological target-focused transcriptomic analysis of native versus cultured human and mouse dorsal root ganglia. Pain 161(7):1497–1517

2. Gemes G, Koopmeiners A, Rigaud M et al (2013) Failure of action potential propagation in sensory neurons: mechanisms and loss of afferent filtering in C-type units after painful nerve injury. J Physiol 591:1111–1131

3. Koerber HR, Woodbury CJ (2002) Comprehensive phenotyping of sensory neurons using an ex vivo somatosensory system. Physiol Behav 77:589–594

4. Cain DM, Khasabov SG, Simone DA (2001) Response properties of mechanoreceptors and nociceptors in mouse glabrous skin: an in vivo study. J Neurophysiol 85:1561–1574

5. Heppner TJ, Tykocki NR, Hill-Eubanks D, Nelson MT (2016) Transient contractions of urinary bladder smooth muscle are drivers of afferent nerve activity during filling. J Gen Physiol 147:323–335

6. Ito H, Aizawa N, Sugiyama R et al (2016) Functional role of the transient receptor potential melastatin 8 (TRPM8) ion channel in the urinary bladder assessed by conscious cystometry and ex vivo measurements of single-unit mechanosensitive bladder afferent activities in the rat. BJU Int 117:484–494

7. Brierley SM, Jones RCW, Gebhart GF, Blackshaw LA (2004) Splanchnic and pelvic mechanosensory afferents signal different qualities of colonic stimuli in mice. Gastroenterology 127:166–178

8. Buckley MM, O'Malley D (2018) Development of an ex vivo method for multi-unit recording of microbiota-colonic-neural signaling in real time. Front Neurosci 12:112

9. Perez-Burgos A, Wang B, Mao Y-K et al (2013) Psychoactive bacteria Lactobacillus rhamnosus (JB-1) elicits rapid frequency facilitation in vagal afferents. Am J Physiol Liver Physiol 304:G211–G220

10. Wenk HN, McCleskey EW (2007) A novel mouse skeletal muscle-nerve preparation and in vitro model of ischemia. J Neurosci Methods 159:244–251

11. Catton WT (1958) Some properties of frog skin mechanoreceptors. J Physiol 141:305–322

12. Catton WT (1957) Responses of frog skin to local mechanical stimulation. J Physiol 137:81–2P

13. González CS, Sánchez JO, Concha JB (1966) Changes in potential difference and short-circuit current produced by electrical stimulation in a nerve-skin preparation of the toad. Biochim Biophys Acta 120:186–188

14. Diamond J, Holmes M, Nurse CA (1986) Are Merkel cell-neurite reciprocal synapses involved in the initiation of tactile responses in salamander skin? J Physiol 376:101–120

15. Weston KM, Foster RW, Weston AH (1984) The application of irritant chemicals selectively to the skin of the leech ganglion/body wall preparation. J Pharmacol Methods 12:285–297

16. Reeh PW (1986) Sensory receptors in mammalian skin in an in vitro preparation. Neurosci Lett 66:141–146

17. Koltzenburg M, Stucky CL, Lewin GR (1997) Receptive properties of mouse sensory neurons innervating hairy skin. J Neurophysiol 78:1841–1850

18. Zimmermann K, Hein A, Hager U et al (2009) Phenotyping sensory nerve endings in vitro in the mouse. Nat Protoc 4:174–196

19. Walcher J, Ojeda-Alonso J, Haseleu J et al (2018) Specialized mechanoreceptor systems in rodent glabrous skin. J Physiol 596:4995–5016

20. Maksimovic S, Nakatani M, Baba Y et al (2014) Epidermal Merkel cells are mechanosensory cells that tune mammalian touch receptors. Nature 509:617–621

21. Maricich SM, Wellnitz SA, Nelson AM et al (2009) Merkel cells are essential for light-touch responses. Science 324:1580–1582

22. Hargreaves K, Dubner R, Brown F et al (1988) A new and sensitive method for measuring thermal nociception in cutaneous hyperalgesia. Pain 32:77–88

23. Yoon C, Young Wook Y, Heung Sik N et al (1994) Behavioral signs of ongoing pain and cold allodynia in a rat model of neuropathic pain. Pain 59:369–376

24. Brenner DS, Golden JP, Gereau RW IV (2012) A novel behavioral assay for measuring cold sensation in mice. PLoS One 7:e39765

25. Abdus-Saboor I, Fried NT, Lay M et al (2019) Development of a mouse pain scale using sub-second behavioral mapping and statistical modeling. Cell Reports 28:1623–1634.e4

26. Rigaud M, Gemes G, Barabas M-E et al (2008) Species and strain differences in rodent sciatic nerve anatomy: implications for studies of neuropathic pain. Pain 136:188–201

27. Vadakkan KI, Jia YH, Zhuo M (2005) A behavioral model of neuropathic pain induced by ligation of the common peroneal nerve in mice. J Pain 6:747–756

28. Swett JE, Woolf CJ (1985) The somatotopic organization of primary afferent terminals in the superficial laminae of the dorsal horn of the rat spinal cord. J Comp Neurol 231:66–77

29. Smith AK, O'Hara CL, Stucky CL (2013) Mechanical sensitization of cutaneous sensory fibers in the spared nerve injury mouse model. Mol Pain 9:61

30. Brennan TJ, Vandermeulen EP, Gebhart GF (1996) Characterization of a rat model of incisional pain. Pain 64:493–502

31. Kang S, Brennan TJ (2009) Chemosensitivity and mechanosensitivity of nociceptors from incised rat hindpaw skin. Anesthesiology 111: 155–164

32. Lennertz RC, Kossyreva EA, Smith AK, Stucky CL (2012) TRPA1 mediates mechanical sensitization in nociceptors during inflammation. PLoS One 7:e43597

33. Lennertz RC, Medler KA, Bain JL et al (2011) Impaired sensory nerve function and axon morphology in mice with diabetic neuropathy. J Neurophysiol 106:905–914

34. Shim HS, Bae C, Wang J et al (2019) Peripheral and central oxidative stress in chemotherapy-induced neuropathic pain. Mol Pain 15:1744806919840098

35. Li L, Rutlin M, Abraira VE et al (2011) The functional organization of cutaneous low-threshold mechanosensory neurons. Cell 147:1615–1627

36. Liu Q, Vrontou S, Rice FL et al (2007) Molecular genetic visualization of a rare subset of unmyelinated sensory neurons that may detect gentle touch. Nat Neurosci 10:946–948

37. Ghitani N, Barik A, Szczot M et al (2017) Specialized mechanosensory nociceptors mediating rapid responses to hair pull. Neuron 95: 944–954.e4

38. Leem JW, Willis WD, Chung JM (1993) Cutaneous sensory receptors in the rat foot. J Neurophysiol 69:1684–1699

39. Knibestöl M (1973) Stimulus—response functions of rapidly adapting mechanoreceptors in the human glabrous skin area. J Physiol 232: 427–452

40. Abdo H, Calvo-Enrique L, Lopez JM et al (2019) Specialized cutaneous schwann cells initiate pain sensation. Science 365:695–699

41. Handwerker HO, Kilo S, Reeh PW (1991) Unresponsive afferent nerve fibres in the sural nerve of the rat. J Physiol 435:229–242

42. Pogatzki EM, Gebhart GF, Brennan TJ (2001) Characterization of Aδ- and C-fibers innervating the plantar rat hindpaw one day after an incision. J Neurophysiol 87:721–731

43. Djouhri L, Lawson SN (2004) Aβ-fiber nociceptive primary afferent neurons: a review of incidence and properties in relation to other afferent A-fiber neurons in mammals. Brain Res Rev 46:131–145

44. Lawson SN, Fang X, Djouhri L (2019) Nociceptor subtypes and their incidence in rat lumbar dorsal root ganglia (DRGs): focussing on C-polymodal nociceptors, Aβ-nociceptors, moderate pressure receptors and their receptive field depths. Curr Opin Physiol 11:125–146

45. Lennertz RC, Tsunozaki M, Bautista DM, Stucky CL (2010) Physiological basis of tingling paresthesia evoked by hydroxy-α- sanshool. J Neurosci 30:4353–4361

46. Kerstein PC, del Camino D, Moran MM, Stucky CL (2009) Pharmacological blockade of TRPA1 inhibits mechanical firing in nociceptors. Mol Pain 5:19

47. Paricio-Montesinos R, Schwaller F, Udhayachandran A et al (2020) The sensory coding of warm perception. Neuron 106(5): 830–841.e3

48. Wong J, Bennett W, Ferguson MWJ, McGrouther DA (2006) Microscopic and histological examination of the mouse hindpaw digit and flexor tendon arrangement with 3D reconstruction. J Anat 209:533–545

Chapter 6

Single-Unit Electrophysiological Recordings of Primary Muscle Sensory Neurons Using a Novel Ex Vivo Preparation

Luis F. Queme and Michael P. Jankowski

Abstract

For decades, multiple electrophysiological techniques have been used to characterize the responses of primary afferents to a variety of stimuli under both normal and pathological conditions. With these strategies, experimenters are able to quantify afferent response properties, such as threshold and firing rates, and determine the functional classifications of individual cells. New research techniques have been developed that allow for both characterization of the afferent responses and correlation of the physiology and neurochemical or molecular identity, providing a comprehensive phenotype of the sensory neurons being analyzed. Dual characterization provides insight about the role of specific markers expressed in primary sensory neurons and how they may influence sensory transduction. In this chapter, we describe a powerful electrophysiological technique that allows for ex vivo recordings from intact primary muscle afferents from two target muscle groups, similar to that previously described for cutaneous afferents.

Key words Electrophysiology, Muscle afferent, Nociceptor, Ex vivo, Dorsal root ganglion

1 Introduction

Sensory transduction involves detection of both non-noxious and noxious stimuli, providing information about the surrounding environment. In the muscle, sensory afferents are similar to their cutaneous counterparts in that they span the range of cell diameters and myelin thickness. Afferents innervating muscles are commonly referred to as group I-IV rather than Aβ-, Aδ-, or C-fibers [1]. Group Ia and II afferents innervate muscle spindles and Group Ib fibers innervate Golgi tendon organs to mainly regulate proprioception and sensory-motor reflexes [2–4]. Detection of noxious and non-noxious stimuli in the muscle is primarily performed by small- and medium-sized afferents, with slower conduction velocities designated as group III (thinly myelinated) and IV (unmyelinated) afferents [1, 5]. These afferents are not known to innervate specific sensory structures and are described as "free" nerve endings throughout the extrafusal muscle fibers [6]. They

Rebecca P. Seal (ed.), *Contemporary Approaches to the Study of Pain: From Molecules to Neural Networks*, Neuromethods, vol. 178, https://doi.org/10.1007/978-1-0716-2039-7_6, © Springer Science+Business Media, LLC, part of Springer Nature 2022

can also localize to small vascular structures in the muscles [7]. Group III and IV afferents have the appearance of beaded structures due to intermittent axonal varicosities [8].

Group III and IV muscle afferents can each respond to one or more mechanical, thermal, or chemical stimuli to detect multiple sensory stimuli. Chemoresponsiveness, in particular, often requires combinations of metabolites such as lactate, protons, and adenosine triphosphate (ATP). The specific sensory modality of group III and IV afferents can be associated with expression of a variety of membrane ion channels, such as the proton sensing channel, transient receptor potential vanilloid 1 (TRPV1) [9], cooling sensing channel, TRP melastatin 8 (TRPM8) [10, 11], and the acid sensing ion channel (ASIC) family to name a few. ASICs in particular have been suggested to play important roles in the perception of fatigue and painful sensations from the muscles [12–19] in many different animal models [13, 19–23]. The P2X family of purinergic receptors has also been implicated in muscle nociception. These receptors perform a variety of functions in muscle afferents, including sensing muscle metabolites [14] by potentially modulating the sensitivity of other co-expressed receptors (including ASIC3) [24]. The expression of these membrane receptors is variable in DRG neurons [25–27], which may explain the diversity in primary afferent subpopulations observed from electrophysiological experimentation [16, 17, 28].

Characterization of muscle primary afferents and their responses has a long history. Muscle afferents were first described by AS Paintal in the late 1950s and early 1960s [5, 29, 30]. Subsequent work by McCloskey and Mitchell on cardiovascular reflexes [31] helped shape the field of muscle afferent research. These investigations included early work on cat group IV afferents and their response to algesic substances by Mense and Schmidt [32]. Mizumura and Kumazawa [33] described muscle polymodal nociceptors in dogs. Further studies implemented recordings from the dorsal roots of decerebrated cats to characterize the responses of group III and IV muscle afferents [34–37]. From that point on, research has moved away from the use of large animals and has focused in rodent models of muscle pain [13, 18, 38–40].

Based on in vitro skin-nerve preparations [41, 42], other groups studied group III and IV muscle afferent responses and potential sensitization in different models of muscle injury using muscle-nerve in vitro recordings of either isolated muscles [38], or a group of muscles attached to the underlying structures [43]. With these preparations, an identified muscle and its corresponding nerve would be isolated. The nerve would then be split repeatedly with small fine tip forceps until a single unit could be isolated to analyze responses to receptive field (RF) stimulation. This strategy provides a plethora of information about individual afferent functions. However, these preparations require that the experimenter record from axotomized neurons, which may change some of the responses of certain cell types. In addition, the cell RF is often localized via mechanical stimulation, which does not allow for mechanically insensitive cells to be characterized. Finally, recording afferent activity from a teased fiber setting does not allow for neurochemical or molecular characterization of the functionally identified sensory neuron subtypes.

With increased availability of immunohistochemical techniques and widespread use of molecular biology strategies to examine the expression patterns of different subpopulations of neurons, there is a need for techniques that pair electrophysiology with detailed profiling of gene expression in primary muscle afferents. To address this technical gap, we recently developed several muscle-nerve-dorsal root ganglion (DRG)-spinal cord preparations based on a cutaneous ex vivo recording preparation originally described by Woodbury and Koerber [44]. This preparation allows us to record from a single muscle afferent, intracellularly label the recorded unit, and then retrieve it for further characterization at the molecular [16] or neurochemical levels [17–19]. This ex vivo preparation uses electrical stimulation of the nerve and muscle to identify the cell RF, allowing an unbiased assessment of primary sensory neuron response properties. This strategy can characterize multiple modalities, including mechanoreception, thermoreception, and chemoreception. Another significant strength of this ex vivo preparation is that the recording setup includes a dual chamber system that separates the muscles from the DRGs and spinal cord. This allows the experimenter to deliver chemical stimuli directly to the muscles without affecting the neurons. The greatest strength of this technique is the ability to match electrophysiological response properties of individual fibers with their gene expression profile or neurochemical content [16–19, 45] (Fig. 1).

The preparation we describe here is based on mouse anatomy; however, similar methodologies could be used for rats. Generation of this preparation has several special considerations. As such, we provide a comprehensive description of the dissecting techniques from two specific variations to address different groups of muscles and describe the important details needed to functionally characterize muscle afferents.

2   Materials

2.1   Recording Solutions

Our ex vivo preparation is dissected and recorded in oxygenated (95% O_2–5% CO_2) artificial cerebrospinal fluid (aCSF). The exact concentration of the components is as follows: 127.0 mM NaCl, 1.9 mM KCl, 1.2 mM KH_2PO_4, 1.3 mM $MgSO_4$, 2.4 nM $CaCl_2$, 26.0 mM $NaHCO_3$, and 10.0 nM D-glucose in 2 L of ultrapure water. Dissection is carried out using one 2 L flask of the aCSF on

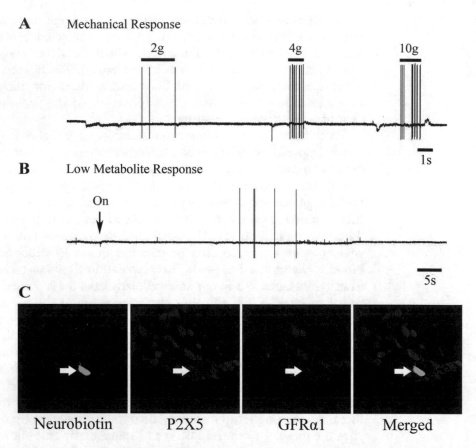

Fig. 1 Example of response properties and neurochemical identity of an individually characterized muscle afferent from the forepaw preparation. A-B, Examples of traces that can be obtained using the muscle-nerve-DRG-spinal cord ex vivo preparation. Responses to a mechanical stimulus (**a**) and "low" metabolite mixtures (15 mM lactate, 1 μM ATP, pH 7.0) (**b**). (**c**) Example of a neuron intracellularly filled with 5% neurobiotin (green) further characterized by immunohistochemistry for the purinergic receptor P2X5 (blue) and the receptor for glial cell derived neurotrophic factor (GDNF) family receptor α1 (GFRα1; red). Arrows indicate the identified cell in each panel

ice. Recordings are carried out with two separate 1 L flasks of aCSF warmed by standard water baths. One flask delivers aCSF to the inner chamber of the recording dish, and the other 1 L aCSF provides the solution for the outer chamber (see below).

2.2 Dissection Dish and Surgical Equipment

To perform the dissection of the preparation, use a 100-mm glass petri dish coated with a layer of silicone mixed with charcoal (to provide adequate contrast while dissecting). The tissues in the dissection dish are held with 0.2 mm wide stainless-steel pins. While standard dissection tools can be used, we recommend using #5 and #3 forceps for dissection and extra-fine Graefe Forceps with 1–2 teeth configuration for handling larger tissues. A Castroviejo needle holder is used to handle pins. Use 2 mm Vannas spring scissors, 6 mm spring scissors and/or 14 mm Noyes spring scissors for blunt and fine dissection. Standard variable speed peristaltic pumps (Masterflex) with appropriate tubing are used to actively remove and recirculate the aCSF used to dissect and record. A similar dish is used for recording, except a 35 mm ring is placed toward the lower part of the 100-mm petri dish and fabricated such that there is an inflow line supplying the aCSF for the inner chamber and another line for supplying outer chamber (see Jankowski et al. 2013 for original description of the preparation).

2.3 Electrical Equipment

The required electronics include an amplifier capable of current-clamp mode with a head stage to hold sharp tip electrodes, a digitizer, at least two stimulus isolation units, an audio monitor, an oscilloscope, and a workstation to store and process the recorded experiments. A Newport micromanipulator controller connected to a series of manipulators that hold the head stage for the electrodes allows for fine step movement when impaling the DRG. For both dissection and recording, Zeiss Stemi 508 dissection stereo microscopes with an independent light source are used. All recording is performed on an air table with surrounding faraday cage. Our specific setup can be found in Table 1.

In order to keep the solutions applied to the inner chamber separated until they are needed during recording, a multivalve electrical controller is attached to an in-line heater that keeps the applied solutions at a temperature matching that of recording solution. For mechanical stimulation, standard von Frey filaments are used. A digitally controlled mechanical stimulator can be used to deliver stepwise forces to receptive fields in the muscles that are localized to the center of the paw (Aurora). Hot and cold stimulations are performed by manually delivering 0.9% saline to the desired receptive field with 10-mL syringes. A suction electrode connected to the stimulus isolation units is used for nerve stimulation. For stimulation of the receptive field in the muscle, use a custom-made concentric bipolar electrode with a rounded tip (125 μm outer diameter/25 μm inner pole).

Table 1
Electronics setup

Device	Model and make
Digitizer	CED micro3 1401
Amplifier	Axoclamp 900A
Headstage	HS-9A x1
Stimulator units	WPI A365
Recording software	Spike 2
Micromanipulator controller	Newport ESP301 with remote

A custom-made acrylic rack with adjustable metal clamps is used to place and hold the recording dish at a working distance from the dissection scope used during recordings. All tubing and perfusion systems are held by standard ring stands and clamps or constructed in the lab to fit with the room and other conditions in the recording setup. The sharp tip electrodes are fabricated from quartz capillaries with inner filament 10 cm long, 1.2 mm outer diameter, and 0.6 mm inner diameter (Sutter Instruments). Quartz is the material of choice as it has greater strength and flexibility to impale DRG neurons. Because of this material choice, a laser-based micropipette puller is required (Sutter Instruments) to make the sharp tip electrodes.

3   Methods

The dissection of the preparation varies depending on the limb or muscles of interest. For outlines of the suggested skin incisions, see Fig. 2. We will describe the general dissecting steps and address the adequate methodological variations where indicated. There are also some special considerations that will improve the yield during recording. These will be discussed at the end of this section. The muscles are extremely sensitive to oxygen deprivation while dissecting and preparing for recording; therefore, speed of dissection is an important consideration.

3.1 Intracardial Perfusion and Extraction of Target Tissues

Mice are first anesthetized with a mixture of ketamine and xylazine (100 mg/kg and 10 mg/kg). Then animals are shaved and intracardially perfused with the ice-cold, oxygenated aCSF. During this step, it is important to achieve continuous perfusion of the mouse and the fastest possible removal of the isolated quadrant from the animal to the dissection dish, where the preparation will then be fully isolated. To start perfusion, the mouse should be secured in a supine position (see **Note 1**). A small amount of heparin (~0.1 mL) is used as an anti-coagulant and injected into the left ventricle prior to perfusion with aCSF. To facilitate a constant perfusion of the animal during the quadrant isolation steps, use a butterfly style needle that can be fixed to the heart by applying a small coat of cyanoacrylate glue. After puncturing the right atrium and initiating cardiac perfusion, the skin on the ventral side of the animal linking the targeted limb to the main body can then be incised so the next step can proceed smoothly. If the targeted limb is the forepaw of the mouse, the skin from the axillar region can be carefully cut, being mindful that the cuts should be as superficial as possible to avoid injuring the brachial plexus. If the target is the hind paw, the skin in the inguinal region and around the tail can be also incised to aid in isolating the lower quadrant. Next, the mouse will be turned to the prone position, leaving the needle intact in order to perform a laminectomy.

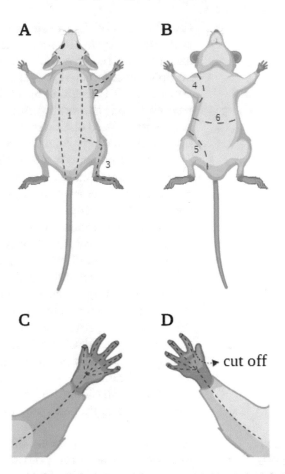

Fig. 2 Incision guideline. Dorsal view of the mouse (**a**). Line 1 depicts the incision pattern to expose the spinal cord, line 2 shows the area where the incision on the forepaw can be made safely. Line 3 shows an incision line to remove the skin of the hind paw. Ventral view of the mouse (**b**). Lines 4 and 5 show the relationship of the ventral skin incisions performed at the time of perfusion in relationship with the incision made to initiate the cardiac perfusion. Dorsal view C and ventral view D of the forepaw. Lines detail areas of hairy (**c**) and glabrous (**d**) skin to cut during dissection. These recommended incision lines are all applicable to the hind paw dissection with the exception of the removal of the first digit as shown in **c**

As soon as the mouse is in the prone position, all limbs can be secured to the dissection tray - taking care not to stretch the limb associated with the quadrant being isolated for further dissection. Make a mid-line skin incision to expose the vertebral column. If the limb of interest is the forepaw, laminectomy is performed from caudal to rostral starting around T10 to all the way up to the base of the skull. In order to dissect the forepaw safely, as soon as the spine is exposed, care should be taken not to overstretch or damage the brachial plexus on the side of interest. The paraspinal muscles should be carefully removed from both sides of the spine. The dorsal region of the ribcage should be completely removed on the contralateral side. The most caudal 1–2 ribs on the side of interest should also be removed. From this point, the spine can be completely transected by separating at the base of the skull and around the T10 region (*see* **Note 2**). The attached viscera can be removed from the ventral side and the forelimb along with spinal segments isolated can be transferred to the dissection dish containing the circulating ice cold aCSF.

If the hind paw is the limb of interest, laminectomy is performed rostral to caudal beginning around T10 and ending around S3. Muscle tissue around the spine is again removed. Do not cut too deep when approaching the hip region, as there is risk of damaging the sciatic nerve. Once the paraspinal muscles have been removed and the dorsal region of the ribcage, latissimus dorsi, and the lumbodorsal fascia have been detached from the spine, the laminectomy should be performed (*see* **Note 3**). Once the laminectomy is completed, the contralateral limb can be detached by completely separating the lateral pelvic bones from the sacral bone. After the contralateral limb is removed, the spine can be transected at the base of the tail and around T10. The remaining tissues attached to the spine can be removed, and this lower quadrant with spine attached should be transferred to the dissecting dish. Duration of these steps is variable but, in ideal conditions, should be approximately 10 min.

3.2 Isolation of the Muscles, Nerves, DRGs, and Spinal Cord

3.2.1 Forepaw Dissection

First, the extracted spine should be pinned, and the forelimb extended in the prone position. The muscles around the scapula can be pinned on the dorsal region using the bone as a shield to prevent the pins from injuring the brachial plexus. This will stabilize the shoulder during the following steps (*see* **Note 4**). The remaining hairy skin of the forelimb is then removed including hairy skin present on each digit. This allows rapid exposure of the forepaw muscles to the circulating aCSF. Next, perform a hemisection of the spinal cord down the dorsal columns on the contralateral side. The brachial plexus is then exposed. This section should be done slowly as it is easy to inadvertently injure the nerves. The nerve bundle containing the median and ulnar nerves will be the most anterior. The radial nerve will project toward the dorsolateral region of the upper forelimb, and it can be safely cut at this point, in addition to the other very thin nerve branches in this region. The median and ulnar nerves will project as a bundle on the ventral surface (*see* **Note 5**). The nerve bundle projecting distally can now be dissected slowly up to the elbow, taking care of removing the surrounding adipose tissue and being especially careful of not stretching the nerves. Once the nerve has been dissected up to the elbow, the split between median and ulnar nerves will be evident.

Once the hemisection of the spinal cord is completed and the nerves are loosened from the upper forelimb, the glabrous skin can be carefully removed from the forepaw and digits (*see* **Note 6**). Next, the median nerve can be dissected up to the carpal tunnel region by following the nerve track. Then the ulnar nerve is isolated. To avoid damage, the ulnar nerve requires extreme care in dissection around the acromial region. Once liberated, the nerve can be easily dissected to the wrist. The final steps include isolating both the median and ulnar nerves at the carpal tunnel region, cleaning them of any remaining adipose tissue and separating the innervated forepaw from the rest of the forelimb. This can be done by cutting the forepaw between the elbow and wrist. The forepaw now can be pinned between the ulna and the radius bones and at the second and fifth digits.

The nerves within the brachial plexus are then further isolated so that the spinal nerves become evident. Finally, the DRGs and bone on the contralateral side can be removed with care to not injure the DRGs of interest (C7-T1) on the ipsilateral side. The ventral roots of each ipsilateral DRG are then transected, and the remaining dura is left intact, so it can be used during the final setup. Excess vertebrae and spinal cord can be trimmed so that C5 to T2 is left intact. Optimal dissection time is <120 min.

3.2.2 Hind Limb

With the preparation placed in the new dissection dish, pin the spine to the silicone with a pin at the most rostral and caudal ends. It is easiest to go through an intervertebral disk—but avoid injuring the spinal cord. Pin out the leg at the digits at approximately a 45-degree angle relative to the spine, gently stretched, with the hairy skin facing upwards. Once the leg is secured, the hairy skin can be removed. Start by making an incision on the lateral face of the leg, and go down all the way to the ankle. Avoid the medial face of the heel, as this is where the tibial nerve will enter the plantar region (*see* **Note 7**).

Next, locate the dorsal midline in the spinal cord and perform a partial hemisection of the cord. The first cut can be performed in the rostral portion of the cord. Then cut slowly toward the sacral region. Extreme care is needed to avoid damaging the dorsal roots of the side of interest and to maintain the integrity of the ipsilateral side of the cord (*see* **Note 8**).

Once the hemisection of the spinal cord is completed, return to the leg and remove the glabrous skin. First unpin all the distal structures, flip the paw around to expose the glabrous skin and re-pin at the digits. For additional stability, another pin can be used in the anterolateral region of the leg, avoiding the tibial nerve near the gastrocnemius muscle. The first incision in the heel should also be on the lateral face of the paw, avoiding the innervation territory of the nerve (*see* **Note 9**). Next, cut through the midline of each digit. Once this is done, remove all remaining skin.

Isolate the sciatic nerve near the biceps femoris muscles (*see* **Note 10**). Slowly expose the nerve, dissecting distally to the popliteal region. Stop dissecting the path of the nerve at this point. Clean the sciatic nerve of all surrounding tissue and axotomize small nerve branches that do not reach the popliteal region (*see* **Note 11**). This is where the sciatic nerve splits into three main branches: the common peroneal, the sural, and the tibial nerves. The tibial nerve can be identified as it moves under the gastrocnemius muscle and next to the tibia. Common peroneal, sural, and gastrocnemius branches can be transected. Isolate the tibial nerve up to the heel of the hind paw. Once this step is completed, the nerve can be cleared of excess adipose or muscle tissue back toward the hip.

Once the nerve is clean and freely moving, detach pins from the leg. Leave anchor points only on the digits, ankle area and spine. Use closed forceps to go under the nerve and gently lift it around the heel region. Using large scissors, cut the bone just above the ankle to free the paw from the rest of the leg. Now the paw will only be attached to the spine by the isolated nerve. Pin the free leg again, away from the paw, and work on liberating the nerve from the pelvic bones. The sacral joint can be separated to release the leg either by cutting through it or by blunt dissection; the latter is the easiest way to separate the leg from the spine, especially in younger animals. Once this is completed, the surrounding muscle is removed to completely detach the leg from the rest of the preparation. The only remaining tissues should be the spine and spinal cord, with the attached sciatic/tibial nerve and the hind paw in continuity. The remaining muscle over the lateral processes of the spine are then removed and the contralateral side of the vertebral column is separated. Avoid cutting the dorsal or ventral roots of the target side. The final step of the dissection is performed in the recording dish.

3.3 Transfer of the Preparation and Setup in the Recording Chamber

The recording chamber should be ready with aCSF circulating prior to transfer. The recording chamber has separated inflows of aCSF to each chamber. We use gravity flow to allow the aCSF to reach the chambers and actively pump it out using custom-made draining tubes placed around the inner and outer chambers. The inflow to the DRG/spinal cord chamber is covered by a thin metal grid to prevent the inflow of any recirculated debris. Mount the paw in the muscle stimulation chamber on top of a custom-made metal crate (~3 mm high). The crate should have a metal grid on top in order to pin through it and secure the paw while allowing for a thorough perfusion with oxygenated aCSF. The aCSF should be chilled on ice during the transfer and setup steps. The aCSF will not be slowly heated until the preparation is ready for recording. Both the inner and the outer chambers are connected by a small opening in the inner chamber, wide enough to allow for the passage of the nerve (s). Seal this area with petrolatum. This opening will eventually be manipulated to allow passage of the nerve(s) from the outer to the inner chamber. It will then be re-sealed once the setup is complete using petrolatum to create a hydrophobic barrier and prevent mixture of the solutions from the two chambers.

In the dissection dish, mount the preparation on a very thin silicone platform (around 2.5 cm by 1.5 cm) that can be pinned to the surface of the dish. This should be done slowly to ensure that the pins in the spine and in the paw do not damage the tissue. Once the preparation is mounted on this new thin platform, we use a small weigh boat containing aCSF to carry the platform with the preparation to the recording dish containing its own circulating aCSF to both inner and outer chambers. Now the preparation can be taken from the platform and pinned to the outer chamber (*see* **Note 12**).

Using a small tool, make a canal through the "petrolatum dam" on the stimulation chamber. This should be only wide enough to let the nerve(s) pass through it. Drain the inner chamber, quickly unpin the paw, and lift it, let the nerve fall into the canal in the dam and place the paw on the metal mesh platform in the inner chamber. Secure the paw to the crate with one pin on the most lateral digits and one in the ankle/wrist area. Replace any lost petroleum jelly over the nerve to seal the chamber and allow the inner chamber to be refilled with its aCSF.

Proceed to pin the spine to the silicone in the outer chamber so that the DRGs targeted for intracellular recording are aligned with the center of the petrolatum dam, but separated from the inner chamber by about 8–9 mm (*see* **Note 13**). There should be enough room from the spine to the inflow grid so that the spinal cord can be taken out and pinned to the recording dish. Next, cut the ventral roots of each DRG. Then slowly insert a pin in the spinal cord and stretch the dorsal roots gently so that the spinal cord is moved off of the vertebrae and pinned to the silicone in the outer chamber. There should be a small amount of tension in the roots but not enough to damage them. Pin back the remaining dura mater over each DRG into the corresponding intervertebral disc. Once these steps are completed, use fine scissors to slowly remove the lateral processes over the DRGs on the target side (*see* **Note 14**).

Once the DRGs are exposed and all the tissues are pinned down, the overlying sheath of the DRGs can be removed. This is the last step of the dissection procedure. Once completed, the preparation will be ready for recording. Using two sets of fine tip #5 forceps, slowly touch the top of the DRG with one set to see the overlying tissue sheath (*see* **Note 15**). Carefully grab only the membrane covering the cells of the DRG, lift it and pierce it with the other set of forceps. From this point on, it will be easy to identify which part of the DRG still has an overlying sheath.

Remove all of the overlying membrane over each DRG to be recorded. Once this is done, the recording setup can be completed (*see* **Note 16**).

Final checklist prior to recording:

- Ensure that there are no gaps between the inner and outer chambers of the recording dish. aCSF can drain between chambers if there are small gaps which will cause overflow of an aCSF flask.
- Verify that both the muscle and the spinal cord are completely covered by the circulating aCSF, flow is constant, and drainage is uninterrupted.
- Place a reference wire connected to the recording head stage into the outer recording chamber.
- Place the suction electrode(s) connected to the stimulus isolation units on the side of the nerve(s) to elicit electrical stimuli. If all of this is working properly, the preparation is ready for recording.

For a final setup of the recording chamber for both preparations, *see* Fig. 3.

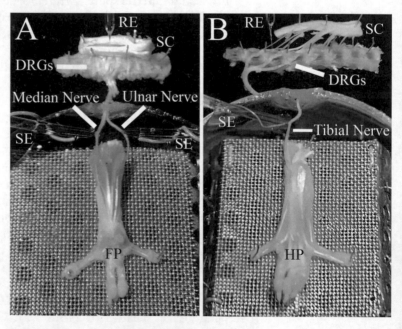

Fig. 3 Ex vivo muscle afferent recording preparations. (**a**) Image of the forepaw muscle ex vivo preparation set up in the recording chamber. (**b**) Hind paw muscle ex vivo preparation in a similar setup. RE: recording electrode, SC: spinal cord, DRGs: dorsal root ganglia, SE: suction electrode, FP: fore paw, HP: hind paw

3.4 Recording Procedures, Stimulation Protocols, and Intracellular Staining

Recordings are performed using Spike 2 software. However, any software package able to receive and track information from the digitized signal can be used. Files can be stored for later analysis offline. To start recording, a sharp-tipped quartz microelectrode filled at the tip with 5% neurobiotin in 1 M Potassium Acetate should be attached to a pipette holder that is attached to the recording head stage. When using this technique to characterize the molecular profiles of functionally identified cells (e.g., Ross et al. 2016), Lucifer Yellow (5%) diluted in 0.1 M LiCl is used in the electrode. This will be directly connected to the head stage, which is connected to the amplifier and the digitizer (as described in Materials). Lower the electrode into the solution and manually place it over of the target DRG using the dissection microscope to get the tip of the electrode as close to the DRG as possible without penetrating the tissue. Once the electrode is placed over the DRG, initiate the recording/stimulation protocol from the computer. Our stimulation protocol gives an electrical stimulation pulse (~0.5–2 mA) through the suction electrode every 2 s followed by 2 square waves (700 ms from the pulse, 10 ms duration, 600 ms apart), to provide enough time to lower the sharp tipped electrode and check electrode impedance. Using the computer-controlled micromanipulator, slowly lower the electrode into the DRG cells. While the recording electrode is being lowered, an orthodromic search stimulus should be delivered to the nerve via the suction electrode. Advance the electrode with the micromanipulator in small (~10 μm) increments and, if available, utilize a remote buzzer connected to the amplifier set to 500 ms to facilitate penetration of the electrode into the cell body. In general, do not penetrate the DRG more than ~300 μm to avoid advancing the electrode too far (*see* **Note 17**).

Once a cell has been impaled, the membrane potential will decrease to approximately −50 to −70 mV, in ideal conditions. If the impaled neuron shows an action potential discharge after the electrical stimulation of the nerve, the receptive field of the neuron can be located by first turning off the suction electrode SIU and stimulating the target muscle tissue with a concentric bipolar electrode. The electrical discharge in the receptive field should be similar to the response observed after nerve stimulation. Once the receptive field has been determined, the stimulation protocols can commence.

If recording from a forepaw preparation, the primary afferents going through the median nerve are usually in C7 and the cephalic side of C8. The afferents from the ulnar nerve are frequently on the caudal side of C8 and in T1. If recording from the mouse hind paw preparation, the target DRGs are L3 and L4 (*see* **Note 18**). Multiple stimuli can then be delivered to the receptive field in the muscle. Mechanical, heat, and cold are the most common, but specific stimuli can be tailored to unique experimental conditions. The described setup allows for a complete change of the bath solution in the inner chamber for the simulation of different environmental conditions (such as innocuous or noxious metabolite mixtures) for the target muscle [14, 17–19]. However, it is recommended to begin with mechanical stimuli so that thermal or chemical stimuli do not influence mechanical responsiveness.

Finally, recorded neurons can be filled by iontophoresis with neurobiotin or Lucifer Yellow for later immunohistochemical or molecular characterization [16–19]. Usual staining protocols create a voltage gradient that iontophoretically injects the neurobiotin or fluorescent dye into the recorded cell after physiological characterization. Iontophoretic injection of 5% neurobiotin can be completed in approximately 3 min at 1.1 nA of current. Ten minutes at -2.1 nA current is required for dyes such as LY [46] (*see* **Note 19**). Be sure to document the location of the stained and characterized cell in the DRG.

3.5 Limitations and Alternatives

As with any procedure, there are some limitations with this methodology. While it allows characterization of response properties of an afferent to a variety of stimuli, the duration of stimulation protocol is limited, as we are impaling cells in a very dynamic setting. Most stimulation and labeling protocols can vary from 5 to 10 min. Longer recordings (hours) from a single unit is typically unsuccessful. To perform recordings for prolonged time periods, a teased fiber technique is a viable alternative. Another limitation of this technique is that it is not in vivo. Thus, it eliminates some of the environmental characteristics of skeletal muscle, such as circulating cells and other blood-related factors, that could influence afferent response. Nevertheless, the ex vivo setting allows us to target receptive fields in muscle specifically without the confound of the overlying skin.

4   Notes

1. Limit the amount of stretch that the targeted limb receives as much as possible. This is of special importance in the case of the forepaw—as the brachial plexus is quite sensitive to stretching. It is important to not overextend the limb in order to avoid damage to the nerves of interest.

2. All anatomical reference points are determined based on the position of the last rib. The T13 DRG can be found immediately below the last rib. This is also a useful reference when dissecting the lumbar region.

3. During the laminectomy step, care should be taken to keep the dura mater covering the spinal cord intact during the laminectomy as this will play a part later in the dissection.

4. An easy way to prevent excessive stretching of the brachial plexus is to gently push and pin the shoulder blade close to the spine.

5. While dissecting the nerve bundles coming out of the brachial plexus, the clavicle will need to be cut. It is recommended to slowly adjust the pins in the arm to let it "roll out" so that the shoulder blade ends up under the nerve bundles and exposes the anterior view of the forelimb.

6. Glabrous skin in the forepaw is closely attached to the underlying muscles. It usually needs to be removed in small increments. Also, the first digit is extremely short, so it must be removed during this step. Caution is needed to avoid injuring the nerves in the carpal tunnel region.

7. Remove all the skin by making small cuts and always pointing the dissection scissors toward the skin-nerve in the direction of the muscles.

8. Performing a hemisection has two purposes: first, it will improve how the aCSF reaches the spinal cord and the DRGs, and second, it will isolate a section of dura mater that will be used to pin back the DRGs for recording. This helps secure the position of the DRG and gain sufficient stability to obtain successful intracellular recordings.

9. During this step, it is easy to injure the plantaris muscles. Therefore, incisions should be small and executed slowly.

10. The white line of fat on the lateral surface of the leg can be used to best determine the location of the sciatic nerve within the muscles.

11. Cleaning the nerves of all surrounding tissue is crucial for the success of this preparation. There is a layer of adipose tissue that frequently pools around nerves at joints and close to points of insertion. Un-dissected adipose tissue will make the nerve float in your dissecting solution—which can affect recording.

12. *Transfer considerations.* The source of oxygen for the dissection dish and the recording dish can be the same, all that is needed is a valve system that allows switching of the gas flow immediately before the transfer (after the preparation is mounted on the silicone). Positioning is not important when initially transferring the preparation, the initial goal is to get the preparation in the recording dish covered by the oxygenated aCSF.

13. Make sure the spine is pinned close enough to the stimulation chamber so that the nerve and connected paw will be able to reach the appropriate area in the inner chamber.

14. During this step, it is very easy to damage either the DRG or the spinal nerves, cut the bone slowly to expose the DRG.

15. The sheath is transparent, and this strategy allows for easier observation of the overlying membrane on the DRGs.

16. The only variation in the transfer procedure for the forepaw is that the DRG exposure can be performed in the dissection dish. There is usually no need to pin the flap of dura at every intervertebral disk in the cervical region and the roots are significantly shorter, requiring more gentle manipulation to avoid causing an avulsion injury.

17. Electrical discharge delivered by the suction electrode should induce a muscle twitch. This is a great indicator of the health of in paw muscles. Lack of a contraction may indicate a need for troubleshooting.

18. Sensory neurons innervating the hind paw are usually (but not exclusively) located in the region of L4 closest to the dorsal root, or in the region of L3 closest to the spinal nerve. This should be used as a guideline to start searching for sensory afferents. This however does not suggest that sensory neurons have a specific anatomical position in the DRG that is related to the tissues that they innervate.

19. The use of these staining molecules are not the only ones available. Different options can be used depending on experimenter needs.

Acknowledgments

This work was supported by grants to MPJ from the NIH (R01AR064551 and R01NS113965).

References

1. Lloyd DPC (1943) Neuron patterns controlling transmission of ipsilateral hind limb reflexes in cat. J Neurophysiol 6(4): 293–315

2. McCord JL, Kaufman MP (2010) Reflex autonomic responses evoked by group III and IV muscle afferents. In: Kruger L, Light AR (eds) Translational pain research: from mouse to man. CRC Press/Taylor & Francis Llc., Boca Raton, FL

3. Moore JC (1984) The Golgi tendon organ: a review and update. Am J Occup Ther 38(4): 227–236. https://doi.org/10.5014/ajot.38.4.227

4. Bewick GS, Banks RW (2015) Mechanotransduction in the muscle spindle. Pflugers Arch 467(1):175–190. https://doi.org/10.1007/s00424-014-1536-9

5. Paintal AS (1960) Functional analysis of group III afferent fibres of mammalian muscles. J Physiol 152(2):250–270

6. Stacey MJ (1969) Free nerve endings in skeletal muscle of the cat. J Anat 105(Pt 2): 231–254

7. von Düring M, Andres KH (1990) Topography and ultrastructure of group II and IV nerve terminals of the Cat's gastrocnemius-soleus muscle. In: Zenker W, Neuhuber WL (eds) The primary afferent neuron. Springer, Boston, MA, pp 35–41. https://doi.org/10.1007/978-1-4613-0579-8_3

8. Messlinger K (1996) Functional morphology of nociceptive and other fine sensory endings (free nerve endings) in different tissues. Prog Brain Res 113:273–298

9. Woodbury CJ, Zwick M, Wang S, Lawson JJ, Caterina MJ, Koltzenburg M, Albers KM, Koerber HR, Davis BM (2004) Nociceptors lacking TRPV1 and TRPV2 have normal heat responses. J Neurosci 24(28):6410–6415. https://doi.org/10.1523/JNEUROSCI.1421-04.2004

10. McCoy DD, Knowlton WM, McKemy DD (2011) Scraping through the ice: uncovering the role of TRPM8 in cold transduction. Am J Physiol Regul Integr Comp Physiol 300(6):R1278–R1287. https://doi.org/10.1152/ajpregu.00631.2010

11. McKemy DD, Neuhausser WM, Julius D (2002) Identification of a cold receptor reveals a general role for TRP channels in thermosensation. Nature 416(6876):52–58. https://doi.org/10.1038/nature719

12. Molliver DC, Immke DC, Fierro L, Pare M, Rice FL, McCleskey EW (2005) ASIC3, an acid-sensing ion channel, is expressed in metaboreceptor sensory neurons. Mol Pain 1:35. https://doi.org/10.1186/1744-8069-1-35

13. Sluka KA, Radhakrishnan R, Benson CJ, Eshcol JO, Price MP, Babinski K, Audette KM, Yeomans DC, Wilson SP (2007) ASIC3 in muscle mediates mechanical, but not heat, hyperalgesia associated with muscle inflammation. Pain 129(1–2):102–112. https://doi.org/10.1016/j.pain.2006.09.038

14. Light AR, Hughen RW, Zhang J, Rainier J, Liu Z, Lee J (2008) Dorsal root ganglion neurons innervating skeletal muscle respond to physiological combinations of protons, ATP, and lactate mediated by ASIC, P2X, and TRPV1. J Neurophysiol 100(3):1184–1201. https://doi.org/10.1152/jn.01344.2007

15. Fujii Y, Ozaki N, Taguchi T, Mizumura K, Furukawa K, Sugiura Y (2008) TRP channels and ASICs mediate mechanical hyperalgesia in models of inflammatory muscle pain and delayed onset muscle soreness. Pain 140(2):292–304. https://doi.org/10.1016/j.pain.2008.08.013

16. Ross JL, Queme LF, Cohen ER, Green KJ, Lu P, Shank AT, An S, Hudgins RC, Jankowski MP (2016) Muscle IL1beta drives ischemic myalgia via ASIC3-mediated sensory neuron sensitization. J Neurosci 36(26):6857–6871. https://doi.org/10.1523/JNEUROSCI.4582-15.2016

17. Jankowski MP, Rau KK, Ekmann KM, Anderson CE, Koerber HR (2013) Comprehensive phenotyping of group III and IV muscle afferents in mouse. J Neurophysiol 109(9):2374–2381. https://doi.org/10.1152/jn.01067.2012

18. Ross JL, Queme LF, Shank AT, Hudgins RC, Jankowski MP (2014) Sensitization of group III and IV muscle afferents in the mouse after ischemia and reperfusion injury. J Pain 15(12):1257–1270. https://doi.org/10.1016/j.jpain.2014.09.003

19. Queme LF, Ross JL, Lu P, Hudgins RC, Jankowski MP (2016) Dual modulation of nociception and cardiovascular reflexes during peripheral ischemia through P2Y1 receptor-dependent sensitization of muscle afferents. J Neurosci 36(1):19–30. https://doi.org/10.1523/jneurosci.2856-15.2016

20. Murase S, Terazawa E, Queme F, Ota H, Matsuda T, Hirate K, Kozaki Y, Katanosaka K, Taguchi T, Urai H, Mizumura K (2010) Bradykinin and nerve growth factor play pivotal roles in muscular mechanical hyperalgesia after exercise (delayed-onset muscle soreness). J Neurosci 30(10):3752–3761. https://doi.org/10.1523/JNEUROSCI.3803-09.2010

21. Sluka KA (1996) Pain mechanisms involved in musculoskeletal disorders. J Orthop Sports Phys Ther 24(4):240–254. https://doi.org/10.2519/jospt.1996.24.4.240

22. Gautam M, Benson CJ, Ranier JD, Light AR, Sluka KA (2012) ASICs do not play a role in maintaining hyperalgesia induced by repeated intramuscular acid injections. Pain Res Treat 2012:817347. https://doi.org/10.1155/2012/817347

23. Gregory NS, Brito RG, Fusaro MC, Sluka KA (2016) ASIC3 is required for development of fatigue-induced hyperalgesia. Mol Neurobiol 53(2):1020–1030. https://doi.org/10.1007/s12035-014-9055-4

24. Birdsong WT, Fierro L, Williams FG, Spelta V, Naves LA, Knowles M, Marsh-Haffner J, Adelman JP, Almers W, Elde RP, McCleskey EW (2010) Sensing muscle ischemia: coincident detection of acid and ATP via interplay of two ion channels. Neuron 68(4):739–749. https://doi.org/10.1016/j.neuron.2010.09.029

25. Flegel C, Schobel N, Altmuller J, Becker C, Tannapfel A, Hatt H, Gisselmann G (2015) RNA-Seq analysis of human trigeminal and dorsal root ganglia with a focus on chemoreceptors. PLoS One 10(6):e0128951. https://doi.org/10.1371/journal.pone.0128951

26. Hu G, Huang K, Hu Y, Du G, Xue Z, Zhu X, Fan G (2016) Single-cell RNA-seq reveals distinct injury responses in different types of DRG sensory neurons. Sci Rep 6:31851. https://doi.org/10.1038/srep31851

27. Ray P, Torck A, Quigley L, Wangzhou A, Neiman M, Rao C, Lam T, Kim JY, Kim TH, Zhang MQ, Dussor G, Price TJ (2018) Comparative transcriptome profiling of the human and mouse dorsal root ganglia: an RNA-seq-based resource for pain and sensory neuroscience research. Pain 159(7):1325–1345. https://doi.org/10.1097/j.pain.0000000000001217

28. Queme LF, Ross JL, Ford Z, Katragadda B, Green K, Hudgins RC, Jankowski MP (2016) Upregulation of GDNF family receptor α1 in the dorsal root ganglia regulates pain-related behaviors and the cardiovascular response to exercise after ischemia with reperfusion injury. Paper presented at the 16th World Congress on Pain, Yokohama, Japan

29. Paintal AS (1959) Intramuscular propagation of sensory impulses. J Physiol 148:240–251. https://doi.org/10.1113/jphysiol.1959.sp006285

30. Paintal AS (1961) Participation by pressure-pain receptors of mammalian muscles in the flexion reflex. J Physiol 156:498–514. https://doi.org/10.1113/jphysiol.1961.sp006689

31. McCloskey DI, Mitchell JH (1972) Reflex cardiovascular and respiratory responses originating in exercising muscle. J Physiol 224(1):173–186

32. Mense S, Schmidt RF (1974) Activation of group IV afferent units from muscle by algesic agents. Brain Res 72(2):305–310

33. Kumazawa T, Mizumura K (1976) The polymodal C-fiber receptor in the muscle of the dog. Brain Res 101(3):589–593

34. Kaufman MP, Rybicki KJ, Waldrop TG, Ordway GA (1984) Effect of ischemia on responses of group III and IV afferents to contraction. J Appl Physiol Respir Environ Exerc Physiol 57(3):644–650

35. Adreani CM, Hill JM, Kaufman MP (1997) Responses of group III and IV muscle afferents to dynamic exercise. J Appl Physiol (1985) 82(6):1811–1817

36. Mense S, Meyer H (1985) Different types of slowly conducting afferent units in cat skeletal muscle and tendon. J Physiol 363:403–417

37. Mense S, Prabhakar NR (1986) Spinal termination of nociceptive afferent fibres from deep tissues in the cat. Neurosci Lett 66(2):169–174

38. Taguchi T, Sato J, Mizumura K (2005) Augmented mechanical response of muscle thin-fiber sensory receptors recorded from rat muscle-nerve preparations in vitro after eccentric contraction. J Neurophysiol 94(4):2822–2831. https://doi.org/10.1152/jn.00470.2005

39. Xu J, Brennan TJ (2009) Comparison of skin incision vs. skin plus deep tissue incision on ongoing pain and spontaneous activity in dorsal horn neurons. Pain 144(3):329–339. https://doi.org/10.1016/j.pain.2009.05.019

40. Xu J, Gu H, Brennan TJ (2010) Increased sensitivity of group III and group IV afferents from incised muscle in vitro. Pain 151(3):744–755. https://doi.org/10.1016/j.pain.2010.09.003

41. Reeh PW (1986) Sensory receptors in mammalian skin in an in vitro preparation. Neurosci Lett 66(2):141–146. https://doi.org/10.1016/0304-3940(86)90180-1

42. Koltzenburg M, Stucky CL, Lewin GR (1997) Receptive properties of mouse sensory neurons innervating hairy skin. J Neurophysiol 78(4):1841–1850. https://doi.org/10.1152/jn.1997.78.4.1841

43. Wenk HN, McCleskey EW (2007) A novel mouse skeletal muscle-nerve preparation and in vitro model of ischemia. J Neurosci Methods 159(2):244–251. https://doi.org/10.1016/j.jneumeth.2006.07.021

44. Koerber H, Woodbury C (2002) Comprehensive phenotyping of sensory neurons using an ex vivo somatosensory system. Physiol Behav 77(4–5):589–594. https://doi.org/10.1016/s0031-9384(02)00904-6

45. Ross JL, Queme LF, Lamb JE, Green KJ, Ford ZK, Jankowski MP (2018) Interleukin 1beta inhibition contributes to the antinociceptive effects of voluntary exercise on ischemia/reperfusion-induced hypersensitivity. Pain 159(2):380–392. https://doi.org/10.1097/j.pain.0000000000001094

46. Hanani M (2012) Lucifer yellow - an angel rather than the devil. J Cell Mol Med 16(1):22–31. https://doi.org/10.1111/j.1582-4934.2011.01378.x

Chapter 7

Electrophysiological Recording Techniques from Human Dorsal Root Ganglion

Jamie K. Moy, Emanuel Loeza-Alcocer, and Michael S. Gold

Abstract

While genetic tools and techniques developed in the mouse have enabled researchers to address questions many thought could never be answered, the persistently high rate at which therapeutic targets fail to translate into novel treatments for human patients has forced some to identify ways to validate preclinical findings prior to initiating expensive and time-consuming clinical trials. Human tissue recovered from patients undergoing surgery and/or from organ donors is readily accessible and. in contrast to heterologous expression systems, enables the study of therapeutic targets in the appropriate species in the native environment. Human dorsal root ganglion (DRG) neurons may be particularly useful in this regard, as they can be maintained in culture and therefore may not only be used for functional analyses, but functional analyses over days, weeks, or months. Because sensory nerves are recovered with DRG, at lengths sufficient for compound action potential recording, this tissue may also be used for functional, as well as biochemical and molecular biological analyses. Described in this chapter are methods used for the recovery and processing of human DRG, to enable the study of isolated human sensory neurons in culture with patch-clamp and/or live-cell imaging techniques.

Key words Human cell culture, Patch clamp, In vitro, Dissociated neurons, Extracellular recordings

1 Introduction

Despite a seemingly endless list of potentially viable therapeutic targets for the treatment of pain almost all of which appear to account for ~100% of the pain behavior assessed in preclinical studies, the rate at which these results are successfully translated to the clinic remains dismally low. There are, of course, many factors that contribute to this failure rate, ranging from the preclinical models used [1, 2] to the ability to identify the patient population likely to be responsive to the intervention being tested [3–5]. We would suggest, however, species differences are likely one of

Supplementary Information The online version of this chapter (https://doi.org/10.1007/978-1-0716-2039-7_7) contains supplementary material, which is available to authorized users.

Rebecca P. Seal (ed.), *Contemporary Approaches to the Study of Pain: From Molecules to Neural Networks*, Neuromethods, vol. 178, https://doi.org/10.1007/978-1-0716-2039-7_7, © Springer Science+Business Media, LLC, part of Springer Nature 2022

the more important factors that are often overlooked. One potential solution to this problem is the use of human tissue. And while there is not enough tissue available to support even low-throughput screening approaches, the refinement of protocols needed to keep human tissue alive combined with increased access to tissue specimens from surgeries as well as from organ donors for research purposes has enabled a number of investigators to perform critical proof of principle studies of target engagement, potency, and selectivity. For example, Bulmer and colleagues have developed a human gut-nerve preparation [6, 7] that has enabled them to confirm the involvement of a K^+ channel in the control of visceral afferent excitability [8]. Similarly, Hildebrand and others have optimized protocols to study physiological properties of neurons in the human spinal cord dorsal horn, enabling them to confirm a pro-nociceptive role of BDNF in spinal cord slices [9]. Human dorsal root ganglion DRG neurons have received the most attention [6, 9–20], likely because of the ease of access and because the physiological properties of these neurons are likely to enable them to survive the procedures needed to study these neurons ex vivo. Because organ donor tissue may be obtained from otherwise healthy individuals, this tissue may be particularly valuable for comparative analyses. Given how extensively rodents are used in biomedical research in general, and pain research in particular, it has been reassuring to learn that there are many similarities between human and rodent sensory neurons with respect to the biophysical and pharmacological properties of ion channels and receptors [14, 16, 18–22]. However, species differences have also be described and include gene expression [17, 23], relative importance of channel subtypes [13, 19, 20], channel kinetics [17–20], and pharmacology [16, 18–20]. These species differences underscore the need to screen human tissue prior to initiating costly and time-consuming clinical trials, as well as to increase the likelihood of the successful translation of preclinical data to more effective therapeutic interventions. To facilitate the wider use of this valuable resource, we describe the methods adapted from [24] that we have employed to recover human DRG and their associated peripheral and central axons and to study this tissue with electrophysiological and imaging techniques.

2   Materials

Bone saw.
Virchow Skull Breaker.
Hemostat.
Scissors.
Autoclaved glass petri dishes 60 mm (PYREX Reusable Petri Dishes: Lids Fisher Scientific (Cat. #- 08-747A) & DWK life

sciences Kimble KIMAX Petri Dishes: Bottom Fisher Scientific (Cat. #- S31471)).

12 mm Round coverslips with Poly-L-Lysine from Fisher Scientific (Cat. #- BD354085).

Prior to use, oxygenate solution for 10–15 min.

Store all aliquots at −20 °C.

Complete Media aliquots.

NGF (2.5S) mouse submaxillary glands from Millipore Sigma (Cat. # NC011):

Make 10 μg in 1 mL of hDRG complete media (CM) (without NGF) for stock solution.

Aliquot 30 μL in each tube and label "NGF."

Freeze at −20 °C.

Use 30 μL in 30 mL of CM for 10 ng/mL.

Glucose:

15 g of glucose in 100 mL of dDiH₂O.

Sterile filter.

Aliquot 900 μL in each tube and label "glu."

Freeze at −20 °C.

Use 1 aliquot for each 30 mL of CM.

L-Glutamine 200 mM (100× Gibco Cat. # 25030–081 for 100 mL):

Comes as a 200 mM stock solution.

Aliquot 300 μL in each tube and label "gluta."

Use 1 aliquot for each 30 mL of CM for final concentration of 10 μL/mL.

Ascorbic acid:

170 mg in 50 mL of dDiH₂O.

Sterile filter.

Aliquot 450 μL in each tube and label "AA."

Use 1 aliquot for each 30 mL of CM.

Glutathione:

8 mg in 50 mL of dDiH₂O.

Sterile filter.

Aliquot 450 μL in each tube and label "G."

Use 1 aliquot for each 30 mL of CM.

5-Fluoro-2-deoxyuridine + uridine (FRDU):

Dissolve 35 mg of uridine +15 mg of 5-fluoro-2-deoxyuridine in 2 mL of $dDiH_2O$.

Sterile filter.

Aliquot 30 μL in each tube and label "FRDU/U."

Sodium bicarbonate:

Weigh out 60 mg of $NaHCO_3$ in each tube.

Use 1 tube for 30 mL of hDRG complete media.

Dissociation solution aliquots.
Collagenase P from Sigma Aldrich, Roche (Cat. # 11213865001):

Record lot # in back of notebook note, start date.

Dissolve in 1× human collection media for 6.25% w/v.

Aliquot 200 μL in each tube and label "CP."

Use 1 aliquot per 5 mL of dissociation media.

Trypsin from Worthington (Cat. # Is003703 for 1 g):

Record lot # in back of notebook note, start date.

Dissolve in 1× human collection media for concentration of 100 mg/1 mL (10% w/v).

Aliquot 50 μL in each tube and label "T."

Use 1 aliquot per 5 mL of dissociation media.

Deoxyribonuclease I (DNAse) from Worthington Biochemical (Cat. # LS002007 for 100 mg):

Record lot # in back of notebook note, start date.

Dissolve in 1× human collection media for concentration of 7.6 mg/0.38 mL (2% w/v).

Aliquot 75 μL in each tube and label "D."

Use 1 aliquot per 5 mL of dissociation media.

"Enzyme stop" solution aliquots.
Store at −20 °C.
Trypsin Inhibitor from *Glycine max* (soybean) from Sigma (Cat. # T9128):

Record lot # in back of notebook note, start date.

Dissolve in 1× human collection media for concentration of 200 mg/1 mL (20% w/v).

Aliquot 70 μL in each tube and label "TI."

Use 1 aliquot per 7 mL of "stop" solution.

Bovine serum albumin (BSA):

Dissolve 500 mg in 10 mL of 1× human collection media for concentration of 50 mg/1 mL (5% w/v).

Aliquot 150 μL in each tube and label "BSA."

Use 140 μL of aliquot per 7 mL of "stop" solution.

$MgSO_4$:

Make 700 mM $MgSO_4$ stock in $dDiH_2O$ (5.55 g in 50 mL) then sterile filter using 0.22 μm low-binding filter.

Aliquot 100 μL in each tube and label "Mg."

Use 1 aliquot for 7 mL of "stop" solution.

Solutions for culture.
Dissociation Solution

5 mL of 1× hDRG collection media.

1 aliquot (200 μL) of coll. P.

*1 aliquot (50 μL) of trypsin

1 aliquot (75 μL) of DNAse.

Trypsin in-series solution

5 mL of HBSS without Mg++ or Ca++ (Gibco Cat # 14170161).

400 μL of 0.02% EDTA (Sigma Cat # E8008).

1 aliquot (50 μL) of trypsin.

Enzyme stop solution

6.3 mL of 1× hDRG collection media.

1 aliquot (70 μL) of trypsin inhibitor.

1 aliquot (140 μL) of BSA.

100 μL of 700 mM $MgSO_4$.

0.7 mL of Heat-inactivated FBS.

Percoll Gradient (Percoll from Millipore Sigma Cat. # P4937):

Create heavy and light Percoll solutions.

For 1 tube in 15 mL.

Heavy in 15 mL tube	1.1 mL Percoll 2.9 mL Human Basal Media
Light in 5 mL tube	0.5 mL Percoll 3.5 mL Human Basal Media

Slowly and carefully layer "light" on top of "heavy."

Look for line of separation.

For 30 mL of complete media

27 mL of Basal media (+ Vit Sol).

900 μL of glucose.

300 μL of glutamine.

450 μL of ascorbic acid.

450 μL of glutathione.

60 mg of sodium bicarb ($NaHCO_3$).

300 μL of Pen/Strep.

30 μL of FRDU/U.

Sterile filter:

Add	3 mL of Heat-inactivated FBS 30 μL of NGF (10 μg/mL)

Adjust pH to 7.35 with 5 N NaOH ~3.5 mL.
Final volume to 1 L in $dDiH_2O$.
Sterile filter and store at 4 °C.
For 1× collection solution, dilute using sterile-filtered or autoclaved $dDiH_2O$.
Adjust pH to 7.35 with 5 N NaOH.
Final volume to 1 L in $dDiH_2O$.
Sterile filter and aliquot 50 mL.
Store at −20 °C.
Basal media (500 mL):

New Leibovitz L-15 w/o glutamine from Sigma Aldrich (Cat. # L5520).

Remove 50 mL.

Add 50 mL of vitamin solution (Table 1).

Note date.

Do not use if older than 1 month.

3 Methods

Organ donation in the US is largely, if not exclusively, managed by nonprofit organizations. These organizations are responsible for managing the organ donation process (coordinating recipient lists, tissue typing, screening of donor, etc.) in largely non-overlapping geographical regions, the size of which is determined by a number of factors. The size of the geographic region covered has implications for the use of human tissue for research purposes, as a longer travel time between the hospital and laboratory may decrease the viability of the tissue recovered. Specific

Table 1
Vitamin solution (Leibovitz L-15 supplement mix)

Solute	mg/L
Imidazole	600
L-aspartic acid	150
L-glutamic acid	150
Cystine	150
ß-alanine	50
Myo-inositol	100
Choline chloride	100
p-Aminobenzoic acid	50
Fumaric acid	250
Vitamin B$_{12}$	20
Lipoic acid[a]	15 μL

[a]Dissolve 1 g in 2.5 mL methanol

policies and procedures around access to, and use of, tissue from organ donors for research purposes are established by these organizations and are also monitored by Institutional Review Boards. The Committee for Oversight of Research and Clinical Training Involving Decedents, or CORID, is the Board that oversees the use of organ donor tissue for research purposes at the University of Pittsburgh. The US Health Resources & Services Administration enables US adults to register to be an organ donor, and in most states in the US, citizens can register to be an organ donor when applying for a driver's license or ID card. This registration process does not cover the use of tissue from organ donors for research purposes. Rather, authorization for the use of tissue from organ donors must be obtained from next of kin or family members.

In addition to the time of travel post-tissue recovery, there are a number of factors that can also influence cell viability. This includes the manner of death, where trauma resulting in blood around the DRG is generally associated with poor viability. Other factors include the time between cross-clamp (time at which systemic circulation has ceased) and tissue recovery, as well as the speed of the tissue recovery process itself.

3.1 Human DRG Recovery

Human DRG are most easily recovered via a ventral approach. We target L4 and L5 ganglia bilaterally because of their size and ease of access and identification. With the abdominal cavity empty of viscera, the spine and associated muscles are prominent. For orientation, the psoas major is the large muscle group running almost

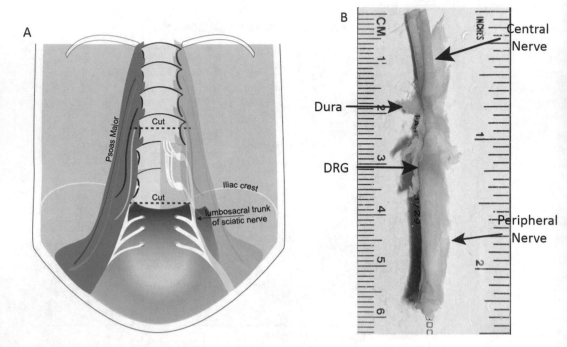

Fig. 1 Dorsal root ganglion dissection from human organ donors. Illustration of the empty abdominal cavity (**a**, left), indicating the psoas major muscle and the nerves connected to L4 and 5 DRG underneath (**a**, right). (**b**) A clean human L5 DRG exposing central nerves, dura, DRG, and peripheral nerve

parallel to the spine (Fig. 1a). Blunt dissection of the muscle from the spine will reveal the sciatic nerve running medial to the muscle body (Fig. 1a, right). The nerve can be well over a 1 cm in diameter and is easily followed rostrally to the point of bifurcation into the L4 and L5 spinal nerves. With a combination of blunt dissection and careful removal of excess muscle, it should be possible to expose the nerves as they enter their respective vertebral foramen. This should be done bilaterally so that once the bone is removed, ganglia can be accessed bilaterally. Once the relevant vertebrae have been identified, a bone saw is used to make four cuts. One is through the middle of the vertebrae rostral to the L4 foramen (Fig. 1a), perpendicular to the spinal column. A second is through the middle of the vertebrae caudal to the L5 foramen, again perpendicular to the spinal column (Fig. 1a). The cuts should be to the full depth of the saw blade with the blade oriented as close to 90° as possible from the plane of the bone. The third and fourth cuts are made on the sides of the vertebrae running between the two perpendicular cuts in a line that runs at the top (ventral) of the foramen. The blade should again be as close to 90° as possible to the lateral surface of the vertebrae. These are the hardest cuts to do well and take a little practice: too deep (dorsal) or angled down, and the blade will cut through the ganglia; too shallow (ventral) or angled up, and there will be too much bone left over the ganglia.

Table 2
Collection solution

For 1 L of 10× collection media dilute 1:10 in sterile H_2O for 1× collection media	
Salts	mM
NaCl	1295
KCl	50
$MgSO_4$	12
$CaCl_2$	10
HEPES free acid	300

The trick is to make the cuts so that when the Virchow Skull Tool is used to pry the vertebrae up, the bones come up en bloc leaving all four ganglia intact and the cauda equine exposed (Fig. 1a, right). Using the hemostat, clamp the spinal (peripheral) nerve of the selected DRG, and then use scissors with a combination of blunt dissection and cutting to free the ganglia from connective tissue in the foramen. Once DRG is freed, the central root should come with the ganglia and can be cut at whatever length is desired (Fig. 1b). If peripheral nerve root recording is to be done, the hemostat should be positioned as far from the ganglia as possible. The ganglion and nerve should be placed immediately into ice cold 1× collection media (Table 2).

3.2 Cleaning up the DRG

Prior to starting, proper PPE should be used including double gloves, gown or lab coat, face mask, and shield. Prepare 7.5 mL of enzyme solution per DRG to be processed, split into 5 mL, i.e., three enzyme solutions for two DRG*. Place the extracted DRG in fresh 1× collection media on ice and identify the peripheral nerve, DRG, and central nerves (Fig. 1b). Isolate the whole central nerves and place it into ice-cold oxygenated Krebs solution. The peripheral and central nerve can be used to record compound action potentials (CAP) identifying changes in Aβ-, Aδ-, and C-waves or conduction velocity (Fig. 2). Gently clean the DRG from the fat, muscle, and dura mater. A clean DRG, absent of dura mater, will have a slightly tan to orange color (Fig. 1b). After cleaning, transfer the DRG into a small 35-mm petri dish with fresh 1× collection media, and cut the DRG into small pieces, approximately the size of ½ rat DRG using a #10 scalpel blade. Separate equally into the prepared enzyme solutions. Because there can a lot of fat surrounding the DRG, it may be necessary to change collection media several times. When doing so, minimize exposure of the ganglia to air.

*For patch experiments, it may be beneficial to treat the cells with trypsin sequentially after collagenase P digestion, which would

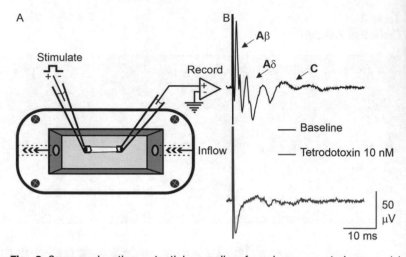

Fig. 2 Compound action potential recording from human central nerve. (**a**) Schematic of the stimulation and recording configuration used to study electrical activity in isolated nerve. Electrically evoked compound action potential in an isolated nerve is measured with a suction electrode. A suction electrode can also be used for stimulating the nerve. The advantage of the suction electrodes is that they provide electrical isolation, avoiding the need for vaseline, oil, agar, or other means to isolate stimulating electrode from the recording electrode, so that it is possible to resolve the propagated activity in the nerve. The nerve is superfused constantly with Krebs solution (at room temperature ~22–25 °C). More physiological temperatures can be used, but it can become difficult to resolve the different fiber types because of the increase in conduction velocity. The nerve is stimulated with a square pulse (0.5 ms duration). (**b**) Fifty times the threshold for Aβ fibers were applied to evoke action potentials in all types of afferents fibers (Aβ, Aδ, and C, black trace). Almost all fiber types are completely and reversibly blocked by tetrodotoxin (10 nM) (red trace), the remaining component of the CAP was eliminated with 100 nM tetrodotoxin (data not shown)

allow the trypsin to work efficiently and leave a cleaner membrane to seal. To do trypsin treatment in series, do not include trypsin in the dissociation solution.

3.3 DRG Dissociation

1. After separating into the enzyme solutions, transfer tissue and enzyme solution into glass petri dishes.

2. Using parafilm, fasten the top and bottom petri dishes then incubate at 37 °C water bath shaking at 70 RPM.

   (a) Water level should be set to ensure contact with the bottom plate, but not so deep that the dishes are submerged.

   (b) Incubation time mainly depends on the age of the donor.

   • <30 years of age, ~1–1.5 h of incubation.

   • Between 30 and 60 years of age, ~1.5–2 h of incubation.

   • ≥60 years of age, may require a second round of digestion using new enzyme solutions.

3. Check on the tissue dissociation ~1 h into incubation time, by using a 1 mL disposable (blue tip) pipette.

   (a) Cut the pipette tip to ~2 mm in diameter.

   (b) Collect a couple of the DRG chunks in the pipette passing them in and out of the pipette tip once or twice.

      • If the tissue begins to break up easily, digestion is complete.

      • If the tissue does not break up, more incubation time is needed. Return tissue to water bath with or without fresh enzyme per above.

   (c) The ganglia may have already started to break up, turning the dissociation media cloudy, which is a good sign the digestion is complete.

4. While tissue is incubating prepare the following solutions:

   (a) The enzyme stop solutions (1 per dish).

   (b) Percoll gradients (1 per DRG processed).

   (c) Human DRG complete media, and place in 37 °C water bath.

   (d) *If needed, Trypsin solutions (1 per dish).

5. *If preparing culture for subsequent patch experiments, spin solution at $1.7 \times g$ for 1 min.

   (a) Remove supernatant and resuspend pellet with "Trypsin In-Series" solution.

   (b) Transfer solution with remaining pieces of ganglion into a glass dish, and place in 37 °C water bath shaking at 70 RPM for no more than 4 min.

   (c) Continue protocol as follows.

6. Triturate remaining chunks of ganglia through 1-mL pipette, starting with a large opening (~2 mm) as described above. Two passes through a series of pipette tips with incrementally decreasing tip diameters should enable close to complete dissociation of ganglia chunks.

7. After triturating, add the dissociated tissue and solution to the enzyme stop solution, invert to mix.

8. Spin tubes at $1.7 \times g$ for 4 min.

9. Remove supernatant without disturbing the pellet.

10. Resuspend pellet in 500 μL of basal media using a cut pipette tip so that the pellet can be gently broken up by passing it through the pipette tip.

11. Carefully load the dissociated solution onto the Percoll gradient.

12. Spin tubes at $3.4 \times g$ for 10 min.

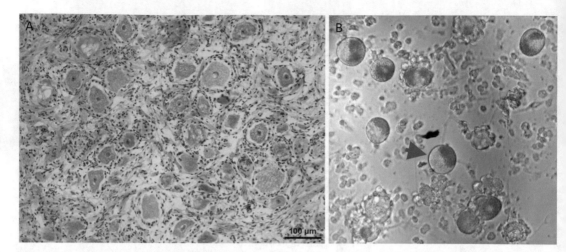

Fig. 3 Example images of human DRG. (**a**) Hematoxylin and eosin staining of intact human DRG. (**b**) After dissociating and plating, DRG neurons are often covered in support cells (*example indicated by orange arrow*), which will migrate off between 4 and 24 h, leaving a neuron readily accessible for patch-clamp recording (*example indicated by purple arrowhead*)

13. At the end of the spin, cells will be at the bottom of the tube.

14. Remove all the Percoll solution without disturbing the pellet.

15. Resuspend pellet into 1 mL of complete media.

16. Using a cut 200 μL pipette tip, ~2 mm in diameter, plate ~50 μL of cells onto poly-L lysine glass coverslips.

    (a) Plate 1 coverslip first to determine density of neurons. For best results, ~30 neurons should be in a field at a 10× magnification.

        • If necessary, add additional media or centrifuge at $1.7 \times g$ for 1 min and resuspend in a lower volume.

    (b) If long-term culture is to be used with microfluidic devices, the chambers commercially available will have to be modified, as human DRG neurons are too large to fit into the space routinely used for plating.

    (c) After plating, neurons will be covered with support cells (Fig. 3b) which will migrate off between 4 and 24 h post plating (Fig. 3c).

17. Allow neurons to attach at in a 37 °C incubator with 5% CO_2 and 95% humidity for ~4 h prior to flooding.

3.4 Electrophysiological Recordings

3.4.1 Cleaning up the Central Nerve

The isolated central nerve from L4/L5 segments are transferred to a dish containing cold Krebs solution (for composition *see* Table 3) oxygenated with carbogen (O_2 95%/CO_2 5%) with a pH ranging 7.2–7.4. The isolated central nerve is carefully desheathed under binocular microscope to get nerve fascicles and then cut it in segments of ~1 cm.

Table 3
Krebs solution for human nerve recordings

	Final concentration (mM)	To prepare 1 L	To prepare 2 L
NaCl	136	7.947 g	15.89 g
KCl	5.6	0.417 g	0.834 g
NaHCO$_3$	14.3	1.2 g	2.4 g
NaH$_2$PO$_4$	1.2	0.143 g	0.286 g
CaCl$_2$	2.2	2.2 mL (stock solution of 1 M)	4.4 mL
MgCl$_2$	1.2	2.4 mL (stock solution of 0.5 M)	4.8 mL
Glucose	11	2 g	4 g

3.4.2 Compound Action Potential Recordings

The conduction of action potentials on afferent fibers can be evaluated by recording the Compound Action Potential (CAP). This technique relies on an electrical stimulus applied to one end of a nerve fascicle to evoke action potentials which are recorded by an electrode placed at the other end of the nerve fascicle (*see* Fig. 2a for example of CAP recordings with suction electrodes). The electrical stimulus simultaneously and synchronously recruits action potentials from numerous afferent fibers and hence the response is referred to as a CAP because it is the sum of the many action potentials recruited.

The shape of the CAP for a peripheral nerve has been determined in several species including humans [25, 26]. Increasing intensity of electrical stimulation produces action potentials in different populations of afferent fibers, so that as the stimulation current is increased, more fibers are recruited to the CAP, and the amplitude grows. As the intensity is increased still further, additional peaks appear at longer latencies which are action potentials evoked in smaller diameter fibers (*see* Fig. 2b). Because the conduction velocity of an axon is dependent on the diameter of the fiber [27], as is also the stimulation threshold (the minimum stimulation current required to elicit an action potential), large diameter fibers are the most rapidly conducting and have a lower threshold for activation than small diameter fibers. The first afferent fibers to be activated are the large diameter Aβ fibers (such as proprioceptors), next Aδ fibers (those related to nociception–warm sensation), followed by C fibers (those generally associated with nociceptors). These three populations of afferents are easily identified in a CAP recording from human central nerve (*see* Fig. 2b).

Recording the CAP from human peripheral nerve in vivo is common in several diagnostic procedures [28]. However, an advantage of studying nerve in isolation is the ability to perform a detailed pharmacological analysis, where the relatively stable CAP can be used to determine potency and efficacy of compounds that may influence action potential propagation [43]. Those that preferentially block the C-fibers, relative to Aδ or Aβ, are more likely to block pain while sparing low threshold fibers needed for proprioception and low threshold touch. Furthermore, because it is possible to control stimulation frequency in addition to intensity, it is possible to assess the impact of compounds that may have use-dependent properties.

3.5 Calcium Imaging

Once human DRG neurons are cultured, there are a number of different ways to evaluate electrophysiological characteristics of the neurons.

If calcium imaging is used, load human DRG neurons with 2.5 μM fura-2 AM or other indicator with 0.025% pluronic F-127 in bath solution for 20 mins in the incubator at 37 °C with 5% CO_2 and 95% humidity (Fig. 4a). Subsequently, neurons will be ready to image. When imaging calcium transients, there are a few tricks that will aid in consistent results.

- Human DRG neurons often contain lipofuscin, a brown-yellow pigment ([29]Fig. 3c), that is auto-fluorescent and could dilute the evoked calcium transient. It is best to avoid lipofuscin when determining a region of interest (Fig. 4a, arrows).

- Due to the large diameter of human DRG neurons, it is best to use a high-density culture when imaging, allowing for multiple neurons to be imaged.

- Calcium transients are often smaller compared to rodent DRG neurons, it may be necessary to use a higher concentration of KCl or electrical stimulation parameters.

  - A successful culture will yield consistent calcium transients in response to repeated stimulation (Fig. 4b).

  - On the other hand, the calcium transients in unhealthy neurons are unstable, with transients that may decrease in response to repeated stimulation (Fig. 4c), and/or transients that recover very slowly, if they recover at all (Fig. 4d).

3.6 Patch-Clamp Recording

If patch-clamp recording is to be employed, investigators should keep a couple of things in mind:

1. These are large cells with large currents. Clamp control can therefore be an issue (*see* Fig. 5). To minimize the impact of a relatively large membrane time constant ($\tau = RC$), R should be reduced as much as possible with a large pipette tip and as much series resistance compensation as possible.

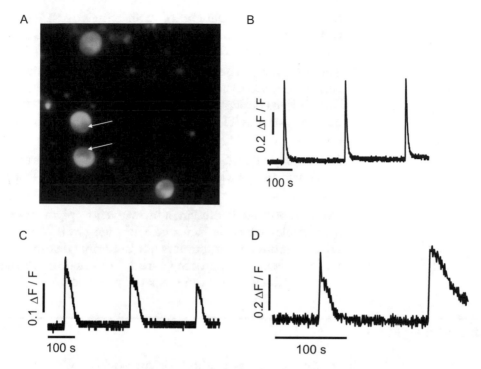

Fig. 4 Calcium imaging in cultured human DRG neurons. Calcium ratio of human DRG neurons loaded with fura-2 AM. Arrows indicate lipofuscin granules that should be avoided when determining a region of interest (**a**). A healthy neuron will produce consistent calcium transients with repeated high-KCl stimulation (**b**), whereas unhealthy neurons will produce run-down (**c**) or non-producible responses (**d**)

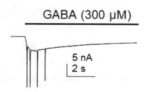

Fig. 5 Poor clamp control of a human DRG neuron. The neuron was studied with whole cell patch with an electrode solution containing 140 mM KCl and a bath solution containing 140 mM NaCl, so that GABAA-evoked currents would have a reversal potential of 0 mV and be inward when the neuron was held at −60 mV. However, the voltage-gated Na + currents were so large in this neuron that the inward current associated with the application of 300 μM GABA in conjunction with the relatively high series resistance (>6 MΩ), large membrane capacitance (>250 pF) resulted in fast transient currents on top of the slower GABA-evoked current

2. To maintain clamp control, it is often not enough to minimize series resistance errors with large pipette tips and series resistance compensation. It may be necessary to minimize current amplitude by manipulating the composition of the recording solutions (e.g., reduce Na^+ concentration in the bath solution

to reduce the amplitude of voltage-gated Na^+ currents), and/or the membrane potential (to reduce the driving force on the current under study).

3. Timing matters. The currents present change over time. Some currents increase with time in culture, such as voltage-gated Na^+ currents or persistent GABA currents [18, 20], and others decrease, such as TRPV1 currents if channels are trafficked to processes.

4. It generally takes 4–72 h for support cells to migrate off of DRG neurons, providing easier access to the neurons for patching (Fig. 3c).

5. Avoid, if possible lipofuscin, a brown-yellow pigment given to fat granules which is increased with age [29]. Most human DRG neurons have it, but it is not uniformly distributed, so it is often possible to approach neurons from a side in which there is less lipofuscin which can block the pipette tip.

4   Discussion

This chapter describes methods or approaches for recovering and processing human DRG as well as different ways electrophysiological properties may be measured from this valuable resource. The potential advantages of using human DRG neurons in the process of translating preclinical observations in sensory neurons into therapy are becoming clear. These include validation of therapeutic targets, pharmacological specificity, conservation of molecular mechanisms, along with identifying potential "off target" effects not found in rodents. This resource could additionally be used in combination with devices such as microfluidics, or tissue chips to reinnervate other cell types such as keratinocytes or tumor cells, or other organs allowing us to study the physiological or pathophysiological interactions. As more investigators have begun to study human neurons, it will be interesting to see whether their promise is able to pay off with a more successful and rapid translation of preclinical targets into better therapies for patients.

For those moving forward with this tissue, however, there are a number of factors to consider. First, there is often an incomplete medical history of the organ donors. Absence of medical history such as cigarette, marijuana usage, alcohol consumption, or recreational opioids may influence data collected. Second, there are variables that may influence cell properties that cannot be controlled when tissue from organ donors is used. These include the duration the donor may have been hypoxic prior to be placed on life support, the duration the donor was a life support, and the time between cross-clamp and tissue recovery. Third, while it is possible to study axons from organ donors, as noted, and groups have developed ex vivo preparations with which to study afferent terminals, and we and others have argued for the utility of the DRG neurons cell body in culture as a model for afferent terminals, it is only a model. Channels are differentially distributed throughout neurons and may not be detectable in the isolated cell body. Other factors such as the relative distribution of other cellular proteins may also be disrupted. Fourth, the neurons have, by definition, been injured, and things change after injury, most notably gene expression.

On the other hand, some labs have been able to combat the incomplete history of organ donors as well as pain models by recovering DRG from patients diagnosed with spinal tumors [11, 22, 30, 31], or as part of a surgical intervention for the treatment of pain (*see* [32, 33]). Other techniques such as differentiating induced human pluripotent stem cells (iPSCs) into nociceptors or sensory neurons not only allow a full and complete patient history and avoid mechanically injuring the cells, but can also model disease and/or genetic mutations [34–42]. However, differentiation protocols are long, it is necessary to maintain differentiated cells in vitro for extended periods of time (weeks), resulting in extensive processes that impact clamp control, and despite considerable effort, the cells generated fail to accurately recapitulate all of the properties of nociceptors observed in vivo.

While many people are registered as organ donors through their driver's license, it is important to mention that this registration is about saving lives. Research has the potential to benefit many more lives still, but the payoff takes considerably longer. Most people do not think about this long-term payoff, let alone the use of their tissue for research purposes. Nevertheless, because the use of tissue for research requires permission from the next of kin, everyone needs to think about this issue now and make sure that friends and family know whether you want your tissues to be used for research after your death.

Acknowledgments

We acknowledge the Center for Organ Recovery and Education (CORE) for their help in coordinating recovery of tissue from organ donors.

References

1. Coderre TJ, Laferriere A (2019) The emergence of animal models of chronic pain and logistical and methodological issues concerning their use. J Neural Transm (Vienna) 127(4):393–406

2. Yezierski RP, Hansson P (2018) Inflammatory and neuropathic pain from bench to bedside: what went wrong? J Pain 19(6):571–588

3. Gewandter JS et al (2020) Improving study conduct and data quality in clinical trials of chronic pain treatments: IMMPACT recommendations. J Pain 21(9-10):931–942

4. Edwards RR et al (2016) Patient phenotyping in clinical trials of chronic pain treatments: IMMPACT recommendations. Pain 157(9):1851–1871

5. Moore RA et al (2015) Systematic review of enriched enrolment, randomised withdrawal trial designs in chronic pain: a new framework for design and reporting. Pain 156(8):1382–1395

6. McGuire C et al (2018) Ex vivo study of human visceral nociceptors. Gut 67(1):86–96

7. Hockley JRF, Smith ESJ, Bulmer DC (2018) Human visceral nociception: findings from translational studies in human tissue. Am J Physiol Gastrointest Liver Physiol 315(4): G464–G472

8. Peiris M et al (2017) Peripheral KV7 channels regulate visceral sensory function in mouse and human colon. Mol Pain 13: 1744806917709371

9. Dedek A et al (2019) Loss of STEP61 couples disinhibition to N-methyl-d-aspartate receptor potentiation in rodent and human spinal pain processing. Brain 142(6):1535–1546

10. Valtcheva MV et al (2016) Surgical extraction of human dorsal root ganglia from organ donors and preparation of primary sensory neuron cultures. Nat Protoc 11(10):1877–1888

11. North RY et al (2019) Electrophysiological and transcriptomic correlates of neuropathic pain in human dorsal root ganglion neurons. Brain 142(5):1215–1226

12. Alexandrou AJ et al (2016) Subtype-selective small molecule inhibitors reveal a fundamental role for Nav1.7 in nociceptor electrogenesis, axonal conduction and presynaptic release. PLoS One 11(4):e0152405

13. Chang W et al (2018) Expression and role of voltage-gated sodium channels in human dorsal root ganglion neurons with special focus on Nav1.7, species differences, and regulation by paclitaxel. Neurosci Bull 34(1):4–12

14. Davidson S et al (2014) Human sensory neurons: membrane properties and sensitization by inflammatory mediators. Pain 155(9):1861–1870

15. Davidson S et al (2016) Group II mGluRs suppress hyperexcitability in mouse and human nociceptors. Pain 157(9):2081–2088

16. Moy JK et al (2020) Distribution of functional opioid receptors in human dorsal root ganglion neurons. Pain 161(7):1636–1649

17. Sheahan TD et al (2018) Metabotropic glutamate receptor 2/3 (mGluR2/3) activation suppresses TRPV1 sensitization in mouse, but not human, sensory neurons. eNeuro 5(2)

18. Zhang XL et al (2015) Inflammatory mediator-induced modulation of GABAA currents in human sensory neurons. Neuroscience 310: 401–409

19. Zhang X et al (2019) Nicotine evoked currents in human primary sensory neurons. J Pain 20(7):810–818

20. Zhang X et al (2017) Voltage-gated Na(+) currents in human dorsal root ganglion neurons. elife 6:e23235

21. Megat S et al (2019) Nociceptor translational profiling reveals the ragulator-rag GTPase complex as a critical generator of neuropathic pain. J Neurosci 39(3):393–411

22. Li Y et al (2015) The cancer chemotherapeutic paclitaxel increases human and rodent sensory neuron responses to TRPV1 by activation of TLR4. J Neurosci 35(39):13487–13500

23. Ray P et al (2018) Comparative transcriptome profiling of the human and mouse dorsal root ganglia: an RNA-seq-based resource for pain and sensory neuroscience research. Pain 159(7):1325–1345

24. Baumann T (1999) Human spinal sensory ganglia. Wiley, New York, pp 398–406

25. Tavee J (2019) Nerve conduction studies: basic concepts. Handb Clin Neurol 160:217–224

26. Stys PK, Ransom BR, Waxman SG (1991) Compound action potential of nerve recorded by suction electrode: a theoretical and experimental analysis. Brain Res 546(1):18–32

27. Waxman SG (1980) Determinants of conduction velocity in myelinated nerve fibers. Muscle Nerve 3(2):141–150

28. Shefner JM, Dawson DM (1990) The use of sensory action potentials in the diagnosis of peripheral nerve disease. Arch Neurol 47(3):341–348

29. Terman A, Brunk UT (1998) Lipofuscin: mechanisms of formation and increase with age. APMIS 106(2):265–276

30. Li Y et al (2018) DRG voltage-gated sodium channel 1.7 is upregulated in paclitaxel-induced neuropathy in rats and in humans with neuropathic pain. J Neurosci 38(5):1124–1136

31. Li Y et al (2017) Dorsal root ganglion neurons become hyperexcitable and increase expression of voltage-gated T-type calcium channels (Cav3.2) in paclitaxel-induced peripheral neuropathy. Pain 158(3):417–429

32. Baumann TK, Chaudhary P, Martenson ME (2004) Background potassium channel block and TRPV1 activation contribute to proton depolarization of sensory neurons from humans with neuropathic pain. Eur J Neurosci 19(5):1343–1351

33. Baumann TK et al (1996) Responses of adult human dorsal root ganglion neurons in culture to capsaicin and low pH. Pain 65(1):31–38

34. Nickolls AR et al (2020) Transcriptional programming of human mechanosensory neuron subtypes from pluripotent stem cells. Cell Rep 30(3):932–946 e7

35. Cao L et al (2016) Pharmacological reversal of a pain phenotype in iPSC-derived sensory neurons and patients with inherited erythromelalgia. Sci Transl Med 8(335):335ra56

36. Eberhardt E et al (2015) Pattern of functional TTX-resistant sodium channels reveals a developmental stage of human iPSC- and ESC-derived nociceptors. Stem Cell Reports 5(3):305–313

37. Young GT et al (2014) Characterizing human stem cell-derived sensory neurons at the single-cell level reveals their ion channel expression and utility in pain research. Mol Ther 22(8):1530–1543

38. Meents JE et al (2019) The role of Nav1.7 in human nociceptors: insights from human induced pluripotent stem cell-derived sensory neurons of erythromelalgia patients. Pain 160(6):1327–1341

39. Guimaraes MZP et al (2018) Generation of iPSC-derived human peripheral sensory neurons releasing substance P elicited by TRPV1 agonists. Front Mol Neurosci 11:277

40. Menendez L et al (2013) Directed differentiation of human pluripotent cells to neural crest stem cells. Nat Protoc 8(1):203–212

41. Chambers SM et al (2012) Combined small-molecule inhibition accelerates developmental timing and converts human pluripotent stem cells into nociceptors. Nat Biotechnol 30(7):715–720

42. Lee KS et al (2012) Human sensory neurons derived from induced pluripotent stem cells support varicella-zoster virus infection. PLoS One 7(12):e53010

43. Pineda-Farias JB et al (2021) Mechanisms underlying the selective therapeutic efficacy of carbamazepine for attenuation of trigeminal nerve injury pain. J Neurosci 41(43):8991–9007

Chapter 8

Human Pluripotent Stem Cell–Derived Sensory Neurons: A New Translational Approach to Study Mechanisms of Sensitization

Katrin Schrenk-Siemens

Abstract

The milestone achievement of reprogramming a human somatic cell into a pluripotent stem cell by Yamanaka and Takahashi in 2007 has changed the stem cell research landscape tremendously. Their discovery opened the unprecedented opportunity to work with human-induced pluripotent stem cells and the differentiated progeny thereof, without major ethical restrictions. Additionally, the new method offers the possibility to generate pluripotent stem cells from patients with various genetic diseases which is of great importance (a) to understand the basic mechanisms of a specific disease in a human cellular context and (b) to help find suitable therapies for the persons concerned. In individual cases, this can even help to develop personalized treatment options. Chronic pain is a disease that affects roughly one in five people worldwide, but its onset is rarely based upon genetic alterations. Nevertheless, the work with sensory-like neurons derived from human pluripotent stem cells has become a more widely used tool also in the field of pain research, as during the past years several differentiation procedures have been published that describe the generation of different types of sensory-like neurons and their useful contribution to studying mechanisms of sensitization. Especially also to complement and verify cellular and molecular mechanisms identified in rodent model systems, the model of choice for decades. Although a sole cellular system is not able to mimic a disease as complex as pain, it is a valid tool to understand basic mechanisms of sensitization in specific subsets of human neurons that might be at the onset of the disease. In addition, the creativity of basic researchers and the more and more advanced available technologies will most likely find ways to implement the derived human cells in more complex networks. In this chapter, I want to introduce a selection of published differentiation strategies that result in the generation of human sensory-like neurons. Additionally, I will point out some studies whose results helped to further understand pain-related mechanisms and which were conducted using the aforementioned differentiation procedures.

Key words Human pluripotent stem cells, Neural crest-like cells, Sensory neurons, Nociceptors, Mechanoreceptors, Sensitization, Pain

1 Introduction

Rodent model systems have been at the center of pain research for decades—and for good reasons: they have helped us to understand the basic cellular, molecular, and mechanistic aspects of

Rebecca P. Seal (ed.), *Contemporary Approaches to the Study of Pain: From Molecules to Neural Networks*, Neuromethods, vol. 178, https://doi.org/10.1007/978-1-0716-2039-7_8, © Springer Science+Business Media, LLC, part of Springer Nature 2022

somatosensation and pain [1]. However, the discrepancy between successful preclinical studies and discontinuation of clinical trials keeps us wondering about the reasons and how we could overcome those, as the need for new and potent analgesic drugs is still unmet [2, 3]. A translational approach that could bridge data from animal models with data derived from human tissue would be desirable. Access to human tissue however is limited or even impossible, especially with respect to chronic pain. The involved neurons are not only located in the brain but also in even less accessible regions of the nervous system, including the spinal cord and dorsal root ganglia. A solution to this conundrum might not be easy to provide, but scientific discoveries in the last 10 years have at least been quite instrumental to overcome the difficulties to some extent. These new developments comprise the work with human pluripotent stem cell (hPSC)-derived neurons. Although human embryonic stem cells (hESCs) have been available for research purposes since the late 1990s [4], it was not until 2012 that the first methods were published on how to derive functional primary sensory nociceptive-like neurons, which are the primary receiver of a potential painful stimuli [5]. Additionally, the milestone achievement by Takahashi and Yamanaka in 2007 [6], the discovery of generating induced pluripotent stem cells (iPSCs) from human somatic cells, has offered an unprecedented opportunity for the broad scientific community to work with human cells, including cells derived from patients suffering from diseases.

2   The Generation of Human Sensory-like Neurons from Pluripotent Stem Cells

The meaningfulness of working with human pluripotent stem cell–derived neurons strongly depends on the availability of efficient and practical procedures describing how to generate them. In the case of acute and inflammatory pain, the first recipients of a potential painful stimuli are the nociceptors. These pain-sensing primary sensory neurons, whose cell bodies reside in the dorsal root or trigeminal ganglia, belong to the peripheral nervous system [7]. Nociceptors comprise a rather heterogeneous population of cells and efforts to classify them have been based on measurements of their electrical, anatomical, and/or molecular properties. Some subtypes do respond to several stimuli such as temperature, mechanical, and chemical stimuli (defined as polymodal), while others only respond to one stimulus [7, 8]. But there seem to be a few commonalities nociceptive neurons share such as the presence of the voltage-gated sodium channel NAV1.8, or the ability to get sensitized by inflammatory molecules.

The possibility to generate human nociceptive-like neurons in vitro would offer the opportunity to study aspects of sensitization that have been found in rodent model systems and to compare similarities and differences on a cellular and molecular level in a human context. This could help to understand potential species-specific differences that could account for difficulties in translating preclinical findings to the clinic.

3   Chambers et al., 2012

In 2012, the research team of Lorenz Studer published the first protocol that described the generation of nociceptor-like cells from human pluripotent stem cells [5]. In contrast to earlier publications that reported the generation of a pan-sensory neuron population from hPSCs [9–12], the protocol by Chambers et al. describes a discrete sensory neuron-like subpopulation with molecular and functional properties similar to what had already been described in mouse nociceptors. The differentiation strategy is based on the combinatorial use of five different small molecules, all of which are inhibiting specific developmental pathways leading to the generation of sensory-like neurons (Fig. 1).

In the first step, human pluripotent stem cells get efficiently neuralized by using dual SMAD inhibition. LDN-193189, an inhibitor of bone morphogenetic protein (BMP), and SB431542, a molecule substituting for Noggin, are used in combination (termed LSB) for the first 5 days. In the second step, 2 days after induction of neuralization, the combination of three more small molecules lead to the generation of neurons from the sensory lineage (termed 3i; Fig. 1). These three molecules are (a) SU5402, a potent inhibitor of VEGF (vascular epithelial growth factor), FGF (fibroblast growth factor), and PDGF (platelet-

Chambers et al. 2012

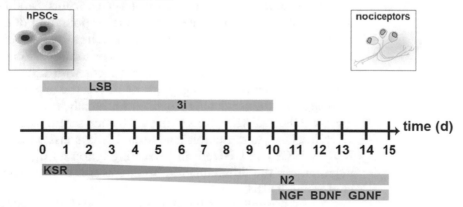

Fig. 1 Simplified schematic drawing of the differentiation method established by Chambers et al. (2012). LSB: name for the medium containing LDN-193189 (inhibitor of bone morphogenetic protein) and SB431542 (Noggin substituent); 3i: synonym for the three small molecules SU5402, CHIR99021, and DAPT; KSR: name for a medium containing Knockout DMEM as well as Knockout Serum Replacement

derived growth factor) tyrosine kinase signaling; (b) CHIR99021, an inhibitor of glycogen synthase kinase-3β (GSK-3β) and thereby acting as a WNT agonist; and (c) DAPT, a γ-secretase inhibitor, blocking notch signaling.

The maturation is further supported by the addition of neuro-trophic factors such as BDNF, GDNF and NGF to the neuronal N2-medium (Fig. 1). The Studer group could show that their approach leads to a rather fast maturation of neurons as opposed to other protocols and the in vivo development. Already 10 days after induction of the differentiation process, >75% of the derived cells showed neuronal markers and 15 days after induction func-tionally mature neurons could be detected. The cells expressed some nociceptive markers, such as NTRK1 (TRKA), Substance P, and calcitonin gene-related peptide (CGRP). Additionally, an array of specific ion channels such as SCN9A (Na$_V$1.7), SCN10A (Na$_V$1.8), and SCN11A (Na$_V$1.9), as well as the purinergic recep-tor P2RX3 or the TRP channels TRPV1 and TRPM8 were present in some of the cells as was shown by gene expression profile. The functionality of some of these channels was shown by electrophysi-ological recordings and Ca-imaging. Albeit, the expression was not homogenous for all the described channels, as capsaicin, the active component of chili peppers and agonist of TRPV1, was only active in a small subset of cells (1–2%).

Based on the functional as well as molecular data, Chambers et al. concluded that the derived sensory neurons resembled mostly a non-peptidergic, IB4-positive nociceptive population, compared to what is known from rodent model systems.

3.1 Modifications and Amendments of Chambers et al.

The protocol by Chambers et al. has been widely used since its publication, and several slight modifications as well as additional characterizations of the derived sensory-like neurons have been performed and published by various research teams. Modifications to the protocol mostly concern the starting point and duration of the treatment with the small inhibitors as well as the composition of the differentiation media used (Fig. 2). For example, Young et al. [13] increased the duration of the treatment with LSB to 7 days, while they delay the starting point of the 3i treatment by 3 days compared to the original protocol. Additionally, they start cultur-ing the cells in DMEM/F12 with fetal bovine serum, BDNF, NGF, GDNF, NT-3, and ascorbic acid after 10 days instead of N2 plus BDNF, NGF, and GDNF in increasing amounts right from the beginning of the differentiation process as was done by Chambers et al., (Fig. 2) [5].

A recent publication by McDermott et al. [14] prolonged the treatment of CHIR for 4 further days. Also, the concentration of the added growth factors varies between publications, from 25 ng/ml [5] to 20 ng/ml [15] to 10 ng/ml [13].

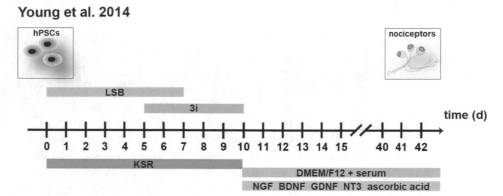

Fig. 2 Simplified schematic drawing of the modifications by Young et al. 2014 done to the differentiation method from Chambers et al. 2012

One criterion that has considerably changed compared to the original publication is the culture period of the cells before using them for experiments. While Chambers et al. emphasized that the use of the small-molecule inhibition results in the generation of functional neurons already after 15 days in culture, most research teams keep the neurons in culture for up to 60 days [14]. Although a short culture period would be advantageous to reduce experimental time, risk of contamination, and costs, it is not surprising that the maturation of a human neuron takes some time.

Although some of the studies compare the characteristics of the derived neurons with the original protocol and claim to not see a major difference [13], in some of the cases, the comparison was only done after short culture periods, and it is very likely that the derived neurons in various publications do show differences at later culture periods in terms of their maturation status.

Although the differentiation procedure published by Chambers et al. is the most widely used in the pain research community, it is not the only protocol that is available for the generation of human sensory-like neurons. In 2015, three different approaches for the generation of different sensory neuron-like subtypes that relied on different strategies were published. While the protocols from Wainger et al. [16] and Blanchard et al. [17] are based on the reprogramming of mouse or human fibroblasts and their direct conversion into neurons of the sensory lineage, the protocol from our group [18] is based on a two-step differentiation, starting with the generation of neural crest-like cells, the in vivo progenitors of sensory neurons and their further differentiation into mechanoreceptor-like cells. Also, in 2015, Boisvert et al. [19] showed the generation of sensory-like neurons using a NGN1-reporter cell line while in 2019, our group developed a NGN1-virus-dependent differentiation strategy using again the hESC-derived neural crest as a starting population [20].

4 Blanchard et al., 2015

For the direct conversion of mouse or human fibroblasts into sensory-like neurons, Blanchard et al. [17] reported the combined use of two doxycycline (dox)-inducible lentiviral vectors harboring the cDNA of two transcription factors (TF), *BRN3A* and *Ngn1* or *BRN3A* and *Ngn2* (Fig. 3). The Ngns are members of a helix-loop-helix transcription factor family and, along with BRN3A (POU4F1), are known to play a role in the development of sensory neurons [21]. The in vitro differentiation is induced by the transient expression of the transcription factor combinations in response to dox treatment. During the first 8 days, the TFs are expressed, but for a remaining minimum of 7 days of maturation, dox is removed and therefore the TF expression is absent (Fig. 3).

The resulting cells show molecular, morphological, and functional characteristics of various sensory neuronal subtypes. Immunocytochemistry for example reveals the expression of certain pan-sensory markers such as ISL1 and BRN3A, as well as more subtype-specific markers such as neurofilament 200 (myelinated fibers), peripherin (nociceptive marker), TRKA, TRKB, and TRKC (peptidergic nociceptors, mechanoreceptors, and proprioceptors, respectively), VGLUT1, VGLUT2 (glutamatergic synaptic transmission), RET (non-peptidergic nociceptors, mechanoreceptors). Functional analysis using Ca-imaging techniques shows in a subset of neurons increased Ca-responses to the TRPV1 agonist capsaicin and in another subset responses to the TRPA1 agonist mustard oil or menthol, the agonist of TRPM8. The functionality of the derived neurons is also supported by electrophysiological recordings, including the presence of the TTX resistant Na_V 1.8 channel, in some of the neurons. Additionally, Ca-imaging

Blanchard et al. 2015

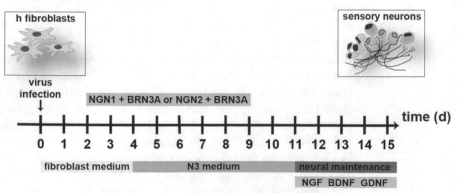

Fig. 3 Simplified schematic drawing of the differentiation method established by Blanchard et al. 2015. Mouse or human fibroblasts are infected with a combination of inducible lentiviruses (BRN3A & Ngn1 or BRN3A & Ngn2)

experiments, using agonists of pruritoceptors, such as histamine, chloroquine, or the peptides BAM 8–22 and SLIGRL reveal a small subset of cells responding selectively to the different compounds. Compared to the Chambers method, it seems the reprogramming strategy of Blanchard et al. works less efficiently (only 1% of human fibroblasts reprogram) and results in a more diverse population of sensory neuron-like subtypes that includes functional nociceptors as shown by the responses to capsaicin.

5   Wainger et al., 2015

The protocol established in the group of Clifford Woolf [16] follows a similar strategy by also converting mouse or human fibroblasts directly into sensory-like neurons, again based on the forced expression of distinct transcription factors. But unlike the inducible system used by Blanchard et al. [17], the lentiviral-transmitted vectors are constitutively active in this case. The procedure requires five different factors for the induction of sensory neurons from fibroblasts (Fig. 4). These factors are Ascl1, a basic helix-loop-helix transcription factor (TF), that has been shown to induce neuronal lineage formation, Myt1L, a Zink finger TF, also important for neuralization; Isl2, a homeodomain TF, Klf7, a Zink finger TF, important for TRKA maintenance and Ngn1, a helix-loop-helix TF, well known for its role during the development of TRKA expressing sensory neurons in mice.

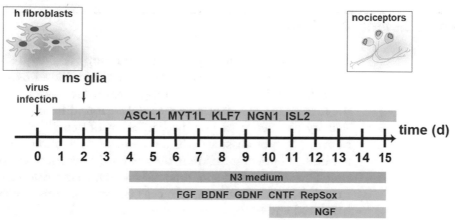

Wainger et al. 2015

Fig. 4 Simplified schematic drawing of the differentiation method established by Wainger et al. 2015. Mouse or human fibroblasts are infected with a combination of five noninducible lentiviruses (ASCL1, MYT1L, KLF7, NGN1, and ISL2). Note the addition of mouse glia to the cultures 2 days after virus infection

The derived sensory-like neurons were mostly characterized with respect to nociceptive properties such as the expression of the Na_V channels 1.8 and 1.9 and their functional resistance to TTX treatment. To show the potential of the derived neurons for disease modeling, fibroblasts from patients with familial dysautonomia (FD) were differentiated and compared to differentiated neurons from healthy donors. FD is a disorder that affects—among other things—the development and survival of sensory neurons, caused by mutations in the IKBKAP gene. Investigating the FD-derived neurons, Wainger et al. could show, using RT-PCR on single cells, the presence of an abnormally spliced transcript of the protein compared to healthy controls. Furthermore, fewer neurons could be differentiated from the FD fibroblasts, and they showed reduced neurite growth and branching.

As most of the characterizations shown in their study have been done with mouse fibroblasts and not human ones, one can only speculate as to what extent the derived human sensory-like neurons show the same functional qualities as those derived from mouse fibroblasts. Additionally, as the characterization of the cells showed a bias toward nociceptive properties, it is difficult to assess the purity of the derived sensory subtypes. As for the efficiency, the generation of human sensory neurons seems less efficient compared to the use of mouse fibroblasts, with only 5% of the fibroblasts showing a signal for the pan-neuronal marker class III beta-tubulin (Tuj1) and from those only 16% showing the presence of peripherin, an intermediate filament protein, expressed in small-sized sensory neurons and therefore considered a marker for nociceptors. It is therefore very likely that this procedure results in a rather heterogeneous population of sensory neuron-like subtypes.

6   Schrenk-Siemens et al. 2015 and 2019

The differentiation strategies from our group [18, 20] are based on a two-step procedure starting with the generation of neural crest-like cells (NCLCs), the in vivo progenitors of sensory neurons, and the further differentiation of the NCLCs into different subtypes of sensory-like neurons. For the derivation of neural crest-like cells, human pluripotent stem cells are cultured as floating spheres in a medium supporting differentiation toward a neuroectodermal lineage. The spheres will spontaneously attach to the non-coated culture dish, and neural crest-like cells will migrate out which can be harvested and even frozen until further use [22].

6.1   Generation of Mechanoreceptors

The generation of mechanoreceptive neurons is based on the addition of various growth factors to the differentiation medium of the NCLCs. The factors BDNF, GDNF, NGF, and NT3 support the further differentiation, but it is especially the addition of retinoic acid which helps to generate functional mechanoreceptor-like neurons. Although the efficiency is not extremely high, the neurons can be easily spotted due to their large size, and therefore, the cells can be used for electrophysiological recordings, as well as picked for sequencing experiments or stained. Characterization of the cells using the aforementioned techniques demonstrate that the cells are functionally resemble low-threshold mechanoreceptors which could be successfully employed to study the role of PIEZO2 in mechanotransduction [18].

6.2 Generation of Nociceptor-like Neurons

As the sole differentiation of the derived NCLCs with growth factors and retinoic acid did not result in any other sensory neuron subtype than mechanoreceptors, we used another approach, similar to what had been used before by others [16, 17, 23]: the lentivirus-mediated induced expression of the transcription factor Neurogenin 1 [20]. A similar approach had already been used by Boisvert et al. [19] who generated a stable NGN1 expressing cell line. Our approach allows the use of the inducible NGN1 virus with any pluripotent stem cell line, including patient-derived iPSC lines. The doxycycline-induced expression of NGN1 for 10 days, followed by a maturation period of at least 2 weeks in differentiation medium containing BDNF, GDNF, and NT3 resulted in the generation of functional nociceptor-like cells (Fig. 5). Due to the presence of a resistance cassette in the NGN1 vector, the addition of puromycin eliminates all cells not infected by the lentivirus, leading to a differentiation rate of almost 100%.

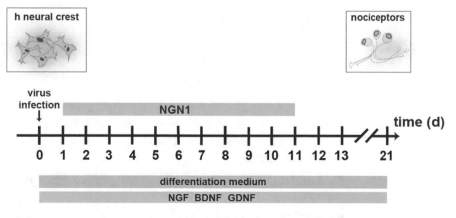

Fig. 5 Simplified schematic drawing of the differentiation method established by Schrenk-Siemens et al. 2019. Human pluripotent stem cell–derived neural crest-like cells are infected with an inducible lentivirus harboring NGN1

Characterization of the neurons using in situ hybridization, deep sequencing, Ca-imaging, and electrophysiology revealed the generation of a homogenous population of TRPV1 expressing neurons, that respond to capsaicin and can be sensitized as expected from nociceptive-like cells.

Deep sequencing experiments support the nociceptive-like phenotype because of the presence of marker transcripts for various Na_V channels, HCN channels, NTRK1, TRPV1, etc.

6.3 Detailed Protocol for the Generation of Nociceptive-like Neurons

1. Generation of neural crest-like cells the in vivo progenitors of sensory neurons (protocol modified from Bajpai et al., 2010 [22]).

6.3.1 Formation of Neuroectodermal Spheres

Materials

10-cm cell culture plates (best experience with cell culture plates with vents, e.g., Greiner #664160)

15-ml tubes

Phosphate-buffered saline (PBS)

0.5 mM EDTA in PBS

Human epidermal growth factor (animal-free; e.g., Peprotech #AF-100-15)

Human fibroblast growth factor basic (e.g., Peprotech #100-18B)

Rho/Rock pathway inhibitor (e.g., StemCell Technologies Y-27632)

Accutase (e.g., Sigma #A6964)

A 80% confluent plate of human pluripotent stem cells (cultured in E8-medium or any other stem cell–specific culture medium; Fig. 6a) is split using the following procedure:

- Wash once with PBS.

- Add 6 ml EDTA (0.5 mM).

- Incubate at 37 °C for 5-6 min.

- Carefully aspirate EDTA solution.

- Harvest cells with 5 ml sphere medium (Table 1), by rinsing the plate several times (! avoid too harsh pipetting).

- Plate cells/colonies on a 10-cm culture plate with vents in 10 ml sphere medium.

- Add hEGF (final conc. 10 ng/ml), hFGF (final conc. 10 ng/ml).

- Add Rho/Rock pathway inhibitor to increase the survival of the cells, (final conc. 1–2.5 μM depending on the cell line).

- Put at 37 °C, 5% CO_2.

! Depending on the pluripotent stem cell line used, the amount of cells taken for sphere formation has to be adjusted. Too many

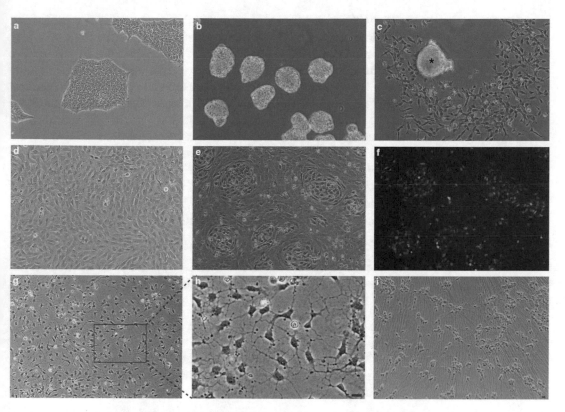

Fig. 6 Representative pictures of subsequent differentiation stages of hPSC-derived nociceptor-like cells. (**a**) Typical morphology of a stem cell colony cultured in E8 medium on a Matrigel-coated dish. (**b**) Neuroectodermal spheres floating on a non-coated cell culture dish. (**c**) Ring of neural crest-like cells, migrated out from a spontaneously attached sphere, whose remnants are still present (asterisk). (**d**) Confluent culture of neural crest-like cells plated on a poly-L-ornithin, laminin, fibronectin-coated dish. (**e**, **f**) Islands of sensory neuron-like progenitors 2 days after induction of NGN1 expression: (**e**) shows a bright-field picture and (**f**) the corresponding fluorescence picture where the green signal indicates the expression of GFP, which is initiated together with the expression of NGN1. Note the GFP-negative cells that refer to NCLCs not having incorporated the virus. (**g**, **h**) Pure sensory neuron-like progenitors 1 day after treatment with puromycin which killed all NGN1/GFP-negative cells. The pink insert in (**g**) is enlarged in (**h**). (**i**) Nociceptor-like neurons after 14 days of differentiation. Scale bar 20 μm

Table 1
Sphere medium

	Final concentration	Company
DMEM/F12	50%	Gibco
Neurobasal	50%	Gibco
l-glutamine	2 mM	Gibco
B27 supplement	0.5%	Gibco
N2 supplement	0.5%	Gibco
Insulin	5 μg/ml	e.g., CSBio Co #CS9212

cells will result in a slow growth of the spheres and a prolonged development of NCLCs; too few cells can result in the same. The first case can be seen by a rather fast depletion of the sphere medium, indicated by the medium color (quite orange already after 1 day), while the second results in the opposite: the medium color stays red over a longer period of time. Too crowded plates can easily be distributed on several plates. As a rule of thumb: one 10-cm plate of confluent hPSCs can be distributed on two 10-cm plates to initiate the formation of neuroectodermal spheres.

First day after sphere induction:

Small floating spheres (Fig. 6b) should be visible and a lot of cell debris. To get rid of the cell debris, the medium needs to be replaced using the following procedure:

- Collect the entire medium plus the spheres into a 15-ml tube.

- Let the spheres sink to the bottom by gravity (depending on the size of the spheres, this can take less than a minute or up to 5 min).

A small pellet of the spheres can be observed at the bottom of the tube.

! When changing the medium on the first day after induction, the pellet might not form as the spheres are very small; the whole medium looks milky because of the cell debris.

- Aspirate the medium but avoid disturbing the pellet (for 1-day--old spheres leave 2 ml in the tube to avoid losing too many small spheres).

- Add 10 ml of fresh medium and plate on a fresh 10-cm culture dish.

- Add fresh EGF and FGF; Rho/Rock pathway inhibitor can be added at the first medium change but not anymore after that.

Change the medium every second day (a routine could be: Monday, Wednesday, Friday, and leave the weekend out).

Especially in the first 2 days of the sphere formation, a lot of debris can be visible, then it might be better to change the medium 2 days in a row.

6.3.2 Harvesting of Neural Crest-like Cells

After 6–9 days, some of the spheres spontaneously start to attach to the plate and cells migrate out (Fig. 6c). They form a nice ring around the sphere and leave part of the sphere in the middle. These migrating cells are neural crest-like cells, the progenitors of sensory neurons.

For harvesting the neural crest-like cells, use the following procedure:

- Remove the medium with the remaining, still floating spheres and put them into a 15-ml falcon tube (do a normal medium change with those).

- Aspirate the remains of the spheres sticking to the plate using the tip of a Pasteur pipette and the vacuum suction device (the remains of the spheres can be identified as big whitish dots).

- Add 1 ml of Accutase and make sure it is evenly distributed over the plate.

- Wait for up to 5 min (in the meantime perform a normal medium change on the remaining spheres you collected in the 15-ml tube).

- Harvest the crest cells by pipetting sphere medium with a 1-ml pipette tip over the plate and collect everything in a 15-ml tube.

- Spin at 1000 rpm for 4 min.

- Aspirate the supernatant.

- The pellet with the NCLCs can be either resuspended in sphere medium and immediately taken for a differentiation or frozen to keep the NCLCs for a later time point.

! The NCLCs can be frozen using standard procedures and stored in liquid N_2 for several months up to years without any trouble. This is quite advantageous as you can create frozen libraries of different cell lines to be able to start differentiations more spontaneously without having to go through the NCLC production process each time.

The spheres will produce NCLCs for up to 12 days. After that, the cells that attach to the plate start to change their morphology and are not NCLCs anymore. One indication is that the spheres become rather flat when attached and no longer show a gap between the migrated cells and the remnant of the sticking sphere.

6.3.3 Differentiation of NCLCs into Nociceptor-like Sensory Neurons

Materials

Culture dishes (e.g., Falcon 3.5 cm plates, #353001)

Round coverslips (use German glass, thickness 1, e.g., VWR #89167–108)

Neubauer counting chamber

Sphere medium (for recipe see *Formation of Neuroectodermal Spheres*)

Lentivirus: two lentiviruses are needed for the successful differentiation of the stem cells: one containing an inducible expression plasmid for Neurogenin1 as well as GFP and a selection cassette for puromycin and the second virus containing an expression plasmid for tTA, whose expression is induced after addition of doxycycline and which then allows the expression of NGN1, GFP, and the puromycin selection cassette. This virus-based induction of differentiation is based on the lentiviruses used by Zhang et al., 2013 [23] with NGN2 exchanged by NGN1.

Doxycycline (e.g., Sigma #D9891)

Laminin (e.g., Sigma #L2020)

Fibronectin (e.g., Gibco #33010018)

Poly-L-ornithin (PORN, e.g., Sigma #P8638)
Mitomycin C (e.g., Sigma #M4287)
Protaminsulfate (e.g., Sigma P3369)

Preparations

Coating of culture dishes

- Culture dishes are coated with poly-L-ornithine (PORN, 15 µg/ml final conc. in sterile water) overnight at 37 °C.
- Wash dishes three times with sterile water.
- UV-treat for 1 h.

!Coated dishes can be stored for several weeks at 4 °C, when sealed with parafilm.

Before plating the cells for differentiation:

- Coat the dishes with laminin (10 µg/ml) and fibronectin (10 µg/ml, in sterile water) overnight at 37 °C.

For some experiments (immunocytochemistry, Ca-imaging, electrophysiology, etc.), the neurons need to grow on glass coverslips. The coverslips need to be pretreated with acid; otherwise, the coating will not work well enough.

- Incubate glass coverslips in 1 M HCl on a vertical rotator, e.g., in a 50-ml tube for at least 1 day.
- Wash glass coverslips extensively with water.
- Put in 100% ethanol.
- Spread out coverslips on a Whatman filter paper and sterilize under the sterile hood using UV light for 1 h.

! It is useful to prepare a good amount of coverslips this way and store them in a sterile container until further use.

The glass coverslips are coated in the same way as already described for culture dishes. To save coating solution, we put the coverslips in special coating chambers, build from a 10-cm culture dish lid wrapped in parafilm (do not stretch the parafilm) and put into a 15-cm culture dish with lid. The coverslips can be placed on the parafilm, and depending on the size of the coverslips, only a small amount of coating solution is needed (e.g., for round coverslips with a diameter of 12 mm, less than 100 µl per coverslip are needed). Coverslips coated with PORN, washed, dried, and sterilized with UV-light can also be kept at 4 °C for several weeks to months.

!The coated cell culture surfaces for the differentiation (containing all three proteins: PORN, laminin, fibronectin) can be stored in the incubator for up to 1 week, then the laminin and fibronectin will start to form precipitates that disturb the proper attachment of the cells.

6.3.4 Differentiation Procedure	• Plate NCLCs on a coated culture dish (PORN, Lam, Fib) in a density of 75,000 cells/cm^2 in sphere medium.

Day -2

  ! The amount of cells needs to be adjusted for each line, as some lines show a higher amount of cells dying; experience also showed that each researcher counts slightly different, and therefore, individual differences in cell densities can be observed. Regarding the size of culture dish to start with, we normally use 3.5 cm plates; one plate results in progenitors enough for 10–20 coverslips (12 mm).

Day -1

- Infect cells with lentivirus harboring *NGN1* and lentivirus harboring tTA in sphere medium containing Hepes (10 mM) and protaminsulfate (8 μg/ml).
- Leave virus on for 6 h.
- Wash three times with PBS.
- Add fresh sphere medium.

  ! Lentivirus requires specific safety standards, make sure to stick to them. The amount of virus needed for a successful infection needs to be tested after each virus production. Therefore, it makes sense to produce large amounts of the virus in the first place.

Day 0

- The morphology of the cells should be pretty much the same as the day before, maybe a bit more dense due to the proliferation ability of the neural crest-like cells (Fig. 6d).
- Change media to differentiation medium (Table 2) containing doxycycline (10 μg/ml).

Day 1

- Cells should start to change slowly their morphology and little islands of a different type of cell should become visible under the microscope (Fig. 6e).
- Change medium to fresh diff+dox.

Table 2
Differentiation medium

	Final concentration	Company
Sphere medium	Fill up to final volume	
Hepes	10 mM	
Anti-anti	1×	Gibco
Human GDNF	10 ng/ml	e.g., Peprotech #450–10
Human BDNF	10 ng/ml	e.g., Peprotech #450–02
Human b-NGF	10 ng/ml	e.g., Peprotech #450–01

! Changes can already be seen at day1, but sometimes, it might take a day longer.

Day 2

- The morphology of the cells should have changed by today: islands of different looking cells should be visible between streaks of other cells (Fig. 6e).

- To eliminate those cells that were not infected by the virus, puromycin (10 μg/ml) is added to the differentiation medium containing doxycycline.

! The lentivirus we are using not only expresses NGN1 but also GFP and contains a selection cassette for puromycin; therefore, cells that did incorporate the virus will express GFP and are resistant against puromycin selection (Fig. 6f).

Day 3

- Due to the addition of puromycin the day before, there will be some cell death visible.

- Change medium to differentiation medium + dox.

- The progenitors should become prominently visible: they look like little spiders.

- Prepare laminin- and fibronectin-coated surfaces, according to the experiments planned (glass coverslips for Ca-imaging, electrophysiology, immunocytochemistry, etc. or larger dishes for isolation of protein or RNA).

Day 4: Splitting of the Progenitors

- A homogenous population of progenitors should be seen: looking all like little spiders (Fig. 6g, h).

- Wash cells once with PBS.

- Add 1 ml Accutase.

- Incubate for up to 5 min at 37 °C.

- Harvest cells by rinsing the plate with sphere medium.

- Collect cells in a 15-ml tube.

- Spin at 1000 rpm for 4 min.

- Aspirate supernatant.

- Resuspend pellet in 1 ml diff + dox medium.

- Count cells.

- Seed 26,000 cells/cm^2 in diff + dox on freshly coated surfaces.

! The amount of cells plated depends again on the cell line used and the person counting. When doing it for the first time with a given cell line, it makes sense to try different densities. The splitting of the cells can also be delayed for 1–2 additional days, in case the cells are slower in their differentiation or another virus infection is needed, for example, to introduce a Ca-sensitive dye such as

GCaMP via virus. After 6 days, the cells already have established a complex network and splitting might result in the loss of a good amount of cells.

Day 5

- Cells should already start to look like neurons with small neurites.
- Change diff + dox.

Day 6–10

- Change diff + dox every day.

  ! Cells look more like neurons with every day, developing an elaborated network of processes. Almost no other cell types should be visible. In case non-neuronal cells come up in the background, another overnight puromycin treatment can be done but only until day 10, when the doxycycline treatment is still done.

Day 11

- Change medium to diff without dox.

Day 13

! To prevent the growth of remaining non-neuronal cells, the cultures can be treated with a mitose-inhibitor, e.g., mitomycin C.

- Add mitomycin C (10 µg/ml) for 45 min to the medium of the cells.
- Exchange medium to fresh diff medium.

Day 14 till end

- Change diff medium 2–3 times a week, do not exchange the whole medium but roughly 75% (Fig. 6i).

Day 26

Neurons respond to capsaicin and show electrical activity.
Can be used for all kinds of experiments.

7 In-Depth Characterization of hPSC-Derived Sensory-like Neurons

A more in-depth characterization of the expression and functionality of ion channels in sensory neuron-like cells was performed by Young et al. [13]. They used whole-genome microarrays to investigate the transcriptome of the sensory-like neurons during maturation from embryonic stem cells up until 39 days in culture and compared the data to adult human DRGs. The data showed a high comparability (84%) of ion channel expression between hDRGs and in vitro derived sensory-like neurons made after 30 days in culture following the protocol by Chambers et al. [5]. Among these commonly expressed ion channels are some associated with sensitization and pain phenotypes such as GABA$_A$R, HCN1, KCNQ2/3, and ASICs, whose functionality was tested by using electrophysiological techniques. Additionally, they performed single-cell qPCR at five different time points during differentiation, with the latest being 16 days in culture. While these experiments showed the expression of neuronal markers in 70% of the cells, it also showed that some of the markers that were present in the bulk sequencing at later time points could not yet be detected, pointing toward a still immature phenotype after 2 weeks in culture.

A study by Eberhardt et al. [15] looked closer at the molecular and functional expression of different voltage-gated sodium channels (Na_V) in the human ESC-derived sensory-like neurons generated based on the protocol by Chambers et al. [5]. Na_V channels are one possibility to identify nociceptors, as the cells express several tetrodotoxin-sensitive and -resistant isoforms, with Na_V 1.7 belonging to the first and Na_V 1.8 and Na_V 1.9 belonging to the second category, and mutations in these Na_V channels have been implicated in some of the rare genetically based forms of pain. Eberhardt et al. [15] investigated the expression and functionality of different Na_V channels over the course of differentiation and found substantial expression of Na_V 1.5, a TTX-resistant channel known to be expressed during rat DRG development, as well as the presence of Na_V 1.7, 1.8, and 1.9. Electrophysiological recordings looking at TTX-resistant currents in the differentiated neurons point to a major contribution of Na_V 1.5 instead of 1.8 and 1.9, suggesting a still more immature phenotype of the derived neurons based on the assumption that Na_V 1.5 is expressed also in immature human sensory neurons as known from rodent organisms.

Our group [20] used RNA sequencing at four different time points to characterize hPSC-derived nociceptor-like neurons in more depth. Because of the almost homogenous response of the derived neurons to capsaicin, bulk analysis rather than single-cell sequencing was performed. The data show an increased presence of transcripts for pan-sensory markers such as PRPH (peripherin), NEFH (neurofilament heavy chain), ISL1, or BRN3A but also for specific nociceptive markers such as NTRK1 (TRKA), TRPV1, SCN10A, SCN11A, RET, and others. At the same time, transcripts for markers such as NTRK2, NTRK3, and MAF could be observed. This could be for several reasons: (1) the derived sensory-like neurons are not yet fully mature, and therefore, the make-up of expressed genes is not entirely refined; (2) the co-expression of specific markers is different in human versus rodents, as was shown for other markers already [24–26]; and/or (3) the presence of the transcript is not equivalent with the presence of the protein, as was shown by our group [20] for the transcription factor MAFA, whose transcript was elevated but the protein could barely be detected by immunostaining.

8 hiPSC-Derived Sensory-like Neurons in Research

HPSC-derived sensory-like neurons mostly generated by using small-molecule inhibitors have become an increasingly used tool in pain-related research, especially to study the impact of Na_V mutations on their functionality and kinetics, either in patient-derived iPSCs or in genetically modified hESCs.

8.1 hiPSC-Derived Neurons with Pain-Related Na_v 1.7 Mutations

More than 1000 disease-related mutations in Na_V channels have been identified including those leading to pain phenotypes, with Na_V 1.7, 1.8, or 1.9 being mostly affected. Loss of function mutations in Na_V 1.7 cause congenital insensitivity to pain (CIP), resulting in the complete loss of pain perception in response to any noxious stimuli in the affected patients. Gain of function mutations in Na_V 1.7 can lead to human pain disorders such as inherited erythromelalgia, paroxysmal extreme pain disorder, small fiber neuropathy, or diabetic neuropathy [27, 28]. Unfortunately, an efficient pharmacological targeting of the sodium channel is not trivial because of the homology of different voltage-gated sodium channels and the nonspecific side effects. Therefore, it is desirable to better understand the mechanisms underlying the different inherited forms of sensitivity to pain or the lack thereof.

A recent study by McDermott et al. [14] used a multi-modal approach to investigate the functional role of Na_V 1.7 in human nociception, including the use of CIP-patient-derived iPSCs and nociceptor-like neurons generated thereof, based on a modified differentiation protocol by Chambers et al. [5] as well as sensory profiling, microneurography (a method to record from peripheral nerves in awake humans), and functional brain imaging in patients. By combining in vivo techniques with in vitro generated data, the researchers could demonstrate that the loss of functional Na_V 1.7 results in not only a loss of pain perception but additional deficits in temperature discrimination and itch. By introducing a tag in the endogenous Na_V 1.7 gene locus, they could further investigate the cellular localization of the channel, which is trafficked to the cell surface, axon, and terminals. Interestingly, CIP patients showed a clear reduction of epidermal innervation by C-nociceptive fibers as could be assessed by microneurography, pointing toward a role for Na_V 1.7 in preserving the structural integrity of the distal nociceptive terminals. Mechanistically, CIP-patient-derived iPSC nociceptor-like neurons were hypoexcitable and required increased current threshold for activation, suggesting a key role for Na_V 1.7 in the regulation of excitability. The iPSC-derived sensory-like neurons from either healthy or CIP patients could also be used to investigate the specificity of two different, supposedly Na_V 1.7-specific drugs. In particular, the use on CIP-patient-derived iPCS sensory-like neurons was instrumental in showing a lack of specificity for Na_V 1.7, as the putative specific effect on Na_V 1.7 was also observed in the absence of this ion channel. This combinatorial approach is a promising example of how the generated sensory-like neurons could help to understand disease-related mechanisms in more detail.

8.2   Patient-Derived iPSCs and their Role for Personalized Pain Therapies

The majority of patients suffering from chronic pain do not show an obvious genetic mutation that would account for the pain phenotype. In these cases, the use of patient-derived pluripotent stem cells might not seem meaningful, as potential epigenetic changes that are also discussed to be causative to pain phenotypes mostly get lost during the process of reprogramming. Nevertheless, a recent study by Namer et al. [29], successfully used iPSCs from a neuropathy patient to find an individual pain-relieving therapy. The 69-year-old Norwegian female was suffering from small fiber neuropathy for more than 10 years with severe continuous burning pain. The pain being especially strong in the late evenings was rated by the patient as a 7.5 on a visual analog scale and had a major impact on the quality of life. Using microneurography, it became obvious that the patient showed increased spontaneous activity in C-fibers compared to healthy age-matched controls. IPSCs derived from the patient and genetically analyzed to assess possible mutations that could account for the pain phenotype, revealed no indication for a monogenetic disorder. The pluripotent stem cells were differentiated into sensory-like neurons, using a modified protocol by Chambers et al., 2012 [5], and the functional fingerprint of the derived neurons was investigated using electrophysiological recordings. A significantly larger number of the patient-derived neurons showed spontaneous activity, compared to the controls, while, e.g., the action potential threshold was unaltered. This observation was corroborated by using multi electrode arrays (MEAs): the patient- as well as control-derived sensory-like neurons were seeded on the MEAs and their spontaneous activity was assessed. The patient-derived cells showed a significant increase in excitability: more neurons fired spontaneously with increased frequency. The authors "treated" the derived neurons with lacosamide, an FDA-approved drug, which had never been used on the patient and could show a strong reduction in the number of spikes in the patient-derived but not the control cells. The patient was finally given an off-labeled treatment with lacosamide which resulted in a tremendous reduction in pain already 5 days later as reflected by a decrease in rating from 7.5 to 1.5. Microneurography on the patient under lacosamide treatment showed a significant reduction of the spontaneously active C-fibers from 26.3% to 7%.

This study showed for the first time the successful use of iPSC-derived sensory-like neurons from a patient without an obvious disease-causing mutation, where the cells could serve as a tool to find an individual therapy that could successfully be implemented. Further studies in this direction will show how efficiently patient-derived sensory-like neurons can be used to find individual treatment options where standard medication is not helping anymore.

9   Challenges and Limits Working with hPSC-Derived Sensory-like Neurons

All of the aforementioned differentiation procedures result in neurons showing more or less hallmark characteristics of their respective in vivo counterparts. But there are two recurrent questions that researchers have to face when working with these cells: (1) how mature are the sensory-like neurons and (2) to what in vivo subtype are they corresponding to?

Although it would be desirable to know which maturation stage and cell type the generated neurons are most closely comparable to, this knowledge might not be necessary in order to take advantage of the differentiated neurons, as examples in sect. 8 demonstrate. Specific—for sensory neurons characteristic—functions are enough to justify their use in order to investigate molecular or functional aspects of somatosensory biology. For a nociceptive neuron for example, the ability to undergo sensitization under inflammatory conditions is well described. An in vitro generated nociceptive-like neuron, being able to get sensitized, is therefore still a valid tool to study aspects of sensitization even if some other characteristic nociceptive markers are not (yet) expressed.

In order to be able to classify the generated cells, we would need a thorough characterization on a molecular, cellular, and functional level of different human sensory neuron subtypes, isolated from human DRGs or trigeminal ganglia. This has been extensively done in the last few years for mouse DRGs [30–34] and the data that has been gathered from single-cell sequencing experiments of sensory neuron subtypes at different time points during development up until adulthood has redefined our perception of the somatosensory system. Unfortunately, studies like this are still missing from human DRGs. But it might just be a matter of time until those data will be available as it has gotten much easier during the last years to get access to human DRGs and TGs. The easier access to this human tissue has already resulted in several studies describing the molecular make-up and function of human sensory neurons. Although these studies performed bulk RNA sequencing or only focused on specific markers using in situ hybridization or immunostainings on slices, they offered some insight as to the distribution of commonly accepted sensory markers in human cells [24–26]. These studies also showed that there are species differences regarding the subtype-specific expression of key sensory markers between rodent and human DRGs, which is of particular interest in relation to translating findings from animal studies into the human organism. One of these studies was conducted in our own lab [24] where we did a comparative analysis between mouse and human sensory ganglia assessing the distribution and co-expression of a defined panel of markers using in situ hybridization or immunocytochemistry. The make-up of the chosen markers was always studied in reference to the neurotrophic receptor *TRKA*, that is largely present in peptidergic nociceptive neurons in rodents. While we could show that the classical correlation between the expression of neurotrophic receptors and somata size of sensory neurons was not different in human versus mouse DRG neurons (*TRKA* mostly found in small, *TRKB* and *TRKC* in larger cells), we found some differences in regard to the distribution of markers that are used to classify sensory subtypes in rodents. One very striking difference is the presence of the heavy neurofilament in every sensory neuron in human DRGs while in rodents this is classically used to define the myelinated Aβ- or Aδ-fibers. Another striking difference could be seen for the distribution of RET, a marker that in rodents is widely used to characterize non-peptidergic nociceptors that have downregulated TRKA expression. We found *RET* expression in roughly half of the *TRKA*-positive cell population in human DRGs. We also found a higher amount of human sensory neurons co-expressing *TRKA* and *TRPV1*, $Na_V1.8$ or $Na_V1.9$, respectively. Some of these findings have been corroborated by a more recent study [25] where the authors used multiplex RNAscope in situ hybridization to investigate the distribution of classic and also unique neuronal markers in human versus mouse DRGs. The distribution of those markers was related to the presence of either calcitonin gene-related peptide (CGRP) or P2X purinergic ion channel type 3 receptor (P2X3R), two markers that—at least in the rodent system—are markers for peptidergic and non-peptidergic neurons, respectively. This exclusive separation was not present in the human DRGs, where a large population of neurons co-expressed CGRP and P2X3R with a higher population of sensory neurons expressing CGRP in general. Also the expression of TRPV1 could be seen in many more nociceptors in human DRGs compared to the mouse, corroborating our results. The broad panel of markers tested by Shiers et al. also included several subunits of nicotinic acetylcholine receptors from which one the CHRNA9 could not even be detected in mouse DRGs. These studies underline the importance of a thorough comparison on a single-cell level between human and mouse sensory neuron subtypes with regard to the translation of data from rodent model systems onto the human organism.

10 Future Perspectives

Animal models have been at the center of pain research for decades, and this might not change in the future as the complexity of pain can never be mimicked in a dish using cells. But, with the discovery of human iPSCs and the ability to generate sensory neuron-like cells in vitro showing molecular and functional characteristic hallmarks of their in vivo counterparts, we have now the unprecedented opportunity to investigate aspects of sensitization in a human cell-based system. By combining these two models, we will create a powerful tool to address and overcome challenges of translational research in the future.

References

1. Mogil JS (2009) Animal models of pain: progress and challenges. Nat Rev Neurosci 10(4): 283–294

2. Percie du Sert N, Rice AS (2014) Improving the translation of analgesic drugs to the clinic: animal models of neuropathic pain. Br J Pharmacol 171(12):2951–2963

3. Burma NE et al (2017) Animal models of chronic pain: advances and challenges for clinical translation. J Neurosci Res 95(6): 1242–1256

4. Thomson JA et al (1998) Embryonic stem cell lines derived from human blastocysts. Science 282(5391):1145–1147

5. Chambers SM et al (2012) Combined small-molecule inhibition accelerates developmental timing and converts human pluripotent stem cells into nociceptors. Nat Biotechnol 30(7): 715–720

6. Takahashi K et al (2007) Induction of pluripotent stem cells from adult human fibroblasts by defined factors. Cell 131(5):861–872

7. Basbaum AI et al (2009) Cellular and molecular mechanisms of pain. Cell 139(2):267–284

8. Le Pichon CE, Chesler AT (2014) The functional and anatomical dissection of somatosensory subpopulations using mouse genetics. Front Neuroanat 8:21

9. Brokhman I et al (2008) Peripheral sensory neurons differentiate from neural precursors derived from human embryonic stem cells. Differentiation 76(2):145–155

10. Goldstein RS et al (2010) Generation of neural crest cells and peripheral sensory neurons from human embryonic stem cells. Methods Mol Biol 584:283–300

11. Pomp O et al (2005) Generation of peripheral sensory and sympathetic neurons and neural crest cells from human embryonic stem cells. Stem Cells 23(7):923–930

12. Pomp O et al (2008) PA6-induced human embryonic stem cell-derived neurospheres: a new source of human peripheral sensory neurons and neural crest. Brain Res 1230:50–60

13. Young GT et al (2014) Characterizing human stem cell-derived sensory neurons at the single-cell level reveals their ion channel expression and utility in pain research. Mol Ther 22(8): 1530–1543

14. McDermott LA et al (2019) Defining the functional role of NaV1.7 in human nociception. Neuron 101(5):905–919 e8

15. Eberhardt E et al (2015) Pattern of functional TTX-resistant sodium channels reveals a developmental stage of human iPSC- and ESC-derived nociceptors. Stem Cell Reports 5(3):305–313

16. Wainger BJ et al (2015) Modeling pain *in vitro* using nociceptor neurons reprogrammed from fibroblasts. Nat Neurosci 18(1):17–24

17. Blanchard JW et al (2015) Selective conversion of fibroblasts into peripheral sensory neurons. Nat Neurosci 18(1):25–35

18. Schrenk-Siemens K et al (2015) PIEZO2 is required for mechanotransduction in human stem cell-derived touch receptors. Nat Neurosci 18(1):10–16

19. Boisvert EM et al (2015) The specification and maturation of nociceptive neurons from human embryonic stem cells. Sci Rep 5:16821

20. Schrenk-Siemens K et al (2019) HESC-derived sensory neurons reveal an unexpected role for PIEZO2 in nociceptor mechanotransduction. bioRxiv

21. Ma Q et al (1999) Neurogenin1 and neurogenin2 control two distinct waves of neurogenesis in developing dorsal root ganglia. Genes Dev 13(13):1717–1728

22. Bajpai R et al (2010) CHD7 cooperates with PBAF to control multipotent neural crest formation. Nature 463(7283):958–962

23. Zhang Y et al (2013) Rapid single-step induction of functional neurons from human pluripotent stem cells. Neuron 78(5):785–798

24. Rostock C et al (2018) Human vs. mouse nociceptors - similarities and differences. Neuroscience 387:13–27

25. Shiers S, Klein RM, Price TJ (2020) Quantitative differences in neuronal subpopulations between mouse and human dorsal root ganglia demonstrated with RNAscope in situ hybridization. Pain 161(10):2410–2424

26. Ray P et al (2018) Comparative transcriptome profiling of the human and mouse dorsal root ganglia: an RNA-seq-based resource for pain and sensory neuroscience research. Pain 159(7):1325–1345

27. DiMario FJ Jr (2016) Inherited pain syndromes and ion channels. Semin Pediatr Neurol 23(3):248–253

28. Kanellopoulos AH, Matsuyama A (2016) Voltage-gated sodium channels and pain-related disorders. Clin Sci (Lond) 130(24):2257–2265

29. Namer B et al (2019) Pain relief in a neuropathy patient by lacosamide: proof of principle of clinical translation from patient-specific iPS cell-derived nociceptors. EBioMedicine 39:401–408

30. Usoskin D et al (2015) Unbiased classification of sensory neuron types by large-scale single-cell RNA sequencing. Nat Neurosci 18(1):145–153

31. Zeisel A et al (2018) Molecular architecture of the mouse nervous system. Cell 174(4):999–1014 e22

32. Faure L et al (2020) Single cell RNA sequencing identifies early diversity of sensory neurons forming via bi-potential intermediates. Nat Commun 11(1):4175

33. Sharma N et al (2020) The emergence of transcriptional identity in somatosensory neurons. Nature 577(7790):392–398

34. Soldatov R et al (2019) Spatiotemporal structure of cell fate decisions in murine neural crest. Science 364(6444):eaas9536

Chapter 9

Intraspinal Transplantation of Precursors of Cortical GABAergic Interneurons to Treat Neuropathic Pain

João M. Braz and Allan I. Basbaum

Abstract

Neuropathic pain arises from peripheral nerve injury-induced loss of GABAergic controls in the spinal cord dorsal horn, resulting in ongoing, often burning pain, and mechanical and thermal hypersensitivity. Here we describe a powerful method for the long-term relief of neuropathic pain based on the rebuilding of lost inhibitory controls after partial peripheral nerve injury. The method, spinal cord transplantation of precursors of embryonic cortical GABAergic interneurons, involves two major steps: first, we harvest and mechanically dissociate inhibitory cells from the embryonic medial ganglionic eminence (MGE) of mice. Second, in adult mice that manifest mechanical hypersensitivity (allodynia) secondary to peripheral nerve injury, we expose the lumbar spinal cord by laminectomy and then stereotaxically inject the dissociated cells into the dorsal horn. Our studies have demonstrated that 2 weeks posttransplantation, MGE cells are fully differentiated into mature GABAergic interneurons and have functionally integrated into the host spinal circuitry.

Key words Transplantation, Spinal cord, Analgesia, Inhibitory interneurons, Chronic neuropathic pain, GABA, Medial ganglionic eminence

1 Introduction

Because dysfunction of the spinal cord inhibitory circuitry is among the major contributors to nerve injury-induced neuropathic pain, most pharmacotherapies developed to manage neuropathic pain aim to increase inhibition in the spinal cord. In many respects, the approach mimics that used to treat seizures that arise from loss of cortical inhibitory controls in epilepsy. Not surprisingly, therefore, anticonvulsants and specifically the gabapentinoids are the first-line therapies for neuropathic pain. However, pain relief with these compounds is only temporary and is often associated with adverse side effects due to their systemic delivery. With some exceptions, pain relief is generally not much better than 30% of the time in 30%

Rebecca P. Seal (ed.), *Contemporary Approaches to the Study of Pain: From Molecules to Neural Networks*, Neuromethods, vol. 178, https://doi.org/10.1007/978-1-0716-2039-7_9, © Springer Science+Business Media, LLC, part of Springer Nature 2022

of patients. Furthermore, and importantly, although these pharmacotherapeutic approaches mitigate the symptoms of neuropathic pain, they do not address the etiology. Building upon previous studies that introduced transplantation of inhibitory cells to overcome some of the shortcomings of traditional seizure pharmacotherapies [1–6], we present here, a method for the intraspinal transplantation of embryonic cortical inhibitory interneuron precursors from the medial ganglionic eminence (MGE). These cortical progenitors integrate into host circuits [7] and locally release the inhibitory neurotransmitter GABA, thereby restoring critical inhibitory tone [8, 9]. We have demonstrated that this approach is not only devoid of systemic side effects, but also provides long-term pain relief in different models of neuropathic pain [10, 11]. Perhaps more importantly, because this cell-based approach aims to repair a dysfunctional system, it has the potential to be disease modifying. Finally, the MGE cells can be genetically modified prior to transplantation [8] so as to deliver other pain-relieving agents [12] that may act synergistically with the MGE neurotransmitter-derived inhibitory controls.

In this procedure, we first harvest precursors of cortical GABAergic interneurons from E13.5 embryonic medial ganglionic eminence (MGE). After mechanical dissociation, we next transplant ~50,000 dissociated MGE cells into the dorsal horn of the spinal cord of mice with peripheral nerve injuries (Fig. 1).

Our studies have shown that 2 weeks after transplantation, the MGE cells disperse throughout the dorsal horn of the spinal cord, but rarely migrate to the contralateral side. MGE cells extend long and ramified processes, making synaptic connections with a wide variety of host neurons [6, 11]. In our studies, we chose the MGE-derived GABA precursor cells because of their proven ability to differentiate into functional GABAergic neurons and integrate

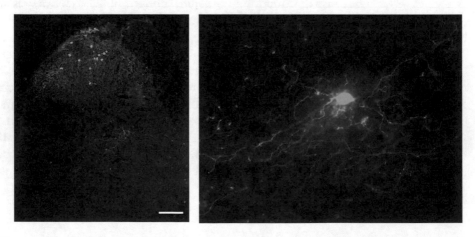

Fig. 1 Two weeks after transplantation, GFP-expressing MGE cell transplants disperse throughout the dorsal horn of the spinal cord (left) and extend long processes (right)

the host circuitry where they can exert strong inhibitory effects [7]. However, as embryonic cells have high intrinsic growth capacity, we believe that the procedure we describe here could be adapted to transplant other subsets of precursor cells. Indeed, in preliminary studies, we used the same approach to transplant human MGE-derived cells as well as human stem cells pre-differentiated in vitro into GABAergic cells, with similar results (data not shown). Finally, and although MGE cells retain their cortical phenotype rather than becoming spinal cord inhibitory interneurons, we have shown that the transplanted MGE cells can adapt to many different environments. Indeed, we achieve good survival rates when we transplanted the MGE cells into the anterior cingular cortex, the amygdala, or intraganglionically in lumbar DRGs.

2   Materials

2.1   Mice

Transplantations are performed in adult, 7- to 10-week-old mice. Donor mice: embryos are harvested from pregnant GAD67-GFP female mice [13]. In this transgenic knock-in mouse line, the green fluorescent protein (GFP) reporter is expressed under the control of the GAD67 promoter. Recipient mice: for transplantation, we use non-transgenic offspring from the GAD67-GFP litters, so that donor and recipient mice have the same genetic background. Alternatively, we purchase C57BL/6 mice (Jackson Labs, strain #664).

2.2   Euthanasia

CO_2 chamber.

Red biohazard bags.

2.3   Dissociation of MGE Cells

We autoclave all surgical tools before use. We buy sterile buffers and cell culture media. To maintain sterility, bottles containing buffers and media are only opened inside a safety cabinet, and keep them at 4 °C until their expiration.

Forceps curved Dumont #7 (Fine Science Tools [FST]; cat #11271-30).

Forceps straight Dumont #5 (FST; cat #11254-20).

Scissors (FST; cat #14084-08).

Microsurgical knife (22.5°, straight, Surgical Specialties Corp, cat #72-2201).

Petri dishes (sterile, 100 × 20 mm; BD Falcon cat#353003).

Fluorescence microscope.

PBS 1× (sterile; Genesee; cat #25-507).

DMEM cell culture medium (Gibco, cat #10565-018).

Leibovitz L-15 cell culture medium (Sigma Aldrich Inc., cat #L5520).

DNAse I (Biolabs, cat #M-0303S).

2.4 Laminectomy

Use anesthetics, antiseptics, and analgesics according to your IACUC protocol.

Sterilize all tools prior to surgery.

Scissors (FST; cat #14084-08).

Scissors, straight (FST; cat #15000-00).

Rongeurs, bone trimmers (FST; cat #16109-14).

Straight Dumont #5 forceps (FST; cat #11254-20).

70% isopropyl alcohol preps (Kendall, cat #6818).

Scalpel (FST; cat #10003-12).

Blades #11 (FST; #10011-00).

Cotton tip applicator, 6-in., sterile.

Sutures, 6-0, black silk, C-22 (Henry Schein; cat #1012636).

Wound clips, 7 mm (Roboz Surgical Instrument Co. Inc.; cat #RS-9258).

Warm lamp or pad.

2.5 MGE Cell Injection

Glass micropipettes, 1.0×0.75 mm, thin-wall, 4″ (A-M Systems; cat #6150).

Micropipette puller.

Mineral oil (Sigma Aldrich Inc.; cat #M5904).

Manual micro-injector (Sutter Instrument Company; cat #MI-10010).

3 Methods

3.1 Pretransplantation Steps

1. Set up breeding pairs or trios to impregnate female mice. *See* **Note 1**.

2. Check female mice every morning to detect sperm plug. When the plug is detected, separate females from the male. The day when a plug is detected is defined as embryonic day 0.5 (E0.5). *See* **Note 2**.

3. Harvest embryos 12–14 days after the plug is detected. *See* **Note 3**.

4. Select adult (8–10 weeks old) recipient mice. *See* **Note 4**.

3.2 Transplantation Day

3.2.1 MGE Isolation and Dissection (~30–60 min)

1. At E13.5, transport pregnant female mice to the laboratory and euthanize in CO_2 chamber.

2. Incise abdomen with sharp scissors to expose ovaries, where embryos can be readily observed. From the exposed ovaries, collect and transfer each embryo to a tissue culture dish containing 15 ml of DMEM cell culture medium; maintain on ice.

3. Dispose of euthanized female mouse in biohazardous red bag.

4. Under a fluorescence microscope, identify GFP^+ embryos. Discard non-GFP embryos into the biohazardous red bag and keep GFP^+ embryos in the dish containing DMEM cell culture medium. *See* **Note 5**.

5. Transfer one GFP^+ embryo to a cell culture dish containing 15 ml of PBS 1× and proceed to dissection under a microscope. *See* **Note 6**.

6. Using sharp forceps and a microsurgical knife, peel off skull to expose the embryonic brain.

7. Separate the brain from the rest of the embryo and discard embryo (Fig. 2).

8. Isolate the telencephalon by cutting and removing the hindbrain.

9. Make a midline incision (sagittal cut) to separate the two hemispheres.

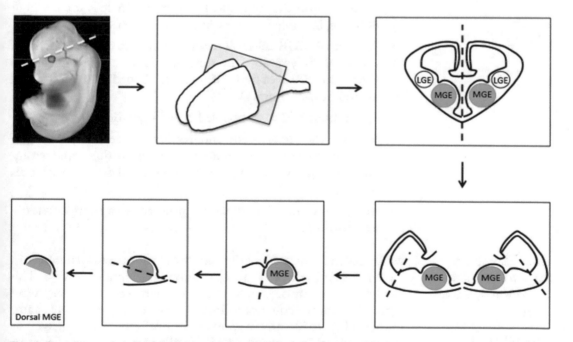

Fig. 2 Precursors of cortical inhibitory interneurons are dissected out from the dorsal medial ganglionic eminence (MGE, green) of E12.5–14.5 embryos. *LGE* lateral ganglionic eminence

10. Expose the ventral telencephalon by dissecting away the dorsally located cortex. The LGE, MGE, and CGE regions will appear as three distinct hillocks.

11. To isolate the MGE, dissect away the LGE and CGE.

12. Rotate the excised MGE 90° to visualize both dorsal and ventral MGE and make incision to separate the dorsal from the ventral MGE.

13. Using a 1-ml pipette, collect and transfer the dorsal MGE to an 1.5-ml Eppendorf tube containing 500 µl of Leibovitz L-15 cell culture medium; store on ice.

14. Repeat **steps 10–13** to isolate dorsal MGE from the second hemisphere.

15. Repeat procedure to collect dorsal MGEs from all remaining embryos and pool with other MGEs in the Eppendorf tube.

16. Proceed with mechanical dissociation.

3.2.2 MGE Cell Dissociation (~15 min)

1. With time, pooled MGEs will fall to the bottom of the tube and form a pellet. Remove medium carefully without disturbing the pooled MGEs.

2. Add 200 µl of fresh Leibovitz L-15 into the Eppendorf tube.

3. Pipette up and down 10 times (with a 200 µl plastic tip) to mechanically dissociate the MGE cells. *See* **Note 7**.

4. Add 1 ml of Leibovitz L-15 medium containing 100 µg/ml DNAse I into the Eppendorf tube and spin cells for 2 min at $800 \times g$, in cooled (4 °C) centrifuge. *See* **Note 8**.

5. A pellet of MGE cells will form on one side of the Eppendorf tube. Carefully remove the supernatant and resuspend the cell pellet with 200 µl of fresh Leibovitz L-15 medium (by pipetting up and down 10 times).

6. Spin cells at $800 \times g$ in cooled centrifuge, for 2 min.

7. Remove the supernatant and resuspend the cell pellet with fresh Leibovitz L-15 medium. Count cells with a hemocytometer and adjust volume to achieve desired cell density. *See* **Notes 9** and **10**.

8. Maintain MGE cells on ice and proceed with transplantation. *See* **Note 11**.

3.2.3 Optional Step: Infection of MGE Cells with Lentivirus to Express cDNA (~1 h)

MGE cells can be genetically modified before transplantation to co-express and release a variety of proteins/peptides in vivo. This can be achieved through infection with a viral vector that expresses a protein coding sequence of interest. For example, in our anatomical studies, to identify the postsynaptic cells that were engaged by the transplants, we generated MGE cells that expressed the transneuronal tracer, wheat germ agglutinin. To genetically modify the MGE cells, we performed the following:

1. After counting, incubate the MGE cells with the viral vector at a multiplicity of infection of 2 (i.e., two viral particles per cell), i.e., add 100,000 viral particles per MGE. *See* **Note 12**.

2. Add DMEM cell culture medium to a final volume of 200 µl and incubate at 37 °C for 45 min to 1 h, in shaker. *See* **Note 13**.

3. After viral infection, spin cells at 2000 rpm (~3000 × *g*) for 3 min.

4. Carefully remove the supernatant and resuspend the cell pellet with 200 µl of DMEM.

5. Repeat **steps 3** and **4** at least three times in order to wash out any unbound viral particles.

6. After the last wash, resuspend the cell pellet in an appropriate volume of Leibovitz L-15 medium, place on ice, and proceed with transplantation.

3.2.4 MGE Transplantation (~30–45 min per mouse)

Presurgical Steps

1. Pull glass micropipettes, bevel the tip, and fill with mineral oil. Insert oil-filled micropipette into the micro-injector.

2. Bring MGE cell suspension to the animal procedure room and maintain on ice throughout the surgery.

3. Follow IACUC guidelines for surgical tool sterilization, anesthesia, and analgesia procedures.

4. Under general anesthesia, shave mouse back with an electric razor, remove fur with sterile 70% alcohol preps and bring the anesthetized mouse to the surgical site.

Surgery

1. Sterilize incision site with antiseptic solution (we use Betadine) and rinse with 70% alcohol prep. Repeat sterilization step twice. Apply local anesthetic (we use 1% lidocaine) at incision site.

2. Using a scalpel blade (#11), make a midline incision over the vertebral column and cut through all overlying muscle layers.

3. In preparation for laminectomy, separate muscle with retractors and remove paraspinal muscle overlying the vertebral column.

4. Identify vertebra overlying the spinal cord lumbar enlargement, and using forceps, stabilize the cord slightly rostral to the targeted spinal level.

5. In the visible space between adjacent vertebra, insert a sharp rongeur under the bone, remove bone, and continue until the laminectomy is complete. *See* **Notes 14** and **15**.

6. Clean exposed dura by removing blood and/or bone debris.

7. Carefully grasp the dura with sharp forceps to separate from the surface of the spinal cord and make a small incision with sharp scissors (a 30-gauge needle can also be used), without injuring the spinal cord. *See* **Note 16**.

8. Using a sterile cotton tip applicator, remove leaked cerebrospinal fluid.

9. Load cell suspension into the glass micropipette. *See* **Note 17**.

10. Bring tip of the micropipette to the surface of the spinal cord (this is your zero) and insert micropipette into the spinal cord to a depth of 0.5 mm, which will target the deep dorsal horn, lamina V. Come back to 0.2 mm, so as to target laminae III–IV. Wait for 1 min and slowly inject cells (400 nl) over 1 min. Wait for another minute before slowly withdrawing the micropipette.

11. Move micropipette to second injection site and repeat **steps 12** and **13** until the desired amount of cells has been injected. *See* **Note 18**.

12. After all injections have been performed, place a piece of gelfoam on top of the exposed dura before suturing the paraspinal and overlying muscles together with 6-0 silk.

13. Staple skin with 7.0-mm wound clips and re-apply antiseptic solution over the incision site.

14. Allow the animal to recover under a warm lamp or on a warm pad before returning to their home cage.

3.3   Troubleshooting

3.3.1   No Cells Detected After Transplantation

Several reasons can account for a lack of cells in the transplanted spinal cord:

1. Cells died before transplantation: Because the dissociation of the MGE cells is mechanical, cells can be damaged during this step. As the MGE tissue is very delicate, triturating 10 times is more than sufficient to dissociate and resuspend the cells. If some tissue is difficult to dissociate (you will see clumps floating in the medium), then it is best to remove the clumps rather than persistently triturate. Before transplantation, one can also perform a viability test to determine whether there has been significant cell death during the dissociation step.

2. Cells were lost during transplantation: The lumbar spinal cord can be challenging to transplant because of breathing-induced movement of the cord. The up-and-down movement of the spinal cord can result in leakage of some medium and cells at the surface of the spinal cord. This can easily be detected under the microscope when it occurs at the cord surface. Breathing can be more problematic depending on the general anesthesia regimen. Isoflurane tends to generate milder breaths than does ketamine/xylazine. In any case, if the breathing makes the injection too challenging, one can use clamps to stabilize the vertebral column to lift the mouse so that it is no longer in contact with the heating pad. This procedure should reduce the breathing-induced movement of the spinal cord. Finally, if the

injection needs to be restricted to the very superficial laminae of the spinal cord (I–II), some medium will invariably leak out during the injection. Slowing down the speed of the injection can help reduce medium leakage.

3. Cells died posttransplantation: MGE cells can adapt to many different environments. For example, we have transplanted MGE cells in different tissues (spinal cord, brainstem, cortex, dorsal root ganglia) and have never found a significant difference in survival rate. Although we have never encountered this problem, it is always possible that some cells will not survive the transplantation. This could be due to bacterial infection if tools were not properly sterilized. Postsurgical antibiotic may help reduce infections (assuming that the antibiotics do not interfere with behavioral endpoints).

3.3.2 MGE Cells Did Not Migrate After Transplantation

Although MGE cells initially form an aggregate at the injection site, they should rapidly migrate from the aggregate and disperse throughout the dorsal horn of the spinal cord. Small cell clumps may remain, but most cells typically leave the injection site. If the cells do not disperse, it is likely that the cells were not properly dissociated before transplantation. If the micropipette is removed too quickly, some MGE cells will be found along the needle tract. If that occurs, we recommend waiting longer before removing the micropipette, as well as reducing the speed at which the pipette is removed.

3.3.3 Behavioral Abnormalities Posttransplantation

If the animals exhibit deficits in motor assessment tests, such as rotarod, hindlimb grip strength, catwalk, etc., it is likely due to tissue damage during the laminectomy and/or transplantation. To determine the extent of spinal cord damage, one can look for the presence of necrotic tissue under a fluorescence microscope; dead cells/debris will be autofluorescent and readily apparent. To prevent and/or minimize spinal cord damage, it is critical to avoid contacting the spinal cord tissue when removing bone during laminectomy or when incising the dura. Slow insertion of the micropipette and slow injection of the cell suspension will also help.

4 Notes

1. Because the GFP coding sequence is knocked into the first codon of the GAD67 locus, homozygous GAD67-GFP mice are deficient for GAD67. It is, therefore, important to pair transgenic mice with wild-type mice, so that knock-out GFP[+] embryos are not selected for transplantation.

2. Plugged females do not always become pregnant. If a subsequent plug leads to pregnancy, staging of the resulting

embryos may not be accurate. For this reason, we recommend separating the females from the male the day the plug is detected.

3. Transplantation success is comparable when using embryos between days E12.5 and E14.5. We found, however, that the MGE region is much easier to identify in E13.5 embryos. At E12.5, the MGE is smaller, and at E14.5 the MGE is elongated (as cells start to migrate), making it more difficult to identify.

4. Both male and female mice can be used as recipients for transplantation. To maximize transplantation success, we use donor and recipient mice with the same genetic background. In our colony, the GAD67-GFP mouse line has been kept on a mixed genetic background (CD1xC57BL6/J). We, therefore, use the nontransgenic offspring in the GAD67-GFP litters as recipients. However, we have also used inbred mice on a C57BL/6 genetic background as recipients and did not observe a significant difference in survival rate of the transplanted cells.

5. The expression level of GFP in the GAD67-GFP mouse line is strong enough to be detected under a microscope (without antibody).

6. At E13.5, the MGE region can easily be identified so this step can be performed under a green fluorescence or a brightfield microscope. However, under green fluorescence, the MGE will be much easier to detect as, at E13.5, GFP$^+$ cells are largely restricted to the three ganglionic eminences (lateral, medial, and caudal GE).

7. Because the MGE tissue is very delicate, cells will be rapidly dissociated and suspended.

8. Centrifugation can also be performed at room temperature, without significant damage to the cells.

9. If cells are too diluted, spin cells down once more and resuspend with appropriate volume of Leibovitz L-15 medium to reach desired concentration.

10. For spinal cord transplantation, we use 2 μl of medium to resuspend 1 MGE (i.e., ~50,000 cells) and transplant per mouse.

11. MGE cells can be stored on ice for up to 4 h without significantly affecting their posttransplantation survival rate.

12. It is critical to use replication incompetent viral vectors to preserve the integrity of the MGE cells. We have used both AAV and lentiviral vectors in these experiments, with ~30% infection yield.

13. Higher yields of infection are achieved with constant shaking. If shaker is not available, resuspend the cells by mixing every 10–15 min. Gently tapping the tube will resuspend them.

14. Hemorrhage can occur during the laminectomy if bone is pulled too forcefully or if the laminectomy extends too laterally. If this occurs, stop the bleeding by applying pressure on the hemorrhage site with a sterile cotton tip applicator or by using gelfoam.

15. Depending on the analysis to be performed, the extent of the laminectomy should be adjusted. For behavioral analyses, we typically remove bone only on one side (hemi-laminectomy), over two lumbar spinal segments. This approach will expose most of the lumbar enlargement where primary afferents that innervate the hindpaw terminate. For anatomical analyses, in which a lower number of transplanted cells is required, laminectomy over one segment is sufficient. If only one injection is to be performed, it is not necessary to perform a laminectomy. In this case, the dura can be accessed and incised exposing the spinal cord through an open space that exists between adjacent vertebra.

16. Because the dura is remarkably firm, even very sharp micropipette tips will not penetrate. For this reason, the dura must be incised. After dural incision, cerebrospinal fluid (CSF) will flow immediately. If there is no CSF leakage, then the dura has not been sufficiently cut.

17. Before loading, make sure to resuspend the cells by gently tapping the tube (or by pipetting two to three times up and down) as cells fall to the bottom of tube with time. For the same reason, do not load cells into the micropipette before the laminectomy is completed and you are ready to inject. Otherwise, the cells will clog the micropipette.

18. As cells will fall to the bottom and clog the micropipette tip, we recommend loading enough cells for only one or two injections.

References

1. Wu HH, Wilcox GL, McLoon SC (1994) Implantation of AtT-20 or genetically modified AtT-20/hENK cells in mouse spinal cord induced antinociception and opioid tolerance. J Neurosci 14:4806–4814

2. Eaton MJ, Plunkett JA, Martinez MA, Lopez T, Karmally S, Cejas P, Whittemore SR (1999) Transplants of neuronal cells bioengineered to synthesize GABA alleviate chronic neuropathic pain. Cell Transplant 8:87–101

3. Cejas PJ, Martinez M, Karmally S, McKillop M, McKillop J, Plunkett JA, Oudega M, Eaton MJ (2000) Lumbar transplant of neurons genetically modified to secrete brain-derived neurotrophic factor attenuates allodynia and hyperalgesia after sciatic nerve constriction. Pain 86:195–210

4. Mukhida K, Mendez I, McLeod M, Kobayashi N, Haughn C, Milne B, Baghbaderani B, Sen A, Behie LA, Hong M (2007) Spinal GABAergic transplants attenuate mechanical allodynia in a rat model of neuropathic pain. Stem Cells 25:2874–8285

5. Basbaum AI, Bráz JM (2016) Cell transplants to treat the "disease" of neuropathic pain and itch. Pain. https://doi.org/10.1097/j.pain.0000000000000441

6. Braz JM, Etlin A, Juarez-Salinas D, Llewellyn-Smith IJ, Basbaum AI (2016) Rebuilding CNS inhibitory circuits to control chronic

neuropathic pain and itch. Prog Brain Res. https://doi.org/10.1016/bs.pbr.2016.10.001

7. Baraban SC, Southwell DG, Estrada RC, Jones DL, Sebe JY, Alfaro-Cervello C, García-Verdugo JM, Rubenstein JL, Alvarez-Buylla A (2009) Reduction of seizures by transplantation of cortical GABAergic interneuron precursors into Kv1.1 mutant mice. Proc Natl Acad Sci 106:15472–15477

8. Bráz JM, Sharif-Naeini R, Vogt D, Kriegstein A, Alvarez-Buylla A, Rubenstein JL, Basbaum AI (2012) Forebrain GABAergic neuron precursors integrate into adult spinal cord and reduce injury-induced neuropathic pain. Neuron 74:663–675

9. Bráz JM, Wang X, Guan Z, Rubenstein JL, Basbaum AI (2015) Transplant-mediated enhancement of spinal cord GABAergic inhibition reverses paclitaxel-induced mechanical and heat hypersensitivity. Pain 156:1084–1091

10. Etlin A, Bráz JM, Kuhn JA, Wang X, Hamel KA, Llewellyn-Smith IJ, Basbaum AI (2016) Functional synaptic integration of forebrain GABAergic precursors into the adult spinal cord. J Neurosci 36:11634–11645

11. Llewellyn-Smith IJ, Basbaum AI, Bráz JM (2018) Long-term, dynamic synaptic reorganization after GABAergic precursor cell transplantation into adult mouse spinal cord. J Comp Neurol 526:480–495

12. Jergova S, Gajavelli S, Varghese MS, Shekane P, Sagen J (2016) Analgesic effect of recombinant GABAergic cells in a model of peripheral neuropathic pain. Cell Transplant 25:629–643

13. Tamamaki N, Yanagawa Y, Tomioka R, Miyazaki J, Obata K, Kaneko T (2003) Green fluorescent protein expression and colocalization with calretinin, parvalbumin, and somatostatin in the GAD67-GFP knock-in mouse. J Comp Neurol 467:60–79

Chapter 10

A Co-culture System for Studying Dorsal Spinal Cord Synaptogenesis

Yanhui Peter Yu and Z. David Luo

Abstract

Studying synapse formation, maintenance, and plasticity in adaptation to developmental and pathological changes is critical in our understanding of cellular mechanisms of biological processes and disease states. However, a major barrier in getting cell-type-specific detail information in these studies is the complexity of in vivo environment in high-density tissue or organs, such as spinal cord, that is packed with different types of cells, connecting tissues and structural components. In this chapter, we describe a co-culture system in which dorsal root ganglion sensory neurons and spinal cord neurons are cultured in separate compartments without culture medium diffusion between compartments, but allowing sensory neuron axons to outgrowth to adjacent chambers to establish synaptic connections with dendrites of spinal cord neurons. This provides an in vitro environment that mimics the in vivo synaptogenic environment between sensory neurons and spinal cord neurons and enables manipulation of specific neuronal populations and studying their detail contribution to synaptic formation, maintenance, and plastic changes.

Key words Pain mechanisms, DRG, Dorsal spinal cord, Neuronal co-culture, Synaptogenesis

1 Introduction

Studying spinal cord synaptogenesis during development and under pathophysiological conditions is critical for our understanding of normal sensory and motor functions as well as relevant disorders, such as chronic pain development and motor function deficits, after peripheral/spinal nerve injury or spinal cord injury. However, it remains a technical challenge to study bio-pathological mechanisms of spinal cord synaptogenesis in vivo due to the complexity of synaptogenic environment and high density of spinal cord tissues. To overcome this barrier, we have fine tuned a primary co-culture system of dorsal root ganglion (DRG) neurons and dorsal spinal cord (DSC) neurons to facilitate studies related to sensory neuron synapse formation in a controlled in vitro environment [1, 2]. Although a cell culture system does not wholly replicate the in vivo milieu, we believe that this simplified system enables

Rebecca P. Seal (ed.), *Contemporary Approaches to the Study of Pain: From Molecules to Neural Networks*, Neuromethods, vol. 178, https://doi.org/10.1007/978-1-0716-2039-7_10, © Springer Science+Business Media, LLC, part of Springer Nature 2022

genetic and biochemical manipulations that are critical for studying the roles of individual genes/factors and signaling pathways important for sensory neuron synaptogenesis in a controlled environment that mimics the interactions between DRG neurons and dorsal spinal cord neurons in vivo. This protocol was derived from primarily a series of published protocols [3–6] along with some method sections from several papers [7–11].

2   Materials

2.1   Cell Culture Supplies

1. Calcium magnesium-free phosphate buffer (CMF) (Sigma-Aldrich, St. Louis, MO).
2. Trypsin (Thermo Fisher Scientific, Waltham, MA).
3. Dulbecco's Modified Eagle Medium (DMEM) (Thermo Fisher Scientific, Waltham, MA).
4. Fetal bovine serum (FBS) (Thermo Fisher Scientific, Waltham, MA).
5. Horse serum (Thermo Fisher Scientific, Waltham, MA).
6. Penicillin-Streptomycin (Pen/Strep, 100X) (Thermo Fisher Scientific, Waltham, MA).
7. Trypan blue (Thermo Fisher Scientific, Waltham, MA).
8. Collagenase (Sigma-Aldrich, St. Louis, MO).
9. Poly-D-lysine (Sigma-Aldrich, St. Louis, MO).
10. Natural Mouse Laminin (Thermo Fisher Scientific, Waltham, MA).
11. NB/B27 media (Thermo Fisher Scientific, Waltham, MA).
12. Uridine (Sigma-Aldrich, St. Louis, MO).
13. 5-Fluorodeoxyuridine (Sigma-Aldrich, St. Louis, MO).
14. Neurobasal media (Thermo Fisher Scientific, Waltham, MA).
15. Methylcellulose (Sigma-Aldrich, St. Louis, MO).
16. Glass coverslips (Thermo Fisher Scientific, Waltham, MA).
17. Cell culture dish (Corning Life Sciences, Tewksbury, MA).
18. Campenot chambers (Tyler Research Corporation, Edmonton, AB, Canada).
19. Epoxy (Home depot).
20. Small insect pins (Carolina Biological Supply Company, Burlington, NC).
21. Silicone grease (Home depot).
22. 5-mL syringe (Thermo Fisher Scientific, Waltham, MA).
23. Glass Pasteur pipette (Thermo Fisher Scientific, Waltham, MA).

24. Filter (40 μM, Thermo Fisher Scientific, Waltham, MA).

25. Blunted 28 or 30 gauge needle (Thermo Fisher Scientific, Waltham, MA, *see* **Note 1**).

26. Hemocytometer (Thermo Fisher Scientific, Waltham, MA).

27. 15 mL conical tube (Corning Life Sciences, Tweksbury, MA).

2.2 Immuno-
fluorescence Staining,
Imaging Acquisition,
and Analysis

1. Methanol (Sigma-Aldrich, St. Louis, MO).

2. Dako antibody diluent (now Agilent Technologies, Inc., Santa Clara, CA).

3. Primary antibodies (various vendors based on experimental needs).

4. Alexafluor secondary antibodies (Thermo Fisher Scientific, Waltham, MA).

5. Glass slides (Thermo Fisher Scientific, Waltham, MA).

6. Vectashield DAPI hard mount media (Vector Laboratories, Inc., Burlingame, CA).

7. Confocal microscope (Zeiss International, Oberkochen, Germany).

8. Imaging analyzing software (*see* **Note 15**).

3   Method

3.1 Spinal Cord
Neuron Isolation

1. Euthanize timed pregnant mice following Institutional Animal Care and Use Committee (IACUC) guidelines.

2. Spinal cord neurons are harvested from E14–E19 mouse embryos by making a lengthwise incision along the ventral side of the spinal vertebrae and collecting the spinal cord in CMF on ice.

3. Dissociate the spinal cord neurons in 5 mL 0.25% trypsin in a 15 mL conical tube for 10 min at 37 °C.

4. Inactivate trypsin by adding 10 mL DMEM + 10% FBS.

5. Triturate the digested spinal cord to separate spinal cord neurons using two flamed glass Pasteur pipettes with gradually constricted opening tips.

6. Spin down the cell suspension at low speed ($200 \times g$) for 5 min.

7. Remove supernatant and resuspend the pellet in fresh DMEM (5 mL) + 10%FBS.

8. Pass the resuspended cells through a 40-μM filter to remove debris and cell clumps.

9. Centrifuge the cell suspension at low speed and resuspend the pellet in 1 mL fresh DMEM + 10%FBS + 10% horse serum + 1% pen/strep.

10. Count cells using a hemocytometer and Trypan blue at 1:10 dilution. A successful isolation should yield a greater than 10:1 live/dead cell ratio.

3.2   DRG Neuron Isolation

1. Euthanize mice following IACUC guidelines. DRG neurons should be promptly isolated from freshly euthanized mice to insure high viability and healthy cells.

2. Remove skin along the back of the animal.

3. Make a lateral cut below the shoulder and above the hip and dissect out the vertebral column. This should roughly include segments T7/8 through L4/5.

4. Adult DRG neurons are harvested by first removing the spinal cord via hydraulic extrusion (*see* **Note 1**), which makes isolating the DRGs easier. Spinal cord can also be removed by dissection after laminectomy if the researcher is careful enough not to pull the attached DRGs while removing the spinal cord.

5. A laminectomy is performed to remove the dorsal half of the vertebrae exposing the spinal cavity.

6. Isolate the DRGs from the segments of interest and store in CMF on ice. We routinely recover DRGs bilaterally from T9 through L5 for our experiments.

7. Dissociate the DRGs in 1 mL of 1.25 mg/mL collagenase solution for 10–15 min at 37 °C in a 15 mL conical tube.

8. Inactivate the collagenase digestion by adding equal or greater volume of DMEM + 10%FBS.

9. Triturate the DRGs using two fire-flamed glass Pasteur pipettes with gradually reduced tip opening.

10. Spin down the cells at low-speed centrifugation ($200 \times g$).

11. Resuspend the pellet in 1 mL DMEM + 10% FBS.

12. Pass the cell suspension through a 100-µM filter. Some DRG neurons have large diameters so a larger filter is used to avoid removing neurons with the cellular debris.

13. Count the cells using Trypan blue at 1:4 dilution.

14. Cell viability should be above 90%. The 100-µM filter should be able to remove some but not all cellular debris (*see* **Note 2**). It is acceptable to have some debris before plating, further filtering through a 40 µM filter is not recommended.

15. Cells are plated onto 0.1 mg/mL poly-D-lysine- and 0.04 mg/mL laminin-coated glass coverslips and grown in NB/B27 media and 100 nM uridine/20 nM 5-fluorodeoxyuridine (U/FrdU).

3.3 Co-culture in Regular Culture Dishes (See Note 3)

In SC/DRG co-cultures, SC neurons are first allowed to grow for 3 days for maturity and neurite sprouting. In our experience, embryonic SC neurons take a few days after plating to sprout neurites while adult DRG neurons are able to sprout neurites overnight. Therefore, DRG neurons are added to the SC neuron cultures at the time of SC culture media change at day 3. From day 3 onwards, neurons are grown in the presence of Uridine/FrdU mitotic inhibitors.

1. Plate the cells onto a 0.1 mg/mL poly-D-lysine and 0.04 mg/mL laminin-coated glass coverslips (*see* **Note 4**).

2. SC neurons are grown in DMEM supplemented with 10% horse serum and 10% fetal bovine serum and 1% pen/strep.

3. After 24 h, the media are replaced with neurobasal media supplemented with B27 supplement (NB/B27) (*see* **Note 5**). After 3 days, DRG neurons are added along with U/FrdU to inhibit non-neuronal cell proliferation (*see* **Note 6**).

4. Cultures are usually ready to use by day 10–14 when neuronal morphology emerges and significant neurite outgrowth is observed.

3.4 Co-culture Using Campenot Chamber Cultures

Campenot chambers are set up as described by Pazyra-Murphy and Segal [12]. The main benefit of such a setup is to isolate the two different neuronal populations (since media exchange between the inner and outer chambers is blocked by the silicone/methylcellulose gel used to affix the chamber to the culture dish lid), so one can selectively treat or manipulate each population of cells in each chamber for desire studies (*see* **Note 7**). There are many designs of Campenot chambers available. We demonstrate here a circular design with a central rectangular chamber dividing the outer two chambers of the apparatus.

1. 35-mm cell culture dish lids are coated with 1 mg/mL collagen (*see* **Note 8**). Grooves are scored on the culture dish lid using an insect pin rake at approximately 200-μm intervals (*see* **Note 9**). The grooves provide a physical barrier preventing axons from growing across them, looping and growing back toward the center chamber, and help to guide the axonal growth toward the outer chambers (Fig. 1). The length of the grooves is usually the width of the center chamber so that axons are allowed to grow freely once they are guided to the outer chambers.

2. Wet grooves with a drop of NB/B27 + 0.6 mg/mL methylcellulose with a sterile pipette.

3. Coat Campenot chamber bottoms with autoclaved silicone grease dispensed from a sterile 5-mL syringe and mount onto

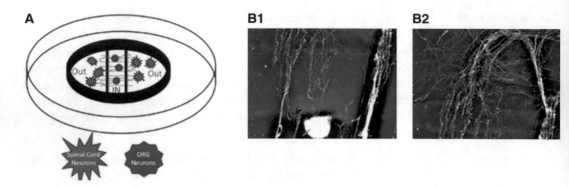

Fig. 1 Campenot chamber setup (modified from Fig. 3a of Yu et al. [2] with permission) and DRG neuron axonal outgrowth. (**a**) Diagram of a Campenot Chamber setup. (**b1**) DRG neuron axons, but not soma, crossing the inner chamber wall. (**b2**) DRG neuron axons crossing into the outer chambers

the scored culture dish lid as shown in Fig. 1a, making sure a proper seal is achieved.

4. Add culture media into the middle chamber leaving the outside chambers empty and store the Campenot chamber overnight in an incubator to check if the silicone grease creates a leak-proof seal (*see* **Note 10**).

5. Plate DRG neurons into the middle (rectangular) chamber and fill the outer chamber with culture media. Culture the DRG neurons in NB/B27 media with U/FrdU as discussed in **step 15** in Subheading 3.2 for around 7 days (*see* **Note 11**).

6. Once DRG axons have started sprouting into the outer chambers (Fig. 1b1, b2), SC neurons are plated into the outer chambers. Chambers are ready to use when sufficient number of DRG axons fully cross the divider between chambers to reach the SC neurons in the outer chambers.

3.5 Immuno-fluorescence Staining

Cells can be stained or imaged directly from the culture plate. For certain microscopes with a short focal length (for certain magnifications) that could not take focused cell images through the bottom of the culture plate, we employed workarounds such as culturing cells on a glass coverslip or looking for a thinner bottom plate such as using a culture dish lid.

3.5.1 For Regular Co-cultures

1. Cultured neurons are fixed in $-20\ °C$ methanol for 15 min (*see* **Note 12**).

2. Samples are then incubated for 24 h with primary antibodies diluted in Dako antibody diluent (Dako). The antibodies that we used include: chicken Microtubule Associated Protein 2 (MAP2) (Abcam), mouse Post Synaptic Density 95 kDa (PSD95) (Pierce), guinea pig Vesicular Glutamate Transporter 2 (VGlut2) (Synaptic Systems) (*see* **Note 13**).

3. Samples are then incubated overnight in corresponding Alexafluor (Thermo Fisher) 488, 594, and 647 secondary antibodies that are species specific to the primary antibodies.

4. Mount samples onto glass slides using Vectashield DAPI hard mount media (Vector Labs). Images are taken from the selected areas using a 63× objective on a confocal microscope (we used the Zeiss LSM700) (*see* **Note 14**) and analyzed using Volocity 6.0 (Perkin Elmer) (*see* **Note 15**).

3.5.2 For Campenot Chamber Co-cultures

1. Remove the Campenot chamber carefully.

2. If possible, wash and carefully remove as much silicon grease as possible.

3. Fix the co-culture cells with methanol as discussed above.

4. Perform immunohistostaining and image acquisition following **steps 3** and **4** for regular co-cultures.

4 Notes

1. Insert the blunted-end tip of a 28- or 30-gauge needle attached to a 10-mL syringe filled with ice-cold PBS into the lambar end opening of the vertebral column.

   The diameter of the opening might vary depending on the size of the animal. Therefore, trying higher and lower gauge size tips may be necessary to find a tip size that fits snuggly into the opening. We found that using pliers to slightly depress the tip to achieve a more oval shape may help the fitting. These tips can be cleaned and reused.

2. Some cell debris remaining after filtration might include myelinated axon bits, which can be washed off in 12–24 h upon changing medium once cells have been attached.

3. Note that this is a multi-day process. Embryonic spinal cord neurons take longer to establish and produce neurite outgrowth. They also need a different medium formulation for the first 24 h to get established. In our experience and referencing protocols published by other laboratories, we have found that SC neurons need to be cultured in DMEM + 10% FBS + 10% horse serum for 24 h before changing to NB/B27 media.

4. The glass coverslips are used to mount the adhered cells to a glass slide for imaging. Alternatively, you can use a plastic culture dish lid.

5. In studying synaptic connectivity, it is important to keep neurons healthy and ensure that they sprout enough neurites to find each other as well as maintain synaptic connectivity. This is

especially critical in low-density cultures, including DRG/SC neuron, and Campenot chamber co-cultures since more consistent neuronal health could improve culture consistency, thus improve the assay window and minimize date variation among different sets of cultures. Therefore, it is worthy in initial experiments to compare (e.g., B27 vs B27Plus, Thermo Fisher Scientific) and select reagents that can better increase culture longevity, improve neurite outgrowth, and enhance synaptic connectivity.

6. U/FrdU is required to control the proliferation of non-neuronal cells in both DRG and SC cultures. Uncontrolled proliferation of non-neuronal cells could overwhelm the culture and reduce neuronal survival rates. Spinal cord neurons are very sensitive to mitotic inhibitors for the first few days in culture. On the other hand, DRG neurons can be cultured directly in NB/B27 + mitotic inhibitors U/FrdU and will adhere and produce neurite outgrowth overnight. Thus, in our hands, allowing spinal cord neurons to establish themselves 3 days prior to addition of DRG neurons provides the best compromise.

   Getting too little or too much mitotic inhibitors could result in glial cell overgrowth or poor neuronal health/excessive cell death, respectively. Newly improved reagents, such as CultureOne (Thermo Fisher Scientific), that can help to suppress glial cell proliferation without affecting neuronal health or accelerating cell death could greatly improve the image analysis and data variation downstream.

7. We also tried the Axis Axon Isolation Device (Millipore-Sigma) but could not get consistent axonal outgrowth across the microgrooves. However, we do believe that it could also be a viable alternative and is worth testing.

8. The cell culture dish bottom is too thick to get focused cell images with our confocal microscope (Zeiss LSM700). We found that the lid is thin enough for downstream imaging applications.

9. To make the rake, use epoxy to glue small insect pins side by side onto a metal stirring spatula to create a handle for the comb tip. Significant force is needed to score the dish so one needs to make sure that the pins are glued firmly together. The grooves are scored directly onto the culture dish lid, thus using a glass culture dish is impractical for this application.

10. In practice, the methylcellulose/silicone layer creates a seal that allows axons to grow through the medium while minimizing medium diffusion from one chamber to another chamber. A proper seal is critical in making sure that the middle chamber is fluidically sealed from the outer chambers. Improper seal will

not only result in medium diffusion across chambers, but also allow SC neurons to flow into the center chamber. If the unsealed gaps are big enough, the whole Campenot chamber can become detached from the dish.

11. After around 7 days in culture, one should observe DRG axons crossing or about to cross from the middle chamber to the outer chambers.

12. The fixation protocol is up to preference. We prefer methanol fixation to paraformaldehyde fixatives.

13. We used DRGs from transgenic mice with a pan DRG-RFP (Red Fluorescent Protein) reporter to identify DRG neurons in co-cultures. Since DRG neurons are pseudo-unipolar neurons without dendrites, immunoreactivity to dendrite specific antibody MAP2 could then be used to identify SC neurons. Alternatively, neuronal subtype-specific markers can also be used.

14. High content analysis instrument: For our synaptogenesis studies, a critical component is acquiring images of the neurons, their neuritis, and associated synapses. Manually finding and taking enough images using a confocal microscope and the subsequent data analysis have proved to be a labor-intensive affair. We would highly recommend using a high content analysis Instrument. Automating data acquisition and analysis takes away user biases and allows the generation of much larger data sets in a timely manner.

15. In addition to Volocity, which is an expensive software that requires a significant learning curve, other software, some of them such as ImageJ (NIH) are free, can also be used.

Acknowledgments

Supported in part by grants from the National Institutes of Health R01NS064341, R01DE021847, R01DE029202 (Z.D.L.).

References

1. Park JF, Yu YP, Gong N, Trinh VN, Luo ZD (2018) The EGF-LIKE domain of thrombospondin-4 is a key determinant in the development of pain states due to increased excitatory synaptogenesis. J Biol Chem 293: 16453–16463

2. Yu YP, Gong N, Kweon TD, Vo B, Luo ZD (2018) Gabapentin prevents synaptogenesis between sensory and spinal cord neurons induced by thrombospondin-4 acting on pre-synaptic Cav alpha2 delta1 subunits and involving T-type Ca(2+) channels. Br J Pharmacol 175:2348–2361

3. Albuquerque C, Joseph DJ, Choudhury P, MacDermott AB (2009) Dissection, plating, and maintenance of dorsal root ganglion neurons for monoculture and for coculture with dorsal horn neurons. Cold Spring Harb Protoc 2009:pdb.prot5275

4. Albuquerque C, Joseph DJ, Choudhury P, MacDermott AB (2009) Dissection, plating, and maintenance of dorsal horn neuron

cultures. Cold Spring Harb Protoc 2009:pdb. prot5274

5. Burkey TH, Hingtgen CM, Vasko MR (2004) Isolation and culture of sensory neurons from the dorsal-root ganglia of embryonic or adult rats. Methods Mol Med 99:189–202

6. Seybold VS, Abrahams LG (2004) Primary cultures of neonatal rat spinal cord. Methods Mol Med 99:203–213

7. Bauer CS, Nieto-Rostro M, Rahman W, Tran-Van-Minh A, Ferron L, Douglas L, Kadurin I, Sri Ranjan Y, Fernandez-Alacid L, Millar NS, Dickenson AH, Lujan R, Dolphin AC (2009) The increased trafficking of the calcium channel subunit alpha2delta-1 to presynaptic terminals in neuropathic pain is inhibited by the alpha2delta ligand pregabalin. J Neurosci 29:4076–4088

8. Joseph DJ, Choudhury P, Macdermott AB (2010) An in vitro assay system for studying synapse formation between nociceptive dorsal root ganglion and dorsal horn neurons. J Neurosci Methods 189:197–204

9. Delree P, Leprince P, Schoenen J, Moonen G (1989) Purification and culture of adult rat dorsal root ganglia neurons. J Neurosci Res 23:198–206

10. Ohshiro H, Ogawa S, Shinjo K (2007) Visualizing sensory transmission between dorsal root ganglion and dorsal horn neurons in co-culture with calcium imaging. J Neurosci Methods 165:49–54

11. Varon S, Raiborn C (1971) Excitability and conduction in neurons of dissociated ganglionic cell cultures. Brain Res 30:83–98

12. Pazyra-Murphy MF, Segal RA (2008) Preparation and maintenance of dorsal root ganglia neurons in compartmented cultures. J Vis Exp 17(20). https://doi.org/10.3791/951

Chapter 11

Visualizing Synaptic Connectivity Using Confocal and Electron Microscopy: Neuroanatomical Approaches to Define Spinal Circuits

David I. Hughes and Andrew J. Todd

Abstract

Visualizing the synaptic connectivity of identified neuronal populations is an essential part of defining neuronal circuits in the central nervous system. In our laboratory, we have optimized several histological approaches that are compatible with confocal and transmission electron microscopy (TEM) to study the anatomical relationships between sensory afferents, interneurons, and projection neurons in the spinal dorsal horn. The aims of this chapter are to provide detailed descriptions of some of these protocols, to highlight both advantages and potential difficulties that may be encountered when using these approaches, and to reference several studies where these have been applied to study spinal circuits in normal and chronic pain states.

Key words Ultrastructure, Synapses, Electron microscopy, Immunohistochemistry, Circuitry, Neurotransmitters, GABA, Glycine, Peptides, Receptors, Enzymes

1 Introduction

Defining synaptic circuits within the central nervous system (CNS) is of critical importance to further our understanding of pain pathways in both normal and pathological states. While electrophysiological recordings, and more recently chemo- and optogenetic approaches, help us study the functional and pharmacological basis of these circuits, determining the anatomical relationship and synaptic connectivity between neurons is equally important. The specificity of immunohistochemical approaches for defining neuronal populations, and the selectivity of the resultant labeling patterns, can be used to define synaptic circuits in anatomical studies using combinations of confocal and transmission electron microscopy (TEM) approaches. Since disinhibition (or increased excitability) of spinal circuits has long been proposed to be an important feature in the development of several chronic pain states [1–3], studying the synaptic connectivity of primary afferent

Rebecca P. Seal (ed.), *Contemporary Approaches to the Study of Pain: From Molecules to Neural Networks*, Neuromethods, vol. 178, https://doi.org/10.1007/978-1-0716-2039-7_11, © Springer Science+Business Media, LLC, part of Springer Nature 2022

terminals, as well as those of excitatory and inhibitory interneurons, in the superficial laminae of the spinal dorsal horn has been of particular interest.

Excitatory transmission at synapses formed by primary afferent terminals and excitatory interneurons is mediated principally through the release of glutamate, whereas inhibitory synaptic transmission in the spinal dorsal horn results from the release of GABA and/or glycine. Immunohistochemical studies have used antibodies raised against glutamic acid decarboxylase (GAD) and its two isoforms (GAD65 and GAD67), to identify axon terminals derived from GABAergic interneurons in the spinal cord [4–7], whereas antibodies raised against the neuronal glycine transporter (GlyT2) have been used to visualize axon terminals from glycinergic neurons [7, 8]. Since virtually all glycinergic neurons in lamina II are also GABAergic [9, 10], antibodies raised against the vesicular GABA transporter (VGAT) are often used to study inhibitory axon terminals in the dorsal horn [11–14]. Glutamatergic axon terminals derived from primary afferents, excitatory interneurons, or descending projections can be labeled using antibodies raised against various vesicular glutamate transporters (VGLUT1, VGLUT2, or VGLUT3). In the spinal dorsal horn of adult rats and mice, VGLUT1 is expressed in axon terminals derived from myelinated afferents or corticospinal projections, and these are localized primarily in laminae II inner, III, and IV [13–15], whereas VGLUT2 is expressed largely in boutons of excitatory interneurons and found in all laminae [12, 15]. Boutons that express VGLUT3 are confined to lamina II and are derived from unmyelinated low-threshold mechanoreceptors (CLTMRs; [16]). Recent transcriptomic studies have identified several subpopulations of cutaneous and visceral primary afferents, spinal dorsal horn interneurons, and projection neurons based on their distinctive genetic signatures [17–21]. These resources allow us to identify permutations of molecular markers that can be used in combination with immunohistochemical approaches to visualize distinct neuronal populations with even greater precision than previously possible, and this will, in turn, help efforts to define discrete spinal circuits [22–24]. Studies adopting such approaches typically show only associations between neurochemically defined axon terminals and possible synaptic targets, and only rarely demonstrate the presence of synapses between immunolabeled profiles [25, 26].

Until relatively recently, when immunolabeling patterns for Homer1 and gephyrin were shown to define excitatory and (most) inhibitory synapses in the spinal dorsal horn, respectively [27, 28], the only way of visualizing synaptic connectivity between cells was using TEM. Excitatory and inhibitory synapses have been defined on the symmetry of pre- and postsynaptic membranes at synaptic active zones, and the shape of synaptic vesicles contained in the presynaptic terminal [29–31], although the reliability of these schemes has been questioned [32]. Immunolabeling studies to detect the expression of glutamate, GABA, and glycine in CNS tissues offer greater precision in identifying excitatory and inhibitory synaptic connections than defining these contacts on morphological features alone. However, this approach is technically demanding, largely because of the challenges of fixing the tissues to preserve the subcellular distribution of these antigens appropriately. The main difficulty is to develop an optimized fixation protocol to ensure that these antigens are preserved in a way that makes them amenable to detection with antibodies. Glutaraldehyde binds free amino acids to nearby proteins; therefore, for optimal retention of these antigens, it is vital that glutaraldehyde-containing fixative is delivered rapidly into tissue. This can be achieved most efficiently through vascular perfusion [33–35]. Furthermore, given that the most reliable antibodies raised against GABA, glycine, and glutamate are directed against glutaraldehyde conjugates of these amino acids [36], it is essential to prepare tissues intended for use with these antibodies with glutaraldehyde-containing fixative.

For optimal retention of tissue antigenicity and integrity of tissue ultrastructure, fixative solutions containing 2.5% glutaraldehyde are typically used in these studies [37, 38]; however, this high concentration is not compatible with immunofluorescence studies since it will restrict the penetration of antibodies into the tissues and increase background fluorescence. Furthermore, certain epitopes are sensitive to over-fixation (e.g., enzymes, neurotransmitter receptors, and intrinsic fluorescent markers such as GFP that may be expressed in transgenic animals), and this forces us to use lower concentrations of glutaraldehyde in primary fixatives for these experiments. Therefore, a balance (or compromise) needs to be reached when developing effective protocols for the use of immunohistochemistry in studying neuronal circuitry, where the retention of antigenicity in the tissue and preservation of its structure are both optimal, while also ensuring that the means of visualizing the immunolabeling is not compromised. Experimenting with optimal fixative combinations for various antibodies and visualization strategies has clear benefits. Where more conventional anatomical studies have prepared tissues for either light or electron microscope studies, establishing protocols for confocal microscopy where low levels of glutaraldehyde are used has allowed us to develop approaches where tissues can be used for correlative light and electron microscopy [39, 40].

The use of antibodies raised against the postsynaptic scaffolding proteins Homer1 and gephyrin have proven to be reliable ways of visualizing excitatory and inhibitory synapses in the spinal cord [27, 28]. One distinct advantage of using antibodies to scaffolding proteins that are associated with excitatory or inhibitory synapses specifically, rather than to neurotransmitter receptors or their subunits, is the ease with which primary antibodies can access their target epitope. Epitopes for transmitter receptors are typically embedded within the postsynaptic density or synaptic cleft. In fixed tissue, these are likely to be inaccessible to primary antibodies due to aggregations of cross-linked proteins and cell adhesion molecules, thereby making it difficult to label synaptic specializations reliably. This can be overcome by targeting specific scaffolding proteins, which anchor receptors and postsynaptic density proteins to postsynaptic sites by adhering them to the underlying cytoskeleton. These epitopes are more accessible to antibodies. A distinct advantage of using these markers is that synapses can be visualized using immunofluorescence and confocal microscopy approaches. These approaches can help define the neurochemical properties of both the presynaptic axon terminal and the postsynaptic target neuron, making it easier to map the incidence of both excitatory and inhibitory synapses from neurochemically distinct axon terminals on to defined neuronal populations [13, 14].

In this chapter, we describe protocols we have developed that allow the study of the circuitry of the spinal dorsal horn using confocal microscopy, TEM, and correlated confocal and TEM. Several variants on these approaches have been used in other laboratories, but we present protocols that work best for our studies.

2   Materials

2.1   Mammalian Ringer's Solution

We use mammalian Ringer's solution to flush blood from terminally anesthetized animals prior to the introduction of fixative solutions during perfusion fixation procedures. To 5 l of distilled water, we add NaCl (45 g), KCl (2 g), $CaCl_2$ (1.25 g), $MgCl_2$ (0.025 g), $NaHCO_3$ (2.5 g), NaH_2PO_4 (0.25 g), glucose (5 g). We also add lignocaine (1 g lidocaine hydrochloride monohydrate, $C_{14}H_{22}N_2O \cdot HCl \cdot H_2O$; Sigma; L5647-100G) to the stock Ringer solution to promote vasodilation during the early stages of perfusion.

2.2   Phosphate Buffer (PB)

We routinely make up stock component solutions of buffer (components A and B) and mix these to produce a working solution of buffer at the correct pH and molarity. For component solution A, we add 18.72 g sodium di-hydrogen phosphate dehydrate ($NaH_2PO_4 \cdot 2H_2O$; EMD Millipore) to 600 ml distilled H_2O, whereas for component solution B, we add 42.45 g di-sodium hydrogen phosphate anhydrous (Na_2HPO_4; EMD Millipore) to 1500 ml distilled water. For a stock solution of 0.2 M of PB, we add 560 ml of component A to 1440 ml of component B, then adjust the pH to 7.4 where necessary. We then mix equal volumes of this buffer and distilled water to make up a working solution of 0.1 M PB.

2.3 Phosphate-Buffered Saline (PBS)

Our phosphate-buffered saline (PBS) is made up to a molarity of 0.3 M sodium chloride. To achieve this, we add 100 ml 0.2 M PB and 36 g of NaCl to 1900 ml distilled water. The relatively high ionic strength is used to reduce nonspecific binding of antibodies during immunostaining.

2.4 Phosphate-Buffered Saline with Triton (PBST)

We use phosphate-buffered saline with the detergent Triton X-100 as a conventional diluent for primary and secondary antibodies. To 1000 ml PBS, we add 3 ml Triton X-100 (Sigma).

2.5 Tris-Buffered Saline with Triton (TBST)

We use 0.05 M Tris (at pH 8.2) with 0.9% NaCl and 0.1% Triton X-100 as the diluent for secondary antibodies in the post-embedding immunogold protocol.

2.6 Fixative Solutions

The composition of the fixative solution is an important consideration when preparing tissues for anatomical studies. All fixative solutions are made up in PB (pH 7.4). We typically use fixative solution containing 4% depolymerized formaldehyde for all conventional immunofluorescence studies, and fixative containing 1% formaldehyde and 2.5% glutaraldehyde for TEM studies (especially if we intend to use post-embedding immunogold approaches, *see* below). Tissues intended for use with a combined confocal-TEM protocol or freeze-substitution Lowicryl embedding will be prepared with fixative solution containing 4% formaldehyde and low levels of glutaraldehyde (between 0.1% and 0.5%) as this minimizes background fluorescence associated with glutaraldehyde-fixed tissues but still affords adequate preservation of the tissue ultrastructure.

2.7 Resins

All tissues prepared for conventional electron microscopy in our laboratory are embedded in Durcupan resin (Sigma), since in our hands it affords us optimal preservation of ultrastructural features and conservation of tissue antigenicity for subsequent post-embedding immunolabeling procedures. We make small batches of fresh resin as/when required, following manufacturer's instructions and the following recipe:

DURCUPAN component A	10 g
DURCUPAN component B: DDSA (hardener)	10 g
DIBUTYL PTHALATE (plasticizer)	0.35 g
DMP 30 (accelerator)	0.15 g

2.8 Reynold's Lead (II) Citrate

Add 1.33 g lead (II) nitrate ($PbNO_3$; Agar) and 1.76 g tri-sodium citrate ($Na_3C_6H_5O_7 \cdot H_2O$; Fischer Scientific) to 30 ml distilled water. Shake vigorously for 60 s, then allow to stand for 30 min. A white precipitate will form on allowing the solution to settle. Add 8 ml 1 M NaOH (CO_2 free) and make up to 50 ml with distilled water. This should be stored at 5 °C and used within 1 month. The pH of this solution should be between 11.5 and 12.5 for optimal staining of grids.

3   Methods

Perfusion fixation of a terminally anaesthetized rats or mice has been approved by the University pf Glasgow's local Ethical Review Panel and is covered by our Home Office Project licence, in accordance with the UK Animals (Scientific Procedures) Act 1986.

3.1   Perfusion Fixation

To ensure optimal preservation of ultrastructural features and antigenicity of spinal cord sections from adult mice (body weight 17–35 g), we routinely fix tissues by vascular perfusion of terminally anaesthetized animals. We administer a lethal dose of sodium pentobarbitone (0.1–0.2 ml; 200 mg/ml Euthatal, i.p.), then quickly expose the heart once we have established that the animal is not displaying either a corneal or withdrawal reflex. A line containing mammalian Ringer solution is then inserted directly into the lumen of the left cardiac ventricle, and the vena cava is punctured using micro scissors at the point where it enters the heart. We run approximately 5 ml of Ringer solution through the animal (at a flow rate of approximately 20 ml/min) to flush out most of the blood before then switching directly to fixative solution (250 ml). A similar approach is used to fix tissues in adult rats (body weight 200–300 g), but we typically run 1 l of fixative through each animal. This transcardial perfusion approach ensures both the rapid clearance of blood from capillary beds and the effective delivery of fixative into the tissue. Tissues are then dissected out of the animal and post-fixed in the same fixative solution for a period between 2 h and 24 h, depending on the intended use of the tissues. We do, on occasion, use immersion fixation for slices from in vitro electrophysiological experiments, but preservation of the ultrastructure of these tissues is often compromised.

3.2   Post-fixation Times

Post-fixation time is an important consideration when planning experiments of this type. In most cases, we keep tissues in fixative solution for a maximum of 4 h (taken from start of perfusion) before transferring it to PB, since this prevents over-fixation of the tissues and lessens the risk of restricted access to certain epitopes for some antibodies. Tissues fixed with high concentrations of glutaraldehyde for TEM studies are usually kept in the same fixative solution for 24 h. Blocks of spinal cord tissue (typically up to three segments in length) can be archived for long-term storage. The most efficient and reliable means of archiving tissues is to cryoprotect the tissues in a 30% sucrose solution made up in PB (weight/volume), then storing the tissues in sealed Eppendorf vials submerged in liquid nitrogen.

3.3 Tissue Preparation

For most of our work, we use free-floating sections for immunohistochemistry. Blocks of tissues (typically lumbar spinal cord) are cut into 60 μm thick sections with a vibrating blade microtome (VT1200 or VT1000S, Leica, Milton Keynes, United Kingdom), with sections then incubated in 50% alcohol for 30 min to permeabilize tissues for subsequent antibody incubations. If sections are to be processed for immunoperoxidase labeling for electron microscopy, they undergo two additional steps, in which they are incubated in 0.3% H_2O_2 in PB for 30 min (to block endogenous peroxidase activity), then in 1% sodium borohydride ($NaBH_4$) in PB for 30 min (to quench free aldehyde groups). Between each step, sections are rinsed thoroughly in PB (9×10 min following incubation in $NaBH_4$, 3×10 min for all other stages). Since the $NaBH_4$ solution effervesces vigorously, it is important to leave the lids off during this stage. At this stage, the tissues can be prepared for (1) immunohistochemistry for confocal microscopy; (2) pre-embedding immunohistochemistry for electron microscopy; (3) immunohistochemistry for correlated confocal and electron microscopy; (4) conventional resin embedding; or (5) freeze-substitution resin embedding. Spare sections can be stored in glycerol at -20 °C and archived for future use.

3.4 Immunohistochemistry for Confocal Microscopy

Free-floating tissue sections are typically incubated in a cocktail of primary antibody solutions for 72 h maintained at 4 °C with gentle agitation. This step is followed by incubation in species-specific secondary antibodies for 12–18 h. Primary antibodies are typically made up in PBS with detergent (PBST), whereas sections are rinsed thoroughly between each stage in PB. Sections are mounted in antifade mounting medium (Vectashield; Vector Laboratories, Peterborough, UK) on to glass slides and cover slipped. Since the medium remains liquid, the coverslips are secured in place with nail varnish. The sections are ready to be viewed and scanned for analysis (*see* Fig. 1) and can then be stored at -20 °C for archiving purposes. *See* Appendix 1 for protocol details.

3.5 Pre-embedding Immunohistochemistry for Electron Microscopy (Peroxidase Labeling)

Free-floating tissue sections are incubated in primary antibody solutions for 72 h maintained at 4 °C, followed by biotinylated species-specific secondary antibodies for 12–18 h. Sections are then incubated in extrAvidin-Peroxidase (Sigma; Cat. No. E2886) made up on PB. Peroxidase activity is then revealed with 3,3'-5,5'diaminobenzidine (DAB; 0.05% weight/volume) in PB in the presence of H_2O_2 (0.01% volume/volume). The resulting reaction product forms an electron-dense precipitate that can be identified in the electron microscope (*see* Fig. 2), and its development should be monitored throughout the procedure. All antibodies are made up in PBS with no detergent. It is critical that tissues intended for use with transmission electron microscopy are not exposed to detergent as this will dissolve the lipid content of cell

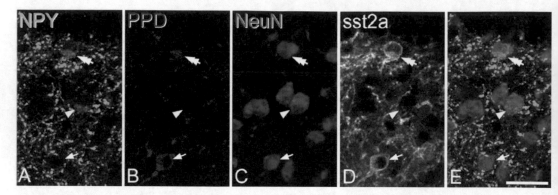

Fig. 1 Neurochemical characterization of interneurons in laminae I and II of the spinal dorsal horn. Using combinations of antibodies such as those against neuropeptide Y (NPY, green), preprodynorphin (PPD, red), the neuronal marker NeuN (blue), and the somatostatin receptor sst_{2A} (gray), we are able to investigate the neurochemical heterogeneity of spinal interneuron populations. Scale bar = 20 μm. (Taken from Boyle et al. [41])

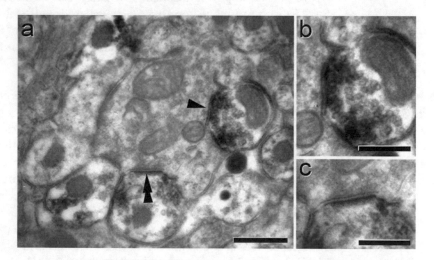

Fig. 2 Synaptic connectivity of a neurochemically distinct population of inhibitory spinal interneurons. Using pre-embedding immunohistochemistry in combination with transmission electron microscopy, we were able to determine the postsynaptic targets of inhibitory interneurons that express the calcium-binding protein parvalbumin. Reaction product is seen as an electron-dense (black) precipitate in immunolabeled profiles (most notable in panel **b**). Scale bars (nm): **a** = 500; **b** and **c** = 250. (Taken from Hughes et al. [25])

membranes and impair the ultrastructure. Tissues at this stage of processing are now ready for resin embedding. *See* Appendix 2 for protocol details.

3.6 Immunohisto-chemistry for Correlated Confocal and Electron Microscopy

This approach allows us to visualize multiple markers in tissue sections using conventional confocal microscopy (*see* Fig. 3), and then prepare the same section for subsequent ultrastructural analysis by converting one of these markers into an electron-dense precipitate (as above). Free-floating tissue sections are typically

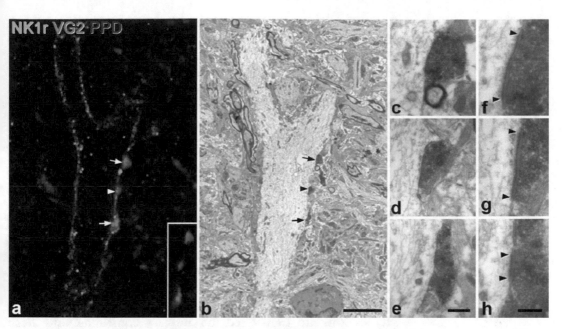

Fig. 3 Correlated confocal and electron microscopic analysis of excitatory synaptic inputs from neurochemically defined axon terminals. (**a**) Confocal microscopy was used to show that cells expressing the neurokinin 1 receptor (NK1r, green) received multiple contacts from axon terminals derived from excitatory dynorphin interneurons, revealed by the coexpression of vesicular glutamate transporter type 2 (VG2, blue), and preprodynorphin (PPD, red). (**b–h**) Subsequent analysis of these appositions (arrowheads) using transmission electron microscopy and peroxidase-labelling of VG2 axon terminals revealed the presence of synaptic specializations at these sites (**f–h**). Scale bars (μm): **a, b** = 5; **c–e** = 0.5; **f–h** = 0.25 (Taken from Baseer et al. [42])

incubated in a cocktail of primary antibody solutions for 72 h maintained at 4 °C, followed by a cocktail of species-specific secondary antibodies tagged with fluorescent markers as well as one (or more) biotinylated secondary antibodies (depending on which marker/markers we wish to visualize at the TEM level) for 12–18 h. Sections can then be incubated in extrAvidin-peroxidase for 12–18 either prior to mounting on to glass slides for viewing and confocal scanning [42], or once all fluorescence images have been collected. As a note of caution, we have found that direct exposure of the peroxidase-labeled tissues to ultraviolet light may lead to quenching of subsequent peroxidase activity in regions of tissues where scanning has taken place [40]. It is therefore up to individual users to determine the most appropriate stage of the protocol to incubate their tissues in extrAvidin-peroxidase. Once all necessary confocal image stacks have been obtained through regions of interest (typically within 10 μm of the surface of the section), the tissues are recovered from the glass slides and processed further to visualize peroxidase activity using 3,3′-5,5′diaminobenzidine (DAB; 0.05% weight/volume) in the presence of H_2O_2 (0.01% volume/

volume), as above. It is critical that tissues intended for use with transmission electron microscopy are not exposed to detergent as this will dissolve the lipid content of cell membranes. Therefore, all antibodies and other reagents are made up in PBS, unless stated otherwise. Tissues at this stage of processing are now ready for resin embedding. *See* Appendix 3 for protocol details.

3.7 Tissue Preparation for Conventional Resin Embedding

Tissue sections intended for ultrastructural studies using the basic resin embedding protocol are first incubated in 1% osmium tetroxide (OsO_4) in PB for 30 min to fix all lipid components and cell membranes, rinsed in distilled water (3×10 min), then incubated in a saturated solution of uranyl acetate in 70% acetone (in distilled water) for 30 min. Uranyl acetate binds to chromatin, nucleic acids, mRNA, and DNA, helping stain the nucleus and cytoplasmic domains. Sections are then incubated in increasing concentrations of acetone as follows: 90% acetone (in distilled water) for 10 min, 100% acetone (3×10 min), followed by incubation in 1 part 100% acetone and 1 part fresh Durcupan resin (overnight). The following day, sections are transferred into bottles containing only fresh Durcupan resin. All incubations are carried out at room temperature with tissue sections placed on a rotator for continuous agitation. Sections are then mounted in Durcupan resin between acetate sheets for flat embedding, with resin cured by placing the sections in an oven at 60 °C for 24/48 h. Embedded sections are then cut from acetate sheets and mounted on to resin stubs. The mounted sections are then trimmed to outline the region of interest, then prepared for serial sectioning using an ultra-microtome. If sections are for correlated confocal-TEM work, it is necessary to consider the orientation of the tissues to ensure that the surface of interest will be uppermost on the block when sections are cut on the ultra-microtome. *See* Appendix 4 for protocol details.

Before viewing, grids are counterstained on droplets of lead (II) citrate to enhance contrast of tissues (and any DAB reaction product) in the ultrathin sections. Grids are incubated on droplets of lead citrate for 5–12 min, then rinsed on droplets of distilled water (5×2 min). Care should be taken not to breathe directly on the grids while on the lead citrate droplets as this will lead to contamination by the formation of lead carbonate precipitate. To reduce the risk of dust or atmospheric CO_2 from contaminating the grids, we carry out this staining protocol on droplets contained within petri dishes. We routinely surround these droplets with pellets of NaOH to reduce levels of atmospheric CO_2 within the dishes.

3.8 Post-embedding Immunogold Labeling

Post-embedding immunolabeling protocols have been developed to allow study of the neurotransmitter content of axon terminals (*see* Fig. 4). To allow for this, the tissues must first be fixed appropriately with an optimal concentration of glutaraldehyde (*see*

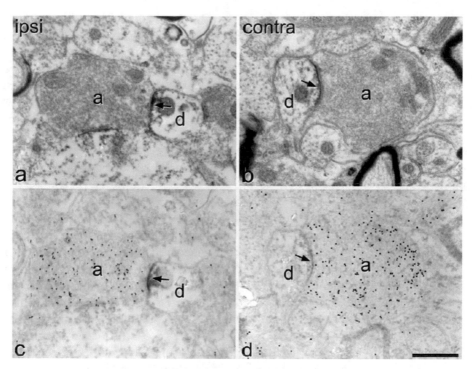

Fig. 4 Combined pre-embedding immunohistochemistry with post-embedding immunogold labeling in serial ultrathin sections. (**a** and **b**) Pre-embedding immunolabeling for the β_3 subunit of the GABA$_A$ receptor is visualized using DAB precipitate (arrows). Adjacent ultrathin sections were processed for post-embedding immunogold labeling for GABA, using 15 nm gold particles to delineate immunolabeled profiles. (**c** and **d**) Both the axon terminals (**a**) illustrated in panels (**a**) and (**b**) show a high density of immunogold particles, confirming the presence of GABA in these boutons. Scale bar = 0.5 μm (Taken from Polgár and Todd [44])

above). Serial ultrathin sections (50–70 nm thick; silver-gold to gold interference color) are mounted on to Formvar-coated nickel grids. This protocol involves the sequential incubation of grids on to droplets (100 μl for buffers and 25 μl for antibodies) dispensed on to a clean strip of Parafilm (Bemis). To ensure that droplets at various stages do not evaporate, and to prevent contamination by airborne particulate matter, these reactions are carried out in humidified chambers (large petri dishes). Small reservoirs (~1 cm diameter) to retain buffers/antibodies that contain detergent for each step in sequence are drawn on the parafilm strip using a liquid-repellent slide marker pen/wax pen/PAP pen prior to starting. Grids are incubated sequentially on droplets of PBST (pH 7.6) with 1% bovine serum albumin (BSA) for 1 h before overnight incubation in primary antibody made up in the same diluent. Grids are then rinsed (3 × 10 min) on droplets of PB, before being transferred to a droplet of Tris-buffered saline (TBST; pH 8.2) for 30 min. Grids are then transferred to a droplet of secondary antibody conjugated to colloidal gold particles (typically

10 or 15 nm goat anti-rabbit from BioCell), diluted 1:25 in TBST (pH 8.2) for 2 h. Grids are rinsed on droplets of TBS (3 × 10 min) and then distilled water prior to viewing [28]. Alternative protocols using only Tris-buffered saline solutions (TBST) have also been described, in which diluent for the primary antibody is at pH 7.6, and that for the secondary antibody is at pH 8.2, and these are equally reliable [38, 43]. *See* Appendix 5 for protocol details.

3.9 Tissue Preparation for Freeze-Substitution Resin Embedding

The conventional resin embedding protocol described above is optimal for analysis of tissue ultrastructure and is compatible with both pre-embedding immunohistochemistry for an array of markers and post-embedding immunogold labeling for the detection of GABA and glycine [25, 28, 37–39]. One important feature of this protocol is the use of osmium tetroxide to fix lipids, allowing cell membranes to be stabilized and visualized. This does, however, obscure several epitopes in the cell membranes including neurotransmitter receptor subunits at synapses. To overcome these difficulties, freeze-substitution resin embedding protocols have been developed to help retain membrane integrity of tissues without the need for osmium stabilization [15, 45]. These procedures require dedicated apparatus, and the protocol we describe is applicable for use with the Leica EM ASF system (Leica Microsystems, Germany). Briefly, 400-μm-thick tissue sections from perfusion-fixed animals are cryoprotected in glycerol then plunge-frozen in liquid propane using a Leica EM CPC unit. Freeze-substitution is carried out with the Leica EM ASF system by immersing sections in 1.5% uranyl acetate in anhydrous methanol for 24 h at −90 °C, followed by impregnation of tissues with Lowicryl HM20 at −45 °C, then flat-embedded and cured using ultraviolet irradiation (48 h at −45 °C, 24 h at 0 °C, and 24 h at 20 °C). Maintaining the tissues at very low temperatures throughout the embedding process (between −45 °C and −170 °C) ensures that the structural integrity of the tissues is maintained, since the lipid component of cell membranes will not be dissolved by the methanol that is required to dehydrate the tissues and impregnate it with resin at these temperatures. By embedding the tissues in resin under these conditions, both the structural integrity and antigenicity of the tissues is preserved, making it more likely that several epitopes can be targeted using post-embedding immunogold labeling protocols. These approaches have been used to detect the subcellular distribution of various epitopes with great precision, including AMPA receptor subunits in postsynaptic densities, and specific peptides to individual dense-core vesicles in axon terminals (*see* Fig. 5). *See* Appendix 6 for protocol details.

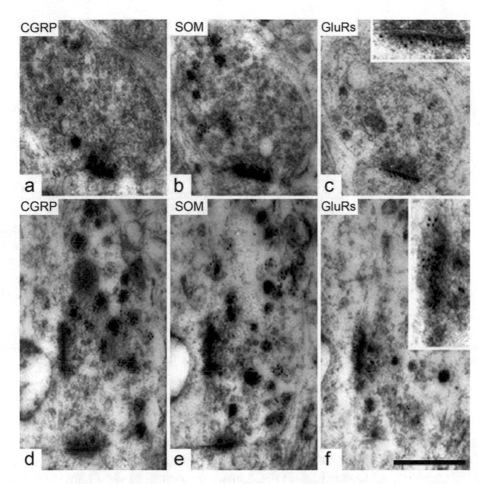

Fig. 5 Post-embedding immunogold labeling of neuropeptides and AMPA receptors in Lowicryl-embedded tissue. (**a–c**) Consecutive serial ultrathin sections showing the absence of immunogold labeling for CGRP (**a**), but the presence of somatostatin (SOM, **b**) in the dense core vesicles of an axon terminal in the superficial dorsal horn. This terminal forms an asymmetrical synapse, and immunogold labeling for GluR1 and GluR2/3 can be visualized in the postsynaptic density. (**d–f**) This series of ultrathin sections demonstrates the coexpression of immunogold labeling for CGRP (**d**) and somatostatin (SOM, **e**) in dense core vesicles of an axon terminal that makes a synapse at which numerous gold particles representing GluR-immunoreactivity are present. Scale bar = 0.5 μm (Modified figure taken from Todd et al. [15])

4 Notes

1. Many of the protocols described here use reagents that are either highly toxic (e.g., osmium tetroxide; glutaraldehyde, formaldehyde; hydrogen peroxide; sodium borohydride; dibutyl pthalate; Lowicryl), carcinogenic (e.g., uranyl acetate; DAB), or flammable (e.g., liquid propane; methanol). Users are urged to familiarize themselves with the potential risks of handling all of the reagents listed prior to use.

2. A common starting point of critical importance in each of the protocols described here is the rapid and effective fixation of tissues to ensure optimal preservation of both tissue structure and antigenicity. If fixation of the tissues is inadequate, all subsequent stages are compromised. Should fixation of the tissues be sub-optimal, this will impact both the quality and reliability of the data generated.

3. The increased use of transgenic animals in neuroscience research is a major technological advancement, offering a means of targeting and manipulating discrete neuronal populations with greater precision than possible previously. In many cases, these populations can be visualized using fluorescent reporter molecules. To validate the selectivity of these transgenic lines, the fidelity of labeling patterns seen for these reporters should be compared to expression patterns of the promoter sequence [46]. One way of assessing these expression patterns is by immunohistochemistry, using protocols as described above.

Acknowledgments

The authors are grateful to the Biotechnology and Biological Sciences Research Council (BBSRC; UK: grants P0079961/1 to DIH and N006119/1 to AJT), the Wellcome Trust (UK: grant 102645 to AJT), the Medical Research Council (MRC; UK: MR/S002987/1), and the National Health and Medical Research Council (NHMRC; Australia: grants 1144638 and 1184974) for supporting research in their laboratories.

Appendix 1: Immunofluorescence Labeling of Sections of Spinal Cord

Tissues are fixed by vascular perfusion and post-fixed in the same fixative for between 2 h and 4 h, depending on antibodies/protocol used. The tissues are then cut into 60 μm thick sections on a vibrating-blade microtome and collected in PB. Free-floating sections processed for immunohistochemistry are maintained on a shaker table to ensure maximal exposure to the medium they are incubated in at each stage.

• Incubate sections in 50% alcohol in distilled water	30 min
• Rinse in PBS	3 × 10 min
• Incubate in primary antibody/cocktail of primary antibodies[a] at 4 °C	48–72 h
• Rinse in PBS	3 × 10 min

(continued)

• Incubate in secondary antibodies[a] tagged with fluorescent molecules at 4 °C	12–18 h
• Rinse in PBS	3 × 10 min
• Mount sections in anti-fade medium on to glass slides.	

[a]All antibodies are made up in PBST, unless the tissues are to be processed further for TEM studies

Appendix 2: Protocol for Pre-embedding Immunohistochemistry (Peroxidase Labeling) of Sections Spinal Cord

Tissues are fixed by vascular perfusion and post-fixed in the same fixative for between 2 h and 24 h, depending on antibodies/protocol used. The tissues are then cut into 60 μm thick sections on a vibrating-blade microtome and collected in PB. Free-floating sections processed for immunohistochemistry are maintained on a shaker table to ensure maximal exposure to the medium they are incubated in at each stage.

• Incubate sections in 0.3% H_2O_2 in PB	30 min
• Rinse in PB	3 × 10 min
• Incubate sections in 1% $NaBH_4$ in PB	30 min
• Rinse in PB	9 × 10 min
• Incubate sections in 50% alcohol in distilled water	30 min
• Rinse in PBS	3 × 10 min
• Incubate in primary antibody antibodies[a] at 4 °C	48–72 h
• Rinse in PBS	3 × 10 min
• Incubate in biotinylated secondary antibody at 4 °C	12–18 h
• Rinse in PBS	3 × 10 min
• Incubate in extrAvidin-Peroxidase (diluted 1:1000 in PB)	12–18 h
• Rinse in PB	3 × 10 min
• Incubate in DAB solution in PB	10 min
• Incubate in DAB solution with 0.01% H_2O_2	4–10 min
• Rinses in PB	3 × 10 min
• Mount on glass slides or process for resin embedding.	

[a]Antibodies and extrAvidin-Peroxidase complex are made up in PBST unless the tissues are to be processed further for TEM studies

Appendix 3: Preparing Tissues for Correlated Confocal and TEM Labeling

Tissues are fixed by vascular perfusion and post-fixed in the same fixative for between 2 h and 24 h, depending on antibodies/protocol used. The tissues are then cut into 60 μm thick sections on a vibrating-blade microtome and collected in PB. Free-floating sections processed for immunohistochemistry are maintained on a shaker table to ensure maximal exposure to the medium they are incubated in at each stage.

• Incubate sections in 0.3% H_2O_2 in PB	30 min
• Rinse in PB	3×10 min
• Incubate sections in 1% $NaBH_4$ in PB	30 min
• Rinse in PB	9×10 min
• Incubate sections in 50% alcohol in distilled water	30 min
• Rinse in PBS	3×10 min
• Incubate in a cocktail of primary antibodies[a] at 4 °C	48–72 h
• Rinse in PBS	3×10 min
• Incubate in secondary antibodies[a] tagged with fluorescent molecules at 4 °C	12–18 h
Note: A biotinylated secondary antibody to the target marker should also be included in this cocktail of antibodies.	
• Rinse in PBS	3×10 min
Mount sections in anti-fade medium on to glass slides and collect image stacks with a confocal microscope. Once all relevant data has been collected, the tissues can be removed from the slides and processed further.	
• Rinse in PBS	3×10 min
• Incubate in extrAvidin-Peroxidase (diluted 1:1000)	12–18 h
• Rinse in PB	3×10 min
• Incubate in DAB solution in PB	10 min
• Incubate in DAB solution with 0.01% H_2O_2	4–10 min
• Rinses in PB	3×10 min
• Process tissues for resin embedding.	

[a]All antibodies and extrAvidin-Peroxidase complex used throughout this protocol are made up in PBS. These tissues should not be incubated in any buffer containing detergent

Appendix 4: Resin Embedding Spinal Cord Tissues for Electron Microscopy

For optimal retention of ultrastructural features, tissue sections (as prepared in earlier protocols listed) must be impregnated with heavy metals and dehydrated prior to infiltration with resin. Tissue samples are processed in clean glass bottles and kept on a rotator at each stage to ensure that the sections are agitated constantly to ensure maximal exposure to the medium they are incubated in at each stage.

• Osmicate sections in 1% osmium tetroxide	30 min
• Rinse in distilled water	3 × 10 min
• Incubate in saturated uranyl acetate in 70% acetone	30 min
• Incubate in 90% acetone	10 min
• Incubate in 100% acetone (i)	10 min
• Incubate in 100% acetone (ii)	10 min
• Incubate in 100% acetone (iii)	10 min
• Incubate in 100% acetone/Durcupan resin mix (1:1)	60 min
• Incubate in Durcupan resin only at room temperature	12–18 h
• Flat embed sections between acetate sheets and incubate at 60 °C for 24–48 h	

Embedded tissues can then be prepared for sectioning on an ultramicrotome

Appendix 5: Post-embedding Immunogold Labeling for Electron Microscopy

Post-embedding immunogold labeling is carried out on ultrathin sections (50–70 nm thick; silver-gold to gold interference color) mounted on to Formvar-coated nickel grids. Grids are incubated on individual droplets dispensed freshly for each step and are not re-used for other grids. The final stages of this protocol involves repeated incubation on droplets of distilled water (×3), then repeated dipping into distilled water (10 times in 30 s). This is necessary to ensure all salts and debris are removed from the surface of the section and supporting film.

• Incubate on PBST (pH 7.6) with 1% bovine serum albumin	1 h
• Incubate on primary antibody made up in PBST (pH 7.6) with 1% BSA	12–18 h
• Incubate on PBST (pH 7.6)	3 × 10 min
• Incubate on TBST (pH 8.2)	30 min
• Incubate on gold-labeled secondary antibody made up in TBST (pH 8.2)	2 h

(continued)

• Incubate on TBST (pH 8.2)	3 × 10 min
• Incubate on distilled water	3 × 10 min
• Dip in distilled water	30 s

Grids are then counterstained in lead (II) citrate before viewing

Appendix 6: Freeze-Substitution Resin Embedding for Electron Microscopy

Tissues are fixed by vascular perfusion and post-fixed in the same fixative for between 2 h and 24 h, depending on antibodies/protocol used. The tissues are then cut into 400-μm-thick sections on a vibrating-blade microtome, then incubated in 4% glucose in PB overnight. Free-floating sections processed for immunohistochemistry are maintained on a shaker table to ensure maximal exposure to the medium they are incubated in at each stage. We use a Leica EM CPC unit for plunging cryoprotected tissues into liquid propane, and a Leica EM ASF system for tissue dehydration, resin infiltration, and resin curing stages. All tissues are embedded in Lowicryl HM20 at −45 °C and cured using ultraviolet irradiation.

• Incubate sections in ascending concentrations of cryoprotectant solution: (10%, 20% then 30% glycerol in PB)	30 min each
• Plunge individual cryoprotected sections into liquid propane for rapid freezing using the Leica EM CPC unit, then transfer these vials in the Leica EM ASF for all subsequent processing steps.	
• Incubate in 1.5% uranyl acetate in absolute methanol (at −90 °C)	24 h
• Incubate in absolute methanol (at −45 °C)	3 × 1 h
• Incubate in 3 parts absolute methanol: 1-part Lowicryl (at −45 °C)	1 h
• Incubate in 1-part absolute methanol: 1-part Lowicryl (at −45 °C)	1 h
• Incubate in 1-part absolute methanol: 3 parts Lowicryl (at −45 °C)	1 h
• Incubate in fresh Lowicryl (at −45 °C)	2 h
• Repeat incubation in fresh Lowicryl (at −45 °C)	2 h
• Repeat incubation in fresh, de-gassed Lowicryl (at −45 °C)	12–18 h
• Incubate in fresh, de-gassed Lowicryl (at −45 °C)	3 h
• Repeat incubation in fresh, de-gassed Lowicryl (at −45 °C)	3 h
• Polymerize the resin using ultraviolet irradiation over 4 days.	

Embedded tissues can then be prepared for sectioning on an ultramicrotome, followed by post-embedding immunogold labeling

References

1. Yaksh TL (1989) Behavioral and autonomic correlates of the tactile evoked allodynia produced by spinal glycine inhibition: effects of modulatory receptor systems and excitatory amino acid antagonists. Pain 37:111–123

2. Sivilotti L, Woolf CJ (1994) The contribution of GABAA and glycine receptors to central sensitization: disinhibition and touch-evoked allodynia in the spinal cord. J Neurophysiol 72:169–179

3. Zeilhofer HU, Wildner H, Yévenes GE (2012) Fast synaptic inhibition in spinal sensory processing and pain control. Physiol Rev 92: 193–235

4. McLaughlin BJ, Barber R, Saito K, Roberts E, Wu JY (1975) Immunocytochemical localization of glutamate decarboxylase in rat spinal cord. J Comp Neurol 164:305–322

5. Feldblum S, Dumoulin A, Anoal M, Sandillon F, Privat A (1995) Comparative distribution of GAD$_{65}$ and GAD$_{67}$ mRNAs and proteins in the rat spinal cord supports a differential regulation of these two glutamate decarboxylases in vivo. J Neurosci Res 42:742–757

6. Tillakaratne NJK, Mouria M, Ziv NB, Roy RR, Edgerton VR, Tobin AJ (2000) Increased expression of glutamate decarboxylase (GAD67) in feline lumbar spinal cord after complete thoracic spinal cord transection. J Neurosci Res 60:219 230

7. Mackie M, Hughes DI, Maxwell DJ, Tillakaratne NJK, Todd AJ (2003) Distribution and colocalisation of glutamate decarboxylase isoforms in the rat spinal cord. Neurosci Lett 119: 461–472

8. Zafra F, Aragón C, Olivares L, Danbolt NC, Giménez C, Storm-Mathisen J (1995) Glycine transporters are differentially expressed among CNS cells. J Neurosci 15:3952–3969

9. Todd AJ, Sullivan AC (1990) Light microscope study of the coexistence of GABA-like and glycine-like immunoreactivities in the spinal cord of the rat. J Comp Neurol 296:496–505

10. Polgár E, Durrieux C, Hughes DI, Todd AJ (2013) Quantitative study of inhibitory interneurons in laminae I-III of the mouse spinal dorsal horn. PLoS One 8:e78309

11. Chaudhry FA, Reimer RJ, Bellocchio EE, Danbolt NC, Osen KK, Edwards RH, Storm-Mathisen J (1998) The vesicular GABA transporter, VGAT, localizes to synaptic vesicles in sets of glycinergic as well as GABAergic neurons. J Neurosci 18:9733–9750

12. Yasaka T, Tiong SY, Hughes DI, Riddell JS, Todd AJ (2010) Populations of inhibitory and excitatory interneurons in lamina II of the adult rat spinal dorsal horn revealed by a combined electrophysiological and anatomical approach. Pain 151:475–488

13. Abraira VE, Kuehn ED, Chirila AM, Springel MW, Toliver AA, Zimmerman AL, Orefice LL, Boyle KA, Bai L, Song BJ, Bashista KA, O'Neill TG, Zhuo J, Tsan C, Hoynoski J, Rutlin M, Kus L, Niederkofler V, Watanabe M, Dymecki SM, Nelson SB, Heintz N, Hughes DI, Ginty DD (2017) The cellular and synaptic architecture of the mechanosensory dorsal horn. Cell 168:295–310

14. Boyle KA, Gradwell MA, Yasaka T, Dickie AC, Polgár E, Ganley RP, Orr DHP, Watanabe M, Abraira VE, Zimmerman AE, Ginty DD, Callister RJ, Graham BA, Hughes DI (2019) Defining a spinal microcircuit that gates myelinated afferent input: implications for tactile allodynia. Cell Rep 28:526–540

15. Todd AJ, Hughes DI, Polgár E, Nagy GG, Mackie M, Ottersen OP, Maxwell DJ (2003) The expression of vesicular glutamate transporters VGLUT1 and VGLUT2 in neurochemically defined axonal populations in the rat spinal cord with emphasis on the dorsal horn. Eur J Neurosci 17:13–27

16. Seal RP, Wang X, Guan Y, Raja SN, Woodbury CJ, Basbaum AI, Edwards RH (2009) Injury-induced mechanical hypersensitivity requires C-low threshold mechanoreceptors. Nature 462:651–655

17. Usoskin D, Furlan A, Islam S, Abdo H, Lönnerberger P, Lou D, Hjerling-Leffler J, Haeggström J, Khaechenko O, Kharchenko PV, Linnarson S, Ernfors P (2015) Unbiased classification of sensory neuron types by large-scale single-cell RNA sequencing. Nat Neurosci 18:145–153

18. Hockley JRF, Taylor TS, Callejo G, Wilbrey AL, Gutteridge A, Bach K, Winchester WJ, Bulmer DC, McMurray G, Smith ESJ (2018) Single-cell RNAseq reveals seven classes of colonic sensory neuron. Gut 68(4):633–644. gutjnl-2017-315631

19. Häring M, Zeisel A, Hochgerner H, Rinwa P, Jakobsson JET, Lönnerberg P, Manno GL, Sharma N, Borgius L, Kiehn O, Lagerstrm MC, Linnarsson S, Ernfors P (2018) Neuronal atlas of the dorsal horn defines its architecture and links sensory input to transcriptional cell types. Nat Neurosci 21:869–880

20. Sathyamurthy A, Johnson KR, Matson KJE, Dobrott CI, Li L, Ryba AR, Bergman TB, Kelly MC, Kelley MW, Levine AJ (2018) Massively parallel single nucleus transcriptional

profiling defines spinal cord neurons and their activity during behaviour. Cell Rep 22: 2216–2225

21. Zeisel A, Hochgerner H, Lönnerberg P, Johnsson A, Memic F, van der Zwan J, Häring M, Braun E, Borm LE, La Manno G, Codeluppi S, Furlan A, Lee K, Skene N, Harris KD, Hjerling-Leffler J, Arenas E, Ernfors P, Marklund U, Linnarsson S (2018) Molecular architecture of the mouse nervous system. Cell 174:999–1014

22. Todd AJ (2017) Identifying functional populations among the interneurons in laminae I-III of the spinal dorsal horn. Mol Pain 13:1–19

23. Moehring F, Halder P, Seal RP, Stucky CL (2018) Uncovering the cells and circuits of touch in normal and pathological settings. Neuron 100:349–360

24. Hughes DI, Todd AJ (2020) Inhibitory interneurons in the spinal dorsal horn. Special issue on pain: aligning new approaches to accelerate the development of analgesic therapies. Neurotherapeutics 17:874–885

25. Hughes DI, Sikander S, Kinnon CM, Boyle KA, Watanabe M, Callister RJ, Graham BA (2012) Morphological, neurochemical and electrophysiological features of parvalbumin-expressing cells: a likely source of axo-axonic inputs in the mouse spinal dorsal horn. J Physiol 590:3927–3951

26. Bell AM, Gutierrez-Mecinas M, Stevenson A, Casas-Benito A, WIldner H, West SJ, Watanabe M, Todd AJ (2020) Sci Rep 10: 13176

27. Gutierrez-Mecinas M, Kuehn ED, Abraira VE, Polgár E, Watanabe M, Todd AJ (2016) Immunostaining for Homer reveals the majority of excitatory synapses in laminae I-III of the mouse spinal dorsal horn. Neuroscience 329: 171–181

28. Todd AJ, Watt C, Spike RC, Sieghart W (1996) Colocalization of GABA, glycine, and their receptors at synapses in the rat spinal cord. J Neurosci 16:974–982

29. Gray EG (1959) Axo-somatic and axo-dendritic synapses of the cerebral cortex: an electron microscope study. J Anat 93: 420–430

30. Uchizono K (1965) Characteristics of excitatory and inhibitory synapses in the central nervous system of the cat. Nature 207: 642–643

31. Colonnier M (1968) Synaptic pattern on different cell types in different laminae of the cat visual cortex. An electron microscope study. Brain Res 9:268–287

32. Klemann CJHM, Roubos EW (2011) The Gray area between synapse structure and function—Gray's synapse Types I and II revisited. Synapse 65:1222–1230

33. Somogyi P, Hodgson AJ, Chubb IW, Penke B, Erdei A (1985) Antisera to γ-aminobutyric acid. II. Immunocytochemical application to the central nervous system. J Histochem Cytochem 33:240–248

34. Ottersen OP, Storm-Mathisen J (1987) Localization of amino acid neurotransmitters by immunocytochemistry. Trends Neurosci 10: 250–255

35. Polgár E, Hughes DI, Riddell JS, Maxwell DJ, Puskár Z, Todd AJ (2003) Selective loss of spinal GABAergic or glycinergic neurons is not necessary for development of thermal hyperalgesia in the chronic constriction injury model of neuropathic pain. Pain 104:229–239

36. Pow DV, Crook DK (1993) Extremely high titre polyclonal antisera against small neurotransmitter molecules: rapid production, characterisation and use in light- and electron-microscopic immunocytochemistry. J Neurosci Methods 48:51–63

37. Todd AJ (1996) GABA and glycine in synaptic glomeruli of the rat spinal dorsal horn. Eur J Neurosci 8:2492–2498

38. Watson AHD, Hughes DI, Bazzaz AA (2002) Synaptic relationships between hair follicle afferents and neurones expressing GABA and glycine-like immunoreactivity in the spinal cord of the rat. J Comp Neurol 452:367–380

39. Todd AJ (1997) A method for combining confocal and electron microscopic examination of sections processed for double- or triple-labelling immunocytochemistry. J Neurosci Methods 73:149–157

40. Hughes DI, Bannister AP, Pawelzik H, Thomson AM (2000) Double immunofluorescence, peroxidase labelling and ultrastructural analysis of interneurones following prolonged electrophysiological recordings in vitro. J Neurosci Methods 101:107–116

41. Boyle KA, Gutierrez-Mecinas M, Polgár E, Mooney N, O'Connor E, Furuta T, Watanabe M, Todd AJ (2017) A quantitative study of neurochemically-defined populations of inhibitory interneurons in the superficial dorsal horn of the mouse spinal cord. Neuroscience 363:120–133

42. Baseer N, Polgár E, Watanabe M, Furuta T, Kaneko T, Todd AJ (2012) Projection neurons in lamina III of the rat spinal cord are selectively innervated by local dynorphin-

containing excitatory neurons. J Neurosci 32: 11854–11863

43. De Zeeuw CI, Holstege JC, Calkoen F, Ruigrok TJ, Voogd J (1988) A new combination of WGA-HRP anterograde tracing and GABA immunocytochemistry applied to afferents of the cat inferior olive at the ultrastructural level. Brain Res 447:369–375

44. Polgár E, Todd AJ (2008) Tactile allodynia can occur in the spared nerve injury model in the rat without the selective loss of GABA or GABAA receptors from synapses in laminae I-II of the ipsilateral spinal dorsal horn. Neuroscience 156:193–202

45. van Lookeren CM, Oestreicher AB, van der Krift TP, Gispen WH, Verkleij AJ (1991) Freeze-substitution and Lowicryl HM20 embedding of fixed rat brain: suitability for immunogold ultrastructural localization of neural antigens. J Histochem Cytochem 39: 1267–1279

46. Graham BA, Hughes DI (2019) Rewards, perils, and pitfalls of untangling spinal pain circuits. Curr Opin Physio 11:35–41

Chapter 12

Using Viral Vectors to Visualize Pain-Related Neural Circuits in Mice

Bin Chen, Jun Takatoh, and Fan Wang

Abstract

Neural circuit tracing methods that take advantage of viral vectors and transgenic mice provide opportunities to gain insight into functional connectivity. Here we describe several protocols for identifying synaptic connections made by projection neurons, specific cell types, and neural ensembles.

Key words Circuit tracing, Capturing activated neuronal ensembles, CANE, FosTRAP AAV, Lentivirus, Retrograde, Anterograde, Monosynaptic, Polysynaptic

1 Introduction

In the past decade, neuroscience has witnessed a massive expansion in the use of viral vectors to label, measure, and manipulate neurons in the mammalian brains. This protocol focuses on using viral vectors to examine neural circuits in the mouse model. While the methods and strategies described here are intended for dissecting nociceptive circuits, they are also generally applicable to studying other circuits in the mouse. Compared to pure genetic approaches that often require crossing two or more transgenic mouse lines to express desired transgene in desired neurons (e.g., crossing a Cre-driver line and a Cre-dependent-transgene line; or crossing Cre-, Flp-, and Cre/Flp co-dependent lines), which have made instrumental impact in advancing our understandings of the peripheral sensory neurons and spinal circuits [1–3], viral vectors offer quicker and more versatile ways to express different transgenes in neurons of interest and can be used in combination with transgenic mouse lines. Here we describe in detail the specific strategies and protocols that use different viral vectors to study neuronal connectivity and functions based on either target projections or cell types or cell ensembles.

Rebecca P. Seal (ed.), *Contemporary Approaches to the Study of Pain: From Molecules to Neural Networks*, Neuromethods, vol. 178, https://doi.org/10.1007/978-1-0716-2039-7_12, © Springer Science+Business Media, LLC, part of Springer Nature 2022

2 Materials and General Strategies

2.1 Mice

For tracing the input-output circuits from specific neuronal cell types, mice expressing Cre recombinase in desired cells are used. For examples, *see* [4–9]. For performing activity-dependent circuits tracing, *Fos*^{dsTVA} knockin mice [10–12] (JAX 027831) or *Fos*$^{-}_{CreERT2}$ mice [13, 14] (JAX 021882) can be used. These mouse lines provide an approach to obtain genetic access to neurons that are activated by defined stimuli, including noxious stimuli.

The optimal age are 8–12 weeks for mice. All experiments using animals should be in accordance with your institutional and governmental biosafety and animal use guidelines.

2.2 Viruses and Strategies

2.2.1 AAV Vectors for Tracing Axonal Projections and Synaptic Terminals and Their Limitations

The tracing of axons and synaptic terminals is sometimes referred to as anterograde tracing. At present, this is done primarily using adeno-associated viruses (AAVs). One can use AAV vectors that drive the expression of fluorescent proteins, generally referred to as XFPs, with a ubiquitous promoter (such as the chicken betaActin promoter, CAG, or the human synapsin promoter, hSyn) to fill the axons, or AAVs that express synaptophysin-fused with XFPs to label synaptic boutons. This can be done either for neurons in a brain region without cell-type specificity (e.g., using AAV-CAG-GFP, Addgene #37825) or for labeling specific class of neurons by using Cre-dependent AAVs (e.g., AAV-CAG-Flex-tdTomato, Addgene #28306) in desired Cre-driver mice. For example, there are many Cre-expressing lines that label different spinal or brainstem neurons implicated in pain-, itch-, or touch-processing [3, 15–20], and in combinations with AAV vectors mentioned above, one can visualize the processes and synapses of these different types of neurons. The Cre-dependent vectors contain either DiO or Flex in their names, which essentially refers to the same mechanism that allows expression of genes in the presence of Cre. One highly useful AAV that can label axons and synaptic boutons with two different colors is AAV-hSyn-Flex-mGFP-2A-synaptophysin-mRuby (Addgene #71760) (Fig. 1) [21, 22]. In this vector, mGFP is a membrane-targeted GFP that labels axons effectively, and the bright synaptophysin-mRuby reveals the locations of synaptic terminals on the labeled axons.

There are some caveats to note. First, AAVs of different serotypes will have different tropisms. Thus, if low or no expression is observed with one serotype, one should test other serotypes. Unfortunately, there is no known universal serotype that works for all neurons. Second, because of the small diameter of AAVs, it can be taken up by axon terminals, thereby retrogradely labeling neurons that project into the injection site. In fact, this forms the

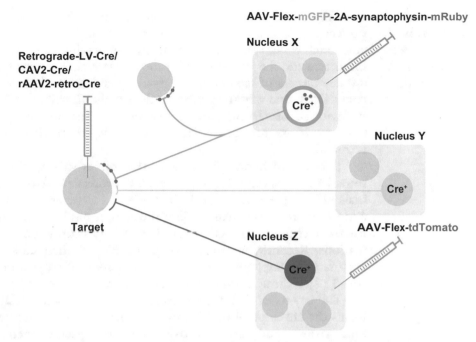

Fig. 1 Example strategy of target-dependent retrograde labeling of neurons. Retrograde virus carrying Cre recombinase, either retrograde-LV-Cre, or CAV2-Cre or rAAV2-retro-Cre, is injected in the target site. Neurons from many nuclei innervating in the target site will take up the retrograde virus from axon terminals and express Cre recombinase in their soma. Neurons innervating the same target but with cell bodies located in different nuclei can be labeled using different Cre-dependent AAV vectors. In the example shown here, neurons projecting from Nucleus X to Target are labeled by AAV-Flex-mGFP-2A-synaptophysin-mRuby, which labels neuronal membrane with green fluorescence and synaptic terminals with red fluorescence. Red dots in the cell body represent synaptophysin-mRuby in the post-Golgi compartment. Neurons projecting from Nucleus Z to Target are labeled by AAV-Flex-tdTomato

basis for one of the retrograde labeling method discussed in Sub-heading 2.2.2. Third, AAVs, after entering into a neuron, can sometimes undergo transcytosis out of the axonal terminals, albeit at a low efficiency. This is the basis for anterograde trans-neuronal tracing described in Subheading 2.2.4.

Finally, note that at present, it is difficult to use AAV-based axon/bouton tracing for labeling only a single neuron. However, it is possible to label a small number of neurons. With such sparse labeling in combination with whole-mount brain-clearing, one can use high-resolution microscopy, such as two-photon microscope to trace and reconstruct the full processes of individual neurons. The Janelia MouseLight project (*MouseLight Neuron Browser: http://ml-neuronbrowser.janelia.org*) used this method to reconstruct many individually labeled cortical neurons [23–26].

In many circuit studies, it is desired to bulk label neurons that project to specific brain centers. For example, in studies of pain and touch circuits, it is known that the parabrachial nucleus in the brainstem and the ventroposterior medial/lateral thalamic neurons are key targets of second-order spinal/medullary projection neurons. Thus, one way to specifically dissect the functions and connectivity of these spinoparabrachial or spinothalamic projection neurons is to retrograde label these cells from their axonal terminals.

It has been known for decades that rabies viruses and pseudorabies viruses can infect mammalian neurons from axonal terminals [27, 28]. However, both the types of viruses are neurotoxic, and therefore, other than labeling the cells, it has been difficult to use these viruses for functional studies such as recording or manipulations. Recently, nontoxic, double-deletion-mutant rabies viral vectors were developed for retrograde targeting of projection neurons [29], although this method has not been widely used.

The key to rabies infecting axons is its glycoprotein (RG) that forms the envelope of the viruses. Since lentiviruses are also envelope viruses and are non-toxic, lentivirus pseudotyped with RG-coat protein was developed (RG-LV, also called HiRet-LV) and was found to infect neurons from axons effectively, and these have been successfully used in labeling neurons based on projections [30–35]. Notably, using the retrograde lentiviruses, it was revealed that corticospinal projection neurons originated from somatosensory cortex can increase the gain of spinal tactile sensory transmission and contribute to touch allodynia after neuropathic pain models [33].

Generally speaking, lentiviral vectors have a low titer and only drive moderate levels of gene expression, and thus, other retrograde viruses that can achieve high- expression levels are desired. To this end, canine adenovirus-2 (CAV-2) was discovered to have high retrograde infection efficiency, stable long-term transgene expression, and minimal cytotoxicity [36–38]. This vector can be purchased from PVM vector core (https://www.pvm.cnrs.fr/plateau-igmm/?cat=17). However, CAV-2 can infect only a narrow range of neurons (i.e., limited tropism) since many neurons do not express the receptor for this virus [39].

Most recently, using a clever direct-evolution and in vivo screening, an AAV vector with the capsid referred to as rAAV2-retro was developed by scientists at the Janelia Farm [40]. rAAV2-retro packaged vectors (Addgene #81070) give rise to high-level and long-term gene expressions, non-toxic, and can retrograde infect many types, although not all types, of neurons [25]. Thus, it is a very versatile tool, but one needs to test its tropism for the desired neurons.

One commonly used strategy for efficient retrograde labeling and manipulating of neurons is injecting the above-mentioned vectors (retrograde-LV, or CAV-2, or rAAV-retro) carrying Cre recombinase at the target site where axons terminate, while in the same animals injecting Cre-dependent AAV vectors carrying desired transgenes at the nucleus or region where neuronal soma are located (Fig. 1). For example, injecting retrograde viral Cre vectors in the parabrachial region and injecting AAV-Flex-XFP (for labeling) or AAV-Flex-opto-genetic/chemicogenetic molecules (for manipulations) in a specific spinal segment. The advantages of this two-viral strategy are twofold: (1) a small amount of Cre is sufficient to induce high-level transgene expression for labeling or manipulation; (2) since there are different populations of neurons projecting into a target, this strategy allows specific labeling/manipulation of one projection population with one set of molecules and another population of projection neurons with a different set of molecules (Fig. 1).

2.2.3   Viral Vectors for Transsynaptic Tracing of Presynaptic Inputs to Desired Neurons

Retrograde transsynaptic tracing of presynaptic input uses the monosynaptic rabies virus system, which has been extensively reviewed previously [41–43]. Therefore, we will simply summarize here and refer readers to the previous reviews for more details and caveats. Currently, there are two strains of monosynaptic rabies viruses can be used for tracing presynaptic cells: the glycoprotein-deleted SAD-B19 vaccine strain of rabies virus (SAD-B19-dG-RV) and the less toxic glycoprotein-deleted CVS-N2c strain (N2c-dG-RV), collectively referred to here as dG-RV. The dG-RVs can be pseudotyped in tissue culture with a foreign coat protein; usually the avian sarcoma virus subtype A (ASLV-A) envelope glycoprotein (EnvA) such that the EnvA-dG-RVs does not infect mouse neurons unless the neurons are made to express the receptor TVA specifically for EnvA. SAD-B19-dG-RV can be acquired from the Salk Institute viral core. EnvA-CVS-N2c-dG-RV needs to be custom made, and all plasmids needed for generating N2c-dG-RV can be found in Addgene.

To enable transsynaptic tracing from desired Cre-expressing neurons (e.g., a type of spinal interneurons labeled by a Cre driver), Cre-dependent expression of two genes is needed first, often in the form of two AAV-helper viruses (Fig. 2a): AAV-CAG-Flex-TVA-XFP (e.g., CAG-Flex-TCB in Addgene #48332) and AAV-CAG-Flex-RG (either AAV-Flex-SAD-B19-Glycoprotein, Addgene #38043; or AAV-DIO-EF1a-Flex-H2B-GFP-P2A-N2c (G) Addgene #73476) to be co-injected into region containing neurons of interest. An optimized version of SAD-B19-Glycoprotein called oG was developed [44]. A single AAV vector driving Cre-dependent expression of both TVA and RG, or TVA and oG, is also available from Addgene (AAV-syn-Flex-splitTVA-EGFP-

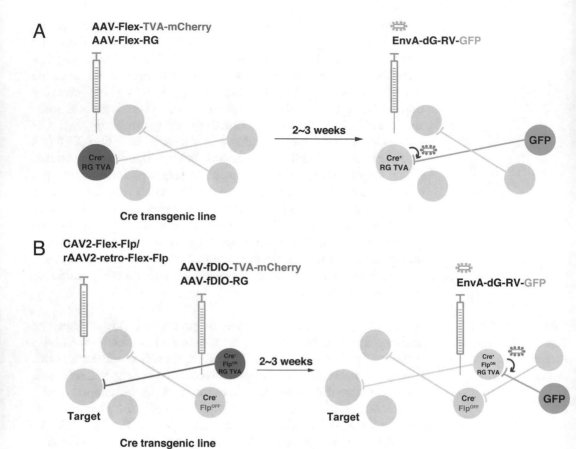

Fig. 2 Strategy of retrograde trans-synaptic tracing of presynaptic inputs to desired Cre-expressing neurons. (**a**) Schematic of standard retrograde monosynaptic tracing strategy. In Cre transgenic mice, Cre-dependent helper viruses AAV-Flex-TVA-mCherry and AAV-Flex-RG are injected into brain regions to label the desired neurons (mCherry). Subsequently, pseudotyped rabies viruses (EnvA-dG-RV-GFP) are injected into the same site. EnvA-dG-RV-GFP infects only TVA-mCherry expressing neurons (yellow represents TVA-mCherry, GFP double-positive cells), and after in vivo complementation using RG expressed in Cre-positive cells (blue dots), the viruses trans-synaptically spread to presynaptic neurons (green). (**b**) Schematic of the TRIO method. In the Cre transgenic animal, retrograde virus carrying Cre-dependent Flp recombinase, either CAV2-Flex-Flp or rAAV2-retro-Flex-Flp, is injected in the target site. In the same animal, Flp-dependent helper viruses AAV-fDIO-TVA-mCherry and AAV-fDIO-RG were injected into a region containing Cre+ cells projecting to the target. In this way, all the neurons innervating the target can take up the retrograde virus from axon terminals, but only the Cre+ neurons can express Flp, which in turn, induces Flp-dependent expression of TVA-mCherry and RG. Subsequently, pseudotyped rabies viruses (EnvA-dG-RV-GFP) are injected into the same region to spread to the presynaptic neurons (green). This method enables tracing the presynaptic inputs (green) of the Cre+ neurons that project to the specific target

B19G, Addgene #52473; AAV-hSyn-Flex-TVA-P2A-EGFP-oG, Addgene #85225). The general strategy is to first inject the helper AAVs into the region containing Cre-expressing neurons, and then after 2–3 weeks to allow sufficient expression of TVA and RG, injecting EnvA-dG-RV-GFP (Salk Viral Vector) into the same region, which will only infect Cre/TVA-expressing neurons, and can be complemented in vivo by the RG and thereby spread into the presynaptic neurons (Fig. 2a).

One can combine the presynaptic transsynaptic tracing with the retrograde labeling of neurons projecting to specific target to simultaneously trace the input and output of a desired type of neurons (expressing Cre), and this method is called TRIO (tracing the relationship between input and output) first described by Liqun Luo's lab [45]. Briefly, a retrograde virus expressing Cre-dependent Flp recombinase (e.g., rAAV2-retro-Flex-Flp) is first injected into the target area in a Cre-driver mouse such that Cre-expressing neurons projecting to the target will also express Flp (Fig. 2b). In the same animals, Flp-dependent helper AAVs (AAV-CAG-fDiO-TVA-mCherry and AAV-CAG-fDiO-RG, Addgene #67827, #67828, respectively) are injected into the cell body area, and finally EnvA-dG-RV is also injected into the same cell body locations to trace the presynaptic inputs of a specific type of neurons innervating a specific target (Fig. 2b).

2.2.4 Anterograde Transneuronal Tracing with High-Titer AAV1-Cre

Anterograde transsynaptic tools for tracing postsynaptic neurons downstream of specific type of neurons are still under development. A recent study reported that high titers of AAV1-Cre can transfer from the axonal terminals of infected neurons into neurons in the downstream target region [46]. Such anterograde transneuronal tracing using AAV1-Cre has been replicated and applied in a few other studies [47, 48]. However, this method cannot be targeted to specific types of neurons, and AAV1-Cre can also be transported in the retrograde direction with a low efficiency [49].

2.2.5 Activity-Dependent Circuit Tracing Using a Viral-Genetic Combination Approach

While strategies based on Cre-drivers and target-dependent retrograde labeling described above have helped reveal connectivity and functions of different populations of neurons in the pain-processing circuits, it is widely acknowledged that such strategies often label multiple types of neurons that may have heterogenous or even opposing functions. Therefore, two methods that based on neuronal activity have been developed to label and manipulate functional relevant neurons and have been applied to study pain-processing circuits. Both strategies rely on the activity-induced expression of the immediate early gene Fos.

One method is called CANE, which stands for capturing activated neuronal ensembles [10], in which engineered pseudotyped lentiviruses (CANE-LV-Cre) is used to infect strongly Fos+ neurons in the Fos^{dsTVA} transgenetic mice (JAX 027831) that express the receptor for CANE-viruses transiently in Fos$^+$ neurons. The second method is called FosTRAP (JAX 021882) [13] or an enhanced version called FosTRAP2 (JAX 030323) [50, 51], in which the tamoxifen-inducible CreERt2, or iCreERt2, is knocked-into the

Fos locus such that when Fos is induced, CreERt2/iCreERt2 is expressed in the activated neurons. In terms of pain-related circuits, the CANE method was used for labeling and connectivity tracing of painful stimuli-activated neurons in the parabrachial nucleus [11], or labeling the general anethesia activated neurons in the central amygadal which have potent pain supression effect [12]. The FosTRAP method was applied to label and manipulate pain-activated neurons in the basolateral amygdala [14].

To use the CANE method, painful stimulus is applied to Fos^{dsTVA} mice, followed by 60–90 min of waiting in the homecage (single-housed), followed by stereotaxic surgery to inject CANE-LV-Cre together with desired Cre-dependent AAV vectors into the brain region of interest. To use the FosTRAP2 method, often Cre-dependent AAV is injected into the region of interest first, weeks later, painful stimuli will be applied as well as 4-hydroxytamoxifen to enable pain/taxomifen co-dependent expression of transgenes. The key to the FosTRAP method is to figure out when to administer 4-hydroxytamoxifen (usually through i.p. injection) to most efficiently activate CreERt2 to label Fos^+ cells activated by painful stimuli, as different neurons and different brain regions have different optimal time points for efficient Cre-expression and labeling. Both methods also have some random labeling which is inevitable due to background Fos expression.

3 Detailed Surgical Procedures for Viral Vector Delivery

3.1 Surgical Apparatus and Tools

1. Kopf Model 940 small animal stereotaxic instruments with digital display console (David Kopf instruments).

2. 10-µL Hamilton syringe with a polyethylene tubing (RN Compression Fitting for Tubing, Hamilton).

3. Aseptico drill (Aseptico lightest duty dental lab motor).

4. Micro4 Micro Syringe Pump controller (World Precision Instruments).

5. Heating pad for small animals (DC temperature controller).

6. VAD veterinary anesthesia machine (VETAMAC anesthesia).

7. Surgical tools: scissors, sharp forceps, scalpel blades (Fine Science Tools), pulled glass capillary pipette, cotton swab, suture wires (surgical specialties), gel foam (Surgifoam).

3.2 Reagents and Drugs for Surgery

1. Ethanol, 70% vol/vol.

2. Sterile saline (e.g., AirLife Unit Dose Saline, 3 mL 0.9% inhalation).

3. Mineral oil (e.g., Fisher BioReagents; Cat.no.BP26291).

4. Povidone-iodine.

5. Meloxicam (nonsteroidal anti-inflammatory drug analgesic, for subcutaneous injection, 5 mg/kg).

6. Dexamethasone (for subcutaneous injection, 2 mg/kg, 4 h injection before surgery).

7. Puralube ophthalmic ointment.

8. Anesthetics: Isoflurane, USP, inhalation anesthetic; or ketamine (intraperitoneal injection, 100 mg/kg) and xylazine (intraperitoneal injection, 10 mg/kg).

3.3 Preparations for Surgery

1. Surgery tools are sterilized (autoclave or other methods), and the surgical work station is cleaned with 70% ethanol.

2. Pull glass capillary pipettes using the micropipette puller (SUTTER Instrument.co, P-87), and sharpen the pipette tip by beveling the tip using a home-made pipette polishing equipment [52]. This beveled pipette tip penetrates dura with minimum damage to the brain surface. After sharpening, connect the glass pipette to a tube with 10-mL syringe, submerge the pipette into 70% EtOH, and wash the pipette for 2–3 times by pushing and pulling the syringe. Keep the sterilized glass pipettes in a pre-sterilized box.

3. Pre-fill mineral oil into the 10-μL Hamilton syringe and the connected polyethylene tubing. Set the Hamilton syringe on the pump.

4. Set the heating pad to 37.5 °C.

5. Prepare a waste container of 10% bleach for disposing of used pipettes that come in contact with the virus and the and tubes that contain viral stocks/aliquots.

6. Get virus aliquots from the −80 °C freezer and keep the tubes on ice.

3.4 Surgical Procedures

1. *Anesthesia*: For isoflurane anesthesia, place the mouse into the anesthesia induction chamber, turn on system with 2–3% isoflurane mixed with 100% oxygen at 1.0–1.2 L/min for induction. As an alternative option, mouse can be anesthetized with an i.p. injection of ketamine (100 mg/kg) and xylazine (10 mg/kg). After the mouse is fully anesthetized (which is characterized by a breathing rate of about 1 time/s, and no responses to toe pinch), carefully remove the mouse from the induction chamber to the stereotaxic frame, place the mouse on the prewarmed heating pad, secure the head of mouse to the ear bars and fix the teeth on the tooth bar in the correct position. Administer the maintenance level of anesthesia at 1–1.5% isoflurane mixed with 100% oxygen at 0.6–0.8 L/min. Inject Meloxicam (5 mg/kg) subcutaneously.

2. *Setting stereotaxic coordinates*: Cover the eyes with Puralube ophthalmic ointment to prevent drying and cataract formation. Wipe the fur on the top of the skull with a cotton swab soaked in 70% ethanol, shave the fur with scalpel blade (F.S.T, 10023-00), and clean the hair with cotton swap. Remove any remaining hair and apply the povidone-iodine on the skin. This decreases the risk of infection at the incision site. Lift a small piece of skin at the midline with tweezers and cut with scissors into 2-cm-long opening. Carefully pull the skin back to expose the skull (to see the bregma and lambda). To avoid bleeding, insert a small piece of surgery foam under the skin around the opening. Place the flat-tip pipette prefilled with ink on the pipette holder. Find the bregma through the microscope, lower the pipette, and make a mark at the bregma. Write down the coordinates of the bregma or set it to all "0" if using a digital reader of stereotaxis. Then move the pipette to the lambda and make another mark. Level the skull using the bregma and lambda, the D-V difference should be less than 0.03 mm. Also, to level the brain along medial-lateral axis, starting from the lambda move to the left for 2 mm, record the height, and then move to the right side of lambda for 2 mm and record the height again. The height difference between the left and right should also be less than 0.03 mm. This procedure is essential for precise targeting of the subsequent injection. After leveling the mouse skull to be flat, move back to the bregma mark and set it as "0" on the stereotaxic device. Next, move to the intended target coordinates (anterior-posterior, and medial-lateral positions relative to bregma) for virus injection, and mark the target location.

3. *Performing craniotomy of the injection site*: Use a dental drill to thin the skull around the target marker carefully and slowly approximately 2 mm in diameter until the skull cracks, and then gently remove the skull without rupturing the meninges. Drilling too fast increases the risk of damaging the surface of the brain and produces too much heat that will induce tissue swelling. Throughout this procedure, keep the skull moist with the application of sterile saline.

4. *Injection procedure*: Fill up a previously prepared, sharpened, and sterilized glass pipette with mineral oil, and place it on the stereotaxic holder, make sure that its orientation is perfectly vertical and straight. Connect the glass pipette to the tubing from the Hamilton syringe using a 2–3 cm tube as connector. Test by pushing the syringe, to see if mineral oil is coming out from the pipette tip. This procedure is critical to make sure there is no leakage or air from the connected parts. Move the pipette to the bregma marker and point the pipette tip exactly on it, read the coordinates or set them all to "0," then move the

pipette to the coordinates (anterior-posterior, medial-lateral, and dorsal-ventral positions relative to bregma and the dura surface). Carefully put a small piece of tissue wipe under the pipette tip to absorb the extra mineral oil coming out from the tip. Lower the pipette tip into the tube containing viruses and withdrawal the virus at the rate of 20 nL/s. Retract the pipette after taking 2 μL of viruses. Start the pump briefly to check if there is virus coming out from the tip of the pipette and absorb the extra virus by using tissue wipe and discard the tissue paper into the prepared bleach bottle.

Next, slowly lower (~1 mm/min) the pipette to the desired depth. Start the pump with desired injection volume at a maximum rate of 50 nL/min. After injection, leave the pipette in the place for 5–10 min before slowly retracting it. Suture the scalp and seal it with tissue glue. Add some disinfection ointment on the skin. Place the animal in the home cage on a heating pad until the mouse recovers. Disinfect the surgical tools and bench.

Tissue collection of anterograde/retrograde tracing can be initiated 3–4 weeks after viral injection. For retrograde trans-synaptic tracing, mouse receive a second time of EnvA-DG-RV virus injection at the same site 2 weeks after the first injection, and samples are collected 6–7 days later. To visualize the labeled neurons, one either directly images brain/spinal cord sections using a fluorescent/confocal microscope or images after immune-fluorescence staining using standard protocols.

4 Concluding Remarks

We have described the detailed strategies and surgical procedures to use viral vectors (sometimes in combination with transgenic mouse lines) to label neurons based on Cre-expression or projection targets and to label their presynaptic partners. Instead of the fluorescent proteins, one can use the same strategy to express optogenetic- or chemical-genetic proteins in the neurons of interest and manipulate the labeled neurons to reveal their functions in somatosensory processing. Similarly, one can express genetically encoded calcium or voltage sensors using the viral strategies described here to monitor the in vivo activity of labeled neurons in response to different sensory stimuli. We foresee that these methods will bring great new insights about neural circuits underlying pain sensation in the near future.

References

1. Abraira VE, Ginty DD (2013) The sensory neurons of touch. Neuron 79(4):618–639. https://doi.org/10.1016/j.neuron.2013.07.051

2. Peirs C, Seal RP (2016) Neural circuits for pain: recent advances and current views. Science 354(6312):578–584. https://doi.org/10.1126/science.aaf8933

3. Duan B, Cheng L, Bourane S, Britz O, Padilla C, Garcia-Campmany L, Krashes M, Knowlton W, Velasquez T, Ren X, Ross S, Lowell BB, Wang Y, Goulding M, Ma Q (2014) Identification of spinal circuits transmitting and gating mechanical pain. Cell 159(6):1417–1432. https://doi.org/10.1016/j.cell.2014.11.003

4. Huang ZJ, Zeng H (2013) Genetic approaches to neural circuits in the mouse. Annu Rev Neurosci 36:183–215. https://doi.org/10.1146/annurev-neuro-062012-170307

5. Zhang S, Lv F, Yuan Y, Fan C, Li J, Sun W, Hu J (2019) Whole-brain mapping of monosynaptic afferent inputs to cortical CRH neurons. Front Neurosci 13:565. https://doi.org/10.3389/fnins.2019.00565

6. Carter ME, Soden ME, Zweifel LS, Palmiter RD (2013) Genetic identification of a neural circuit that suppresses appetite. Nature 503(7474):111–114. https://doi.org/10.1038/nature12596

7. Han L, Ma C, Liu Q, Weng HJ, Cui Y, Tang Z, Kim Y, Nie H, Qu L, Patel KN, Li Z, McNeil B, He S, Guan Y, Xiao B, Lamotte RH, Dong X (2013) A subpopulation of nociceptors specifically linked to itch. Nat Neurosci 16(2):174–182. https://doi.org/10.1038/nn.3289

8. Pan H, Fatima M, Li A, Lee H, Cai W, Horwitz L, Hor CC, Zaher N, Cin M, Slade H, Huang T, Xu XZS, Duan B (2019) Identification of a spinal circuit for mechanical and persistent spontaneous itch. Neuron 103(6):1135–1149.e1136. https://doi.org/10.1016/j.neuron.2019.06.016

9. Acton D, Ren X, Di Costanzo S, Dalet A, Bourane S, Bertocchi I, Eva C, Goulding M (2019) Spinal neuropeptide Y1 receptor-expressing neurons form an essential excitatory pathway for mechanical itch. Cell Rep 28(3): 625–639.e626. https://doi.org/10.1016/j.celrep.2019.06.033

10. Sakurai K, Zhao S, Takatoh J, Rodriguez E, Lu J, Leavitt AD, Fu M, Han BX, Wang F (2016) Capturing and manipulating activated neuronal ensembles with CANE delineates a hypothalamic social-fear circuit. Neuron 92(4):739–753. https://doi.org/10.1016/j.neuron.2016.10.015

11. Rodriguez E, Sakurai K, Xu J, Chen Y, Toda K, Zhao S, Han BX, Ryu D, Yin H, Liedtke W, Wang F (2017) A craniofacial-specific monosynaptic circuit enables heightened affective pain. Nat Neurosci 20(12):1734–1743. https://doi.org/10.1038/s41593-017-0012-1

12. Hua, T., Chen, B., Lu, D. et al. (2020). General anesthetics activate a potent central pain-suppression circuit in the amygdala. Nat Neurosci 23, 854–868. https://doi.org/10.1038/s41593-020-0632-8

13. Guenthner CJ, Miyamichi K, Yang HH, Heller HC, Luo L (2013) Permanent genetic access to transiently active neurons via TRAP: targeted recombination in active populations. Neuron 78(5):773–784. https://doi.org/10.1016/j.neuron.2013.03.025

14. Corder G, Ahanonu B, Grewe BF, Wang D, Schnitzer MJ, Scherrer G (2019) An amygdalar neural ensemble that encodes the unpleasantness of pain. Science 363(6424):276–281. https://doi.org/10.1126/science.aap8586

15. Sun S, Xu Q, Guo C, Guan Y, Liu Q, Dong X (2017) Leaky gate model: intensity-dependent coding of pain and itch in the spinal cord. Neuron 93(4):840–853.e845. https://doi.org/10.1016/j.neuron.2017.01.012

16. Gao ZR, Chen WZ, Liu MZ, Chen XJ, Wan L, Zhang XY, Yuan L, Lin JK, Wang M, Zhou L, Xu XH, Sun YG (2019) Tac1-expressing neurons in the periaqueductal gray facilitate the itch-scratching cycle via descending regulation. Neuron 101(1):45–59.e49. https://doi.org/10.1016/j.neuron.2018.11.010

17. Peirs C, Williams SP, Zhao X, Walsh CE, Gedeon JY, Cagle NE, Goldring AC, Hioki H, Liu Z, Marell PS, Seal RP (2015) Dorsal horn circuits for persistent mechanical pain. Neuron 87(4):797–812. https://doi.org/10.1016/j.neuron.2015.07.029

18. Cui L, Miao X, Liang L, Abdus-Saboor I, Olson W, Fleming MS, Ma M, Tao YX, Luo W (2016) Identification of early RET+ deep dorsal spinal cord interneurons in gating pain. Neuron 91(5):1137–1153. https://doi.org/10.1016/j.neuron.2016.07.038

19. Barik A, Thompson JH, Seltzer M, Ghitani N, Chesler AT (2018) A brainstem-spinal circuit controlling nocifensive behavior. Neuron 100(6):1491–1503.e1493. https://doi.org/10.1016/j.neuron.2018.10.037

20. Gatto G, Smith KM, Ross SE, Goulding M (2019) Neuronal diversity in the somatosensory system: bridging the gap between cell type and function. Curr Opin Neurobiol 56: 167–174. https://doi.org/10.1016/j.conb. 2019.03.002

21. Beier KT, Steinberg EE, DeLoach KE, Xie S, Miyamichi K, Schwarz L, Gao XJ, Kremer EJ, Malenka RC, Luo L (2015) Circuit architecture of vta dopamine neurons revealed by systematic input-output mapping. Cell 162(3): 622–634. https://doi.org/10.1016/j.cell. 2015.07.015

22. Zhang S, Xu M, Chang WC, Ma C, Hoang Do JP, Jeong D, Lei T, Fan JL, Dan Y (2016) Organization of long-range inputs and outputs of frontal cortex for top-down control. Nat Neurosci 19(12):1733–1742. https://doi. org/10.1038/nn.4417

23. Economo MN, Clack NG, Lavis LD, Gerfen CR, Svoboda K, Myers EW, Chandrashekar J (2016) A platform for brain-wide imaging and reconstruction of individual neurons. eLife 5: e10566. https://doi.org/10.7554/eLife. 10566

24. Economo MN, Viswanathan S, Tasic B, Bas E, Winnubst J, Menon V, Graybuck LT, Nguyen TN, Smith KA, Yao Z, Wang L, Gerfen CR, Chandrashekar J, Zeng H, Looger LL, Svoboda K (2018) Distinct descending motor cortex pathways and their roles in movement. Nature 563(7729):79–84. https://doi.org/ 10.1038/s41586-018-0642-9

25. Winnubst J, Bas E, Ferreira TA, Wu Z, Economo MN, Edson P, Arthur BJ, Bruns C, Rokicki K, Schauder D, Olbris DJ, Murphy SD, Ackerman DG, Arshadi C, Baldwin P, Blake R, Elsayed A, Hasan M, Ramirez D, Dos Santos B, Weldon M, Zafar A, Dudman JT, Gerfen CR, Hantman AW, Korff W, Sternson SM, Spruston N, Svoboda K, Chandrashekar J (2019) Reconstruction of 1000 projection neurons reveals new cell types and organization of long-range connectivity in the mouse brain. Cell 179(1):268–281.e213. https://doi.org/10.1016/j.cell.2019.07.042

26. Economo MN, Winnubst J, Bas E, Ferreira TA, Chandrashekar J (2019) Single-neuron axonal reconstruction: the search for a wiring diagram of the brain. J Comp Neurol 527(13): 2190–2199. https://doi.org/10.1002/cne. 24674

27. Ugolini G (2011) Rabies virus as a transneuronal tracer of neuronal connections. Adv Virus Res 79:165–202. https://doi.org/10.1016/ B978-0-12-387040-7.00010-X

28. Song CK, Enquist LW, Bartness TJ (2005) New developments in tracing neural circuits with herpesviruses. Virus Res 111(2): 235–249. https://doi.org/10.1016/j. virusres.2005.04.012

29. Chatterjee S, Sullivan HA, MacLennan BJ, Xu R, Hou Y, Lavin TK, Lea NE, Michalski JE, Babcock KR, Dietrich S, Matthews GA, Beyeler A, Calhoon GG, Glober G, Whitesell JD, Yao S, Cetin A, Harris JA, Zeng H, Tye KM, Reid RC, Wickersham IR (2018) Nontoxic, double-deletion-mutant rabies viral vectors for retrograde targeting of projection neurons. Nat Neurosci 21(4):638–646. https://doi.org/10.1038/s41593-018-0091-7

30. Hirano M, Kato S, Kobayashi K, Okada T, Yaginuma H, Kobayashi K (2013) Highly efficient retrograde gene transfer into motor neurons by a lentiviral vector pseudotyped with fusion glycoprotein. PLoS One 8(9):e75896. https://doi.org/10.1371/journal.pone. 0075896

31. Nelson A, Schneider DM, Takatoh J, Sakurai K, Wang F, Mooney R (2013) A circuit for motor cortical modulation of auditory cortical activity. J Neurosci 33(36):14342–14353. https://doi.org/10.1523/JNEUROSCI. 2275-13.2013

32. Kato S, Kobayashi K, Inoue K, Kuramochi M, Okada T, Yaginuma H, Morimoto K, Shimada T, Takada M, Kobayashi K (2011) A lentiviral strategy for highly efficient retrograde gene transfer by pseudotyping with fusion envelope glycoprotein. Hum Gene Ther 22(2):197–206. https://doi.org/10.1089/ hum.2009.179

33. Liu Y, Latremoliere A, Li X, Zhang Z, Chen M, Wang X, Fang C, Zhu J, Alexandre C, Gao Z, Chen B, Ding X, Zhou JY, Zhang Y, Chen C, Wang KH, Woolf CJ, He Z (2018) Touch and tactile neuropathic pain sensitivity are set by corticospinal projections. Nature 561(7724): 547–550. https://doi.org/10.1038/s41586-018-0515-2

34. Wang X, Liu Y, Li X, Zhang Z, Yang H, Zhang Y, Williams PR, Alwahab NSA, Kapur K, Yu B, Zhang Y, Chen M, Ding H, Gerfen CR, Wang KH, He Z (2017) Deconstruction of corticospinal circuits for goal-directed motor skills. Cell 171(2):440–455. e414. https://doi.org/10.1016/j.cell.2017. 08.014

35. Stanek E, Rodriguez E, Zhao S, Han BX, Wang F (2016) Supratrigeminal bilaterally projecting neurons maintain basal tone and enable bilateral phasic activation of jaw-closing muscles. J Neurosci 36(29):7663–7675. https://doi. org/10.1523/JNEUROSCI.0839-16.2016

36. Boender AJ, de Jong JW, Boekhoudt L, Luijendijk MC, van der Plasse G, Adan RA (2014) Combined use of the canine adenovirus-2 and DREADD-technology to activate specific neural pathways in vivo. PLoS One 9(4):e95392. https://doi.org/10.1371/journal.pone.0095392

37. Ekstrand MI, Nectow AR, Knight ZA, Latcha KN, Pomeranz LE, Friedman JM (2014) Molecular profiling of neurons based on connectivity. Cell 157(5):1230–1242. https://doi.org/10.1016/j.cell.2014.03.059

38. Junyent F, Kremer EJ (2015) CAV-2—why a canine virus is a neurobiologist's best friend. Curr Opin Pharmacol 24:86–93. https://doi.org/10.1016/j.coph.2015.08.004

39. Li SJ, Vaughan A, Sturgill JF, Kepecs A (2018) A viral receptor complementation strategy to overcome CAV-2 tropism for efficient retrograde targeting of neurons. Neuron 98(5):905–917.e905. https://doi.org/10.1016/j.neuron.2018.05.028

40. Tervo DG, Hwang BY, Viswanathan S, Gaj T, Lavzin M, Ritola KD, Lindo S, Michael S, Kuleshova E, Ojala D, Huang CC, Gerfen CR, Schiller J, Dudman JT, Hantman AW, Looger LL, Schaffer DV, Karpova AY (2016) A designer AAV variant permits efficient retrograde access to projection neurons. Neuron 92(2):372–382. https://doi.org/10.1016/j.neuron.2016.09.021

41. Callaway EM, Luo L (2015) Monosynaptic circuit tracing with glycoprotein-deleted rabies viruses. J Neurosci 35(24):8979–8985. https://doi.org/10.1523/JNEUROSCI.0409-15.2015

42. Luo L, Callaway EM, Svoboda K (2018) Genetic dissection of neural circuits: a decade of progress. Neuron 98(2):256–281. https://doi.org/10.1016/j.neuron.2018.03.040

43. Reardon TR, Murray AJ, Turi GF, Wirblich C, Croce KR, Schnell MJ, Jessell TM, Losonczy A (2016) Rabies virus CVS-N2c(DeltaG) strain enhances retrograde synaptic transfer and neuronal viability. Neuron 89(4):711–724. https://doi.org/10.1016/j.neuron.2016.01.004

44. Kim EJ, Jacobs MW, Ito-Cole T, Callaway EM (2016) Improved monosynaptic neural circuit tracing using engineered rabies virus glycoproteins. Cell Rep 15(4):692–699. https://doi.org/10.1016/j.celrep.2016.03.067

45. Schwarz LA, Miyamichi K, Gao XJ, Beier KT, Weissbourd B, DeLoach KE, Ren J, Ibanes S, Malenka RC, Kremer EJ, Luo L (2015) Viral-genetic tracing of the input-output organization of a central noradrenaline circuit. Nature 524(7563):88–92. https://doi.org/10.1038/nature14600

46. Harris JA, Oh SW, Zeng H (2012) Adeno-associated viral vectors for anterograde axonal tracing with fluorescent proteins in nontransgenic and cre driver mice. Curr Protoc Neurosci Chapter 1:Unit 1.20.1–Unit 1.20.18. https://doi.org/10.1002/0471142301.ns0120s59

47. Zingg B, Chou XL, Zhang ZG, Mesik L, Liang F, Tao HW, Zhang LI (2017) AAV-mediated anterograde transsynaptic tagging: mapping corticocollicular input-defined neural pathways for defense behaviors. Neuron 93(1):33–47. https://doi.org/10.1016/j.neuron.2016.11.045

48. Oh SW, Harris JA, Ng L, Winslow B, Cain N, Mihalas S, Wang Q, Lau C, Kuan L, Henry AM, Mortrud MT, Ouellette B, Nguyen TN, Sorensen SA, Slaughterbeck CR, Wakeman W, Li Y, Feng D, Ho A, Nicholas E, Hirokawa KE, Bohn P, Joines KM, Peng H, Hawrylycz MJ, Phillips JW, Hohmann JG, Wohnoutka P, Gerfen CR, Koch C, Bernard A, Dang C, Jones AR, Zeng H (2014) A mesoscale connectome of the mouse brain. Nature 508(7495):207–214. https://doi.org/10.1038/nature13186

49. Castle MJ, Gershenson ZT, Giles AR, Holzbaur EL, Wolfe JH (2014) Adeno-associated virus serotypes 1, 8, and 9 share conserved mechanisms for anterograde and retrograde axonal transport. Hum Gene Ther 25(8):705–720. https://doi.org/10.1089/hum.2013.189

50. Allen WE, DeNardo LA, Chen MZ, Liu CD, Loh KM, Fenno LE, Ramakrishnan C, Deisseroth K, Luo L (2017) Thirst-associated preoptic neurons encode an aversive motivational drive. Science 357(6356):1149–1155. https://doi.org/10.1126/science.aan6747

51. DeNardo LA, Liu CD, Allen WE, Adams EL, Friedmann D, Fu L, Guenthner CJ, Tessier-Lavigne M, Luo L (2019) Temporal evolution of cortical ensembles promoting remote memory retrieval. Nat Neurosci 22(3):460–469. https://doi.org/10.1038/s41593-018-0318-7

52. Canfield JG (2006) Dry beveling micropipettes using a computer hard drive. J Neurosci Methods 158(1):19–21. https://doi.org/10.1016/j.jneumeth.2006.05.009

Chapter 13

Recording Pain-Related Brain Activity in Behaving Animals Using Calcium Imaging and Miniature Microscopes

Biafra Ahanonu and Gregory Corder

Abstract

Pain is a multifaceted percept formed by information processing in the brain of ascending signals from the periphery and spinal cord. Numerous studies in humans and animals, using technologies such as fMRI, have demonstrated that noxious stimuli activate a distributed network consisting of multiple brain regions. These human and preclinical studies suggest that the nervous system relays nociceptive information through a vast network of high-order cognitive, motivational, and motor-planning brain regions to generate the perception of pain and resulting nocifensive behavior. While these previous studies have improved our understanding of brain network function in pain, they present limitations due to low-resolution, static snapshots of neural activity, or a difficulty tracking the same cells longitudinally across extended periods of time ranging from weeks to months. Here we present a protocol that uses recent advances in in vivo microscopy and computational techniques to address these questions. Miniaturized fluorescence microscopes (miniscopes) using microendoscopy allow for imaging of intracellular Ca^{2+} transients, which function as a proxy for neural activity. This innovative technology permits high-resolution imaging of large neuronal populations (up to 1000+ neurons in a single animal) located in deep brain regions of freely behaving mice over a time scale of months. This technology puts researchers in a position to answer many fundamental questions regarding the coding principles of nociceptive information and to identify pain-specific neural pathways in the brain. Furthermore, it is now possible to determine how brain neuronal networks evolve their activity dynamics over several months, before, during, and after chronic pain has developed while also understanding how existing and novel analgesics restore both behavior and neural activity to alleviate pain.

Key words Calcium imaging, Pain, Imaging analysis, Behavior analysis

1 Introduction

Pain is a perception. Like all perceptions, it is computed by complex, highly interconnected, and dynamic networks consisting of tens of thousands, if not millions, of neurons in the brain receiving sensory information from the periphery [1]. Different aspects of our perception of pain, for example, the unpleasantness or the sense intensity, can shift or become more salient depending on the environment, time of day, social situations, or our general mood. Such

Rebecca P. Seal (ed.), *Contemporary Approaches to the Study of Pain: From Molecules to Neural Networks*, Neuromethods, vol. 178, https://doi.org/10.1007/978-1-0716-2039-7_13, © Springer Science+Business Media, LLC, part of Springer Nature 2022

malleable intensities of how we experience pain indicate that these neural computations fluctuate on a rapid time scale and integrate more than just noxious sensory information into our conscious experiences [2]. This is painfully obviously in pathological pain conditions, where normally protective and necessary pain perceptions arise under completely unnecessary situations and without noxious stimulation [3]. Indeed, we know a great deal about how chronic pain, which affects nearly a third of the population with many patients suffering due to inadequate therapies, can reshape entire regions of the spinal cord and brain as pain becomes less of an evolutionary advantage signal and more of an outright disease of the nervous system [4, 5].

There are many approaches to discovering how the dynamic nature of nociception, or pain-related neural information, is processed in the central nervous system of human pain patients. These range from electrical methods with high temporal resolution, to temporally imprecision but anatomically enlightening whole-brain techniques, such as functional magnetic resonance imaging (fMRI) [6, 7]. Indeed, human physiology studies have been invaluable in providing a detailed roadmap for what regions of the brain are likely to be involved in the different dimensions of our perceptions of pain. However, the resolution of fMRI studies limits interpretation to voxels of activity that represents 100,000+ neurons; thus, the next phase of biomedical and translational pain research will need to improve this functional anatomy by an order of magnitude if one's goal is to identify and disentangle the specific individual neural circuits in the brain that are most critical to generate acute and chronic pain experiences.

There are a variety of existing techniques that allow dissection of neural circuits, but each comes with limitations. Existing approaches to measure activity of individual neurons are limited by only allowing static snapshots of neuronal activity, for example immediate early gene expression [8], or present challenges when attempting to understand the spatial organization of neural activity or track activity, of genetically identified cells, longitudinally over chronic pain relevant time scale of months, such as with in vivo electrophysiological recordings [9]. Further, while head-fixed two-photon calcium imaging circumvents many of these limitations [10], it places the animal is a stressful, constrained environment that may alter pain processing and prevents observation of many ethologically relevant nocifensive behaviors in response to noxious stimuli [11]. Studies aiming to identify functional brain regions—using lesions, chemogenetic, or optogenetic manipulations—likely modulate activity in neural networks unrelated to pain (e.g., those that assign positive valence to stimuli). Similarly, pharmacological approaches often impact functionally heterogeneous populations of neurons. Consequently, how distinct pain modalities are represented in the brain and how neural activity evolves during the development of chronic pain requires additional precision using new technologies available for preclinical studies in animals.

The advantage of trying to understand nociceptive brain processes in nonhumans, especially rodent subjects, is that an entire new range of invasive imaging technologies have rapidly become available that can provide single-neuron resolution of large networks of neurons across the brain in freely behaving animals. Recent advances in in vivo microscopy and microendoscopy, which take advantage of implantable optics, have led to the development of miniaturized fluorescent microscopes (here referred to as miniscopes [12, 13]) that use microendoscopes to image Ca^{2+} transients in neurons and astrocytes (i.e., elevation of intracellular Ca^{2+} that reports cellular activity). These one-photon microscopes permits high-resolution imaging of large neuron populations, up to 1000+ neurons simultaneously [14–16], located in deep brain regions of freely behaving mice over time scales of months. This will prove invaluable in allowing researchers to make conclusions about what type of dynamic network computations contribute to the acute-to-chronic pain transition period along with how neural networks are functionally restored as pain is relieved by existing and novel analgesics. Even more relevant to the pain field, this methodology permits concurrent recording of sensory and affective nocifensive behavior in freely moving animals to uncover the precise neural network coding features that lead to specific behaviors.

Recently, we applied miniscope calcium imaging to one of the first large-scale studies of nociceptive processing in the brain with the aim of elucidating the network computations related to the affective component of pain. In this chapter, we will expand on the technical details for utilizing miniscope imaging and concurrent behavioral analysis for determining the dynamic neural circuit mechanisms of pain in the brain, in a step-by-step protocol with critical considerations for improving this approach in future pain studies to more precisely identify potential nociceptive-specific neural circuits. Many of the procedures, analyses, and techniques described herein can also be applied to non-pain studies of neural activity in freely moving animals.

2   Materials

Materials table located in Table 1. For Ca^{2+} imaging analysis, we recommend users download MATLAB and our calcium imaging analysis software package CIAtah (pronounced cheetah) or calciumImagingAnalysis (CIAPKG) [17], which can be downloaded at https://github.com/bahanonu/calciumImagingAnalysis.

Table 1 lists the main materials and equipment that are used to prepare Ca^{2+} indicator expressing mice along with acquiring and analyzing Ca^{2+} imaging movies, behavior videos, accelerometer data, and other experimental variables related to imaging neural activity in mice experience noxious stimuli.

Table 1
Materials used for in vivo imaging of the mouse brain during pain

Items	Product number, vendor, and comments
Animals	
Mice	• Wild-type mice. C57BL6/J. Jackson.
Viral injections and implantation surgery	
Cabinets for surgical tool organization	• Plastic Small-Parts Cabinet. 9619T61. McMaster.
Waste scavenging, suction, and filtration	• Patterson Scientific: EVAC 2—78918181. • Fisher Scientific: Filter Units (50 mm dia.), Millex Inlet and outlet, Pore Size: 0.22 μm, SLFG05010.
Surgery supplies	• Cotton tips. • Alcohol, 70% in water. • Betadine (D1415 Povidone Iodine). Dynarex. • 1-mL syringe. • 27G needle. VWR International. • 30G needle. VWR International. • Mammalian Ringers. 50-980-246. Fisher Scientific • Lens Tissues. MC-5. Thorlabs • Biological tape Tegaderm Transparent Dressing. NC9033794. 3M. • Nail polish, black.
Metal cannulas	• Custom order of 304S/S Hypo Tube 18X GA. 0495/.0505″ OD × .0410/.0430″ ID × 4.3 mm long; cut and deburred. Ziggy's Tubes and Wires, Inc.
Microendoscopic lenses	• 1-mm-diameter gradient refractive index (GRIN) lens. Grintech GmBH. • Lens Probe 1.0 mm diameter, ~9.0 mm length. ID: 1050-004596. Inscopix. • Lens Probe 0.6 mm diameter, ~7.3 mm length. ID: 1050-004597. Inscopix.
Cover glass	• Custom order of 2.0-mm diameter cover glass. TLC International (http://www.tlcinternational.com). • Alternative: Small round cover glass, #0 thickness, 3 mm, 100 pack. 64-0726 (CS-3R-0). Warner Instruments.
Surgical tools	• Delicate Bone Scraper. 10075-16. Fine Science Tools. • Bonn Micro Probes. 10033-13. Fine Science Tools. • Micro Points. 10066-15. Fine Science Tools. • Bonn Micro Probes. 10031-13. Fine Science Tools. • Dental mirror. B07NZMK31Y. Amazon. • Slotted Screwdriver Set. 5714A4. McMaster.
Headbars	• Get headbars custom laser cut or CNC machined, for example from Protolabs, Laser Alliance (San Jose), and other companies. *See* design in Fig. 3j.
Headbar clamp	• C-clamp. CC-2. Siskiyou.
Screws	• S/S Machine Screw #000-120 × 1/16″ Flat Head, Slotted Drive. MX-000120-01SFL. Component Supply.

(continued)

Table 1
(continued)

Items	Product number, vendor, and comments
Cement	• S380 - C&B METABOND® Quick Adhesive Cement System • S371 - "C" Universal 4-META Catalyst 0.7 mL • S398 "B" Quick Base For C&B METABOND®
Dental cement	• Hygenic® Perm, Powder and Liquid Kits. 379-8840. Coltene. • https://www.pattersondental.com/Supplies/ProductFamilyDetails/PIF_52528
GRINjector parts	• 3D printed parts, files at https://github.com/bahanonu/GRINjector. • Vernier micrometer. SM-13. Newport. • Bearing stage. 9066-COM. Newport. • Model 1770, Kopf Instruments. For attaching GRINjector to stereotaxic arm. • 2× - 8–32 × 1″ cap screw. • 1× - 4-40 × 3/16″ cap screw. • 1× - 4-40 × 1/4″ cap screw. • 4× - 4-40 × 5/16″ cap screw. • Luer lock needle matched to the size of the GRIN lens probe. • 0.5-mm diameter needle that is ~80-mm long.
Drill	• Drill. EXL-M40 (http://www.osadausa.com/exlm40.html). Osada. • PS-SC Optical pedestal sliding clamp. Newport.
Micro drill burrs	http://www.finescience.com/Special-Pages/Products.aspx?ProductId=268&CategoryId=126 • 0.5-mm burrs. 19007-05. Fine Science Tools. For skull screws. • 1.4-mm burrs. 19007-14. Fine Science Tools. For 1-mm GRIN hole.
Micro drill trephines	• 1.8-mm trephines. 18004-18. Fine Science Tools.
Optical glue	• Optical Adhesive 81. NOA81. Norland Products.
Optical glue curing gun	• oGeee 5W Dental Wireless Cordless LED Curing Light Lamp Cure.
UV glue	• Light-Activated Adhesive #4305, 1 oz. Bottle. 303389–30769. Loctite.
UV gun	• Edmund Optics LED UV Curing GunNT 59-270 • Alternative: UV Mini Flashlight. 1159N2. McMaster.
Verifying calcium indicator expression	
Peristaltic pump	
Needle	• 25G needle.
Curved forceps	• Dumont #7 Forceps. 11272-40. Fine Science Tools.

(continued)

Table 1
(continued)

Items	Product number, vendor, and comments
Blunt forceps	• Dumont #5 Forceps. 11295-00. Fine Science Tools.
1x PBS	• Dilute 10× PBS 1:10 in Milli-Q water.
4% formaldehyde	• Dilute formaldehyde 1:10 in 1× PBS
Avertin	• Dilute stock Avertin 1:40 in saline
Histology solutions	• 30% sucrose in 1× PBS • 1x PBS with 0.3% Triton X-100
Serum	• Donkey Serum. 017-000-121. Jackson ImmunoResearch.
Anti-GFP antibody	• a-GFP rabbit at 1:1000 dilution. A11122. Invitrogen.
Anti-rabbit secondary	• DyLight 550 Donkey anti-rabbit 1:500. ab96892. Abcam.
Nuclear stain	• DAPI 50 ng/mL in 1× PBS.
Mounting Medium	• Fluoromount-G™. 00-4958-02. Thermo Fisher Scientific
Checking implant quality and baseplate mounting	
Parts to set up holder for awake animal fluorescence activity checking.	All the below parts are from Thorlabs. • BA1S Mounting Base, 1″ × 2.3″ × 3/8″ • MB4 Aluminum Breadboard 4″ × 6″ × 1/2″, 1/4″-20 Taps • PH1.5 Post Holder with Spring-Loaded Hex-Locking Thumbscrew, L = 1.50″ • SWC Swivel Post Clamp, 360° Continuously Adjustable • TR4 Ø1/2″ × 4" Stainless Steel Optical Post, 8-32 Stud, 1/4″-20 Tapped Hole • TR3 Ø1/2″ × 3″ Stainless Steel Optical Post, 8-32 Stud, 1/4″-20 Tapped Hole • TR6 Ø1/2″ × 6″ Stainless Steel Optical Post, 8-32 Stud, 1/4″-20 Tapped Hole • GN2—Small Dual-Axis Goniometer
Rotary encoder.	• BQLZR 600P/R Incremental Rotary Encoder. N04452. Signswise.
Running wheel	• InnoWheel, Catalog No.14-726-577. Fisher Scientific.
Micromanipulator	• XYZ Linear Stage, Compact, Dovetail, 0.375 in. Travel, 3 lb., 8–32. MT-XYZ. Newport. • Alternative: XYZ Linear Stage, 0.55 in. Travel, 8-32 & 1/4-20, Triple Divide. 9065-XYZ. Newport.
Miniature microscope holder	• Gripper Part. 1050-002199. Inscopix.
Miniature microscope system	• nVista 2.0 or 3.0. Inscopix
Ca^{2+} imaging experiments	
Miniature microscope system	• nVista 2.0 or 3.0. Inscopix.
Camera lens	• Inesun 6-60 mm 1/3″ CS Lens CCTV. B07BTP96RL. Amazon.

(continued)

Table 1
(continued)

Items	Product number, vendor, and comments
Camera	• USB 3.0 color industrial camera. DFK 23UP1300. The Imaging Source • Alternative: Guppy Pro F-125 1/3" CCD Monochrome Camera. 68-567. AVT.
Camera trigger cable	• Hirose 12-pin female to tinned leads I/O/power cable, 2 m. CB-I/O-02M. 1STVISION INC.
Data acquisition laptops	• Aspire 5 Slim Laptop. A515-43-R19L. Acer.
Sensory testing rack	All the below items and catalog numbers are from McMaster: • 1× - High-Flow Perforated Sheet Steel, Staggered Holes, 0.03" Thick, 36" Wide × 40" Long. Cut to 12" × 24". 92725T22. • 8× - Aluminum Inch T-Slotted Framing System Four-Slot Single, 1" Solid Extrusion, 2' Length. 47065T101-47065T209. • 4× - Aluminum Inch T-Slotted Framing System Four-Slot Single, 1" Solid Extrusion, 1' Length. 47065T101-47065T411. • 16 - Aluminum Inch T-Slotted Framing System 90 Degree Bracket, Single, 2-Hole, for 1" Extrusion. 47065T236. • 32× - Compact Head End-Feed Fastener, for 1" & 2" W Aluminum Inch T-Slotted Framing System. 47065T142.
Sensory testing equipment	• Touch Test (von Frey). 95060-230. Stoelting.
Needle	• 25G needle. VWR.
Noise generation	• Piezo Buzzer. TDK PS1240. Digi-Key.
Hot stimulus	• Hot plate that can reach at least 70 °C.
Cold stimulus	• Cooling device that can reach at least 0 °C.
Valve for sucrose	• Tube Normally Closed Pinch Valve. 161P011. NResearch.
Lick spout	• Reusable Small Animal Feeding Needles: Straight. 7922. Cadence Science.
Aversive odor	• Isopentylamine. SKU #126810, CAS #107-85-7. Sigma-Aldrich. • Tissue paper. #05511. Kimtech.
Odor and air delivery	• Blood serum tube. 02685A. Fisher Scientific. • Valve. MB202-VB30-L203. Gems Sensors and Controls. • Medical-grade compressed air. UN1002.
Open field (circular)	• Diameter: 24 in, Thickness: 1/2 in High density VHMW M/M White Opaque. TAP Plastics. • Custom order white Polystyrene sheet. 030" thick, 16" × 96". Mr. Plastics (San Leandro, CA).
Open field (square)	• HDPE (High Density Polyethylene) Sheet, Opaque White, 1/4" Thickness, 24" Width, 24" Length. • HDPE (High Density Polyethylene) Sheet, Opaque Off-White, 0.250" Thickness, 12" Width, 24" Length.

(continued)

Table 1
(continued)

Items	Product number, vendor, and comments
Electronics	• 3× Arduino Uno. Adafruit. • 1× Arduino Mega. Adafruit. • 3 × 4 Phone-style Matrix Keypad. 1824. Adafruit. • LCD Shield Kit w/ 16 × 2 Character Display. 772. Adafruit. • Electronics Semiconductor Kit. Joe Knows. • Electronics 1/4 W Resistor Starter Kit. Joe Knows. • Breadboard 830 Point Solderless Prototype PCB Board. • Multiwire cable. 2714/5. Daburn Electronics & Cable. Ask if they have left over from normal spool to save money. • Alternative multiwire cable: NMUF5/36-2550SJ. Cooner Wire Company.
Counter-balance arm	• Multi-Axis Lever Arms. SMCLA. Instech labs.
Air Purifier	• HEPA Filter Air Purifier. 895916000851. Germ Gaudian.
Light meter	• Digital Lux Meter. LX1010B. Dr. Meter.
Logic analyzer.	• Logic 8. Saleae. Ask for the academic discount.
Stage micrometer	• Micrometer. 94 W 9910. WARD's Natural Science.
Power meter	• PM100D and S120C. Thorlabs
Imaging and behavior analysis software and hardware	
Imaging analysis	• CIAtah or *calciumImagingAnalysis* (CIAPKG), a calcium imaging analysis software suite. Biafra Ahanonu. • Code and instructions for use can be found at https://github.com/bahanonu/calciumImagingAnalysis.
MATLAB	MATLAB 2018a or above. MathWorks.
ImageJ or Fiji	Download at https://imagej.net/Fiji/Downloads.
Python	2.7 or 3.7
Logic analyzer software	Saleae Logic 1.2.xx data collection software.
Shock software	Freeze Frame, Actimetrics
Analysis workstation	i7 or better CPU with ≥64 GB of RAM, ≥128 GB SSD, and ≥1 TB hard drive.
10 Gb networking	http://www.amazon.com/Intel-E10G41BFLR-Ethernet-Adapter-X520-LR1/dp/B002IYDGMU/ http://www.newegg.com/Product/Product.aspx?Item=N82E16833106066 http://www.amazon.com/Lynn-Electronics-LCLCDUPSM-3M-Yellow-Single-Mode/dp/B008BQ1L2G/
Local server or network attached storage.	Not required, but makes large-scale analysis easier.

3 Methods

This method for imaging and analyzing neural responses to noxious stimuli (Fig. 1) is based on protocols and techniques developed by the Schnitzer and Scherrer groups [18]. The goal of these methods is to enable identification of neurons that encode nociceptive information (Fig. 2) or play a role in nocifensive behaviors. To this end, the procedures here cover the entire protocol from expression of calcium indicators to analysis of neural and behavioral data. In addition, we will highlight the many control experiments and stimuli that allow researchers to gain greater confidence when claiming specificity of neural responses to noxious (painful) stimuli. We will also explain key decision points and reasoning either in the Methods or in the associated Notes.

Prior to starting any calcium imaging analysis experiment, it is prudent to investigate whether prior publications have performed Ca^{2+} imaging in the brain region(s) you wish to investigate. This will be particularly useful for determining which gene delivery strategy, stereotaxic coordinates, and control stimuli to use and delivery.

Throughout the methods, we give coordinates as anterior-posterior (AP), medial-lateral (ML), and dorsal-ventral (DV) from bregma, unless specified otherwise (e.g., we sometimes measure DV from dura). For all coordinates, negative AP, ML, and DV indicates left, caudal, and ventral from bregma. These coordinates can be verified by checking the widely used Paxinos and Franklin reference mouse atlas [19].

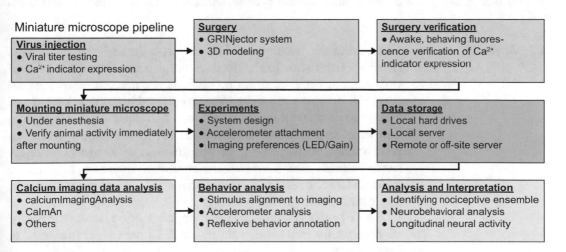

Fig. 1 Overview of calcium imaging pain experiments. Outline of the major steps to conduct pain-related brain calcium imaging experiments. Grey, steps to express virus and verify imaging. Green, pain-related brain imaging steps. Yellow, post-experiment data analysis and interpretation steps

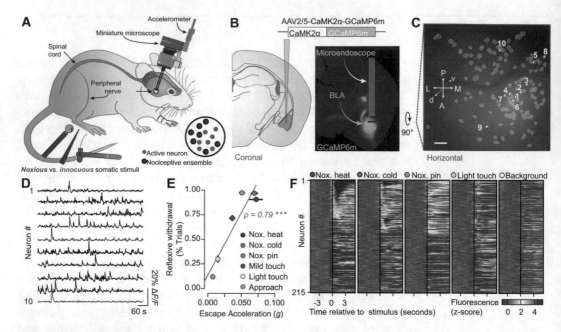

Fig. 2 Overview of calcium imaging and analysis of pain. (**a**) Neural activity is imaged in freely behaving mice with a microendoscope and the virally expressed fluorescent Ca^{2+} indicator, GCaMP6m. Noxious mechanical (pin prick) and thermal (55 °C H_2O and 5 °C H_2O or acetone) stimuli are delivered to the left hind paw, while reflexive and affective-motivational behavior are monitored via a scope-mounted accelerometer. (**b**) Micro-endoscope placement and GCaMP6m expression in the right, contralateral BLA. (**c**, **d**) Map of active BLA neurons ($n = 131$ neurons) with numbers in (**c**) matching PCA-ICA derived neuron activity traces in (**d**). Scale bar, 100 μm. (**e**) Spearman's correlation between reflexive withdrawal and affective-motivational escape acceleration. (**f**) Mean Ca^{2+} response (Z-scored $\Delta F/F$ per trial) across all trials for all BLA neurons imaged during a single session ($n = 215$ neurons). Neurons are aligned from high to low Ca^{2+} responses in the noxious heat trials. Individual neuron identifications between different stimuli are consistent across the trial rows

3.1 Animals

Ensure that you have approval for the procedures described in these methods from your institutional Administrative Panel on Laboratory Animal Care, as these procedures required multiple survival surgeries over several weeks. We recommend housing mice 1–5 per cage and maintaining them on a 12 h light–dark cycle in a temperature-controlled environment with ad libitum access to food and water. To protect the imaging site, we suggest you singly house animals that have had the miniature microscope baseplate mounted and/or are undergoing active Ca^{2+} imaging experiments. For experiments involving imaging at a single brain location, use C57Bl/6J mice (Jackson Laboratory, stock #664, male), or strains of comparable size and weight, at 8–12 weeks at the start of experiments.

3.2 Viral Injection and Expression of Calcium Indicators in Specific Brain Regions

The below are general steps to take to express a Ca^{2+} indicator in rodents (Fig. 2b). Please follow specific protocols as detailed in your lab's Administrative Panel on Laboratory Animal Care (APLAC) or Institutional Animal Care and Use Committees (IACUC) approved protocols.

1. Conduct all surgeries under aseptic conditions with glass bead sterilized surgical tools (Dent-Eq, BS500) after autoclave sterilization (follow institutional guidelines). Prior to surgery, clean and disinfect the surfaces upon which you will perform the surgery or place sterile surgical tools. Wash hands with disinfectant and don sterile gloves.

2. Anaesthetize mice with gaseous isoflurane in O_2 (2–5% induction) until they no longer respond to a squeeze of forelimb or hind limb. Alternatively, use a mixture of ketamine and xylazine per institutional guidelines.

3. In a separate area, place a laboratory absorbent sheet and place the animal then clean the scalp and remove scalp hair with either a depilatory cream (Nair) or an electric trimmer.

4. Transfer the animal to a digital small animal stereotaxic instrument (David Kopf Instruments). Anesthetize the animal with isoflurane (1–2% maintenance, do not go above 2% isoflurane unless it is absolutely necessary as it increases the risk of adverse events).

5. For the entire surgery maintain body temperature using a closed-loop rectal-probe heating pad (FHC, DC Temperature Regulation System). Remember to continually monitor the breathing rate and temperature of the animal while keeping the eyes lubricated by a drop of ophthalmic ointment.

6. Inject nonsteroidal anti-inflammatory drug (NSAID) (e.g., carprofen [5 mg/kg] or meloxicam, subcutaneously) as directed by institutional guidelines.

7. Apply lidocaine (0.5%, subcutaneous injection, max 5.0 mg/kg) to the incision site.

    (a) You can apply a mixture of 2% lidocaine (max 5 mg/kg)—1:100,000 epinephrine (approx. 2.5 mcg/kg) subcutaneously to the incision site, which minimizes bleeding.

8. Clean incision site with 70% ethanol followed by Betadine. Wait a minute then clean with a final 70% ethanol rinse.

9. Make an incision through the scalp at the level of the eyes along the midline for a length of ~1 cm. Retract the skin to expose the surface of the skull and clear the periosteum.

10. Use sterile absorbent eye-spears, or similar sterile absorbent material, to resolve any bleeding at the skull surface. It is also possible to use a blunt tipped needle attached to a vacuum system to dry the skull surface as well.

11. Measure and record the bregma to lambda distance, this can be used to later adjust coordinates for mouse size should you need to improve targeting.

12. Ensure that the brain is level in the dorsoventral land medio-lateral axes. This is a *critical* step as tilt in either axes can lead to mistargeting of the final point of reagent delivery. *See* **Note 1**.

   (a) Note 1: It is preferable for users to use a dye, such as Fast Green or Methylene Blue, initially to determine that injection coordinates hit their region(s) of interest. This can give results in the same day and will help determine whether the chosen leveling procedure and coordinates will allow you to target the correct brain location. You can test multiple coordinates in the same mouse for structures that are not near the midline by injecting the left and right hemispheres then sacrificing the mice per the procedure outlined in Subheading 3.5.

13. Use a burred drill bit to make a ~0.5-mm craniotomy at your preferred AP-MV coordinates. Set the micro drill to the highest speed and make several successive up-down applications of the drill to the skull surface to gradually remove the skull and to ensure a clean hole is formed with minimal bone fragments. Else make a small incision of the dura over the foramen magnum.

14. Zero needle on bregma, move to AP-MV coordinates, measure and record DV dura distance, and lower needle to DV coordinates. Make sure the slowly lower the needle as this will minimize the chance that the brain tissue under the needle is compressed as it is lowered into the brain.

15. Inject 100–500 nL (recommend 250 nL/min, this will depend on volume and specific compound being injected) of material into the brain. After 5 min from start of injection, raise the needle 100 μm (modify as needed for brain region and anatomy) for an additional 5–10 min to allow the virus to diffuse at the injection site, and then slowly withdraw the needle over an additional minute, *see* **Note 2** [19].

   (a) Material refers to experimental reagents, for example adeno-associated virus, lentivirus, retrovirus, rabies/vsv-g-lenti, or of a fluorescent dye (e.g., Texas red dextran; Calcium Green).

   (b) We injected using a beveled 33G needle (WPI, NF33FBV-2), facing medially, attached to a 10-μL microsyringe (Nanofil, WPI) using a microsyringe pump (UMP3, WPI) and its controller (Micro4, WPI).

   (c) For Ca^{2+} imaging using GCaMP6m (*18*) in BLA *Camk2a* + principal neurons, we intracranially injected 500 nL of AAV2/5-*Camk2a*-GCaMP6m-WPRE. Care must be taken to ensure that the brain is level in both the dorsoventral and mediolateral axes.

(d) Note 2: Raising the needle 100 μm will create a small pocket where the injected material can accumulate and minimize backflow. It is very important that you slowly remove the needle from the brain to minimize backflow of reagent along the injection path. This can include moving the needle half-way out of the brain and waiting another several minutes before moving it again slowly out of the brain.

16. If you have no additional viral injections planned, rinse the needle used for virus injection in 70% ethanol. Soak in 10% bleach any instruments and material that contacted the virus or the infected tissue for virus inactivation prior to disposal or usual cleaning procedures.

17. Suture the skin overlying the injection site (suture size 5-0, single interrupted pattern) or glue with Vetbond as per institutional guidelines. If needed, also suture any neck muscles cut during the procedure (sterile absorbable 5-0 suture, single interrupted pattern).

18. Administer long-lasting analgesic (e.g., sustained-release buprenorphine [1.5 mg/kg, subcutaneous injection]) postoperatively, following your institutional guidelines. *See* **Note 3**.

(a) Note 3: If your experiments involve investigations of the endogenous opioid system or related signaling mechanisms, alternative use of a nonopioid perioperative analgesic should be considered (e.g., carprofen or meloxicam) as the opioid may alter endogenous signaling or expression. This can be critical for studies that investigate neural activity before and after chronic pain develops.

19. After surgery, let animals recover from anesthesia on a heating pad to maintain body temperature. Provide food pellets in the recovery chamber and inject mice with warm saline (1 mL, intraperitoneal) to improve recovery.

3.3 GRINjector: Fabrication of Lens Probe Injection Device

The GRINjector (Fig. 3d–f) is a device for injecting microendoscopes (microendoscopic gradient refractive index (GRIN) lens probe) into animals that can be used in Subheading 3.4. The advantages are: (1) the injection needle displaces brain tissue, improving targeting and minimizing tissue compression; (2) the GRIN is held in place during retraction of the placement needle from the GRIN and the GRINjector device; and (3) the guide needle can be easily swapped for other guide needles depending on the diameter and length of the GRIN lens probe. We have created a GitHub repository containing computer-assisted design (CAD) files to assist in 3D print necessary parts along with instructions, located at https://github.com/bahanonu/GRINjector.

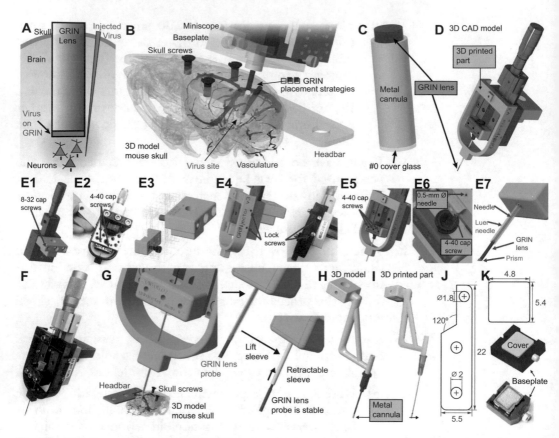

Fig. 3 Implantation and verification of GRIN lens in brain tissue. (**a**) Diagram of GRIN lens probe placement and virus expression. (**b**) 3D CAD model of surgery and miniature microscope mounting. Different colored cylinders indicate several potential GRIN lens probe placements to avoid vasculature while still reaching a deep brain region. (**c**) GRIN lens probe inside a stainless-steel cannula with #0 cover glass glued to the bottom. (**d**) A custom GRINjector (green, 3D printed parts) was designed in PTC Creo 3.0 around a base (gray) consisting of a Newport Vernier micrometer (SM-13) and bearing stage (9066-COM). A thin metal rod (red) is guided through a blunt needle (yellow) where the 0.6 mm GRIN lens (blue) is held via friction. (**e**) Step-by-step (1–7) instructions for assembling the GRINjector device. (**f**) Image of functional GRINjector with 3D printed parts (black) attached to a Kopf Instruments stereotaxic holder arm. (**g**) Same as in (**d**). After placement of the GRIN in the correct brain location (here the locus coeruleus), the blunt needle (yellow) is retracted and the fixed in place thin metal rod (red) helps ensure the GRIN does not move from implant location. (**h**) 3D CAD model of device that holds metal cannula [as in (**c**)]. (**i**) Image of actual 3D printed part. (**j**) Technical drawing (units, mm) of headbar [as in (**b**, **g**)]. Units in mm. (**k**) Technical drawing (units, mm) of laser-cut baseplate cover along with picture of cover in a miniscope baseplate. Units in mm

We provide STEP and STL files for the main parts to allow creation on commonly used 3D printers (Straysys, uPrint, etc.). We recommend printing with a high-resolution (sub-30-micron layer thickness) printer to ensure optimal fit and one with easy to remove support material. We have used VeroBlackPlus on a Stratasys Objet30 and found that to work quite well. Below are step-by-step by step instructions for assembling the GRINjector after printing and obtaining the necessary parts (*see* Table 1).

1. Connect the SM-13 and 9066-COM to form a functional micrometer stage. Remove the locking plate from the 9066-COM as you do not need it as part of the final assembled GRINjector. For help with this, *see* the user manual for more, for example https://web.archive.org/web/202011160230 59/https://www.newport.com/medias/sys_master/images/images/he5/h5b/9366526427166/906806B-9066-9067-User-Manual.pdf.

2. Using 8–32 cap screws, connect the Stereotaxic holder attachment to the back of the 9066-COM, connecting to the top-most holes near the micrometer (Fig. 3e1).

3. Next, pull down the bearing stage so that you expose the two 4–40 holes in the back then place 4–40 × 5/16″ cap screws in each hole (Fig. 3e2).

4. Tap the "GRINjector main needle bottom" 3D printed part holes with a 4–40 tap then place it onto 9066-COM as in Fig. 3e3 and screw in the 4–40 × 5/16″ cap screws (red in Fig. 3e3).

5. Align the lock screws that comes with 9066-COM to each of the four holes on each side of the "GRINjector bottom connector" (Fig. 3e4).

6. Next, place a blunt or syringe tip Luer needle whose inner diameter matches or is barely larger than the diameter of your GRIN lens probe into the groove of the "GRINjector main needle bottom" piece then place the "GRINjector main needle top" piece on top of the Luer needle (Fig. 3e5). Screw the "GRINjector main needle bottom" piece in place with 4–40 × 5/16″ cap screws then secure the Luer needle with a 4–40 × 1/4″ cap screw. A syringe (beveled) tip Luer needle allows easier parting of the underlying tissue as you lower the GRINjector into the brain or spinal cord tissue. Note, this setup allows fast swapping of a different Luer needle into the GRINjector, thus enabling use for GRIN lens probes of various diameters.

7. Next, push the Luer needle through the bottom hole in the "GRINjector bottom connector" then obtain a long, 0.5-mm diameter needle and thread it through the small opening at the top of "GRINjector bottom connector" piece (Fig. 3e6). The needle will then go through the Luer needle. Lock the needle in place with a 4–40 × 3/16″ cap screw as indicated in Fig. 3e6.

8. Lastly, verify that the chosen Luer needle fits the GRIN lens probe tightly and that you can push the GRIN lens probe out of the Luer needle smoothly by twisting the SM-13 micrometer (Fig. 3e7, g).

A quick guide to using the GRINjector in animals, detailed more in Subheading 3.4:

1. Attach the GRINjector onto a stereotaxic arm then perform your normal GRIN lens probe crainotomy surgery.

2. Lower the GRIN into the brain using the GRINjector (Fig. 3g, left).

3. After placement at desired coordinates in tissue, raise the blunt tip needle by twisting the micrometer until the needle is clear of the GRIN lens probe. The red needle secures the GRIN in place during retraction of the device (Fig. 3g, middle and right).

4. Secure the GRIN in place with dental cement, UV glue, or your adhesive of choice then after it as set, slowly raise the GRINjector using the stereotaxic arm.

5. If your stereotaxic arm allows angles, you can use the GRIN-jector to target the same site at different angles to avoid vasculature or vital brain regions (Fig. 3b).

3.4 Microendoscope Implantation

This procedure aims to implant either a microendoscope (micro-endoscopic gradient refractive index (GRIN) lens probe) into the brain or to implant a cannula (Fig. 3a–i). In the case of the cannula, during later animal experiments you will place a GRIN lens probe inside to image neural activity.

1. Fabricate 1.06-mm-diameter stainless steel cannulas (we custom cut 18G McMaster's 89935K66 to 4.2-mm length pieces at Stanford Varian Physics Machine Shop or ordered 304S/S Hypodermic Tubing 18G custom cut to 4.3-mm length pieces from Ziggy's Tubes and Wires). *See* Fig. 3c and **Note 4**.

   (a) Note 4: We strongly encourage researchers new to this procedure to attempt the metal cannula implants as the first set of procedures should there be cost concerns. This allows the researcher to conduct dozens of surgeries for a very low cost compared to the thousands of dollars when attempting direct GRIN lens probe implants. All the metal cannula implanted animals can be checked with a single GRIN lens and those that do not meet prespecified criteria for activity in an awake animal can be retired early from the experiment, saving time and money.

2. Attached a 2-mm-diameter 0.1-mm-thick Schott Glass (TLC International, custom order) onto one end of the cannula using optical adhesive (Norland Optical Adhesive No. 81, NC9586074).

3. We ground down the excess glass using a polisher (Ultra Tec ULTRAPOL End & Edge Polisher, #6390) and film (Ultra

Tec, M.8228.1), washed the surface with deionized water, and then placed the completed cannula in a sealed scintillation vial until use during implantation surgeries.

4. We performed stereotaxic implantation of a stainless-steel cannula or microendoscope 1–2 weeks after AAV viral injections. *See* **Note 5**.

   (a) Note 5: It is possible to conduct viral injections and implantation procedures in the same surgery, but we advise against this as it leads to a prolonged procedure for the animal, is less efficient due to needing to switch out different equipment. Further, the implantation procedure likely leads to a heightened immune response that may reduce efficiency of viral uptake. Newer approaches allow coating of GRIN lens probes with viral proteins that can allow for single virus plus implant surgery [20].

5. Follow **steps 1–12** in Subheading 3.2 to prepare the animal for surgery and level the skull.

   (a) For implantation surgeries, anaesthetize mice with isoflurane (2–5% induction, 1–2% maintenance, both in oxygen) and maintain their body temperature using a heating pad (FHC, DC Temperature Regulation System).

6. Remove head hair (Nair, Church and Dwight Co. NRSL-22339-05) and open the mouse skin using scissors or a scalpel.

7. Using a 0.5-mm burr drill bit, perform small craniotomies in three locations. For implant locations caudal to bregma, we often use the following coordinates (ML, AP) unless they are near an implant location: (−0.7, 5.2), (2.1, −3.6), and (−3.1, −3.6) mm. *See* **Note 6**.

   (a) Note 6: We often use the drill at maximum speed and do several up down vertical motions to slowly grind away the skull. This minimizes downward force and reduces the amount of skull that will be pushed into the brain. After placing the screws, this leads to two anchor points for the cement on each side of the skull once you implant the GRIN lens probe or metal cannula. Adjust coordinates to achieve the same result should you be implanting in a different area than described here. Alternatively, you can prepare the skull with dentin activator to increase adhesion of cement onto the skull.

8. Screw three stainless steel screws (Component Supply Company, MX-000120-01SF) into the skull until they touch dura. Do not push the screws all the way into the brain.

9. Perform a craniotomy using a drill (Osada Model EXL-M40) and 1.4-mm round burr drill bit (FST, 19007-14). Clean away bone fragments and other detritus from the opening using

sterilized forceps (Fine Science Tools, Dumont #5 Forceps, 11252-20) or curved micro probe (FST, Hook Bonn Micro Probes, 10033-13). Note that some skull bits will have bent themselves into the brain nearly parallel to the surface, make sure all skull bits are removed to reduce chance that the area will become chronically inflamed.

10. Continuously apply mammalian Ringers (Fisher Scientific, 50-980-246) to the surgical area when necessary for the remainder of the craniotomy and implant portions of the surgery. However, after implanting the cannula or microendoscope, it is critical that the skull and surrounding skin is kept dry to improve adhesion of dental cement.

11. Attached a 1.06-mm stainless steel cannula onto a custom designed 3D printed cannula holder (Stratasys Objet30 printer, VeroBlackPlus material, Fig. 3h–i) or place the microendoscope (GRIN) into the GRINjector device (Fig. 3d–g). Both devices are designed to attach into stereotaxic surgery holders (Model 1770, Kopf Instruments). Make sure for the microendoscope that there is enough clearance between the bottom of the probe and the outer metal sheath to be able to lower the entire probe down to the DV coordinates for your brain region.

    (a) CAD design files (in STL and STEP formats) and instructions for 3D printing and assembling the GRINjector can be found at https://github.com/bahanonu/GRINjector.

12. Dip some lens tissue paper in water and set aside to later wipe the lens probe.

13. Zero the bottom-center of the cannula or microendoscope at bregma then move along AP axis until at AP position where implant will take place. This allows the user to be ready for a fast implant after the tissue aspiration and related steps.

14. Prepare two 27-G needles for aspiration by using a digital caliper or ruler to mark the distance you plan to aspirate down to on the needle with a permeant marker. As a secondary check again aspirating past the designated target, bend the needle at the location previously marked. The second needle is a backup in case the first needle becomes clogged during the procedure, which can often occur.

15. To prevent increased intracranial pressure and improve quality of the imaging site, aspirate all overlying tissue down to ~300 μm (0.3 mm) above final implant site using a with a 27-G needle (Sai-Infusion, B27-50-27G or VWR Cat. No. 89134-172). As you aspirate the tissue, conduct continuous circular motions as you slowly move downward to efficiently remove the tissue and minimize clogging and sudden suction of large amounts of tissue. In our recent study

investigating the basolateral amygdala (BLA) [18], we aspirated down to −4.20 mm (BLA mice) or −2.10 mm (dorsomedial striatum [DMS] mice) from bregma. *See* **Note 7**.

(a) Note 7: Determining how deep to aspirate is a function of both anatomical and optical considerations. We advise before deciding on the distance above the injection site to aspirate, that you consult the technical specifications for the microscope system and GRIN lens probe that you plan to use along with a mouse reference atlas. For example, many of the Inscopix nVista miniscopes and GRINTECH lens probes they sell have a working distance of 100–300 µm. Due to this, it is advisable to plan to implant the lens probe greater than 200–300 µm above the desired imaging site. This allows you to account for error in placement and it is in general preferred to be slightly high then implant too deep and risk destroying the desired imaging location.

16. After tissue aspiration, move the cannula or microendoscope to proper AP and ML coordinates then lower until right at dura. Record the DV coordinates then quickly lower down to final DV coordinates. Wait about 10 s then retract the cannula or microendoscope from the craniotomy site, take the wet lens tissue and wipe the bottom of the cannula or microendoscope, and move out of the way of the implant hole. Quickly aspirate away any additional debris or blood that may have been pushed down during the initial implant then rapidly move the cannula or microendoscope back into position and lower down into the brain a final time.

    (a) This step allows you to remove blood and debris (e.g., bits of skull that were not properly removed) from the bottom of the aspirated hole and thus can improve final image quality and reduce inflammation at implant site. It is possible for very deep brain regions, such as the hypothalamus, that you will want to avoid doing this since it is more difficult to see the bottom of the aspiration hole and reimplantation will likely lead to worse results.

    (b) For BLA-implanted mice, we lower the cannula to AP: −1.70 mm, ML: +3.30 mm (right BLA) or −3.30 mm (left BLA), DV: −4.50 mm. For DMS-implanted mice, we lowered the cannula to AP: −0.80 mm, ML: +1.50 mm, DV: −2.35 mm. This placed the cannula ~100–300 µm above the imaging plan based on the specifications of the GRIN lens microendoscope's imaging side working distance.

17. It is critical at this stage to dry the entire exposed skull and surrounding skin. Ensure that there is no blood leaking out of the implant site. If you cannot stop bleeding from implant site,

surround with a small amount of UV curable glue then cure to seal the implant site. Score the surface of the skull with a bone scraper and apply dentin activator per instructions on your kit.

18. Cover the cannula or microendoscope with adhesive cement (C&B, S380 Metabond Quick Adhesive Cement System), and allowed the cement to set for 2–3 min. Monitor for any blood pooling under the cement as this can prevent it from properly adhering to the skull surface.

19. Place custom designed laser cut headbars (Fig. 3j, LaserAlliance, 18–24G thickness stainless steel) over the left posterior skull screw (Fig. 3b) and applied a layer of dental cement (Coltene Whaledent, Hygenic Perm) to affix both the headbar and cannula to the skull. The cement should evenly cover the entire skull surface and a thin layer of skin near the edge of the incision site.

20. Let the cement dry for 7–10 min before covering the cannula with bio tape (NC9033794 Tegaderm Transparent Dressing), fixing the tape to the cement with ultraviolet (UV) glue (Loctite(R) Light-Activated Adhesive #4305), and allowing the animal to recover from anesthesia on a heated pad. Alternatively, if directly implanting a GRIN lens, cover with the cut bottom end of a PCR tube then cover with bio tape, apply UV glue around the edge of the tape, and cure the UV glue.

21. Monitor the animal daily after the surgery for signs of weight loss or other signs of pain and distress as laid out in your institutional protocols and guidelines.

3.5 Verification of GCaMP Expression in Fixed Tissue

Check viral expression several weeks after injection to confirm both targeting. This involves perfusing the animal, sectioning the brain, and staining for your calcium indicator to enhance sensitivity (Fig. 2b). Skip over this section if your research environment already has established protocols for verifying viral protein expression.

1. Assemble and sterilize tools (e.g., autoclave) according to institutional guidelines.

2. Prime the peristaltic pump with $1\times$ PBS and 3–4% paraformaldehyde (PFA). Then run $1\times$ PBS through the entire line until a small amount comes out of a 25G beveled needle attached at the end.

3. Inject the mouse with avertin (IP, ~1 mL). Alternatively, place an isoflurane-soaked paper towel in an empty mouse cage and place the mouse inside. In both cases, wait for the animal to lose withdrawal reflexes with breathing near cessation.

4. Hold mouse in place by taping or clipping hind paws and forepaws onto the injection apparatus. Cut open the abdomen

and the diaphragm, take care not to damage any organs and especially the heart and lungs.

5. Cut along the ribcage until you almost reach the armpit then clip a hemostat onto the ribcage and pull back to expose the heart. The hemostat helps keep the ribcage from moving back and blocking your view of the heart.

6. Make a small incision along the right atrium of the heart, a small amount of dark red blood should come out.

7. Place a 25G needle into the left ventricle and start the pump to begin perfusing animal with PBS. The fluid exiting from the right atrium should become progressively clearer. Perfuse animal with 20–30 mL of PBS then switch the line to PFA and perfuse the same amount. During perfusion with PFA the animal's tail should twitch and may rise into the air, this is often the sign of a good perfusion procedure.

8. After you have finished perfusing, remove the brain from the skull and fix in 4% PFA overnight on a shaker. Depending on preferred procedure for brain slicing, a day or two after PFA fixation switch the brain to 30% sucrose until the brain sinks to the bottom of the tube.

9. After brains have fixed, place them in 2–4% agarose then cut on vibratome or place them in an OCT compound then cut on a microtome per normal lab procedures.

10. Place brains in a 24 well plate and perform immunohistochemistry as described below.

11. Perform the below immunohistochemistry steps to check expression of GCaMP in your brain region of interest. Using an anti-GFP antibody will help boost detection of low fluorescence GCaMP that might be either have low expression or not be in an active state for a variety of reasons yet would still be visible when imaging. This reduces the chance of false negatives, that is, that you conclude targeting or expression is not sufficient when it in fact is.

    (a) Wash brain slices three times for 5 min at 25 °C in PBS and 3% Triton X-100.

    (b) Leave brains in 10% Serum in PBS and 3% Triton X-100 for 60 min. Serum should match species of the secondary antibody.

    (c) Shake overnight at 4 °C in primary antibody in PBS (Invitrogen a-GFP A11122 rabbit at 1:1000 dilution) with 0.3% Triton X-100 and 1% Serum.

    (d) Wash brain slices three times for 5 min at 25 °C PBS + Triton X-100 (0.3%).

(e) Incubate brains for 1–2 h at 25 °C in secondary antibody (e.g., DyLight 549 donkey anti-rabbit at 1:500 dilution) in PBS with 3% Triton X-100 and 1% Serum.

(f) Incubate brains with DAPI (50 ng/mL in 1× PBS) for 10–20 min.

(g) Wash brain three times for 5 min at 25 °C in 1× PBS.

12. Mount brain sections in PBS on a glass slide then allow the sections to dry. Then wash with distilled water (to prevent crystal formation) and allow to dry again for several minutes. Subsequently, put a layer of mounting medium (Fluoromount-G) over the brains. Use the pipette tip to spread the mounting medium evenly over the brains before placing the glass cover-slip on top to reduce the number of bubbles. Store at 4 °C in a dark container until imaging under a confocal microscope.

3.6 Verification of Microendoscope Implantation and GCaMP Expression in Awake, Behaving Mice

Several weeks after implantation, check awake animals for GCaMP fluorescence and Ca^{2+} transient activity on a custom designed apparatus (Fig. 4a, b). Checking in awake animals increases the probability that you will be able to identify animals with appropriate amounts of neural activity, as compared to under anesthesia when many brain regions are silent (Fig. 4c, d).

1. Build the head-fix apparatus, *see* Fig. 4a, uses the parts outlined in Table 1.

2. Head-fix mice by clamping (Siskiyou, CC-1 or CC-2) their headbar and allowed them to run on the running wheel (Fisher Scientific, InnoWheel, Catalog No.14-726-577), which is attached via a custom designed 3D printed part (Stratasys Objet30 printer, VeroBlackPlus material) to a rotary encoder (Signswise 600P/R Incremental Rotary Encoder). *See* **Note 8**.

(a) Note 8: We strongly recommend to avoid using anesthesia either during checking or for any experiments after base-plate mounting as this causes many structures in the brain, such as the amygdala, prefrontal, striatum, etc., to exhibit reduced activity or become silent, which may potentially lead you to classify animals incorrectly as unusable due to lack of neural activity even though their neurons may become active if the animal had been awake. Recent min-iscope imaging from Dr. Fan Wang and colleagues has shown that isoflurane and ketamine activates an endoge-nous analgesia circuit [21]. Thus, anesthetics might repress signaling in your target area, thereby presenting a false-negative wherein an animal may be inappropriately discarded at this step. Further, repeated anesthesia is reported to alter long term neural activity, both passively

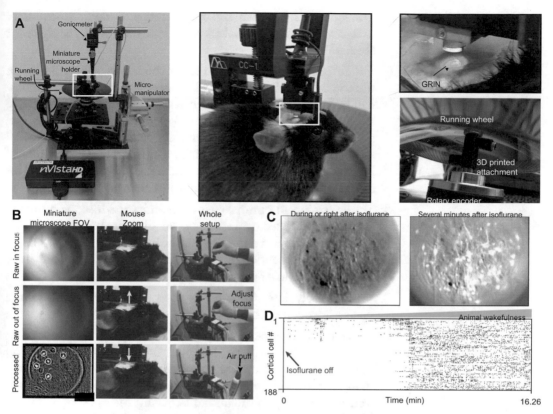

Fig. 4 Verifying indicator expression and experimental arenas. (**a**) Picture of setup to allow awake verification of neural activity in implanted animals. Insets show zoomed in views of the miniature microscope over the GRIN lens probe and an underside view of the running wheel system. (**b**) Example of the miniature microscope field-of-view (FOV) as cells are moved in (top row) and out (middle row) of focus. Bottom row, single frame from an in-focus FOV after calculating $\Delta F/F0$ showing active cells. (**c**) Example average image (500 frames each) showing activity of frontal cortex neurons during or right after isoflurane anesthesia and several minutes after anesthesia when the animal is more awake. Note the marked difference in neural activity. (**d**) Same animal as in (**c**), showing neural activity rasters from Ca^{2+} transients extracted from the entire over 16-min (4878 frames at 5 Hz, downsampled $10\times$ for visualization) movie. As can be seen, there is little activity right after isoflurane is turned off, which mirrors activity during anesthesia. This hinders evaluation of whether an animal has sufficient cells to proceed to imaging

and in behavioral tasks; thus, keeping the number of times an animal has to be anesthetized to a minimum is paramount for obtaining accurate and reproducible data.

3. If you implanted a cannula, use a forceps, or a 27G needle attached to a vacuum line, to lower a custom-designed microendoscope probe (1.0-mm-diameter gradient refractive index (GRIN) lens Grintech GmBH) into the stainless-steel cannula.

4. Attach the miniature microscope onto a holder (Inscopix, Gripper Part, ID: 1050-002199) connected to a goniometer (Thorlabs, GN1). The latter allows you to tilt the miniature microscope in x–z and y–z planes to correct for any

misalignment between the implanted and miniscope GRIN surfaces. Connect the holder to a three-axis micromanipulator (Fig. 4a).

5. Use the micromanipulator to lower the miniscope until you see the top of the GRIN surface. To ensure the entire field-of-view will be in focus, adjust the miniature microscope's tilt relative to the microendoscope. Do this by checking that you are level to the GRIN by moving the miniscope in the x–y plane to all four corners of the GRIN surface, if any corners are out of focus make the proper adjustments using the goniometer. This method is easier than trying to adjust by eye.

6. To determine an optimal part of the microendoscope to image neural activity, make minor position adjustments of the miniature microscope in the x–y plane using the micromanipulator until the microendoscope is in the center of the miniature microscope field of view.

7. Begin slowly lowering the microscope until you are in the microendoscope's focal plane.

8. If your microscope allows, use the imaging software (e.g., Inscopix, nVista 2.0 or 3.0) to display incoming imaging frames in units of relative fluorescence changes ($\Delta F/F$) rather than just the raw image. This allows you to observe Ca^{2+} transient activity in the awake behaving mice, especially weak transients that might be hard to see when looking at the raw video if there is high background fluorescence.

9. Check for time-locked responses to both auditory (e.g., hand clap) and sensory (e.g., tail pinch) stimuli. In the raw video, check for any signs of indicator overexpression (i.e., brightly fluorescent neurons lacking Ca^{2+} transient activity). You should mount mice passing both tests with a miniscope baseplate and used for experiments. *See* **Note 9**.

(a) Note 9: Remember to record several videos at a couple of different focal planes that appear to have a fair number of cells. After checking the animal, analyze the videos with CIAtah to determine which focal plan has the greatest number of cells or those with the highest signal-to-noise ratio, whichever is more important for your experiments. Generally, high SNR cells will be easier to find with automated cell-extraction algorithms (such as PCA-ICA [22]) and you will be more likely to track them across days. Save a single frame from this preferred focal plan as you will use this later as a reference when mounting the miniature microscope.

3.7 Miniature Microscope Baseplate Mounting

1. Conduct miniature microscope mounting on anesthetized (2% isoflurane in oxygen) mice that met the criteria described in Subheading 3.6.

2. If you implanted a cannula, carefully drop a microendoscope into the metal cannula. Lower the miniature microscope until you obtain a similar field of view as seen previously with awake verification then remove the miniature microscope and fix the microendoscope in place with UV curable epoxy (Loctite (R) Light-Activated Adhesive #4305).

3. Next, attach the miniature microscope in a manner similar to **step 4** in Subheading 3.6 but instead do so on a stereotaxic animal setup then stereotaxically lowered the miniature microscope, with the baseplate attached, toward the top of the microendoscope until the brain tissue is in focus. Use the imaging session videos taken previously for that animal to obtain the same focal plane where you saw the maximum number of cells or amount of activity.

4. To ensure that the entire field-of-view is in focus, use the goniometer (Thorlabs, GN1) attached to the holder to adjust the orientation of the miniature microscope until it is parallel to that of the microendoscope using a similar strategy as before, that is, focus the FOV on the top of the GRIN lens probe then move the miniscope to the center each of the four corners of the microendoscope. They should all be in focus, if not, use the goniometer to adjust the miniature microscope until all four corners are in focus.

5. Refocus on the desired focal plane where you have previously determined cells are located. During the rest of the mounting procedure, reduce the LED power so you can just barely see the microendoscope or fluorescence, this will ensure that you do not lose the focal plane due to accidental contact with the miniature microscope, animal, or other parts of the stereotaxic stage that can cause the FOV to shift.

6. To fix the baseplate onto the skull, first build a layer of blue-light curable composite (Pentron, Flow-It N11VI) from the dental cement on the mouse's skull toward, but not touching, the baseplate. Try to get it as close to the baseplate as possible.

7. Follow this by placing a layer of UV-curable epoxy (Loctite (R) Light-Activated Adhesive #4305) to affix the baseplate to the composite. Shine UV light on the epoxy for at least 20 s to allow it to cure. To prevent external light from contaminating the imaging field-of-view, coat the outer layer of the composite and UV glue with black nail polish (OPI, Black Onyx NL T02).

8. Before ceasing the procedure, start recording an imaging video of neural activity then remove the mouse from the stereotaxic setup and place them in their homecage with a heating pad

(FHC, DC Temperature Regulation System) underneath the cage. Continue to record activity until the animal has completely recovered then use the system outlined in Subheading 3.6 to remove the miniature microscope. The video collected can be analyzed in CIAtah, which will allow you to obtain an initial sense of how stable the recording is and how many cells you might expect to obtain during experiments.

9. Attach either a provided or custom-designed baseplate cover (LaserAlliance, 16G thickness mild steel, magnetic to allow to snap into baseplate, *see* Fig. 3k) to the baseplate to protect the microendoscope and return the animal to its home cage.

3.8 Miniature Microscope Ca^{2+} Imaging of Pain

We provide a diagram of the experimental setup (Fig. 5) and general outline of the two main experimental imaging sessions and overall pain imaging timeline (Fig. 6). The experimenter should stay in the testing environment throughout the habituation portion of each session to limit variations related to stress and stress-induced analgesia [23]. The main protocol consists of three or four imaging sessions performed on non-consecutive days (e.g., 7, 5, 3, and 1 day(s) before nerve injury or other chronic pain manipulation) to allow time for animals to recover after each imaging session and to reduce photobleaching resulting from long imaging sessions.

3.8.1 Miniature Microscope Behavior Recording Hardware

This section will describe the setup (Fig. 5) to collect and synchronize data from Ca^{2+} imaging, behavior, stimulus timing, accelerometer, and lick detection circuits for animals run through the procedure in Fig. 6. The miniscope acts as the master controller of event timing as we consider time locking to Ca^{2+} activity the most critical aspect of the experiment. We describe two hardware setups for collecting all relevant behavior videos, stimulus delivery times, and accelerometer data: one relies on a set of Arduino microcontrollers and the other one uses two Saleae Logic 8 (SL8 or logic analyzer) along with helper Arduinos. Both use two cameras to record animal behavior (Fig. 5e). It is critical that you test both behavior cameras before running actual experiments by running them for 20–60 min and testing for whether there are any dropped frames relative to the miniature microscope that they are synced to; if that is the case then lower the resolution being recorded at, close unnecessary programs, or remove as many peripherals from the USB hub to minimize possibility that data rate is being interrupted (this is not an issue for IEEE 1394 [FireWire] interfaces).

1. In the first setup (Fig. 5a), which is slightly more expensive but we recommend due to ease of use and reliability, the first logic analyzer measures analog outputs from the miniature microscope attached accelerometer (100 Hz sample rate, *see* Fig. 8),

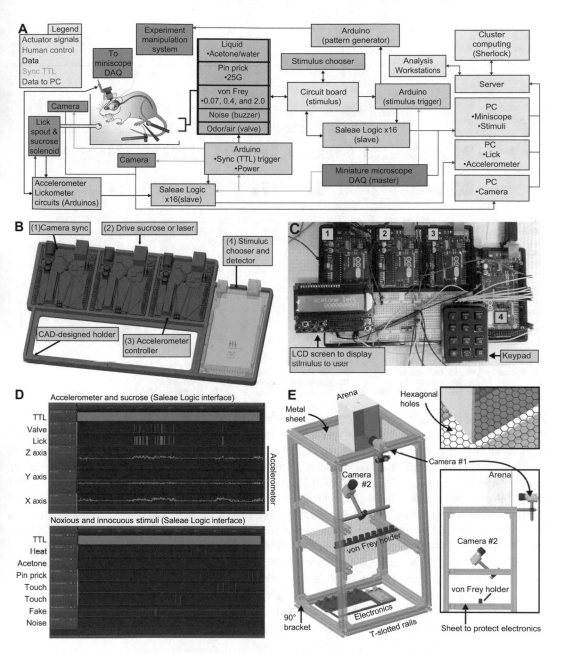

Fig. 5 Experimental setup for synchronous pain imaging and behavior. (**a**) Schematic showing connections between various elements needed to conduct pain imaging experiments with synchronized calcium imaging, accelerometer, lick detection, stimulus application, behavior cameras, and (optionally) experimental manipulation device (e.g., laser for optogenetics). (**b**) 3D CAD model of a holder (red) for several Arduinos (blue and green) used to control various experimental devices [*see* (**a**)]. (**c**) Example image of Arduinos and stimulus chooser, numbers match those in (**b**). (**d**) Interface for digital acquisition device (Saleae Logic) collecting experimental variables indicated. TTL is the synchronization trigger pulse from the miniature microscope. (**e**) 3D CAD model of setup used for delivering noxious stimuli while imaging neural activity with the miniscope and recording behavioral responses. Note the second rack with the von Frey holder, by putting an absorbent sheet here you can also protect sensitive electronics or equipment below

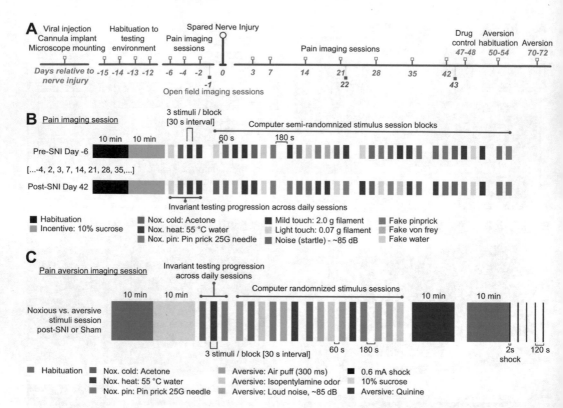

Fig. 6 Outline of pain imaging experiments. (**a**) Experimental timeline and daily imaging protocol for miniature microscope imaged BLA-implanted mice before and after Spared Nerve Injury surgery (SNI). "Drug control" is to check whether neural activity is affected by drug of interest. "Aversion" tested responses of BLA neurons to noxious and aversive stimuli. "Habituation" involved mice being habituated to the fear conditioning chambers used in "Aversion" imaging sessions. (**b**) Each session begins with imaging basal neural activity in the BLA without explicit sensory stimulation (Habituation). Next, mice have free access to a lick port delivering ad libitum 10% sucrose (Incentive), which was removed after 15 min. We then applied somatosensory stimuli to the hind paw. *See* text for details and explanation of experimental design. During "Approach" trials either a von Frey filament, water droplet, pin, or noise device was moved toward the animal similar to other trials but with no actual contact or stimulus delivery. Open field imaging sessions consist of 30 min sessions of animals exploring a 2 m-diameter circle or 2 m² open field apparatus. (**c**) To compare BLA neuron responses to noxious and aversive stimuli, we study the response of animals to noxious (noxious heat, cold, and pin), five commonly used aversive (air puff [to the face], isopentylamine [odor], loud noise [~85 dB], quinine [bitter taste], and 0.6 mA footshock), and a positive valence (10% sucrose) stimuli

connected via a four- (Daburn Electronics & Cable, #2714/4) or a five- (Daburn Electronics & Cable, #2714/5 or Cooner Wire Company NMUF5/36-2550SJ) conductor wire.

2. Simultaneously, in the same logic analyzer, record timestamps for the onsets of licks and control signals for sucrose delivery (triggered on the rise of each signal pulse to obtain exact timing information). To do this, create a separate circuit consisting of two Arduinos (Uno and Mega) and a custom lick detector [24] that measures the mouse's licks (signal #1) and sends a control

signal (signal #2) to turn on a solenoid. The first logic analyzer should record both signals #1 and #2, in both cases use secondary wires to route the signals to the logic analyzer.

3. Drive both behavior cameras with TTLs from the miniscope's DAQ "sync out" line. Use the Image Acquisition Toolbox in MATLAB (MathWorks) to collect TTL triggered video frames from each camera.

    (a) Within the Toolbox's interface, remember to set Frames Per Trigger to 1 and to allow an infinite number of triggers. We recommend saving as a grayscale, uncompressed AVI as this allows easiest use by downstream analysis software. If hard drive space is an issue, make sure that you conduct a test recording and run it through downstream software before the first experiment to ensure you do not have codec compatibility issues. Run each camera on a separate computer if possible, to minimize the chance of dropped frames.

4. Use the second logic analyzer (SL8) to collect stimulus-onset times from a custom-designed circuit that allows you to select a stimulus and press a button to timestamp when you deliver each stimulus, this delivers a signal to the SL8 (Fig. 5d). Use the Saleae software (Logic 1.2.xx) to record and save all data from each SL8. We wrote custom Python and MATLAB scripts to extract the data from each Logic data file for use in subsequent analysis, *see* the below software:

    (a) Use the CIAtah saleae_stimulusFileConvertToMatlab.py function after installing https://github.com/ppannuto/python-saleae.

5. After each imaging session, manually check each session's annotated stimulus-onset times. We use a custom MATLAB program interface to scroll manually through videos recorded from the camera positioned below the mouse. Your program should automatically find all times you pressed a button and move the video to that timestamp. This saves time from scrolling through the entire video and allows you to look in a defined time window around stimulus onset then correct instances in which the annotation did not match the actual onset time of stimulus delivery. *See* **Note 10**.

    (a) Note 10: Determining stimulus onset time is critical for obtaining reliable and consistent data. Variability in this parameter can reduce sensitivity when trying to determine which neurons are part of the nociceptive ensemble. We recommend delivering the full stimulus set to a test mouse then looking through the recorded videos at stimulus onset times and come to a consensus on what you consider a successful stimulus delivery. These same criteria should be used for all subsequent animals. For example, the first frame in which you see an indentation of the plantar surface of the hind paw is annotated as pin prick onset time.

In the second, alternative setup, use a series of Arduinos (three Unos and one Mega) to acquire the data and record the data using COM device recording software on a computer, for example in MATLAB or Python.

1. Make sure that your Arduino code outputs data to a computer with internal session times based off of each Arduino's internal clock as well as a frame counter from the miniature microscope frame number (via the TTL). Use the miniscope frame numbers to do the final synchronization in later analyses with the miniscope Ca^{2+} imaging data. Each data-collecting Arduino should receive a synchronizing TTL signal from the miniature microscope's DAQ.

2. Stream data to a PC and save it using a custom MATLAB script. We provide MATLAB code to read serial data from Arduinos, *see* the "saveArduinoCOMPort.m" function in CIAtah.

3. The miniature microscope DAQ should output a TTL that triggers an interrupt pin on an Arduino Uno, which then produces a 5 V output pulse on a separate port, driving each behavior camera and allowing you to synchronize each camera video frame with the Ca^{2+} imaging data. Alternatively, a BNC splitter can be used to directly send the miniature microscope output TTL to each camera, but this can at times lead to dropped frames.

4. Use a separate Arduino Mega to collect information on stimulus delivery times via a custom circuit that allows you to select the current stimuli, using a keyboard (Adafruit, Product ID #1824) and LCD display (Adafruit, Product ID #772), and to click a button upon stimulus application to record the delivery time for later analysis. Setup the keyboard and LCD according to guides on the Adafruit website. The button allows current to flow from a wire attached to a 3.3 V or 5 V line on one of the Arduinos to an interrupt pin, this allows precise detection of the button press without needing to use while loops in the code. In the code for this Arduino, output the timestamps of each miniscope frame along with timestamps whenever a stimulus is pressed.

5. Use a third Arduino Uno to measure the analog voltage signal from the accelerometer (Sparkfun ADXL335 or ADXL345) attached to the miniscope with a custom 3D printed attachment. Output a timestamp, and associated session frame number, of x, y, and z channel values at 100 Hz.

6. Use the last Arduino to measure the onset of licks and send control signals to open a solenoid (NResearch, 161P011) and release sucrose (10% w/v in water).

7. Conduct stimulus onset time correction identical to the alternative setup.

3.9 Behavioral Apparatus for Noxious Stimuli Delivery

We designed the behavioral apparatus to allow experimenters to easily gain access to the animal's paws while also being able to record neural and behavior activity to aide in downstream analysis. We recommend creating a behavioral apparatus that utilizes T-slotted framing and Thorlabs parts or similar construction materials, as detailed in Fig. 5e. This design and parts have several benefits: they are sturdy and thus less susceptible to movement when the experimenter bumps into the setup, allowing the animal to remain less disturbed; they allow for multiple points at which to anchor experimental devices; they are easy to disassemble and clean; and they allow flexibility to adjust positioning of supplemental devices like cameras and counter-balance arms.

1. Use either steel or aluminum perforated sheets and place on top of the T-slotted setup. We recommend aluminum as it does not rust and thus will be easier to clean over time.

   (a) Do not use chicken wire or similar types of holed metal sheets as they contain numerous pressure points that might be painful to animals in certain experimental conditions; for example, those in which the animal has mechanical hypersensitivity, such as spared nerve injury.

   (b) If you plan to have the electronics (detailed in the next section) or other items that you do not want contaminated by animal waste, make a 2nd metal sheet layer below the bottom behavior camera and place an absorbent sheet on that 2nd layer. This layer can also function as a place to put von Frey filaments and other stimuli for more rapid access.

2. Create plastic containers, where animals will be during experiments, that are either all transparent red or is white on all sides except for the front, which if possible should be red to reduce the ability for the animal to see the experimenter [25] while still allowing behavioral recording of the animal. Drill a small hole onto the left side of the container, this will be used to place the sucrose lick spout through.

3. While the miniscope only weighs a couple of grams, during a long experiment this can place unnecessary strain on the animal. We recommend using a counter-balance system to reduce the weight the animal carries, which will also allow you to get a better measure of nocifensive behaviors. Attach a counter-

balance arm (SMCLA, Instech labs) to the T-slotted system using Thorlabs posts then position the above the animal and container.

(a) Before starting any experiment, loop the miniscope through the arm and test that the counter-balance arm's weight or spring is set to allow some give, but keeps the miniscope suspended, as this will reduce weight on the animals head while still allowing it to move around freely.

4. Use two cameras to record mouse behavior and position one below the mouse, to capture stimulus delivery and reflexive responses, and another facing the test chamber, to capture the mouse's affective-motivational nocifensive behaviors. Follow the design in Fig. 5e.

(a) Take care that you focus the bottom camera on the animal's paw, as this will be crucial during analysis to help determine when stimuli have contacted in animal's paw.

(b) For low-light conditions or if the cameras do not have high sensitivity, purchase an infrared light and shine it on the paw area; this will allow you to better visualize the paw in the videos but will not affect animal behavior as rodents do not see in infrared.

(c) If there is a worry about animal moving out of focus, purchase a lens with (or adjust existing one to have) a high F-stop to allow more parts of the apparatus to remain in focus during experiments.

3.10 Recording Neural Activity During Delivery of Noxious Stimuli

The goal of these experiments is to identify neurons that are responsive to noxious (painful) stimuli. We term these noxious-stimuli responsive neurons the "nociceptive ensemble." A second aim of these experiments is to have a set of control stimuli that reveal what percent of the nociceptive ensemble is unique to pain and which also respond to any (e.g., regardless of valence) salient, aversive, and other stimuli. This battery of tests will allow researchers to better understand what types of stimuli their brain region of interest responds to and whether using computational tools, they are able to disentangle what aspect of neural activity is pain versus nonpain related and how this activity changes in a chronic pain state.

1. After you mount the miniature microscope onto a mouse, check that the animal still has adequate GCaMP expression, either by using CIAtah to analyze the Ca^{2+} imaging movie recorded as the animal woke up from mounting or by running a 10-min open field imaging session. *See* Subheading 3.14 for details on analyzing imaging movies.

2. Habituated each mouse to the testing environment for at least three days prior to imaging (Fig. 6a). Each habituation session should last 20–30 min and take place in the same chamber as the final experiment. Use the light meter to ensure that the lux in the room is similar across days. *See* **Note 11**.

    (a) Note 11: To preclude any emotional contagion between mice [26], we brought only one mouse into the isolated, light-, sound-, and temperature-controlled testing environment. Further, we housed mice individually after mounting to reduce probability that animals would chew on the baseplate, resulting either in a need to remount the animal or removing it from the experiment. Follow institutional guidelines and modify protocols accordingly if this is not possible. Isolating the animals may also be critical during chronic pain experiments, where housing naïve mice with those in chronic pain may lead to alteration of naïve mice's behavior and neural activity.

3. Before starting the pain imaging sessions design a randomized stimulus delivery protocol for each session, subject to the following constraints: give a defined order to light touch, noxious cold, mild touch, and innocuous liquid or noxious heat at the beginning of each session; the same stimuli should not have adjacent stimuli blocks; and "Approach/No contact" stimuli blocks should occur during the first 3 main stimuli superblocks (Fig. 6b). *See* **Note 12**.

    (a) Note 12: Having the first four stimulus blocks always occur in the same order across days allows you to track daily pain behaviors and the development of chronic neuropathic pain. All subsequent stimulus blocks are semirandomized computer-generated sequences within and across days with the condition that the same stimulus block does not occur twice in sequence, nor does the same daily protocol repeat on any given day. This ensures that neural responses are due to stimulus delivery rather than the animal expecting said stimulus or being put in a specific, stimulus-induced state over time. We designed this protocol to be less than 2.5 h for each animal's imaging session; to give enough stimuli to have sufficient statistical power to identify stimulus-responsive neurons; and to incorporate sufficient "down time" between stimuli, in order to avoid potential photobleaching of imaging area or animal exhaustion. During "Approach" trials you will move either a von Frey filament, water droplet, pin, or noise device toward the animal similar to other trials but with no actual contact or stimulus delivery.

4. Turn on your cooling and heating devices and place a beaker of water on each. Allow each to reach 0 °C and 65 °C, respectively, and immerse the 1-mL syringes in each to equilibrate the temperature. This last step ensures that there will be minimal change of the water temperature toward baseline after removing the syringe from the water. If you are using acetone instead of cold water for cold stimuli, instead keep a vial of acetone ready next to the testing setup. Another alternative is to use dry ice [27].

5. At the start of each imaging session, head-fix the mouse (Siskiyou CC-1 clamp), mount the miniature microscope, check for GCaMP fluorescence, align the field of view (FOV) to the previous session FOV (*see* **Note 13**), take a picture of the current day's FOV, and place the mouse within the test chamber.

   (a) Note 13: It is *very important* that you spend the extra minute or two to get the FOV aligned as closely as possible to the previous days FOV, as that will be critical later on when attempting to align cells across imaging sessions. Alternatively, you can always align to the first day's FOV, this will minimize the chance of a slow drift of the FOV if you only align to the previous day's FOV; we prefer this method to ease later computational analysis. Further, we strongly recommend that you determine the size of the frame you will need to include the entire relevant FOV (e.g., 1000 pixels × 950 pixels) and *keep this fixed* for all experiments with that particular animal. This will greatly simplify downstream cross-session analysis.

6. Clean the chamber and metal grating with 70% ethanol then place the mouse in the chamber and run the miniature microscope and accelerometer cable through a counter-balance arm placed above the chamber. This reduces the weight of the miniature microscope on the animal's head and reduces the probability that the cable will drop into range of the animals mouse (if that happens, animals may chew the cables).

7. Start the behavior video cameras and logic analyzers (Fig. 5a) so that they are waiting for a synchronization signal from the miniature microscope. At the start of each session, begin the miniature microscope recording and make sure that all other software and hardware is collecting correctly (Fig. 5b, c). Discard this data then setup all devices for the real experiment.

8. Before sensory stimulation, measure spontaneous neural activity by recording Ca^{2+} activity for 10 min while the mouse habituates to, and freely moves within, the testing box. The mouse should receive no explicit experimenter-delivered sensory stimuli during this period. However, in each session block,

record all stimuli, behavior camera, accelerometer, lick detection, and sucrose delivery signals (Fig. 5d) as this will simplify data analysis later.

9. After baseline recording, give the mouse 15 min ad libitum access to an incentive (sucrose) to capture neural responses to positive valence stimuli. To induce mice to lick without needing prior water deprivation, use a 10% sucrose solution. *See* **Note 14**.

   (a) You can detect licks and delivered sucrose using a custom-built circuit based on previous designs [24] and Fig. 5a. You will need to build a custom electronic circuit (built using Arduino elements) to collect lick data and synchronize all incoming data using output TTL pulses (5 V signals) from the miniature microscope DAQ (*see* Fig. 5a). Control signals from this circuit will drive a solenoid (NResearch, 161P011) that delivers 10% sucrose instantly after the 1st lick in a bout. Program a 5-s cooldown period between liquid deliveries. Thus, even if the mouse licks the port continuously, this approach provides enough time between incentive deliveries to relate the evoked neural activity with specific sucrose delivery time points.

   (b) Note 14: It is critical that the mouse should not be water- or food-restricted during any part of this protocol unless that is an intended experimental manipulation. In part, we want to minimize long-term water restriction as this protocol can take up to 2–3 months when including the chronic pain and negative valence studies. Further, as it is known that thirst and stress alter brain-wide neural dynamics [28, 29] and pain processing, the goal of this experiment is to minimize animal stress as much as possible to reduce confounds in downstream analysis and interpretation of the data.

10. Next, begin the main sensory testing protocol. Here you will deliver a battery of stimuli to the mouse's hind paw: 0.07-g and 1.4- or 2.0-g von Frey hairs (light and mild touch); 25-G needle (noxious pin prick); water drops at ~5 °C or acetone (noxious cold), 30 °C (innocuous liquid), and ~55 °C (noxious heat) delivered via applying a small drop from a 1-mL syringe; fake-out stimuli where no contact is made with the mouse after a stimulus device is placed near their hind paw ("Approach/No contact"); and noise (startle response control, delivered near the face). Deliver all stimuli at least 15 times per session, except "Approach/No contact" and noise, which you can deliver 9 times each. *See* Fig. 6 for detailed timing information. *See* **Note 15**.

(a) Note 15: The number of stimuli to give for each stimulus should be determined based on the reliability of the neural activity seen in pilot studies. We used neural power analysis (run in a similar manner to common behavior power analyses) to determine how many stimuli we would need in order to observe significant, stimulus-responsive neurons at a specific power (e.g., 0.8). We give less of the control stimuli to reduce the length of the imaging session and in cases, like loud noise, where the neural activity is reliable.

(b) For von Frey and pin prick stimulation, attempt to hit the same lateral portion of the animal's hind paw to reduce variability and have a location that is still sensitive to mechanical stimulation after spared nerve injury (depending on which branches of the sciatic nerve are cut).

(c) For water or acetone delivery, we normally applied a 50–100 μL droplet onto the hind paw. Another option is to spray 100 μL onto the hind paw; however, this can lead to simultaneous mechanical stimulation.

(d) Include "Approach/No contact" trials to detect possible neural responses related to expectation of stimulus delivery and error-prediction. To conduct these "Approach/No contact" imaging trials, bring either an 0.07-g von Frey hair, a 25G needle, 1-mL syringe, or an 85-dB noise delivery device toward the animal but neither making contact nor turning on the noise. Randomly intersperse "Approach/No contact" trials between other stimuli blocks to ensure you capture any changes in anticipatory behavior that occurs as the session progresses.

(e) To control for the possibility that the neural responses to hind paw stimuli are startle-induced, use a loud tone (~80–85 dB) as an aversive, but nonnociceptive, sensory stimuli. Deliver the tone (centered around 4 kHz in our experiments but can vary depending on the needs of your experiment and equipment availability) for 300 ms by triggering an Arduino, loaded with custom code, to drive a TDK PS1240 Piezo Buzzer. *See* many online tutorials for producing sound using Arduinos and Piezo buzzers.

11. While imaging neural activity, also measure withdrawal reflexes and affective-motivational behaviors (attending and escape) using high-speed cameras (AVT Guppy Pro F-125 1/3″ CCD Monochrome Camera #68-567 or The Imaging Source DMK 23FM021) and accelerometers (Sparkfun ADXL335 or ADXL345, with data collected using an Arduino Uno or Saleae Logic 8). Wire them together as in Fig. 5a.

(a) The accelerometer can be assembled using guides available on commercial websites for the products in Subheading 2. Connect the x, y, and z accelerometer channels to individual channels on the logic analyzer and in the acquisition software, set to acquire at 100 S/s. This ensures that you oversample behavior relative to miniscope data acquisition rate, which allows for more precise timing of behavior changes relative to neural activity.

12. We recommend between each superblock that researchers save data from the logic analyzers, miniature microscope, and behavior cameras. This prevents loss of a large amount of data in the case of electronic failure or accidents during the experiments. You can do this during the 3 min break periods between each super block (Fig. 6b).

13. At the end of each imaging session, remove the miniature microscope and return the animal to its home cage.

3.11 Miniature Microscope Recording Parameters

We recommend recording all miniscope videos at a frame rate of 20 or 30 Hz, especially if using GCaMP6m or a Ca^{2+} indicator with similar rise and decay kinetics. We typically use between 213 ± 3 and 390 ± 7 µW LED light intensity (measured from miniature microscope GRIN with a Thorlabs PM100D and S120C). Each frame from the Inscopix miniature microscope CMOS camera is 12-bit of varying video dimensions—we typically ran analysis on videos of size 250–275 × 250–270 pixels after downsampling in each spatial dimension by a factor of 4 from the raw data. Use a stage micrometer (Ward's Natural Science, 94W 9910) to empirically calculate the real-world dimensions of each video pixel by placing the miniscope on the holder from Subheading 3.4 and lower down until the micrometer is in focus then take a picture. In an image editing software, open the resulting image and measure the number of pixels between markers of known width, for example 100 µm. This will then allow you to calculate the pixels per micron for each miniature microscope used. In the case of the Inscopix nVista 2.0 miniature microscopes, this was typically 2.51 µm × 2.51 µm per pixel.

3.12 Noxious and Aversive Stimuli Experiments

These experiments are similar to those in Subheading 3.8 and use the protocol outlined in Fig. 6c. The goal of these experiments is to determine how specific the nociceptive ensemble activity is to noxious (painful) stimuli as compared to other aversive stimuli.

1. Habituated mice to a fear conditioning test chamber, similar to prior setups [30], for 30 min on four consecutive days prior to conducting experiments.

2. Prepare the animals for imaging as in Subheading 3.8. Use 70% ethanol to clean all chambers and surfaces that animals will contact before each experimental imaging session.

3. After you mount the miniature microscope, habituate mice to the test chamber for 10 min.

4. After habituation, give mice 10 min of ad libitum access to 10% sucrose, record accelerometer, miniature microscope, stimulus delivery time, and behavior camera data during this time to facilitate easier downstream analysis.

5. Follow the protocol outlined in Fig. 6c and used the same data collection hardware as in Subheading 3.8.1. You will deliver a range of noxious, aversive, and appetitive stimuli to animals: noxious cold (acetone), noxious heat (~55 °C water), noxious pin (25G needle), air puff (300 ms), isopentylamine (~85 mM in H_2O, delivered via 300-ms air puff), loud noise (~85 dB for 300 ms, same as previously described), electric footshock (0.6 mA for 2 s), quinine (0.06 mM), and 10% sucrose.

(a) Deliver quinine last in the main stimulus chamber (before moving to the fear conditioning chamber) to reduce confounds by exposing the animal to a bitter, aversive stimuli before testing noxious and aversive stimuli. Give mice quinine (0.06 mM in deionized water [31]) after they lick a metal tube in an identical manner as 10% sucrose but through a different tube to avoid contamination by sucrose.

(b) Because the main behavior chamber and the fear conditioning chambers were in separate rooms, we allotted time for the mouse to rehabituate to the fear conditioning chamber for 10 min after the ad libitum quinine access.

(c) Deliver noxious cold (acetone), noxious heat (~55 °C water), noxious pin (25G needle), and loud noise (~85 dB) as described in Subheading 3.8.

(d) Isopentylamine (Sigma-Aldrich SKU #126810, CAS #107–85-7) has an odor shown to be aversive in multiple previous studies [32–34]. Other odors are possible to use, just avoid using fox urine or any odors that might cause long-term changes in the room's odor and thus the animal's level of stress.

(e) To conduct isopentylamine and air puff experiments, place 50 μL of isopentylamine onto a small piece of tissue paper (Kimtech, #05511) and place it immediately into a 10-mL blood serum tube (Fisher # 02685A) and re-cap. Then insert two 16G needles through the tube cap and attach one of them using plastic tubing to a valve (Gems Sensors and Controls, MB202-VB30-L203) controlling air delivery. Deliver air by attaching the other needle to flexible tubing that leads to a metal tube or 16G blunt needle, which you will use to manually direct odorant to animals in the test chamber. Deliver air puffs through a blunt, 16G needle, but bypass the odorant-filled serum tube; the best way to do this is to use a three-way stopcock to redirect airflow to a separate, empty tube with a needle at the end. We recommend delivering both isopentylamine and air puff for 300 ms with medical-grade compressed air (UN1002) at between 20 and 30 PSI. Aim the isopentylamine and air puff stimuli during delivery at the nose and front half of the animal's face, respectively.

(f) For footshock trials, habituate mice for 10 min then deliver five 0.6-mA electric footshocks with 2 min between each stimulation (Fig. 8c, d). To synchronize the onset time of each footshock with Ca^{2+} imaging data and each behavior cameras' videos, collect TTLs output by the miniature microscope DAQ and footshock software (Freeze Frame, Actimetrics) on a logic analyzer (Saleae Logic 8), which allows you to determine the specific image frames of the Ca^{2+} video that were synchronous with each footshock.

6. Collect all data in this procedure, process the Ca^{2+} videos, and perform analyses as in the main protocol in Subheading 3.8.

3.13 Analysis of Locomotor Behavior in the Open-Field Assay

The open-field procedure servers two purposes: as a quantitative measure to ensure that the animals exhibit normal locomotor behavior after each surgery and as a control to determine how correlated with general locomotion a given brain area is, as this can affect interpretation of stimulus-evoked activity in Subheading 3.16.

1. Create an open field chamber that is either circular (60.96-cm D, 38.1-cm H, opaque white polyethylene walls and floor) or square (24″ × 24″ × 12″, white opaque HDPE) using tape.

2. Habituate animals in the testing room for 5 min before starting the experiment. We recommend only having one mouse present in the room during all imaging sessions.

3. Conduct open field experiments either at 20–25 or 102 lux (measured with light meter) or at the same lux as the experiments in Subheading 3.8 using a diffuse overhead fluorescent light to illuminate the arena. Ambient room temperature should be ~26 °C. Place a camera above the chamber, load the Image Acquisition Toolbox in MATLAB, and position the chamber in the center of the camera frame. Crop any regions outside the chamber from the camera frame to save storage space. Using a BNC cable to connect the miniature microscope DAQ "sync out" TTL port directly to the camera (e.g., via a HIROSE cable if using The Imaging Source cameras).

4. Mount the miniature microscope on the animal as in Subheading 3.8. Start the video camera so that it is waiting on a trigger from the miniature microscope.

5. Wipe all surfaces down with 70% ethanol before each session then place individual mice in a black plastic container. Place this container in the center of the arena. This allows you to release the animal into the arena without having to touch them and gives you time to setup the imaging parameters while the animal calms down after mounting.

6. Release the animal facing the peripheral zone (e.g., the chamber wall) and immediately start the miniature microscope recording. Record for 15–30 min depending on your experimental needs. At the end of the session, remove the miniscope from the animal and return to the home cage.

7. Analysis is done as in [18, 35, 36]; the most important analysis is to compare the average population firing rate vs. specific velocity bins and look at the average population firing rate at the onset and offset of locomotion. *See* **Note 16**, the code and instructions for tracking animal position at https://bahanonu.github.io/calciumImagingAnalysis/help_animal_tracking.

   (a) Note 16: Some brain regions are known to be highly correlation with locomotion, such as the striatum [35]. For these types of regions, additional care must be taken when interpreting stimulus-evoked activity as it is likely to be nonspecific and occur in response to any salient stimuli. Use control and aversive stimuli described in the main- and aversion-pain experimental protocols to help determine whether most stimulus-responsive neurons are in fact just locomotion related. Further, using the accelerometer, you can analyze portions of the pain imaging experimental sessions when animals exhibit movements that are of equal magnitude as those seen in response to experimenter-delivered stimuli and look at whether the activity of the neurons is on average above baseline during those periods. There are several ways to get around the issue of locomotion-induced activity masking the response to noxious stimuli. One is to perform the experiments under light anesthesia, which is not ideal as detailed in Subheading 3.6. Another possibility is to body restrain the animals, but this can cause stress and alter neural responses to noxious stimuli.

3.14 Preprocessing of Ca²⁺ Imaging Data

In this step we detail analysis of the Ca^{2+} imaging data collected in prior experiments. You can perform all the analysis steps in our open source *CIAtah* (pronounced cheetah, also known as calciumImagingAnalysis [*CIAPKG*]) software package. The code is located at https://github.com/bahanonu/calciumImagingAnalysis while and instructions for use can be found at https://bahanonu.github.io/calciumImagingAnalysis. We refer to CIAtah modules that correspond to each step using italics. We processed all Ca^{2+} imaging data in the MATLAB software environment using methods similar to previous studies [13, 18, 30, 35]. You can perform many of the steps described below (semi)-automatically using CIAtah; we describe the algorithms and relevant details here in case the reader would like to replicate them using their own code or software environment.

1. To reduce computational processing times and boost signal-to-noise, downsample imaging movies collected from the miniature microscope in both x and y lateral spatial dimensions using 4×4 (or 2×2 if you prefer to retain spatial details) bilinear interpolation.

   (a) Within CIAtah, run the *modelDownsampleRawMovies* module or perform spatial downsampling within the *modelPreprocessMovie* module.

2. To remove motion artifacts, register all frames in each imaging session to a chosen reference frame using *Turboreg* [37]. We often register to the 100th frame to avoid any LED warm-up or other artifacts that can occur in earlier frames in an imaging session. Select and register a sub-region of the field-of-view rather than use the entire frame for registration. This allows you to choose a region with stable, high-contrast features (e.g., blood vessels) and without artifacts (e.g., dust particles on the optics) that could impede registration. *See* **Note 17**.

   (a) Note 17: To improve the performance of motion correction, we make a duplicate movie to obtain translation coordinates. First, normalize the movie by subtracting the mean value from each frame. Then, spatially bandpass-filtered each frame of the movie; we use a cutoff frequency of ~0.10 to 0.16 cycle/μm using a Gaussian cutoff filter, which highlighted spatial features at the ~6 to 10 μm scale. Next, perform an image complement operation on each frame by subtracting each pixel value from the maximum pixel value in that frame (i.e., dark areas became light, and vice versa); this inverts the image and generally makes blood vessels and other dark static features appear more prominently, which benefits image registration. Then obtain the two-dimensional spatial translation, skew, and rotation coordinates from *Turboreg* by having the algorithm compare each processed frame to a reference frame (e.g., the 100th movie frame). Discard the duplicate movie used in this step to save memory.

3. To facilitate cell extraction by enhancing cell signals and diminishing neuropil and other background fluctuations, divide each frame of the raw Ca^{2+} movie by a low-frequency bandpass-filtered version of itself (cutoff frequency: ~0.0014 to 0.0063 or ~0.0014 to 0.01 cycle/μm using a Gaussian cutoff filter).

(a) We recommend users test their movies with the CIAtah *viewMovieRegistrationTest* module to determine what filtering parameters best reduce background fluorescence while maximizing cellular signals.

(b) It is important to conduct spatial filtering *before* transforming each image. The reason for this is that transformation often leads to 0 s or NaNs near the edges of the movie and this will often lead to errors or sub-optimal results when trying to apply a spatial filter (e.g., after a fast Fourier transform) to the movie.

4. Conduct an affine or projective transformation on the resulting image frames using the two-dimensional spatial translation, skew, and rotation coordinates obtained in **step 2**, but on the Ca^{2+} movie preprocessed in the manner described in **step 3**.

(a) In certain cases, if users perform multiple rounds of motion correction to improve results, the absolute transformation from initial to final image can be obtained by using the translation (T), skew (S), and rotation (R) transformation matrices to compute $D = \Pi_1^N T_i * S_i * R_i$, where i is the iteration, N is the number of registration rounds, and D is the absolute transformation matrix. This can also be used when transforming cell ROIs during cross-session alignment.

5. Since motion correction can cause the movie edges to take on inconsistent borders filled with 0 s or NaNs due to variable translation distances, you should determine the maximum amount all frames were translated during the motion correction procedure in each dimension (t_{max}) and then add a border of size t_{max} pixels extending from the edge of each frame toward the middle of the frame. Set a maximum border size (t_{max}) to ensure a single frame does not cause most of the movie to become a border, for example 14 pixels (~35 μm with Inscopix nVista 2.0).

6. Covert each movie frame to relative changes in fluorescence using the following formula: $\frac{\Delta F(t)}{F_0} = \frac{F(t) - F_0}{F_0}$, where $F(t)$ is frame at time t and F_0 is the mean image over the entire movie.

7. Next, temporally smooth each movie by downsampling from the original 20 or 30 Hz to 5 Hz. Specifically, for a $x \times y \times t$ movie, bilinearly downsample in $x \times t$ to reduce computational processing times, which is equivalent to performing a 1D linear interpolation in time of the intensity values at each pixel. Skip this step if you are in a brain region or performing an experiment in which higher temporal resolution is of paramount importance.

8. As an optional step, you can manually crop out regions of the movie that correspond to large artifacts (e.g., dust particles) or

regions outside the microendoscope. This will improve cell extraction and save manual cell sorting time by reducing sources of high variance noise within the movie.

9. You can perform these steps within the CIAtah *modelPreprocessMovie* and *modelModifyMovies* modules.

3.15 Extraction of Neuron Shapes, Locations, and Activity Traces

After processing each session's Ca^{2+} imaging videos, you want to computationally extract individual neurons and their activity traces. There are several options for extracting cell images and fluorescence activity traces from Ca^{2+} imaging movies. There is the widely used principle component analysis followed by independent component analysis (PCA-ICA) algorithm [22]. In addition there are several constrained nonnegative matrix factorization methods (e.g., CNMF and CNMF-E), which are part of the CalmAn software package [38]. You can use these and other cell-extraction algorithms using the CIAtah *modelExtractSignalsFromMovie* module, *see* **Note 18**. We will focus here on PCA-ICA, as it is a fast and well-established method for analyzing miniscope data.

1. Note 18: We recommend users process their data with multiple cell-extraction algorithms to ensure that their results are independent of whichever algorithm they used. Our CIAtah software package provides seamless support for many commonly used cell-extraction algorithms along with several new techniques (e.g., CELLMax and EXTRACT) that we and others will be releasing in the future. Further, CIAtah supports the Neurodata Without Borders: Neurophysiology (NWB:N) data standard for optical neurophysiology, which facilitates the use of your cell extraction data cross software pipelines and languages.

2. Run PCA-ICA cell extraction using the following parameters: $\mu = 0.1$, a maximum of 750 iterations, and request ~1.2 to 1.5× the estimated true number of active cells in the field of view.

   (a) The parameter μ is the relative weight of temporal information in ICA, and $\mu = 0.1$ indicates we performed a spatiotemporal ICA with greater weight given to the spatial than to the temporal skewness.

   (b) If running PCA-ICA on an animal for the first time, we recommend running with 1.5–2.0× the number of estimated cells. This will increase the likelihood that you detect most of the cells. Since our implementation of PCA-ICA orders outputs roughly in order of their signal-to-noise ratio (SNR), you can then plot various cell shape (e.g., size) and activity trace (e.g., SNR) features and determine by eye, or using a threshold, where these parameters start to change. This change point is likely around the true number of cells in the movie. Rerun analysis for that movie with the new estimate and run all subsequent analysis for that particular animal with the new estimate, unless you have reason to believe based on observations in the movie that the number of active cells has gone significantly up or down.

3. PCA-ICA outputs a series of candidate spatial filters ($x \times y \times n$) and temporal traces ($n \times t$)—where n is the number of neurons, t is the frame, and (x, y) are spatial dimensions—associated with temporally varying sources, which you then need to manually verify as neurons.

4. An optional step is to use the CIAtah *viewCellExtractionOn-Movie* to view the cell images overlaid on the imaging movie. This allows you to visualize whether the cell-extraction algorithms are missing any cells and adjust the number of requested cells or cell elimination parameters accordingly.

3.15.1 Manual Neuron Identification

After performing cell extraction, a human scorer needs to manually check algorithm outputs for accuracy and removal of false positives. You can perform these steps in CIAtah using the *computeManualSortSignals* module GUI.

For each imaging session, load the CIAtah GUI that displays the spatial filter and activity trace of each candidate cell, along with the candidate cell's average Ca^{2+} transient waveform (Fig. 7a). The GUI also shows a maximum projection image of all output spatial filters (Fig. 2c), on which the GUI highlights the currently selected candidate cell. Example PCA-ICA neuron extraction outputs from a single mouse (same as in Fig. 2c, d) showing accepted and rejected ICA outputs (Fig. 7a). *See* **Note 19**.

1. Note 19: We manually classify all neurons from nonneuron cell-extraction candidates obtained from imaging data based on a variety of parameters, such as the filter shape, the event triggered movie activity (e.g., whether it conformed to prior expectation of one-photon neuron morphology and GCaMP activity), location within the imaging field of view (e.g., not within a blood vessel), and the shape of the transient having characteristic GCaMP dynamics. We did not use automated heuristics to further remove accepted neurons in our studies, but these can be used should you want to save time. In Fig. 7a, "Spatial filters" are the PCA-ICA output filters, "Activity in movie" is a 31×31 pixel square region cropped from the movie around the candidate neuron's centroid location during that candidate neuron's transients (black outlines are "Spatial filter" derived neuron contours), and "Activity traces" shows the mean (black) and per transient (gray) PCA-ICA activity of a candidate neuron from the imaging session.

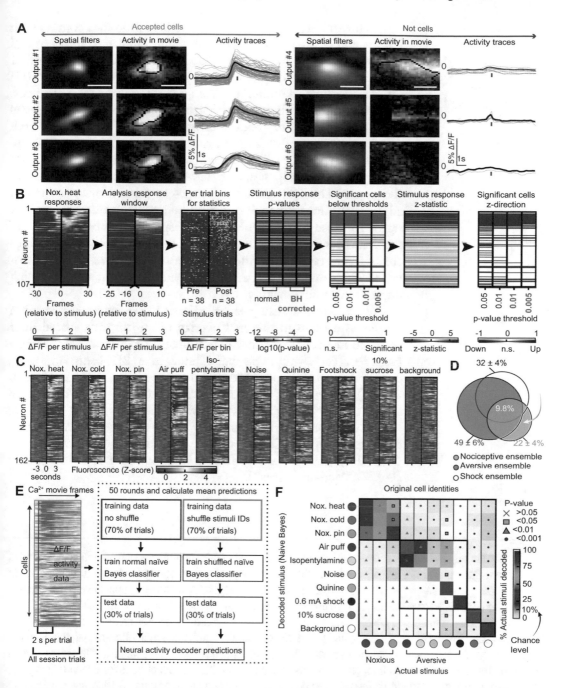

Fig. 7 Calcium imaging analysis of nociceptive ensemble. (**a**) Example PCA-ICA neuron extraction outputs from a single mouse showing accepted and rejected ICA outputs. We manually classified all neurons used in imaging-related aspects of this study based on a variety of parameters. *See* text for details. Scale bars, 25 μm. (**b**) Steps to identify cells that significantly respond to a stimulus. In the right-most graph, for certain brain regions we only consider cells that increased activity (red). (**c**) Mean stimulus response across all trials for all BLA neurons during a single imaging session in an uninjured mouse ($n = 162$ neurons). Neuron identifications across different stimuli are consistent, demonstrating that some neurons encode multiple different types of

Cell-extraction algorithms often yielded candidate sources with images and activity traces that look highly similar to those of real neurons but that are actually associated with neuropil or other sources of contamination in the movie. Thus, the CIAtah GUI interface includes a section to avoid including these false positives. Specifically, the GUI crops the movie to a 31 pixel × 31 pixel (~78 μm × ~78 μm) region centered on the centroid of each candidate cell then creates movies containing 10 frames before and after the onset of an individual peak in the candidate Ca^{2+} activity trace to help visualize actual transient-related activity in the movie. The GUI creates up to 24 of these movies (users can adjust this value) for each output based on each output's highest signal-to-noise (SNR) peaks. It then spatially concatenates all of these movies associated with a specific cell extraction output to create a montage movie that allows you to view movie data associated with peaks in the activity trace for each output at once, which improves decision-making.

We recommend using several criteria to classify a cell extraction output as a neuron: minimal overlap of an output's spatial filter with blood vessels or other contaminating signal sources, resemblance of each output's spatial filter to a 2D Gaussian or an expected neuron shape based on prior knowledge (Fig. 7a, "spatial filters"), similarity of the spatial filter to activity within the movie and proximity of output's centroid to movie activity (Fig. 7a, "activity in movie"), and similarity of the average transient waveform to a typical Ca^{2+} transient waveform as observed using GCaMP6, such as a fast rise time followed by a slow decay (Fig. 7a, "activity traces").

Fig. 7 (continued) noxious and aversive stimuli, while a separate neuron population uniquely encodes nociception. Background is a negative control showing average response during random trial time points at least 10 s away from any defined stimuli. (**d**) Divergent neural populations encoding appetitive (10% sucrose consumption) stimuli versus the nociceptive ensemble. (**e**) To test out the specificity of the neuronal ensemble dynamics between stimuli, we constructed a nine-way naïve Bayes decoder. For cross-validation, we split data each round 70:30 between training and test datasets using 2 s from each trial. After training the decoder, it was run on the test neuron activity data and the predicted stimuli state compared to the actual stimuli delivered. The decoder was run through 50 rounds subsampling different sets of trials for use in training and test datasets. (**f**) We constructed a naïve Bayes decoder, as described in Subsection 3.16.2, and applied it to the noxious vs. aversive stimuli experiments ($n = 6$ mice, 1 session each). The decoder was then run on neural activity data for a new subset of stimuli and the actual stimuli at those frames compared to those predicted. We then normalized each actual stimuli column by the number of total actual stimuli to allow comparisons of how accurate the de-coder was. Better performance (red/orange) occurred on noise and 10% sucrose than on innocuous (light purple) or noxious (blue and dark purple) stimuli. Symbols in the off-diagonal indicate whether prediction of correct stimuli was significantly higher than prediction of that stimuli (Wilcoxon sign-rank, Benjamini–Hochberg)

3.16 Basic Analysis of Noxious-Stimuli Responsive Populations

This next section will cover basic analysis of noxious-stimuli encoding cells (Fig. 7b–d). A critical aspect of this analysis involves analysis of neural responses to nonnoxious stimuli (Fig. 7c, d). To determine which neurons significantly responded to a given stimuli, the philosophy taken here is to identify which neurons encode for a given stimulus [39] as well as using decoding analysis [40] to demonstrate this neural activity is predictive of which stimuli animals received at a given time in the session.

1. Load the stimulus event times and neural activity traces from Subheadings 3.8.1 and 3.15.

2. Conduct **steps 3–5** below for each stimulus given in an imaging session.

3. Take the neuronal activity data (e.g., PCA-ICA output traces) from a 2-s-post-stimulus interval for all trials. This involves creating a $n \times t \times f$ matrix, where $n =$ number of neurons, $t =$ number of trials, and $f =$ number of frames per trial. Then convert the matrix into 1-s bins by taking the mean of each bin's $\Delta F/F$ activity.

4. For each cell, compare the binned stimulus-response activity values to those in an identically binned 2-s window from -5 s to -3 s before the stimuli, *see* **Note 20**.

   (a) Note 20: We recommend this range to minimize the effect that anticipatory behavior or stimulus onset timing errors have on the baseline. If you observe large neural responses to "Approach, no contact," then either rerun imaging session being more careful to approach the animal from an angle that is most likely to minimize anticipatory behavior or shift the prestimulus comparison window closer to stimulus onset to include that neural activity in the analysis. The later would essentially be looking for added responses beyond anticipatory that are induced by the stimulus.

5. Pool this activity across all presentations (or trials) of a specific stimulus and calculate a p-value for each neuron using a Wilcoxon rank-sum. Designate any neurons for which $P < 0.01$ as being significantly responsive to a given stimulus. Adjust this value depending on the SNR of cells in your experiment or if after checking whether identified cells are marked significant, they have stimulus-evoked responses in the range that you expect (Fig. 7b). *See* **Note 21**.

   (a) Note 21: Consider using a one-tailed Wilcoxon rank-sum, which has more statistical power, if prior literature indicates that neurons in your area of interest have low firing rates and that you are unlikely to be able to observe true *decreases* in Ca^{2+} activity in response to a stimulus. Make sure to justify this *before* starting your statistical analyses and state this in any methods [41].

6. We define the *BLA nociceptive ensemble* consists of neurons significantly responsive to noxious pin (25G needle), noxious heat (55 °C water), *or* noxious cold (5 °C water or acetone). Separately assign stimuli responsive neurons to the nociceptive ensemble for each animal's imaging sessions using the above definition.

3.16.1 Analysis of the Overlap in Neural Ensembles Responsive to Different Stimuli

Whether the overlap in neural populations active to separate stimuli are significant can be determined by looking at whether the neuronal ensembles responsive to two different stimuli are consistent with a hypothesis of statistical independent coding channels. This is important, as just looking at the percent overlap is not informative absent knowledge about the stimulus-responsive and total active neural populations. To test this hypothesis, you need to compute the likelihood that statistically independent assignments of cells' coding identities would yield the observed level of overlap in the two coding ensembles. There are two ways to calculate the expected level of overlap under an assumption of independence: using bootstrapping to estimate an empirical null distribution and compared the actual overlap to that or a newer, faster method we introduced to calculate an alternative, exact solution [18].

The idea behind the exact solution is that you calculate the extent to which the observed overlap was unexpected by chance. This is a specific instance of the classic statistics thought experiment of drawing without replacement balls from an urn containing black and white balls. In these studies, we have a population of N neurons and are seeking the probability, p, of having k successes (number of significant neurons for stimulus #2) in a population with predefined K successes (number of significant neurons for stimulus #1) in n drawings (number of significant neurons for stimulus #2).

To conduct the analysis, use the *hygecdf* and *hygestat* functions in MATLAB (or the equivalent in your language of choice) to calculate p and the expected number of overlap neurons given the actual number of significantly responsive neurons observed for stimuli #1 and #2 (say, noxious heat and cold). You can validate this method obtains accurate results by comparing to shuffle tests based on the same parameters and using 1000 rounds of 1,000,000 shuffles to construct bootstrapped distributions. Specifically, each shuffle consists of randomly rearranging the vectors containing 1's and 0's, with the number of 1 s matching the number of stimulus-responsive neurons, indicating which cells are significant for each stimulus then comparing them to get the overlap. Do this 1,000,000 times to get a distribution then calculate the fraction of values that are greater than the actual overlap value. Do this 1000 times to get a distribution of p-values. We demonstrated [18] that the two methods attained nearly identical results and thus we strongly recommend using the hypergeometric distribution instead of shuffle tests to reduce computational processing times and to obtain an exact *p*-value.

To determine whether the overlap in coding ensembles becomes more expected than chance, either before or after spared nerve injury, perform a Wilcoxon rank-sum tests using a Benjamini–Hochberg multiple comparisons correction [42] to identify whether the overlap differed significantly that expected by chance. You can obtain the expected overlap by using *hygestat*.

3.16.2 Decoding of Stimuli Based on BLA Neuron Activity

To test the specificity of the neural code in response to various stimuli delivered to animals, you can implement one of many classifiers. We have previously used naïve Bayes classifiers [18] and general linear model (GLM) decoders [35] to predict stimuli or behavior based on neural activity. You can implement both in the MATLAB programming environment using standard MATLAB libraries and toolboxes. In the below procedure, we will describe the use of Naïve Bayes classifiers, which assume statistical independence between predictors (i.e., neurons), but also work well when this is not strictly the case [13]. The goal of a naïve Bayes classifier is to predict the response, y (stimuli given at time point t) based on predictors, x_{1-n} (neurons), where n is the number of neurons in our case. Formally,

$$P\big(y(t), |x(t)_1, \ldots, x(t)_n\big) = \frac{P(y(t))P\big(x(t)_1, \ldots, x(t)_n | y(t)\big)}{P\big(x(t)_1, \ldots, x(t)_n\big)}$$

where t is a particular poststimulus frame during the session, $P(y(t))$ is the probability of a stimulus at a given time point, $P(x|y)$ is the probability of activity in neurons (x) given a stimulus (y) was present, and $P(x)$ is the probability of a neuron being activated at time t. Thus $P(y|x)$ gives you the probability of stimulus y given you observe x activity pattern within the entire neural ensemble. Then use the following classification rule under the assumption of statistical independence between predictors (e.g., neurons or x):

$$\hat{y}(t) = \arg\max_{y} P(y(t)) \ \Pi_i^n P(x_i(t)|y(t))$$

This rule allows you to predict at each time point, t, the most likely stimulus (class of $\hat{y}(t)$) given the observed response of all n neurons. Procedurally, the steps are as follows, *see* Fig. 7e.

1. Split the neural data into testing and training sets on a per trial basis with 70% of trials for training and 30% for testing.

2. Perform a 50-fold cross validation by training a new decoder using randomly chosen set of training trials and testing that decoder on a nonoverlapping set of test trials sampled from the entire set of stimuli trials. See **Note 22**.

(a) Note 22: Since you are creating decoders that must simultaneously predict either nine (Fig. 6b) or ten (Fig. 6c) stimuli, you need to avoid introducing biases in the test and training sets due to some stimuli having more trials than others. To correct for this, limit the number of each stimuli's trials used for testing and training to whichever stimuli in that session has the minimum number of trials. However, still allow sampling from the full range of trials for each stimulus during each test round.

3. For the training set, construct a $n \times f$ predictor matrix consisting of all n neurons in a session and f frames composed of all ten frames in a 2 s window after stimulus delivery for all trials and stimuli. Construct the response as a $1 \times f$ vector composed of the same 2-s window after the stimulus for all chosen stimuli trials, marked with 1 s to indicate which stimuli trial each frame was associated with. For example, frame 450–455 mark with a 2 (for noxious heat) and frame 900–905 mark with a 3 (for noxious cold).

4. Next, use the predictor matrix and response vector to train a naïve Bayes classifier using the *NaiveBayes* MATLAB class with a Gaussian distribution assumed for $P(x(t)_1, \ldots, x(t)_n | y(t))$, seen as below, where μ_y and σ_y are the mean and standard deviation estimates of neuron i response to a particular class (e.g., stimuli) in y.

$$P(x(t)_i | y(t)) = \frac{1}{\sqrt{2\pi\sigma_y^2}} \; exp\left(-\frac{\left(x(t)_i - \mu_y\right)^2}{2\sigma_y^2}\right)$$

5. Next, run each trained classifier on another $n \times f$ matrix containing the test set neural activity data, this will give you an output of naïve Bayes classifier predictions of which stimulus trial is associated with each time point f.

6. Take the resulting predictions and compare to the actual given during those test set time points. One way to visualize this is to construct a confusion matrix from the predicted and actual stimuli then normalize each column (corresponding to each actual stimuli) by the number of actual stimuli given to allow comparison of the decoder accuracy for each stimulus compared to others, *see* Fig. 7f.

7. To ensure that the decoding specificity is due to an individual neurons' specific activity in response to each stimulus, run another 50 rounds where you shuffle the stimulus identities (e.g., $1 \times f$ response vector in **step 3**) used to train the decoder but keep the testing set stimuli unchanged. This should remove much of the predicted stimuli specificity.

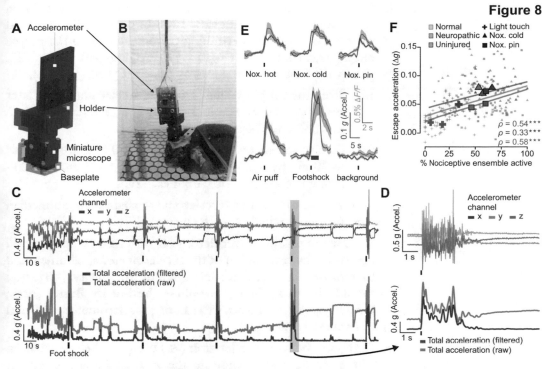

Fig. 8 Accelerometer-based recording of nocifensive behaviors. (**a**) 3D CAD model of the accelerometer (red) and 3D printed holder (blue) attached to the miniature microscope. (**b**) Picture of (**a**) in an experiment. (**c**) Example raw accelerometer traces for *x*, *y*, and *z* channels from an aversion experiment where a foot shock was delivered. Bottom traces show the raw acceleration and after filtering to remove artifacts (e.g., static component of gravity). (**d**) Zoomed in view of shaded region in (**c**). (**e**) Temporal dynamics of the mean $\Delta F/F$ of neurons within the nociceptive ensemble (cyan) and mean affective escape acceleration (red) for all imaging sessions and mice ($n = 6$ mice, 1 session each). (**f**) Correlation between % of nociceptive ensemble activated and escape acceleration per imaging session (light colored points) and across animal groups and conditions (dark, larger points) show significant correlation (Spearman's $\rho = 0.54$ [Normal], 0.33 [Neuropathic], and 0.58 [Uninjured])

3.16.3 *Analysis of Accelerometer Traces*

This section describes collecting and analyzing quantitative measurement of animal movement in response to stimuli using a triaxial accelerometer (Sparkfun, ADXL345 or ADXL335 accelerometer). The accelerometer is attached to the miniature microscope body by a custom 3D-printed part (Stratasys Objet30 printer, VeroBlack-Plus material, Fig. 8a). The *computeAccelerometerOutput* CIAtah function helps with processing the accelerometer data.

1. Using a soldering iron, connect the accelerometer *x*, *y*, *z*, power, and ground lines to a multiwire cable. If using an Arduino, connect the *x*, *y*, and *z* lines to analog inputs 1–3, respectively. If using a Saleae logic analyzer, connect to inputs 0–2. In both cases, connect the accelerometer to a 3.3 V power line and ground on an Arduino. Slot the accelerometer into the 3D-printed attachment and secure in place with setscrews before sliding onto the miniature microscope (Fig. 8a, b).

2. Collect accelerometer x, y, and z channel data using either a Saleae logic analyzer (Logic 8, 100 Hz sample rate, preferred method) or an Arduino running custom code (100 Hz sample rate). Synchronize each with Ca^{2+} imaging data using TTL pulses from the miniature microscope DAQ box (Fig. 5a).

   (a) Store the output voltage from the accelerometer for later analysis at 10-bit resolution (1024 value range, for both Arduino and Saleae setups).

   (b) For accelerometer data recorded at 100 Hz, downsample to 20 Hz by slicing the 100 Hz data into 50 ms bins and taking the mean value within each bin.

3. To remove noise in each accelerometer channel, use a median filter (4 frames wide) on each channel.

4. Since without using an accelerometer with a gyroscope you will not know the orientation of the accelerometer at all times, and to remove the static acceleration due to gravity [43], you should high-pass filter (zero-phase third-order Butterworth, 0.5-Hz cutoff frequency) the entire accelerometer x, y, and z channel digital signal.

5. Compute the total acceleration (A_t) using the below equation where a_x, a_y, and a_x indicate the x, y, and z accelerometer output channels (Fig. 8c, d).

$$A_t = \sqrt{(a_x)^2 + (a_y)^2 + (a_z)^2}$$

6. To further reduce noise, zero-phase lowpass filter the total acceleration (butterworth, 1 Hz cutoff, 3rd order).

7. Downsample the resulting total acceleration from 20 Hz to 5 Hz by binning the 20 Hz data into 200 ms sections and calculating the mean acceleration within each section.

8. For displaying the magnitude of change in behavior from baseline after a stimulus is given (Fig. 8e), it is at times preferable to use Z-scored acceleration instead of in units of g. In this case, calculate the mean (μ_{pre}) and standard deviation (σ_{pre}) response for 3–5 s before the stimuli (using the stimulus vectors constructed in Subheading 3.8.1) and compute the Z-score for the 5 s before and after the stimuli using the below equation where t is the frame relative to stimulus onset.

$$Z_{score}(t) = \frac{A_t(t) - \mu_{pre}}{\sigma_{pre}}$$

9. To determine whether animals have moved statistically more than chance, either before or after spared nerve injury, calculate the mean session response per animal in a 2-s window after each stimulus and compared it to a baseline from 3 to 5 s before each

stimulus using a Wilcoxon rank-sum test with Benjamini–Hochberg multiple comparison correction for the multiple stimuli being tested.

10. To ensure accuracy of the accelerometer recordings, compare accelerometer data for each stimulus to human scored data (*see* Fig. 2e) by calculating the % of that stimuli's trials in a session the animal scores as responding to a stimulus and compare this to the mean acceleration for that stimuli across all trials in the same session.

   (a) We have found that the two measures are highly correlated (Spearman's $\rho = 0.79$, p-value < 0.001) and that the accelerometer offered a greater dynamic range to separate noxious stimuli that otherwise saturate near 100% of trials responsive with the binary measure of behavior used by humans.

11. After calculating the mean acceleration for each animal's session across all stimuli, compare the result to the percentage of neurons for a given stimulus that make up the nociceptive ensemble to see whether the degree of activation of "pain" neurons correlates with increased vigor in locomotion (Fig. 8f). Remember to compare these results to control locomotion experiments as outlined in Subheading 3.13, to reduce the possibility that these are really locomotion-related signals.

3.17 Cross-Day Analysis of Neuronal Activity

There is much interest both within the broader neuroscience community as well as within the pain field as to how stable neurons are in the stimuli they respond to and information they encode. With respect to pain, this is of interest both in terms of understanding how the nervous system relays nociceptive information to the brain (e.g., using labeled line, population coding, or another Schemes [44]) and has therapeutic implications: if the nociceptive ensemble is stable across time that would likely lead to an easier therapeutic target than an ensemble that is constantly changing (Fig. 10d, right). Further characterization of a stable ensemble might thus reveal a genetic component that leads to targeting of specific receptors (e.g., GPCRs) expressed by that stable population to help alleviate pain. Thus, to match neurons across days we will describe a multistep algorithm (Fig. 9) that expands on prior work [13, 18, 30]. Run the method on your data using the CIAtah *compute-MatchObjBtwnTrials* module or *matchObjBtwnTrials* function. Below we describe the procedural steps for users who want to better understand the method or recreate using their own routines or programming environment.

1. Load the cell extraction spatial filters and threshold them by setting to zero any values below 40% the maximum for each spatial filter and use these thresholded filters to calculate each

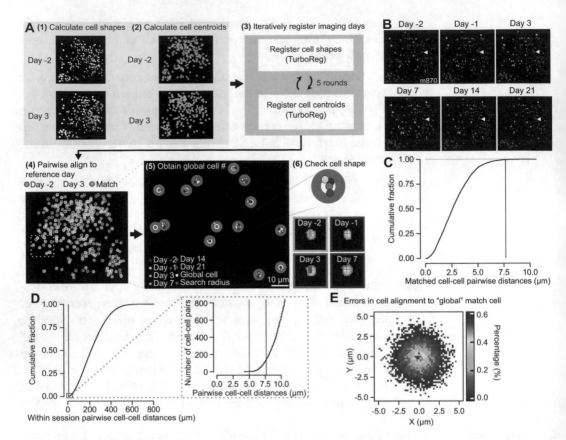

Fig. 9 Cross-session alignment of cells. (**a**) Method for cross-day alignment of neural ensembles using real data from an example mouse. Day −2 and 3 are with respect to nerve injury surgery day. After neurons had been matched (steps 4 and 5), they were associated with a global cell that was then used to analyze their responses across days. *See* text for detailed procedure. Red crosses in (step 6) are the global neuron centroid location overlaid on the cell filters. (**b**) Example neuron spatial filter maps showing cross-day alignment for an example mouse's imaging sessions. Global cells matched across at least 70% of the imaging sessions are coded by a unique color. White arrow points to a neuron active across all aligned days for that animal. Scale bars, 100 μm. (**c**) Pairwise centroid Euclidean distances restricted to neuron–neuron pairs within the same global cell for all mice ($n = 17$), demonstrating the majority of neuron matches assigned to the same global cell are less than 5 μm apart. Red line is at the same location as the 99.99th percentile in (**d**) inset. (**d**) Same calculation as in (**d**) for all imaging sessions across mice ($n = 17$) showing that the vast majority of neurons are >10 μm apart. Inset, zoomed in view showing the absolute number of neuron pairs within 10 μm of one another. Red line indicates 0.01th percentile. Grey line indicates threshold used to group neurons in (**a**) into a global cell. (**e**) Individual neuron distances from their respective global cell centroid location if they were matched to another neuron on at least one other session ($n = 13,558$ session neurons)

neuron's centroid location. You can calculate the centroid using the *regionprops* function in MATLAB. *Do not* round each neuron's centroid coordinates to the nearest pixel value as this would reduce accuracy of cross-day alignment.

2. Next, create simplified spatial filters that contained a 10-pixel-radius circle centered on each neuron's centroid location. This

allows you to register different days while ignoring any slight day-to-day differences in the cell extraction algorithm's estimate of each neuron's shape even if the centroid locations are similar.

3. For each animal, we recommend that if you have N sessions to align that you choose the $N/2$ session (rounded down to the nearest whole number) to align to (*align session*) in order to compensate for any drift that may have occurred during the course of the imaging protocol.

4. For all imaging sessions create two neuron maps based on the thresholded spatial (Fig. 9a, step 1, "thresholded neuron maps") and 10-pixel-radius circle (Fig. 9a, step 2, "circle neuron maps") filters by taking a maximum projection across all x and y pixels and spatial filters (e.g., a max operation in the third dimension on a $x \times y \times n$ neuron spatial filter matrix, where n = neuron number).

5. You then need to register these neuron maps to the *align session* using *Turboreg* [37] with rotation enabled for all animals and isometric scaling enabled for a subset of animals in cases where that improves results. The registration steps are as follows (Fig. 9a, step 3):

   (a) Register the thresholded neuron map for a given session to the *align session* threshold neuron map.

   (b) Use the output 2D spatial transformation coordinates to also register the circle neuron maps.

   (c) Then register the circle neuron map with that animal's *align session* circle neuron map.

   (d) Apply the resulting 2D spatial transformation coordinates to the thresholded neuron map.

   (e) Repeat this procedure at least five times.

   (f) Lastly, use the final registration coordinates to transform all spatial filters from that session so they matched the *align session*'s spatial filters and repeat this process for all sessions for each animal individually.

6. After registering all sessions to the *align session*, recalculate all the centroid locations (Fig. 9a, step 4).

7. Set the *align session* centroids as the initial seed for all *global cells* (Fig. 9a, step 5). Global cells are a tag to identify neurons that you match across imaging sessions.

   (a) For example, global cell #1 might be associated with neurons that are at index number 1, 22, 300, 42, and 240 within the cell extraction analysis matrices across each of the first five imaging sessions, respectively.

8. Starting with the *align session* for an animal, calculate the pairwise Euclidean distance between all global cells' and the selected session's (likely 1st) neurons' centroids.

9. Then identify any cases in which a global cell is within 5 μm (nominally ~2 pixels in our data) of a selected session's neurons. This distance depends on the density of cells in your imaging sessions, a stricter cutoff should be set for more dense brain areas, *see* Fig. 9d. When you find a match, then check that the spatial filter is correlated (e.g., with 2D correlation coefficient) above a set threshold (e.g., $r > 0.4$) with all other neurons associated with that global cell (Fig. 9a, step 6).

10. If a neuron passes the above criteria, add that neuron to that global cell's pool of neurons then recalculate the global cell's centroid as the mean location between all associated session neurons' centroid locations and annotate any unmatched neurons in that session as new candidate global cells.

11. Repeat this process for all sessions associated with a given animal.

12. After assigning all neurons across all animal's imaging sessions to a global cell, conduct a manual visual inspection of each animal's cross-day registration results. We recommend removing imaging sessions that did not align well with other sessions associated with a particular animal; while this leads to a loss of data the mis-aligned session can have detrimental effects on the cross-session algorithm's performance.

13. To quantify alignment accuracy, you should calculate the pairwise distance between all session neurons' centroid locations that are associated with a common global cell. Most alignments should be below 5 μm displacement from the global cell centroid (Fig. 9c, e).

    (a) Use the CIAtah *computeCrossDayDistancesAlignment* module after running *computeMatchObjBtwnTrials* to obtain these numbers in a CSV table.

14. To calculate the number of sessions a global cell responded to specific stimuli (Fig. 10a), use the classification of significantly coding neurons found in Subheading 3.16. Create a $n \times s$ matrix where $n =$ number of global cells and $s =$ session, in which each matrix cell indicates 1 if a cell significantly responds on that session or a 0 if not.

15. Then check for each global cell the number of sessions it responds to a given stimuli while ignoring any global cells who only had activity on a single session.

16. To calculate maximum duration of time that a cell maintains responses to a given stimulus, use the actual date (since imaging sessions might not have been run on the exact dates

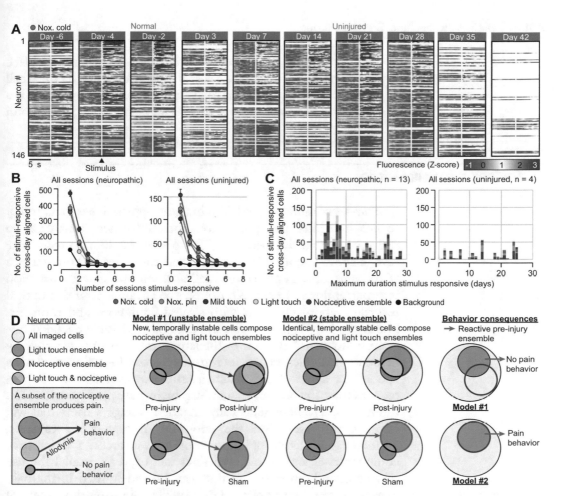

Fig. 10 Longitudinal analysis of the nociceptive ensemble. (**a**) Example animal showing all global cells ($n = 146$) that were active during greater than half of that animal's imaging sessions. A subset of neurons (top rows) are stimulus responsive to noxious cold (acetone) across multiple imaging sessions and days to weeks of time. Black sections indicate sessions in which no associated neuron was found for that global cell. (**b**) Indicates number of global cells that significantly coded for indicated stimuli (*see* Fig. 7b) across either one or more imaging sessions irrespective of temporal distance separating imaging sessions. Gray line is 150 global cells and is common across the neuropathic (top row) and uninjured (bottom row) imaging groups. "Nociceptive ensemble" stimuli are a global cell that responded to either noxious pin and/or noxious cold on any given imaging session. (**c**) To determine how long global cells coded for specific stimuli (color-coded), actual imaging session dates were used to calculate the maximum duration a global cells was found to be stimulus responsive. Of the 3223 global cells matched across two or more imaging sessions, ~11% (350 global cells) responded to noxious stimuli with at least a week separating their first and final noxious stimuli responses. (**d**) Two potential models of BLA nociceptive ensemble activity evolution following injury. In model #1, if the nociceptive ensemble changed cell identities across time and was identifiable on any given day, the cross-day analysis would show that the cells responding to light touch in a neuropathic state are different from the cells constituting the nociceptive ensemble cells preinjury. Regardless, if the neuropathic light touch ensemble is more similar to the neuropathic nociceptive ensemble (*see* orange shaded area), this would suggest that when observing each ensembles' activation, they is more overlapped between ensembles, suggesting that the animal may abnormally perceive the light touch stimulus as more aversive, or painful, than is warranted based on the nonnoxious character of the physical stimuli. In contrast, in model #2, the same effect would be observed except that the cells would be stable across time. This can be tested by using techniques to reactivate the preinjury ensemble and observe whether pain behavior is observed when those same cells are reactivated on later days (rightmost column)

specified in the Fig. 5 protocol) the imaging session took place on to calculate both the earliest and latest date that a global cell significantly responded to each stimuli. Take this difference as a measure of how long a neuron stably coded for said stimuli (Fig. 10b, c).

4   Outlook and Conclusions

The power of miniscope calcium imaging approach lies in its ability to link individual and population nociceptive neural activities to real-time nocifensive behavior. This opens the door to reevaluate several classic theories of how the brain computes and integrates a wide-range of sensory, contextual, and internal-state information into perceptions of pain. No doubt, the advantages of this optical recording approach come with several limitations and room for improvement. While the current class of genetically encoded calcium sensors has been one of the largest catalysts of contemporary neuroscience, calcium activity remains a proxy for action potentials and subthreshold electrical activity that carries immensely important information for understanding brain network dynamics. On the horizon are new genetically encoded voltage sensors that will improve this area, but the adoption of high-speed cameras, improved optics, multiphoton laser light sources, and improve computational and storage pipelines will need to parallel the use of voltage indicators before they can be widely used in the pain neuroscience community. Additionally, light-sensitive neural activity-indicators can be combined with light-sensitive opsins to turn pain-relevant circuits on or off at will. Using a closed-loop all-optical optogenetic-imaging approach could provide even further control for the reading and writing of pain-associated neural activity with tight anatomic resolution and subsecond precision. Nevertheless, the caveats and limitations of one-photon miniscope calcium imaging, as with any emerging technology, should be carefully considered and incorporated into your unique experimental designs and data analysis needs. Investigations into the neural circuits of pain-processing in the brain can now leverage this powerful tool to reverse translate the remarkable depth of key human brain imaging's results by drilling into specific brain regions to visualize the coding principles of nociceptive information and to identify, potentially, pain-specific neural pathways in the brain.

References

1. Basbaum AI, Bautista DM, Scherrer G, Julius D (2009) Cellular and molecular mechanisms of pain. Cell 139:267–284

2. Baliki MN, Apkarian AV (2015) Nociception, pain, negative moods, and behavior selection. Neuron 87:474–491

3. Devor M (2013) Neuropathic pain: pathophysiological response of nerves to injury. In: Wall and Melzack's textbook of pain. Elsevier Saunders, pp 861–888

4. Apkarian AV, Baliki MN, Geha PY (2009) Towards a theory of chronic pain. Prog Neurobiol 87:81–97

5. Pizzo PA, Clark NM, Carter Pokras O (2011) Relieving pain in America: a blueprint for transforming prevention, care, education, and research. In: Institute of Medicine. National Academies Press, p 382

6. Baliki MN, Geha PY, Fields HL, Apkarian AV (2010) Predicting value of pain and analgesia: nucleus accumbens response to noxious stimuli changes in the presence of chronic pain. Neuron 66:149–160

7. Baliki MN et al (2012) Corticostriatal functional connectivity predicts transition to chronic back pain. Nat Neurosci 15:1117–1119

8. Guenthner C, Miyamichi K, Yang HH, Heller HC, Luo L (2013) Permanent genetic access to transiently active neurons via TRAP: targeted recombination in active populations. Neuron 78:773–784

9. Dhawale AK et al (2017) Automated long-term recording and analysis of neural activity in behaving animals. eLife 6:e27702

10. Rumyantsev OI et al (2020) Fundamental bounds on the fidelity of sensory cortical coding. Nature 580:100–105

11. Corder G et al (2017) Loss of μ-opioid receptor signaling in nociceptors, and not spinal microglia, abrogates morphine tolerance without disrupting analgesic efficacy. Nat Med 23:164–173

12. Ghosh KK et al (2011) Miniaturized integration of a fluorescence microscope. Nat Methods 8:871–878

13. Ziv Y et al (2013) Long-term dynamics of CA1 hippocampal place codes. Nat Neurosci 16:264–266

14. Kim TH et al (2016) Long-term optical access to an estimated one million neurons in the live mouse cortex. Cell Rep 17:3385–3394

15. Stringer C, Pachitariu M (2019) Computational processing of neural recordings from calcium imaging data. Curr Opin Neurobiol 55:22–31

16. Stringer C et al (2019) Spontaneous behaviors drive multidimensional, brainwide activity. Science 364:255–255

17. Ahanonu B (2018) calciumImagingAnalysis: a software package for analyzing one- and two-photon calcium imaging datasets. GitHub

18. Corder G et al (2019) An amygdalar neural ensemble that encodes the unpleasantness of pain. Science 363:276–281

19. Paxinos G, Franklin KB (2019) Paxinos and Franklin's the mouse brain in stereotaxic coordinates. Academic

20. Jackman SL et al (2018) Silk fibroin films facilitate single-step targeted expression of optogenetic proteins. Cell Rep 22:3351–3361

21. Hua T et al (2020) General anesthetics activate a potent central pain-suppression circuit in the amygdala. Nat Neurosci 23:1–15

22. Mukamel EA, Nimmerjahn A, Schnitzer MJ (2009) Automated analysis of cellular signals from large-scale calcium imaging data. Neuron 63:747–760

23. Sorge RE et al (2014) Olfactory exposure to males, including men, causes stress and related analgesia in rodents. Nat Methods 11:629–632

24. Slotnick B (2009) A simple 2-transistor touch or lick detector circuit. J Exp Anal Behav 91:253

25. Peirson SN, Brown LA, Pothecary CA, Benson LA, Fisk AS (2018) Light and the laboratory mouse. J Neurosci Methods 300:26–36

26. Langford DJ et al (2006) Social modulation of pain as evidence for empathy in mice. Science 312:1967–1970

27. Brenner DS, Golden JP, Gereau RW IV (2012) A novel behavioral assay for measuring cold sensation in mice. PLoS One 7:e39765

28. Allen WE et al (2019) Thirst regulates motivated behavior through modulation of brainwide neural population dynamics. Science 364:253–253

29. Livneh Y et al (2020) Estimation of current and future physiological states in insular cortex. Neuron 105(6):1094–1111.e10

30. Grewe BF et al (2017) Neural ensemble dynamics underlying a long-term associative memory. Nature 543:670–675

31. Blednov YA, Cravatt BF, Boehm SL, Walker D, Harris RA (2007) Role of endocannabinoids in alcohol consumption and intoxication: studies of mice lacking fatty acid amide hydrolase. Neuropsychopharmacology 32:1570–1582

32. Kobayakawa K et al (2007) Innate versus learned odour processing in the mouse olfactory bulb. Nature 450:503–508

33. Root CM, Denny CA, Hen R, Axel R (2014) The participation of cortical amygdala in innate, odour-driven behaviour. Nature 515:269–273

34. Saraiva LR et al (2016) Combinatorial effects of odorants on mouse behavior. Proc Natl Acad Sci 113:E3300–E3306

35. Parker JG et al (2018) Diametric neural ensemble dynamics in parkinsonian and dyskinetic states. Nature 557:177–182

36. Li Y et al (2017) Neuronal representation of social information in the medial amygdala of awake behaving mice. Cell 171:1176–1190. e1117

37. Thévenaz P, Ruttimann UE, Unser M (1998) A pyramid approach to subpixel registration based on intensity. IEEE Trans Image Process 7:27–41

38. Giovannucci A et al (2018) CaImAn: an open source tool for scalable calcium imaging data analysis. *bioRxiv*, 339564

39. Quiroga RQ, Panzeri S (2009) Extracting information from neuronal populations: information theory and decoding approaches. Nat Rev Neurosci 10:173–185

40. Rolls ET, Treves A (2011) The neuronal encoding of information in the brain. Prog Neurobiol 95:448–490

41. Ruxton GD, Neuhäuser M (2010) When should we use one-tailed hypothesis testing? Methods Ecol Evol 1:114–117

42. Benjamini Y, Hochberg Y (1995) Controlling the false discovery rate—a practical and powerful approach to multiple testing. J Royal Stat Soc B 57:289–300

43. Mathie MJ, Lovell NH, Coster ACF, Celler BG (2002) Determining activity using a triaxial accelerometer. In: Proceedings of the second joint 24th annual conference and the annual fall meeting of the biomedical engineering society. IEEE, pp 2481–2482

44. Craig A (2003) Pain mechanisms: labeled lines versus convergence in central processing. Annu Rev Neurosci 26:1–30

Chapter 14

Optical Imaging of the Spinal Cord for the Study of Pain: From Molecules to Neural Networks

Kim I. Chisholm and Stephen B. McMahon

Abstract

A deeper understanding of somatosensory processing within the spinal dorsal horn is emerging with newly developed methods to monitor the activity of cell populations in response to sensory stimuli in vivo. Here, we describe methods that have been developed to gain access to superficial dorsal horn laminae for acute and chronic imaging of calcium signals as surrogates of neural activity. Microglia, oligodendrocytes, and astrocytes also can be visualized in these preparations as can the inputs from primary sensory neurons. We also include a detailed protocol used by our laboratory to monitor cells in laminae I and II outer in mice as well as a discussion of the limitations of cell imaging in the spinal cord.

Key words In vivo imaging, Microscopy, Calcium indicators, Spinal cord, Pain, Fluorescence, Multi-photon, Intravital microscopy

---

1 Introduction

The spinal cord, especially the dorsal horn, is an important relay center for sensation in general and pain in particular. Primary sensory neurones innervating the peripheral tissues of the body, enter the spinal cord via dorsal roots. Large diameter proprioceptive and mechanoreceptive afferents project directly into the dorsal columns but all sensory fibers have local branches terminating in spinal segments at and near the segment of entry.

The dorsal column contains axons of primary sensory neurons carrying tactile and proprioceptive information. Currently we only have a limited understanding of the circuitry that is critical for sensory processing [1], including the action of essential analgesic drugs, including morphine and related compounds [2]. The complexity of the dorsal horn with multiple (at least 15) subtypes of inhibitory and excitatory neurones [3] and limited anatomical order has made studies of circuitry very difficult.

One important traditional method of studying the spinal cord is single cell electrophysiology. This technique is, however, very

Rebecca P. Seal (ed.), *Contemporary Approaches to the Study of Pain: From Molecules to Neural Networks*, Neuromethods, vol. 178, https://doi.org/10.1007/978-1-0716-2039-7_14, © Springer Science+Business Media, LLC, part of Springer Nature 2022

labor intensive with low throughput and as a result cannot easily portray the complex interplay between different cell types and networks. In fact, even the most sophisticated applications of traditional methods, for instance ex vivo slice preparations in which two neurons are patch clamped and recorded from simultaneously [4] are not only labor intensive but provide only a very small and incomplete picture of the entire network, with only around 10% of recorded pairs being connected [5, 6]. Additionally, these studies involve the nonphysiological settings of a slice preparation. The difficulty of culturing adult spinal cord neurons means that this approach also harbors numerous potential confounds and difficulties as does the use of organotypic slice cultures. Another traditional approach has relied on anatomy. Unfortunately, the spinal cord does not show the elegant anatomical organization that is seen for instance in the cerebellum, and this, coupled with the number of cell subtypes, has limited the power of anatomical studies.

More recently an alternative method to study functional responses of multiple neurones simultaneously has emerged and been optimized to the point where it has considerable resolution. This is in vivo optical imaging and it now provides a viable alternative to these traditional methods and overcomes some of their main limitations. Of course, it simultaneously brings its own limitations.

In this chapter, we will start with a brief overview of the principles of in vivo optical imaging of the spinal cord in rats and mice and then discuss some of the benefits and pitfalls of the method. We also provide some more practical information on the options currently available for implementing the method. For a nonexhaustive selection of publications relevant to in vivo spinal cord imaging within the field of pain and sensation *see* Table 1.

1.1 Principles of In Vivo Optical Imaging of the Spinal Cord

In vivo imaging of the spinal cord involves the use of light to visualize the spinal cord (Fig. 1). It can be used to study structures such as blood vessels, neurons, mitochondria or it could be used to visualize their functions, such as blood flow, calcium transients, or membrane potential. This can be done after the introduction of fluorescent molecules, for example GFP or Oregon Green BAPTA-1, into the spinal cord or it can also be achieved with endogenous signals already present, such as intrinsic optical signals or higher harmonic generations (discussed in more detail below). To visualize structures in the spinal cord it is essential to provide optical access to the tissue of choice. For the vast majority of published work, this pertains to the dorsal aspect of the spinal cord, either the dorsal horn [8–15] or the dorsal columns [15–23]. But the ventral aspects have also been successfully exposed and visualized optically [24]. Most work is focused on the more easily accessible lumbar spinal segments, though cervical imaging has also been achieved [25–30]. The protocols to expose the spinal cord for optical access vary considerably between groups but can be broadly divided into

Table 1
A nonexhaustive selection of publications relevant to in vivo spinal cord imaging within the field of pain and sensation. Inclusion in this list does not necessarily represent an endorsement

Paper	Description	Relevance
Bélanger et al. (2012) Live animal myelin histomorphometry of the spinal cord with video-rate multimodal nonlinear microendoscopy. J. of Biomedical Optics	In this paper a commercially available endoscope (GRINTECH) is used to visualize myelin through CARS imaging together with two-photon imaging of glia and axons signal.	Demonstrates the use of commercially available endoscopes as well as CARS imaging
Cartarozzi et al. (2018) In vivo two-photon imaging of motoneurons and adjacent glia in the ventral spinal cord. J Neurosci Methods	This paper shows a technique to surgically expose the ventral horn of the mouse spinal cord and visualize motoneurons as well as glia more than 200 μm deep inside the ventral horn with a two-photon microscope	Demonstrates in vivo imaging of the ventral horn
Chisholm et al (2021) Encoding of cutaneous stimuli by lamina I projection neurons. Pain	This paper describes a visualisation of lamina I projections neurons and their encoding of cutaneous stimuli, with a focus on thermal stimulation. The data reveal a strong sensitivity to cooling in lamina I projection neurons, as well as their quick adaptation to cold stimuli. Responses to heating stimuli were surpirsingly reliant on baseline temperatures	Relevant to the field of pain and sensation
Davalos and Akassoglou (2012) In vivo imaging of the mouse spinal cord using two photon microscopy. J Vis Exp.	This article describes the use of a spinal cord stabilizations setup together with deep anesthesia to reduce the movement artefact associated with in vivo imaging of the spinal cord. The authors show the applicability of their technique to repeat imaging without a spinal window. They suture and reopen the same incision site for repeat imaging.	A video protocol of an acute and chronic in vivo spinal cord imaging preparation
Farrar and Schaffer (2014) A procedure for implanting a spinal chamber for longitudinal in vivo imaging of the mouse spinal cord.	The authors describe a method to implant a spinal cord window in mice for repeated spinal cord imaging over at least 8 weeks.	A video protocol for the implantation of a spinal cord window for in vivo imaging

(continued)

Table 1
(continued)

Paper	Description	Relevance
Fenrich et al. (2013). Implanting glass spinal cord windows in adult mice with experimental autoimmune encephalomyelitis. J Vis Exp.	This article describes a protocol for implantation of a spinal cord window for repetitive in vivo imaging. This method uses inexpensive and readily accessible materials, including paper clips and staples.	A video protocol for the implantation of a spinal cord window for in vivo imaging
Figley et al. (2013) A spinal cord window chamber model for in vivo longitudinal multimodal optical and acoustic imaging in a murine model. PLoS One	This paper uses a multimodal approach to the study of the spinal cord and vasculature by combining fluorescence and bioluminescence microscope with optical coherence tomography, Doppler ultrasound (for vascular structures) and photoacoustics (for hemoglobin oxygen saturation). They conducted repeat imaging using a spinal window which was fixed to the muscle and skin of the back of the mouse.	A video protocol for the implantation of a spinal cord window for in vivo imaging
Ikeda et al. (2006) Synaptic amplifier of inflammatory pain in the spinal dorsal horn. Science	They used Oregon Green 488 BAPTA-1AM pressure injected into the rat spinal cord combined with DiI injections into the Periaqueductal gray, to identify projection neurons. Here they showed that peripheral stimulation can lead to increases in intracellular calcium in lamina I neurons, among them projection neurons.	Relevant to the field of pain and sensation
Johannssen and Helmchen (2010). In vivo Ca2+ imaging of dorsal horn neuronal populations in mouse spinal cord. J Physiol.	Here the authors show the technique of multicell bolus loading of Oregon Green BAPTA-1AM into the mouse spinal cord to visualize neurons in the dorsal horn. They were able to visualize spontaneous as well as evoked activity using two-photon microscopy.	Description of the multicell bolus loading technique in the spinal cord
Nikić et al. (2011) A reversible form of axon damage in experimental autoimmune encephalomyelitis and multiple sclerosis. Nature Medicine	This paper shows the use of multiple vital dyes for in vivo spinal cord imaging, including dyes for mitochondrial membrane potential, myelination, hydrogen peroxide and nitric oxide.	Includes a description and use of multiple vital dyes

(continued)

Table 1
(continued)

Paper	Description	Relevance
Nishida et al. (2014) Three-dimensional distribution of sensory stimulation-evoked neuronal activity of spinal dorsal horn neurons analyzed by in vivo calcium imaging. PLoS One	Calcium imaging in the dorsal horn in mice using the FRET based calcium indicator Yellow Cameleon loaded through in utero electroporation. The responses of dorsal horn neurons to cutaneous stimulation, including pinch, brush and heat was assessed.	Relevant to the field of pain and sensation and demonstrates the use of a FRET based calcium indicator
Ran et al. (2016) The coding of cutaneous temperature in the spinal cord. Nat Neurosci.	In this paper the authors bulk-loaded Oregon Green 488 BAPTA-1AM into the spinal cord and spinal cells were visualized using two-photon microscopy. Experiments showed that neurons sensitive to peripheral heating responded to absolute temperatures while neurons sensitive to peripheral cooling responded to relative temperatures, or temperature changes and adapted readily to cold stimuli.	Relevant to the field of pain and sensation
Sekiguchi et al. (2016) Imaging large-scale cellular activity in spinal cord of freely behaving mice. Nat Commun.	Here the authors describe two protocols to visualize the spinal cord in awake and behaving mice. Either a miniature microscope mounted on a spinal window or a head and spinal cord restrained preparation with a spherical treadmill were used to visualize activity of dorsal horn neurons and astrocytes in response to peripheral stimulation.	Demonstrate two approaches to in vivo imaging in awake, behaving mice with a miniature microscope and a head and spinal cord fixed setup

two approaches, the acute approach or the chronic approach (Fig. 1b). The acute approach is arguably the easier of the two and can be combined with dyes, which would leak and fade over more chronic imaging experiments. Essentials for an acute protocol include anaesthetizing the animal, exposing a part of the spinal cord and then stabilizing the vertebral column. This is usually done with a clamp which stabilizes the spinal vertebrae on either side of the exposure but can also include pins which are pushed against the column, providing stabilization via the vertebral bones. To provide additional stability some groups use a coverslip applied onto the

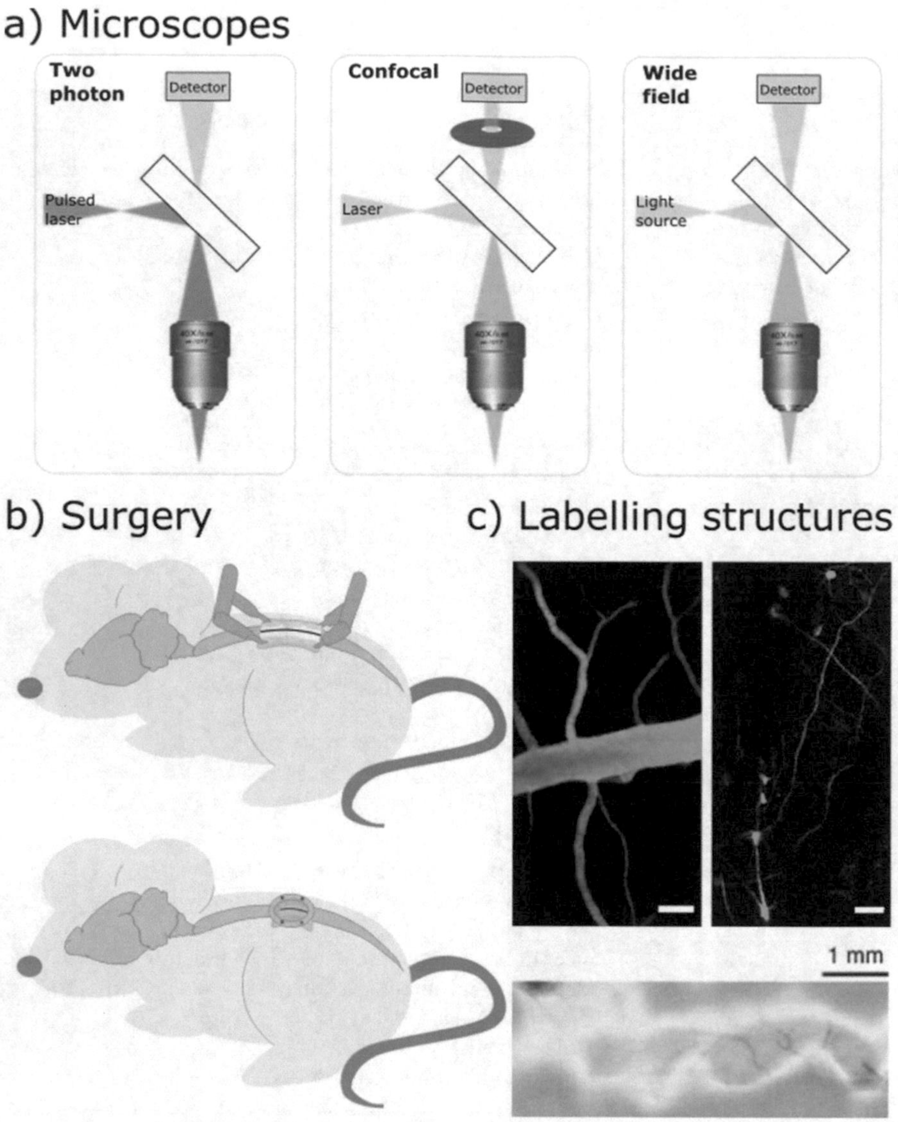

a) Microscopes

b) Surgery

c) Labelling structures

1 mm

Fig. 1 Basic in vivo spinal cord imaging setups. In order to conduct in vivo spinal cord imaging it is necessary to choose an appropriate microscopy setup. (**a**) Most commonly used microscopes in in vivo spinal cord imaging: Two photon microscopes provide increased penetration depth as well as automatic optical sectioning. As the probability of absorption of two photons near simultaneously is only significant at the focal plane, all emitted light is known to originate from a single focal plane. Therefore, all light can be collected, even light that was scattered in its path through the tissue. Confocal microscope uses higher energy excitation light. As fluorophore excitation is not limited to the focal plane, optical sectioning is guaranteed by the addition of a pinhole which rejects light originating outside the focal plane, as well as scattered light from the focal plane. Some signal light is therefore lost when using confocal microscope. Wide field microscope does not require scanning of a laser beam and can therefore acquire images quicker. However, no optical sectioning is provided. (**b**) Different types of surgical techniques have been used to access spinal cord tissue. These divide into acute and chronic surgical preparations. Acute preparations are easier and usually involve a spinal cord clamp which needs to fit onto a microscope stage. Chronic preparations usually involve the implantation of a coverslip as well as a stabilization system usually attached to the vertebrae. (**c**) Different kinds of labels can be added to the spinal cord. These can be broadly divided into structural and functional labels. Top left: a structural label of blood vessels of the spinal cord using intravenous Alexa Fluor 488–conjugated Dextran.

spinal cord, ensuring minimal compression. It is also possible to use a superfusion system in which a well (e.g., made of agarose) is formed around the exposed spinal segment and is filled with artificial cerebrospinal fluid or saline. In the latter case, a dipping objective is used. Additional stability can also be achieved by elevating the rodent off the heating mat. This will provide an opportunity for the breathing movement to be translated downward, rather than upward. In addition, the tail and head can be stretched and stabilized respectively, providing additional suppression of movement [31]. Tracheotomizing the rodent can help for acute procedures as it reduced head and thereby spine movement. If tracheotomy is chosen it is also possible to artificially ventilate the rodent which means breathing can be briefly halted during short acquisitions or it can be synchronized with image acquisition, eliminating all breathing related movement artifacts [32, 33]. To reduce the movement associated with heartbeat, cardiopulmonary bypass surgery can also be used [34]. Additional postprocessing [35] and adaptive movement compensation techniques have also been applied to the spinal cord [36].

Chronic in vivo imaging of the spinal cord, on the other hand, requires repetitive optical access to spinal cord tissues. Surgery is again undertaken under anesthesia and the animals are typically anesthetized during recording sessions. In some cases the spinal cord is surgically reexposed [15, 18, 22, 30, 33, 37–42]. In others cases optical access is maintained, through the installation of a spinal "window" [43–48] or through the use of miniature microscopes to record from freely behaving animals [8].

The protocols involving reexposure of the spinal cord have a lot in common with acute imaging protocols. The stabilization is usually through clamps or pins onto the vertebral column but procedures such as tracheotomies and cardiopulmonary bypass surgeries must be avoided, to improve recovery and reimaging. To allow the animals to recover with minimal scar tissue formation and impact on the spinal tissue, the cord can be covered before closing the surgery site, with for example agarose [18, 42], artificial dura [32, 33, 40, 41], and/or synthetic matrix membrane [41].

In order to reduce the impact of repeated surgical interventions, including the development of scar tissue, chronic spinal windows have been designed. These spinal windows include a coverslip and a stabilization structure, attached to the spinal column or muscle surrounding the exposed spinal cord (Fig. 2). The original spinal cord window, purpose-built for in vivo imaging of

Fig. 1 (continued) Scale bar 100 µm. Top right: a genetically encoded calcium indicator (GCaMP6) labeling spinal cord projection neurons for functional studies. Scale bar 100 µm. Bottom: It is also possible to visualize the intrinsic optical signal of the spinal cord which does not require the addition of a label and provides a readout of spinal cord function (reproduced with permission from [7])

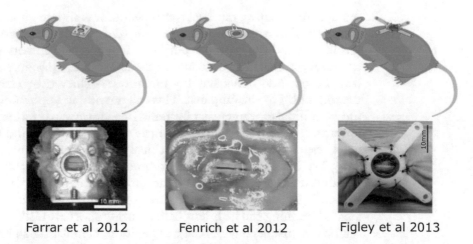

Farrar et al 2012   Fenrich et al 2012   Figley et al 2013

Fig. 2 Chronic spinal cord imaging preparations. Reproduced with permission from [49]. Representations of three chronic in vivo spinal cord imaging preparations. Left most protocol is by Farrar et al. 2012 and involves a custom designed metal top plate with window and two notched bars hugging the vertebral columns. Middle protocol was published by Fenrich et al. 2012 and involves modified staples and paperclips attached to the vertebral columns by dental cement and cyanoacrylate glue. The right most protocol is by Figley at al 2013 and includes a spinal window composed of polycarbonate material sutured to skin and muscle

the spinal cord of mice, was developed by Farrar et al. This involves two miniature bars with notched grooves, fitted onto either side of three vertebral segments. A top plate with a circular opening covers the spinal cord exposed after laminectomy of one vertebral bone. A coverslip is placed onto the circular opening and the space between the spinal cord and window is filled with transparent silicone elastomer. The entire spinal window is held in place with cyanoacrylate glue and dental cement [47, 48]. The authors detected no behavioral changes or vertebral damage after implantation of the spinal window. Despite some degradation of the signal over time, due to the growth of fibrous tissue, the authors were able to collect usable data for more than 5 weeks postimplantation for 50% of mice. The authors do, however, report a change in microglial density and morphology, suggestive of an inflammatory response to spinal window implantation [48].

Shortly after the publication of the above protocol Fenrich et al. proposed a low-cost solution to a spinal window. Using staples and paperclips they were able to fashion a stable spinal cord window which allowed visualization over a maximum of 350 days and a 77% success rate at 5 weeks postimplantation. The paperclips are bent into a modified, open diamond shape and attached below the transverse processes of the vertebrae on either side of the exposure. Using these as the anchors, a modified paperclip, bent into a bracket shape, is inserted below the paperclips to provide an anchor for external stabilization. The exposed spinal cord is covered with transparent silicone elastomer and a coverslip. These pieces are glued together and onto the spinal column using cyanoacrylate glue and dental cement [43, 44]. Using this spinal window the authors noted an increase in inflammatory cells in the spinal cord with partial resolution by 40 days postimplantation [44].

To overcome the limitations of metallic spinal windows, which are incompatible with some other imaging techniques, Figley et al. developed a spinal window which used polycarbonate materials to stabilize and visualize the surgically exposed spinal cord. This window was then used for multimodal imaging, including in vivo fluorescence and acoustic imaging in mice and rats. The window consists of a cross structure, with four arms to provide stability during imaging and a central hole for the coverslip. While the previous designs by Farrar, Fenrich and colleagues, involved windows attached to the vertebral column, the window by Figley et al. is sutured to the dorsal skin and muscle, likely resulting in more movement artifacts related to respiration and heartbeat. Figley et al., however, report low levels of spinal cord inflammation and no macroscopic damage up to 72 h after implantation. They were able to visualize the spinal cord for 29 days after implantation by clearing mild tissue accumulation from the coverslip which is easily removable [46].

Overall, the use of chronic spinal windows requires a high levels of surgical expertise. However, once mastered these windows can allow for repeated analysis of spinal cord structures and events, providing powerful longitudinal methodology with the potential to reveal causal relationships.

1.2 Example Preparation for In Vivo Spinal Cord Imaging

To better describe the practicalities of in vivo optical imaging we will outline a sample preparation which is frequently used in our group [50] to visualize cell bodies of the spinal cord in lamina I and lamina II outer. We describe the preparation in mice, though other rodents, including rats, can be used. It should be noted, however, that increased myelin thickness reduces the penetration depth of the excitation and emission light, limiting imaging depth.

2   Materials

- Mouse with cells labeled in the spinal cord. This can be a transgenic mouse, a mouse in which the spinal cord has been directly labeled with dyes or viral vectors or a mouse in which a projection target has been injected with a retrograde traceable dye or viral vector. Bear in mind that only the upper laminae are visible using confocal and even two photon microscope (for more details *see* Subheading 3.2.1).

- An epifluorescence, confocal or multiphoton microscope with a microscope stage low enough to fit the animal, and with stabilizing equipment.
- An imaging stage (this is where the animal, the spinal cord stabilization equipment and heating mat will be fitted).
- A homeothermic heating mat controlled with a rectal thermometer.
- Spinal clamps securely attached to the imaging stage.
- A dissection microscope under which the mouse and the imaging stage can be placed.
- Anesthesia (we use injectable anesthesia—urethane—but gaseous anesthesia is also useable; however, a delivery system with a mask and extraction vent needs to be included).
- Normal saline (0.9%).
- A tracheal catheter.
- Hair removal cream or hair clippers.
- Hemostatic sponge (e.g., Spongostan).
- Cotton buds.
- Silicone elastomer (for example from World Precision Instruments).
- Fluorescent dye to label the blood vessels, for example dextran tetramethylrhodamine (Thermo-Fisher Scientific)—optional.
- Surgical tools.
  - Fine forceps.
  - Surgical scissors.
  - Rongeurs.

3   Method

The surgical area should be prepared before anesthetizing the rodent. The imaging stage should be placed underneath the dissection microscope and the spinal clamps should be safely attached to the imaging stage. Now the mouse can be anesthetized using gaseous or injectable anesthesia. We use injectable urethane (12.5% w/v) at an initial dose of 37.5 mg (in a volume of 0.3 ml), injected IP. Further doses are administered at an interval of around 20 min until surgical depth is achieved. Anesthetic depth should be assessed throughout the experiment through limb and corneal reflexes. Once full anesthesia is achieved the animal can be hydrated with a subcutaneous injection of 0.5 ml of normal saline (0.9%).

After anesthesia the mouse should be placed onto a homeo-static controlled heating mat and a rectal probe inserted. The core body temperature should be maintained at around 37 °C. Next a tracheotomy can help to reduce movement artifacts. To perform a tracheotomy the mice are placed in a supine position and the paws taped to the imaging stage in an extended position. Next an incision is made over the thyroid gland, just below the jaw and the lobes of the thyroid gland are separated to reveal the underlying trachea. Using fine forceps, a small hole is made between two tracheal rings, just big enough to fit the tapered end of a tracheal catheter. The catheter is secured using sutures and superglue (sparingly). The incision site can then be sutured and secured with superglue, before the animal is returned to a prone position on the heating mat. If gaseous anesthesia is being used it is possible to attach the tracheal tube to a ventilator pump. When using injectable anesthesia the animal can breathe freely.

Now the surgical area over the spinal cord can be cleared of hair, using hair clippers or hair removal cream. An incision in the skin over the lumbar enlargement can expose the underlying spinal column, muscle, and connective tissues. The muscle and connective tissues along the lumbar enlargement are removed using fine scissors and rongeurs. Once the underlying spinal column is exposed a laminectomy can be performed using rongeurs. To achieve this the tip of the rongeurs needs to be placed below the spinal column, facing rostrally. The bone can then carefully be chipped away until the underlying cord is exposed. Ideally no more than two vertebrae should be removed to maintain stability. Any bleeding should be stopped using cotton buds, saline, and hemostatic agent.

Once the cord is exposed it can be covered with cotton wool and saline to prevent drying during the following steps. The spinal column can be secured using spinal clamps placed on the vertebrae immediately adjacent to the exposed segments. To visualize dorsal columns the prone position is ideal but if the dorsal horn is to be visualized the cord may need to be clamped in a position between prone and lateral recumbent (i.e., slightly rotated), for better visual access. The clamps need to be adequately tight to ensure stability but not overtightened as not to compromise blood flow or damage tissue. Additionally, one should take care not to accidentally tighten the spinal clamps around the intervertebral space as this would damage the underlying cord. Once the cord is exposed and cleaned (the dura should remain intact, unless removal of the dura is required for dye absorption [51]) a thin layer of silicone elastomer can be applied over the spinal cord. Any bubbles should be carefully removed before the elastomer hardens.

In order to easily visualize the cord and find an appropriate focal plane it can be useful to label anatomical structures such as blood vessels. To this end an injection of a fluorescent dye (e.g., dextran tetramethylrhodamine) should be made into the tail vein of the mouse. If the vein is difficult to find it may be helpful to warm the tail using a heating lamp or warm water to dilate the vessel.

The preparation can now be transferred to the microscope stage. To visualize cell bodies a 10× dry objective can be used. The fluorescent excitation and emission signal should initially be tuned to the vascular dye but once the area of interest and focal plane have been found the acquisition settings can be changed to the fluorophore of interest. Recording should be made with a trade-off between image quality and temporal resolution in mind. To increase temporal resolution, one may reduce the pixel dwell time/use faster scan speeds and/or reduce the area of interest. Instead, if temporal resolution is of no concern one may use slower acquisition speeds/pixel dwell times and larger areas of interest, possibly including acquisition of several images in the z plane (when using a microscope with optical sectioning capabilities). Unless absolutely necessary, averaging of the signal in real time is not recommended as this can easily be achieved postprocessing.

3.1 Advantages of Spinal Cord Imaging Techniques

3.1.1 The Scope of Information Collected

In vivo imaging in general provides a high-throughput approach in which a variety of neuronal responses can be investigated on a larger, more integrated scale. This means that several cell types can be studied at the same time and their ensemble activity easily visualized. This is especially important in a complex system such as the dorsal horn of the spinal cord [52]. The spinal cord dorsal horn contains a multitude of different neurons, including local circuits of interneurons as well as projection neurons, which transmit pain-related signals from the spinal cord to the brain stem. These neuronal types are often classified by function, physiology, morphology, location, and more recently by their transcriptional profiles [3]. Many of these characteristics are readily and simultaneously assessed when using optical imaging approaches, providing an immediate breadth of information not readily achieved with traditional techniques such as electrophysiology. The function and physiology can be assessed with the use of calcium indicators in which single action potentials can often be detected [53]. Morphology is also provided due to the optical nature of the technique, which also provides three dimensional information about location when techniques with an optical slice are used (e.g., confocal or two-photon microscope). Finally, genetically identified subclasses of cells can be easily studied using selective labeling techniques via transgenic mice or selective viruses.

3.1.2 The Range of Cell Types Studied

In addition to neuronal networks, nonneuronal cells can readily be studied, both functionally and structurally. Nonneuronal cells, including for example astrocytes [54–58], oligodendrocytes [59, 60], and microglia [61–68], have been reported to play a major role in certain pain conditions. Different transgenic mouse lines have made it possible to study these cells using in vivo

(a)   (b)

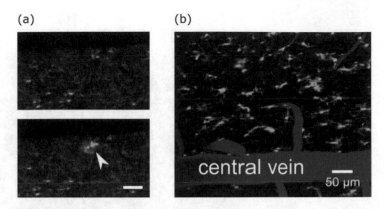

Fig. 3 In vivo imaging allows for the visualization of a range of cell types. (**a**) Astrocytic calcium imaging in the spinal cord. Arrowhead shows spontaneous calcium transients in an astrocyte near the dorsal vein. Scale bar 100 μm. (**b**) Image of spinal cord microglia labeled with GFP in a CX3CR1-EGFP transgenic mouse with the spinal cord vasculature labeled with Texas Red-dextran. Reproduced with permission from [69]

imaging, both functionally and structurally. For example mice in which Cx3cr1, CCR2, LysM, CD2, or CD11c positive cells are transgenically labeled (e.g., Fig. 3) have been used in spinal cord imaging to visualize immune cells and their interaction with other spinal cord tissues [15, 16, 21, 23, 24, 31, 35, 38, 39, 44, 45, 47, 48, 69–75]. Astrocytes (Fig. 3a) have also been visualized and studied in the spinal cord using GFAP-XFP transgenic mice [15, 24, 48, 76]. Alternatively, viral vectors can also be used to label dorsal horn astrocytes which were reported to be responsive mostly to noxious, rather than innocuous peripheral inputs [9, 77]. In vivo imaging of spinal cord endothelial junctions, on the other hand, is limited to a small number of manuscripts, including those in which Connexin43, VE-cadherin, and Claudin5 have been labeled with XFP [25, 44, 78].

In addition to the transgenic labeling of various cell types, in vivo spinal cord imaging can easily be used to study vascular structure (Figs. 1c and 3b), responsiveness, and leakiness as well as blood flow in the spinal cord [17, 26, 29, 69–71, 79–83]. The labeling of blood vessels using intravenous injections of fluorescent compounds, including dextrans or quantum dots, also makes it possible to study the interactions of various cell types, including immune cells and neurons, with the spinal cord vasculature [18, 25, 28, 31, 84–87]. Vascular labeling can also provide a useful landmark with which to evaluate and correct changes in the focal plane or image drift during longer imaging sessions or to relocate the same imaging site during chronic experiments. When the vasculature is labeled repeatedly Qdots can help to avoid the accumulation of dextran fluorophores in perivascular phagocytic cells with repeated administration [44].

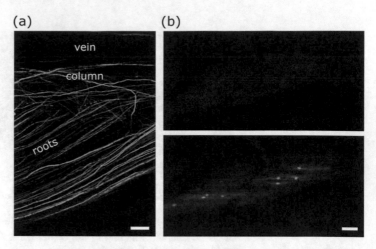

Fig. 4 In vivo imaging allows for a range of structures to be studied simultaneously, including components of the peripheral and central nervous system as well as grey and white matter. (**a**) Labeled afferents entering the spinal cord through the dorsal roots (roots) and running up and down the spinal cord in the dorsal column (column), adjacent to the dorsal vein (vein). With permission from Dimitra Schiza and Kenneth Smith. (**b**) A genetically identified subset of lamina II interneurons, labeled with a calcium indicator at baseline (top) and during noxious peripheral stimulation (bottom) (in collaboration with David Hughes and Andrew Todd). Scale bar 100 μm

3.1.3 The Range of Structures Studied Simultaneously

A further advantage to in vivo imaging is particular to the spinal cord. Here, the complexity of spinal cord sensory circuits is not only due to the many different types of cells relevant to pain and sensation but is further complicated by the presence of primary afferent processes from the dorsal root ganglion (Fig. 4a), which synapse onto and communicate with spinal cord interneurons and projection neurons. This system therefore provides the first processing station of primary sensory input as well as the relay center of this information to higher centers in the brain [15, 30, 33, 40, 41, 88]. Similarly, both the gray and white matter are optically accessible in the spinal cord (Fig. 4). In vivo imaging therefore, allows for the study of the peripheral and central components as well as white and gray matter of the nervous system simultaneously in situ (Fig. 4)—a feat not easily achieved in any other part of the mammalian anatomy.

The study of neuronal cell bodies in the spinal cord for example, has revealed interesting insights into pain and sensation. Nishida et al. were able to visualize the calcium transients of around 100 dorsal horn neurons simultaneously across 150 μm in depth, studying their responsiveness to different types of peripheral stimulation [8]. Similarly Ran and colleagues visualized dorsal horn neuron's responsiveness to thermal stimuli and found differential coding for hot and cold input in the dorsal horn [10]. In addition to studying stimulus evoked responses it is also possible to study spontaneous activity in dorsal horn neuronal cell bodies [88]. Alternatively, defined subsets of cell bodies in the dorsal horn grey matter can also be visualized, for example by anatomical labeling [14] or by using transgenic mouse lines [9].

Axons in the white matter, on the other hand, are part of the peripheral nervous system and have mostly been studied in relation to spinal cord injury. These studies often require sparse labeling of axons in order to distinguish individual structures [18, 20, 22, 33] and can benefit from the potential of highly localized laser injuries to replace more large scale mechanical damage [19, 22, 48, 69]. Though this may seem to have limited applicability to the study of pain and sensation, these experiments have revealed interesting inflammatory mechanisms that may be relevant to the pain community, for example, the role of nitric oxide in the dynamics of microglial activation [69]. With optical imaging it is possible to visualize the interaction of different components of the CNS, for example tracking the interaction between microglia/macrophages and axons during injury [23]. Arguably, the involvement of white matter pathology in various conditions, including pain conditions, is most readily studied in the spinal cord compared to other parts of the CNS due to the superficial location of the white matter. And the ability to visualize afferent terminals, especially with functional indicators, provides great potential for the study of sensory pathways.

3.1.4 A Flexible Scale of the Recording

Not only does in vivo imaging provide an overview of different cell types, tissues and networks but it also allows a more detailed view into subcellular compartments or an overview over larger areas. Indeed, in vivo imaging is well suited for a relatively easy change in scale. A large-scale overview may be particularly useful in for example the assessment of the time course of a model [89] or when outlining the spatial distribution of a response area to peripheral stimulation [7], which could then be followed up with a more detailed view at an appropriate time and location.

Smaller scale views of axonal or somal compartments are very common in the spinal cord and have been discussed in more detail above (*see* Subheading 3.1.3). It is also possible to focus on parts of these structures. For example, cell membranes in the spinal cord have been visualized using the mt/mG mouse line [90], which highlights nodes of Ranvier in the spinal cord [74]. It is also possible to study even smaller scales, for example individual dendritic spines [13].

On an even smaller scale it is possible to study organelles and their role in various pain related conditions. Mitochondria have for example been studied both structurally and functionally (using membrane potential sensitive dyes) in the spinal cord (Fig. 5), during mouse models of multiple sclerosis [38, 91]. As

a)

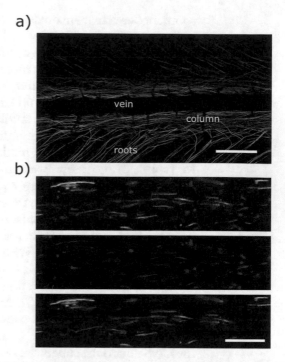

b)

Fig. 5 In vivo imaging of the spinal cord is well suited for the study of structures and functions at different scales. A flexible scale allows the visualization of (**a**) the mesoscopic organization of the spinal cord (shown using a Thy1-YFP transgenic mouse: the clearly labeled axons of sensory neurons delineate the dorsal vein (vein) and show the organization of dorsal columns (column) and dorsal roots (roots), scale bar 500 μm) and (**b**) small subcellular structures within neurons. Axonal mitochondria (blue) and mitochondrial membrane potential (red, tetramethylrhodamine, methyl ester, perchlorate) are visualized using high magnification imaging. Scale bar 10 μm. With permission from Dimitra Schiza and Kenneth Smith.

mitochondria have been variably implicated in some pain conditions [92] this line of research could provide an interesting avenue also in the field of pain and sensation.

3.2 Disadvantages of Spinal Cord Imaging Techniques

3.2.1 The Penetration Depth of Optical Signals

In vivo imaging of the spinal cord does not only come with advantages but also has its own unique set of pitfalls. One of the most commonly sighted problems is the limited penetration of light into biological tissues. This is particularly the case in the spinal cord, in which white matter is covering most of the exposed surface. Myelin has a higher refractive index and shorter scattering length compared to the underlying grey matter and together with increased aberration due to the rounded shape of the spinal cord it becomes harder to probe the spinal cord optically at significant depths. However, luckily, some major components of the pain and sensation pathways lie very superficially on the dorsal horn and the dorsal column. For example, a major pathway to relay pain signals to

(a)

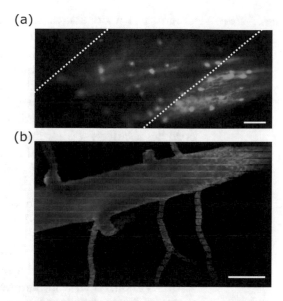

(b)

Fig. 6 In vivo imaging suffers from limitations in penetration depth and movement artifacts. (**a**) A genetically identified subset of lamina II interneurons. The signal originating from these cells is blurred and eventually lost under the scattering tissue of the dorsal roots (in collaboration with David Hughes and Andrew Todd). Dotted lines indicate the approximate presence of obscuring dorsal root. Optimal optical access for grey matter imaging is therefore between roots entering the spinal cord. (**b**) Dorsal vein in the spinal cord labeled with an intravenous injection of Fluorescein Dextran. If the spinal cord is not stabilized optimally and the imaging parameters are not adjusted, movement artifacts can obscure the image and show up as horizontal lines across the image. Scale bar 100 μm

higher brain centers involve lamina I projection neurons which have been multiply implicated in chronic pain conditions [93–95]. Due to their superficial location these projection neurons, as well as other pain-related interneurons, are optically accessible [50], especially if viewed in myelin free areas, for example between roots (Fig. 6a). It has been shown that imaging as deeps as lamina II outer is possible with two-photon and even one-photon imaging [9]. Deeper imaging can also be achieved with two-photon microscopy coupled with fluorescent probes shifted toward longer wavelengths, as redder light penetrates biological tissues more easily and red probes have less spectral overlap with autofluorescence (supporting the slogan: "redder is better"). With currently available nonlinear imaging techniques, cellular resolution in the highly scattering spinal cord is realistically limited to around 100–200 μm depth [49]. Of course, a significant remaining proportion of the spinal cord dorsal horn circuitry is not superficial enough and techniques such as three photon microscopy may be necessary to study deeper layers of the dorsal horn optically.

3.2.2 Movement Artifacts

One of the major issues surrounding any in vivo optical imaging technique is the movement inherent to live preparations. Movement has already been an issue for cortical imaging, but the spinal cord is significantly more difficult to stabilize effectively. Not only is it necessary to contend with heart beat artifacts as blood vessels pulsate, but due to the proximity of the lungs, breathing artifacts are significantly more difficult to combat. Additionally, the spinal cord lies relatively loosely inside the spinal column. As stabilization is achieved along the spinal column and not the cord itself, the movement issues remain significant (Fig. 6b). However, several ways to overcome these issues have been suggested and tested. These include, the addition of agar or silicone elastomer and a coverslip onto the exposed cord [88] and/or suspending the rodent with its chest and abdomen off the stage to allow breathing movements to transmit downward, rather than upward [31]. Nadrigny et al. showed that dorsal aspects of the cord can also be visualized without laminectomy, directly through the intervertebral space, which reduced movement artifacts [15]. In addition, rodents can be tracheotomized. They can either breath spontaneously, reducing the movement originating from the head during breathing and gasping, or alternatively, they can be artificially ventilated. In this case image acquisition can be synchronized with the breathing rate or breathing could be temporarily suspended during acquisition [32, 33].

Johannssen and Helmchen suggest that the choice of rodent can also facilitate stabilization, with mice being easier to stabilize than rats and that 6–8 week old mice are optimal [96]. Our experience supports this to some extent, as mice have less force behind their breaths, making it easier to counteract the movement with spinal stabilization. On the other hand, in our experience it is easier to stably anesthetize rats, limiting the amount of gasping which can occur with longer imaging sessions in mice, although this can be reduced by regularly clearing mucus from their airways [96]. Younger mice tend to breath more steadily in our experience and can be kept more evenly anesthetized across longer experiments.

Additional discussion of movement artifacts and compensation can be found under Subheading 1.1.

3.2.3 To Observe Is to Change

Another issue of in vivo optical imaging surrounds the effect of instrumentation and labeling on the area studied. Laminectomies are arguably more invasive than craniotomies and even the latter has been shown to significantly impact on local circuitry [97, 98].

The relatively severe surgery used to expose the spinal cord is particularly problematic when considering chronic visualization of spinal circuits. The removal of vertebrae and implantation of spinal windows is significantly more difficult compared to cortical windows as the spinal cord is much more mobile and flexible compared to the cranium. This chronic instrumentation has been shown to lead to some spinal cord inflammation [44, 48]. Although gross behavioral abnormalities have not been reported it is likely that the implantation of spinal windows does lead to more nuanced changes in behavior. Repeated exposure of the cord is also possible [30–33] but repeated surgeries, even if the cord is viewed through the intervertebral space [15], provide an added risk of infection and are likely to cause more inflammation and damage to the system studied.

A further complication is introduced by the injection of dyes or expression of large amount of fluorescent proteins, in particular calcium sensing proteins/dyes, which do not only indicate intracellular calcium but also act as exogenous buffers [99–102]. The buffering capacity of an indicator depends on its affinity, concentration and the amount of endogenous buffer inside the cell, against which the exogenous buffer is competing [103] and therefore varies considerably between experiments. Even though calcium indicators should be used at their minimum concentration possible, they usually provide significant buffering at any concentration sufficient to be optically detectable, especially in vivo where autofluorescence demands higher expression levels for signal detection [103]. In addition to problems associated with buffering other considerations, including dye toxicity and interferences with intracellular signaling, need to be considered when conducting in vivo imaging in the spinal cord [51, 104]. This leaves the experimenter with a fundamental trade-off between physiologically accurate and sensitive measurements of calcium transients.

Advances in protein engineering have reduced this trade-off to a certain extent by creating a series of ratiometric sensors with fewer calcium binding sites, while maintaining high calcium affinity [105]. In addition, some fast and easy rules-of-thumb can be implemented to avoid obvious damage, including avoidance of nuclear filling [106, 107], though more subtle effects on cellular physiology should not be ruled-out when interpreting any data from in vivo calcium imaging.

Along the same line, excessive use of lasers can cause significant tissue damage, especially when higher laser powers are needed for deeper penetration with one-photon systems. Some of this damage can be reduced by using two-photon excitation in which out-of-focus tissue is unaffected. However, also here careful calibration of laser power is necessary to avoid phototoxicity.

On the other hand, the damage capabilities of laser light can of course also be used experimentally. Indeed lasers can act as lesion instruments [108, 109] and can therefore be used as a very precise tool to cause highly localized damage in the spinal cord [19, 22, 48, 69] (Box 1).

Box 1 The Early Choices

If you decide that in vivo spinal cord imaging is suitable for you and your lab, here are some of the first few considerations and choices you have to make:

Firstly, it is important to decide whether chronic or acute imaging is more appropriate. Certainly, for the first few experiments acute imaging preparations are likely to be more manageable. This decision would influence your choice of labeling: if a chronic preparation is to be used, it is unlikely that a dye would be suitable (unless used to label blood vessels). Instead genetic labeling would be more stable. Then a structural vs functional label (or both) needs to be chosen. One may choose to use for example quantum dots, injected intravenously, to label blood vessels, which would not only provide information about the vascular network but also an anchor for the stability of the preparation. Then for example a calcium indicator could be used in neurons or nonneuronal cells and a structural indicator, such as RFP could be applied to get better morphology and to visualise also those cells which do not exhibit calcium transients.

The microscope could be a simple fluorescence microscope or a point scanning system which may be slower but has higher sensitivity and greater penetration depth. Add-ons to the point scanning systems can also increase the acquisition frequency, for example the addition of a resonant scanner or a piezo objective scanner for faster z scanning.

The details of the surgical preparation depend on the type of experiment, chronic vs acute, but will always require stabilization via the spinal column, for example through spinal column clamps or a spinal window. When using acute imaging it is also possible to choose between the use of a coverslip or a superfusion system. The coverslip would provide added stability while the superfusion system would provide an access port for the application of drugs. Additional stabilization can be added through a tooth bar and/or tail clamp and suspension of the rodent off the mat. The latter, however, should only be chosen if body temperature can still be maintained, for example through ambient temperature. During acute procedures it is also possible to add a tracheotomy which involves increased instrumentation but can reduce breathing artifacts, especially if coupled with artificial ventilation (in which case breathing can be controlled or briefly stopped during image acquisition).

Table 2
Labeling techniques used in the spinal cord with examples and selected pros and cons

Type	Examples	Pros	Cons
Endogenous signal	Intrinsic optical signal [7] Endogenous autofluorescence [110] Higher harmonic generation [111] Anti-Stokes Raman scattering [35] Spectral confocal reflectance [74]	Does not require the application of labels and therefore reduces the perturbation of the system	Does not provide high resolution imaging in the case of intrinsic optical signal and endogenous autofluorescence Only allows for the visualization of myelinated structures in the case of higher harmonic, anti-Stokes Raman scattering and spectral reflectance microscopy
Dyes	Calcium imaging with Oregon Green 488 BAPTA-1 [88] Visualization of mitochondrial membrane potential [38] Studying vascular dynamics associated with spinal cord injury [26]	Can be used in any model organism Provides flexibility Reduced cost	Cannot label specific subsets of cells Reduced signal to noise because of background labeling Fades over time Requires injection or removal of dura
Proteins	GCaMP imaging in astrocytes using GFAP as a promoter [77] Visualizing axonal calcium levels after spinal cord injury [112] Using Cx3cr1-GFP//Thy1-CFP transgenic animals to visualise microglia/macrophages and spinal cord axons during a model of multiple sclerosis [71]	Specific labeling of genetically identified cells Stable expression over months and years is possible Good signal to noise	Requires breeding of transgenic animals or the use of viral vectors Can be more costly, both financially and with greater time commitments

3.3 The Choice of Optical Labels

The problems associated with in vivo labeling of spinal cord cells, touched upon above, make the choice of labeling technique extremely important. It is necessary to carefully consider a trade-off between sensitivity, selectivity, intensity and the effect on the system studied. Table 2 provides an overview of the labeling techniques used in the spinal cord.

One of the arguably least invasive ways to visualize the spinal cord is through the intrinsic properties of the spinal cord tissue. This has broadly been achieved through two approaches in the spinal cord: the overview visualization of intrinsic optical signal or endogenous fluorescence and the mapping of myelination in the dorsal spinal cord through higher harmonic generation, coherent anti-Stokes Raman scattering or spectral confocal reflectance microscopy. None of these approaches require the application of dyes and therefore provide some of the most physiological protocols for spinal cord in vivo imaging.

Intrinsic optical signal involves the visualization of changes in tissue reflectance and is known to vary with neuronal activity [113]. The signal change is believed to be related to changes in at least three factors (the relative importance of each of these factors in the signal vary depending on the wavelength employed): changes in oxygen saturation of hemoglobin, changes in blood volume, and changes in light scattering associated with variations in extracellular volume and capillaries, ion and water levels, and/or release of neurotransmitter [114–117]. Although intrinsic optical signal is predominantly used in the study of functional cortical maps, Sasaki et al. were the first to visualize intrinsic optical signal in vivo in the spinal cord. They were able to elicit graded signal changes with increasing electrical stimulation and were able to differentiate termination zones of peripheral nerves [7].

As it does not involve the expression of foreign protein this approach can be relatively noninvasive and indeed has been used to study human cortical activity [118–122]. However, intrinsic optical signal provides relatively low spatial and temporal resolution. Indeed, the strongest signal in the spinal cord is likely related to oxygen delivery and is therefore not applicable to single cell resolution or high temporal resolution, with events spaced less than 80 ms apart being indistinguishable [123].

In addition to reflectance, it is also possible to use fluorescence to detect further changes in endogenous signals. In the spinal cord and the cortex the endogenous green fluorescence of oxidized flavoproteins is believed to be related to neuronal metabolic demand and therefore postsynaptic neuronal activation [110, 124, 125]. Electrical stimulation was found to lead to a graded increase in green fluorescence in the ipsilateral (and to a less extent contralateral) spinal cord, while lidocaine application to the sciatic nerve prevented this [110]. Sensitizing agents, like capsaicin, or nerve injury increase the flavoprotein signal change in response to peripheral stimulation [110, 124], demonstrating the potential application of this technique to the field of chronic pain.

Thus, intrinsic reflectance and fluorescence can be used to study large scale changes in the spinal cord under physiological conditions but with low spatial and temporal sensitivity.

Label-free imaging can also reveal details about spinal cord myelination. This has been achieved through three different approaches: third harmonic generation, coherent anti-Stokes Raman scattering, and spectral confocal reflectance microscopy.

Third harmonic generation imaging requires the excitation of a molecule by three photons of a certain wavelength which results in the emission of a single photon of a shorter wavelength. The sum of the frequency of the original photons gives the frequency of the emitted one (third harmonic). In biological samples, third harmonic generation picks up signals where the refractive index within a sample is variable, for example where lipid-rich structures meet aqueous tissues [126]. In the spinal cord this is provided by the layering of cytoplasm and myelin in white matter fiber tracts. Just like two-photon fluorescence microscope, third harmonic generation only occurs at the focal plane, so it is ideally suited for the highly scattering spinal cord tissues. However, unlike fluorescence microscope (where emission is equal in all directions), the emission resulting from third harmonic generation is predominantly in the forward direction making it less suitable for in vivo imaging, where laser induced sample damage limits this approach to 50–70 µm in depth [111]. Nevertheless, Farrar and colleagues were able to demonstrate the feasibility of third harmonic generation imaging in the spinal cord in vivo, to detect changes in myelination during pathological states [111]. However, the laser required makes this an unlikely approach for a standard research laboratory.

Instead coherent anti-Stokes Raman Scattering (CARS) imaging has been more widely applied to the label-free visualization of myelin structures in the spinal cord in vivo. CARS works by visualizing molecular bonds by virtue of their vibrational energy. In the case of myelin it picks up carbon-hydrogen bonds in lipids. It requires excitation by lasers with different wavelengths. The difference of the wavelengths needs to parallel the vibrational frequency of the carbon-hydrogen bonds in the myelin. Just like multiphoton microscopy and third harmonic generation CARS is a nonlinear process and thereby provides automatic optical sectioning. As CARS relies on pulsed lasers it is compatible with in vivo imaging and provides good in-plane and depth resolution. Shi et al. were able to visualize myelin in the rat spinal cord longitudinally in vivo using CARS and were able to detect individual nodes of Ranvier. They saw no behavioral or structural damage as a result of repetitive imaging [42]. Bélanger et al. used a commercially available endoscope to visualize CARS signal in the spinal cord. They simultaneously used the two-photon imaging capabilities of the endoscope and were therefore able to visualize both myelin morphology and microglial activation [35].

Spectral confocal reflectance microscopy, on the other hand, can be achieved with commercially available confocal scanning microscopes, without modification. This technique uses multiple confocal lasers (488 nm, 561 nm, and 633 nm lasers were sufficient to detect continuous bands of myelination) and combines their reflection signal to generate a complete view of myelination. By detecting a narrow band of reflected light from each excitation laser, Schain and colleagues were able to visualize spinal cord myelination in vivo [74].

However, especially in the field of pain and sensation, experimental questions are often not limited to the study of myelination. For the study of spinal cord structure or function at the cellular level therefore it is often essential to add specific labels to the structures of interest. The available options are essentially divided into two camps, the use of dyes or genetically encoded proteins. The benefit of using dyes comes from the relatively easy application and the ability to use these in many species of animal. Dyes are often loaded into the nervous system through bolus loading techniques [127], which is capable of loading multiple cells simultaneously up to around 150 μm from the injection site [127]. Alternatively some dyes can also be applied topically to the spinal cord after removal of the dura [51].

The study of neuronal function has significantly benefit from the availability of calcium indicator dyes. Calcium is a vital intracellular messenger in neurons with a very intimate relationship to action potential (AP) firing. Although variable, most resting neurons have an intracellular calcium concentration of between 50 and 100 nM. This can increase 10–1000 times during electrical activity. The advantages of calcium imaging also provide some of the limitations. Calcium transients are amplified and prolonged relative to action potentials, both in time (calcium transients are slower than AP) and in space (calcium transients spread away from the localized membrane changes associated with APs). This means that calcium transients provide a more easily detectable neuronal event, but at the trade-off of reduce temporal sensitivity. Additionally, of course calcium measures do not report on subthreshold voltage-changes.

Johannssen and Helmchen bolus-loaded an acetoxymethyl (AM) ester form of a calcium indicator (Oregon Green 488 BAPTA-1), diluted in DMSO and Pluronic F-127, into the dorsal horn of the spinal cord [88]. It is important to use AM dyes [128], which have masked carboxylates and are therefore capable of crossing cell membranes. Once inside the cell cytosolic esterases will cleave the ester groups, leaving the dye trapped inside the cell. The pressure injection of the dye mixture 100–200 μm below the spinal cord surface resulted in surface labeling of the spinal cord dorsal horn. This technique was used similarly by Ran et al. to study the coding of cutaneous temperature in the dorsal horn [10]. As this technique labels all cells in the spinal cord, an injection of a more specific dye can aid in untangling the network of cells being visualized. One approach for example is the injection of sulforhodamine 101 which seems to show preferential labeling of glia-like cells [88], but is not as specific to astrocytes in the spinal cord as in the cortex [129]. In addition to neurons and astrocytes other structures can also be labeled in the spinal cord in vivo. In an elegant paper by Nikić et al. they were able to label mitochondrial membrane potential (using tetramethylrhodamine, methyl ester, perchlorate), myelination (using Cell Trace BODIPY TR methyl ester dye), hydrogen peroxide (using Amplex Ultra Red), and nitric oxide (using DAF-FM diacetate) [38]. Additionally an Alexa Fluor–conjugated IB4-isolectin can be used to label IB4+ unmyelinated fibers and activated microglia/macrophages [51, 130]. These dyes were simply incubated on the exposed spinal cord surface for around 30 min to 1 h before the spinal cord was washed and imaged. In addition to labeling the spinal cord parenchyma, labeling of the blood vessels using dyes is straight forward. An intravenous injection of a fluorescent marker, for example a fluorescent dextran, can provide not only a reference point for imaging, for example for repeat imaging or to reduce the effects of movement and drift, but can also provide information about the functioning of the vasculature. For example, differently sized dyes can be used to investigate the leakiness of blood vessels during inflammatory pain models. By using one dye before and a differently colored dye after an injury, it is even possible to visualize vascular dynamics across spinal cord trauma [26].

However, the application of dyes comes with a string of considerations. Firstly, as with all labeling techniques the presence of nonphysiological factors means that the imaged network is somewhat altered. In fact, dye toxicity needs to be considered and limited by running concentration trials and using dyes at the minimum dilution feasible (see Subheading 3.2.3). Additional damage is provided through the administration rout, for example through injection or through the removal of the dura, which will occur at or close to the imaging site. Injecting into and imaging through the intervertebral space can avoid the damage associated with laminectomies [15, 131]. In addition to any damage incurred, it should be noted that dyes are not usually specific to cell types. Therefore, it is only possible to label all cells in the spinal cord through bolus injections or topical application, or individual cells through patch pipette injections. This former approach usually also results in significantly reduced signal-to-noise ratios as the background labeling can obscure areas of interest. Furthermore, the labeling spread with dyes is limited. For example after topical application, most dyes do not penetrate beyond 30 μm [51]. This means that only surface structures can be visualized or if bolus loading is used, imaging needs to occur relatively close to the injection site (with a labeling diameter of around 300 μm one can image a maximum of 150 μm away from the injection site) [127]. Finally, dyes do not last. Not only can they potentially leak/fade throughout a long imaging session but they also do not provide consistency for chronic imaging. If repeated imaging is required one would potentially cause repeated damage through injection. Additionally, repeated application of dyes inevitably leads to added variability, although at least some dyes seem to have good labeling reliability across repeat application [51].

Some of these issues are counteracted by the expression of fluorescent proteins which allow for less invasive, longitudinal, and highly specific labeling in identified cells. Structural labeling is easily achieved with a wide range of differently colored fluorescent proteins expressed in different cell types in an ever-increasing number of transgenic mouse lines. Advancements in red shifted proteins offer particular advantages to spinal cord imaging where penetration depth is hampered due to the overlying, scattering myelin. With longer wavelengths better penetration is achieved. For a review on the evolution of red shifted fluorescent proteins, *see* for example [132].

Structural labeling is not the only avenue offered by fluorescent proteins. Various functional indicators are now commercially available and have been used for spinal cord imaging, including the well-known genetically encoded calcium indicators (GECIs). GECIs are able to signal an increase in intracellular calcium through changes in fluorescence properties upon calcium binding and now have sensitivity sufficient to detect calcium changes in response to individual action potentials [53, 107, 133, 134], while providing stable expression over months [135].

Genetically encoded calcium indicators come in essentially two flavors: single wavelength indicators like the very well-established GCaMP family, or ratiometric calcium indicators, including the Cameleon family.

Among the most popular GECIs for spinal cord imaging, is the rapidly evolving, single wavelength indicator family, GCaMP [136]. Their high signal to noise ratio provides better signal for the highly scattering tissues of the spinal cord. They have been expressed in the spinal cord through microinjections [11, 50] including promoter driven expression to target astrocytes [77] as well as through transgenic mouse lines [17] including a cre-recombinase system [24].

Currently a very popular GCaMP iteration is the GCaMP6 group which is sensitive enough to detect changes in intracellular calcium in response to single action potentials with up to 99% detection rate of single spikes [107]. GCaMP6f for example has a dynamic range of around 1300% and a rise time of 50–80 ms. The signal of GCaMP6s and m is stronger but exhibits longer decay times [107].

Ratiometric indicators, on the other hand, can offer advantages in biological preparations with movement and other artifacts, including the spinal cord. Such ratiometric probes are based on Förster resonance energy transfer (FRET) and are therefore composed of two fluorescent molecules. In the presence of calcium, a conformational change in the indicator results in the two fluorescent molecules coming into close enough contact for energy transfer to occur between them through nonradiative dipole–dipole coupling. This means, the excited molecule emits a so-called virtual photon (these are undetectable, making this a nonradiative interaction). This is then instantaneously absorbed by the acceptor molecule, which becomes excited and releases energy in the form of a photon. Therefore, a ratio of donor vs acceptor emission provides a readout of the levels of calcium inside cells. Such a ratiometric approach has indeed found application in the spinal cord [8, 112] where it could compensate for elevated movement artifacts, which affect both the donor and acceptor molecule equally. However, the reduced sensitivity of these indicators with inherently higher noise levels (due to combination of two independent shot noise components), often provides practical limitations to in vivo spinal cord imaging [137]. Instead, nonratiometric indicators suffer from variable cellular expression, limiting the precision of these tools. This can at least partially be compensated for by examining postmortem fluorescence where it is assumed that intracellular calcium equilibrates sufficiently across cells. This should theoretically allow assessments of expression levels.

Of course, genetically encoded indicators are not limited to calcium indicators and instead come in a multitude of flavors, including for example peroxide sensors [138], pH sensors [139], often used as indicators or vesicular release during presynaptic activity, glutamate sensors [140, 141] and voltage indicators [142–144]. However, to our knowledge these have not yet been used in the spinal cord and many suffer from limitations including signal strength and photostability.

All genetically encoded indictors need to be expressed in cells of interest with sufficient specificity. In the spinal cord this has been achieved through the use of transgenic animals, in utero electroporation and viral vectors. The transgenic approach has gained huge traction with the ever-increasing number of cre lines available, which can "unstop," just in cells of interest, a ubiquitously expressed transgene, for example a fluorescent protein. Particularly in the spinal cord, in which the cellular composition is highly complex [52], the huge array of available cre lines provide a great advantage. However, only a few publications so far have taken advantage of cre lines in the spinal cord and these are limited to visualize astrocytic calcium [24] and structural labels for somatostatin positive interneurons [9]. Problems associated with cre lines can originate from off target effects due to some possible leakiness, but more importantly due to expression of genes during development. However, the latter problem can be combatted by using inducible cre lines.

Transgenic mouse lines often require significant investment, both financially and through time spent breeding. Viral vectors and in utero electroporation provide alternative. This (arguably more flexible) approach, however, suffers from some labeling variability and predictable outcomes are harder to achieve. Viral vectors can provide specificity, through the tropism of the virus, the promoter being used as well as the injection site. For example the labeling of primary afferents can be achieved through intrathecal injections of AAV9 [133], while injections into projection neuron targets will label them selectively [50, 145]. Unfortunately, this approach cannot be applied to the majority of the spinal cord dorsal horn circuitry which is predominantly local. Instead, the viral vector would have to be injected directly into the spinal cord and therefore involves similar limitations as dye injections. Further, both viral vectors and *in utero* electroporation can only carry constructs of certain sizes, imposing limits on the promoters that can feasibly be used.

The ability to apply multiple labels simultaneously is advantageous to functional indicators with low baseline fluorescence, including GCaMP. The addition of a supporting, structural label allows the visualization of cells even if they do not show calcium transient and reduces reliance on search stimuli. Overall, this will provide a far more balanced view of the spinal cord neuronal network with fewer biases toward certain stimuli or responses, and adds the possibility of comparing the number of responding cells to the total number of cells labeled.

3.4 What Promises Does In Vivo Imaging of the Spinal Cord Hold

In vivo imaging of the spinal cord has been made possible by advances in two key fields: the physics/engineering of multiphoton microscopy and the development of new labeling strategies to visualize the structure and function of selected parts of the spinal cord. Further advances in these fields continue to increase the available tools to optically probe the spinal cord. Here, we will take a brief look at selected recent developments that we predict will have a significant impact on in vivo spinal cord imaging in the near future.

Advancements in miniaturization of electronic and optical elements has made it possible to generate microscopes small enough to be carried by awake, behaving mice. The further commercialization of these microscopes makes them accessible to laboratories focused on pain and sensation, even with little experience in optical techniques (e.g., https://www.inscopix.com/). While already widely used in the brain, such techniques have not yet been applied much to the spinal cord, although the beginning has already shown great promise [9]. We expect the ability to record from the spinal cord of unanesthetized animals to provide a huge advance to the field of pain and sensation where anesthetics have traditionally provided a significant barrier.

In addition to miniaturization, the improvements in laser technology has allowed researchers to apply longer wavelength imaging, providing a window to move from two-photon microscopy to three-photon microscopy. Although two-photon microscopy has already allowed visualization of surface structures of the dorsal column and dorsal horn of the spinal cord with penetration of up to 100–200 μm [49], the use of three-photon microscopy, promises to allow even deeper penetration and the ability to visualize ever more parts of the pain circuitry, including across myelinated structures. Indeed, three-photon microscopy has already shown great potential in the study of subcortical areas [146, 147] and a proof of concept visualization of spinal cord structures up to 500 μm below the surface has been shown in a recent review [49]. Such imaging depths could potentially allow the visualization of the entirety of the mouse dorsal horn using only optical techniques. This could mean that alternative techniques, including for example the use of implanted endoscopic systems, often used in the brain to reach deeper structures [148, 149], may not be necessary for mouse spinal cord imaging in the field of pain and sensation.

Not only do we need to look deeper into the living spinal cord, we also need to be able to probe the spinal circuitry in different ways. The development of different forms of functional indicators is invaluable in our search to understand spinal cord circuitry. One such advancement comes from voltage indicators [150].

To our knowledge these have not been used for spinal cord imaging. However, they provide potential advantages over calcium indicators, not only because they offer a more direct readout of neural activity with much faster kinetics, but also because of the potential to detect subthreshold activity. Advances have provided some very promising indicators including a genetically encoded voltage sensor (GEVI) which is capable of capturing single action potentials and subthreshold activity in neurons both in vivo and in vitro [151] and a hybrid approach using dye capture protein that can bind a brighter and more stable synthetic voltage indicator dye [152]. A recent comparison of GEVIs, however, suggests that available probes, though promising, are still lacking essential characteristics to provide a one-size-fits-all probe [153].

Even with drastic improvements in voltage probes, calcium indicators are unlikely to be supplanted. Instead these two systems offer complementary strengths and weaknesses [154, 155]. Calcium transients cover more space and time than voltage changes. This limits the spatial and temporal sensitivity of calcium indicators, compared to voltage indicators but also makes it significantly easier to detect change in signal with standard one or two-photon setups. Alone the increased speed at which microscope systems need to operate in order to detect voltage changes provides significant barriers, especially for laser scanning systems. In addition, due to their membrane localization, most signal from voltage indicators will originate from the membranes of axons and dendrites and therefore neuropil [154, 156]. This is significantly less easy to differentiate into cellular origins compared to signal from calcium indicators, which is preferentially located in the cytoplasm, including in the soma. Voltage indicators are therefore more likely to provide excellent temporal resolution with less spatial precision [156–160].

Thus, we expect the future of functional imaging in the spinal cord to be bright but also colorful with new indicators supplementing existing tools, rather than replacing them.

These advances in technology will hopefully provide a wealth of new insights in the near future. However, even without any new technologies it seems that many new questions in the pain and sensation field could be explored. This includes the characterization of a large number of dorsal horn interneurons, both structurally and functionally. A quickly expanding number of cre driver lines expressed in different, genetically distinct interneurons, is a huge benefit to this endeavor.

Finally, it is possible that functional in vivo imaging of the spinal cord could be applied to human subjects, as has been done in the skin and brain [118–122, 161]. Here, label free approaches are the most immediately promising tools, including intrinsic optical signal and the visualization of metabolic states through flavoproteins and pyridine nucleotides.

4 Conclusion

In vivo imaging of the spinal cord has many advantages, including the ability to visualize many diverse cellular and subcellular structures, while its current limitations, including penetration depth, movement issues, and physiological impact, are continuously being addressed and improved. As such this technique has generated a wealth of information relevant to the field of pain and sensation. There is much reason to believe that spinal cord in vivo imaging will continue to provide vital insight into the (patho)-physiology of the spinal cord.

References

1. McMahon SB, Koltzenburg M, Tracey I, Turk DC (2013) Wall and Melzack's textbook of pain. Elsevier

2. Besse D, Lombard MC, Zajac JM, Roques BP, Besson JM (1990) Pre- and postsynaptic distribution of μ, δ and κ opioid receptors in the superficial layers of the cervical dorsal horn of the rat spinal cord. Brain Res 521(1–2):15–22

3. Häring M et al (2018) Neuronal atlas of the dorsal horn defines its architecture and links sensory input to transcriptional cell types. Nat Neurosci 21(6):869–880

4. Graham BA, Brichta AM, Callister RJ (2007) Moving from an averaged to specific view of spinal cord pain processing circuits. J Neurophysiol 98(3):1057–1063

5. Lu Y, Perl ER (2003) A specific inhibitory pathway between substantia gelatinosa neurons receiving direct c-fiber input. J Neurosci 23(25):8752–8758

6. Lu Y, Perl ER (2005) Modular organization of excitatory circuits between neurons of the spinal superficial dorsal horn (laminae I and II). J Neurosci 25(15):3900–3907

7. Sasaki S et al (2002) Optical imaging of intrinsic signals induced by peripheral nerve stimulation in the in vivo rat spinal cord. NeuroImage 17(3):1240–1255

8. Nishida K, Matsumura S, Taniguchi W, Uta D, Furue H, Ito S (2014) Three-dimensional distribution of sensory stimulation-evoked neuronal activity of spinal

dorsal horn neurons analyzed by in vivo calcium imaging. PLoS One 9(8):e103321

9. Sekiguchi KJ et al (2016) Imaging large-scale cellular activity in spinal cord of freely behaving mice. Nat Commun 7:11450

10. Ran C, Hoon MA, Chen X (2016) The coding of cutaneous temperature in the spinal cord. Nat Neurosci 19(9):1201–1209

11. Chen T et al (2018) Top-down descending facilitation of spinal sensory excitatory transmission from the anterior cingulate cortex. Nat Commun 9(1):1886

12. Cirillo G, De Luca D, Papa M (2012) Calcium imaging of living astrocytes in the mouse spinal cord following sensory stimulation. Neural Plast 2012:425818

13. Matsumura S, Taniguchi W, Nishida K, Nakatsuka T, Ito S (2015) In vivo two-photon imaging of structural dynamics in the spinal dorsal horn in an inflammatory pain model. Eur J Neurosci 41(7):987–995

14. Ikeda H et al (2006) Synaptic amplifier of inflammatory pain in the spinal dorsal horn. Science 312(5780):1659–1662

15. Nadrigny F, Le Meur K, Schomburg ED, Safavi-Abbasi S, Dibaj P (2017) Two-photon laser-scanning microscopy for single and repetitive imaging of dorsal and lateral spinal white matter in vivo. Physiol Res 66:531–537

16. Evans TA, Barkauskas DS, Myers JT, Huang AY (2014) Intravital imaging of axonal interactions with microglia and macrophages in a mouse dorsal column crush injury. J Vis Exp 93:e52228

17. Tang P et al (2015) In vivo two-photon imaging of axonal dieback, blood flow, and calcium influx with methylprednisolone therapy after spinal cord injury. Sci Rep 5:9691

18. Dray C, Rougon G, Debarbieux F (2009) Quantitative analysis by in vivo imaging of the dynamics of vascular and axonal networks in injured mouse spinal cord. Proc Natl Acad Sci U S A 106(23):9459–9464

19. Ylera B et al (2009) Chronically CNS-injured adult sensory neurons gain regenerative competence upon a lesion of their peripheral axon. Curr Biol 19(11):930–936

20. Ertürk A, Hellal F, Enes J, Bradke F (2007) Disorganized microtubules underlie the formation of retraction bulbs and the failure of axonal regeneration. J Neurosci 27(34):9169–9180

21. Dibaj P, Steffens H, Nadrigny F, Neusch C, Kirchhoff F, Schomburg ED (2010) Long-lasting post-mortem activity of spinal microglia in situ in mice. J Neurosci Res 88(11):2431–2440

22. Lorenzana AO, Lee JK, Mui M, Chang A, Zheng B (2015) A surviving intact branch stabilizes remaining axon architecture after injury as revealed by invivo imaging in the mouse spinal cord. Neuron 86(4):947–954

23. Evans TA et al (2014) High-resolution intravital imaging reveals that blood-derived macrophages but not resident microglia facilitate secondary axonal dieback in traumatic spinal cord injury. Exp Neurol 254:109–120

24. Cartarozzi LP, Rieder P, Bai X, Scheller A, de Oliveira ALR, Kirchhoff F (2018) In vivo two-photon imaging of motoneurons and adjacent glia in the ventral spinal cord. J Neurosci Methods 299:8–15

25. Jahromi NH et al (2017) A novel cervical spinal cord window preparation allows for two-photon imaging of T-cell interactions with the cervical spinal cord microvasculature during experimental autoimmune encephalomyelitis. Front Immunol 8:406

26. Chen C et al (2017) An in vivo duo-color method for imaging vascular dynamics following contusive spinal cord injury. J Vis Exp 130:2017

27. Vajkoczy P, Laschinger M, Engelhardt B (2001) α4-integrin-VCAM-1 binding mediates G protein-independent capture of encephalitogenic T cell blasts to CNS white matter microvessels. J Clin Invest 108(4):557–565

28. Laschinger M, Vjakoczy P, Engelhardt B (2002) Encephalitogenic T cells use LFA-1 for transendothelial migration but not during capture and initial adhesion strengthening in healthy spinal cord microvessels in vivo. Eur J Immunol 32(12):3598–3606

29. Ishikawa M et al (1999) In vivo rat closed spinal window for spinal microcirculation: observation of pial vessels, leukocyte adhesion, and red blood cell velocity. Neurosurgery 44(1):156–162

30. Bareyre FM, Garzorz N, Lang C, Misgeld T, Büning H, Kerschensteiner M (2011) In vivo imaging reveals a phase-specific role of stat3 during central and peripheral nervous system axon regeneration. Proc Natl Acad Sci U S A 108(15):6282–6287

31. Davalos D et al (2008) Stable in vivo imaging of densely populated glia, axons and blood vessels in the mouse spinal cord using two-photon microscopy. J Neurosci Methods 169(1):1–7

32. Misgeld T, Nikic I, Kerschensteiner M (2007) In vivo imaging of single axons in the mouse spinal cord. Nat Protoc 2(2):263–268

33. Kerschensteiner M, Schwab ME, Lichtman JW, Misgeld T (2005) In vivo imaging of axonal degeneration and regeneration in the injured spinal cord. Nat Med 11(5):572–577

34. Drdla R, Gassner M, Gingl E, Sandkühler J (2009) Induction of synaptic long-term potentiation after opioid withdrawal. Science 325(5937):207–210

35. Bélanger E, Crépeau J, Laffray S, Vallée R, De Koninck Y, Côté D (2012) Live animal myelin histomorphometry of the spinal cord with video-rate multimodal nonlinear microendoscopy. J Biomed Opt 17(2):021107

36. Laffray S, Pagès S, Dufour H, de Koninck P, de Koninck Y, Côté D (2011) Adaptive movement compensation for in vivo imaging of fast cellular dynamics within a moving tissue. PLoS One 6(5):e19928

37. Davalos D, Akassoglou K (2012) In vivo imaging of the mouse spinal cord using two-photon microscopy. J Vis Exp 59:1–5

38. Nikić I et al (2011) A reversible form of axon damage in experimental autoimmune encephalomyelitis and multiple sclerosis. Nat Med 17(4):495–499

39. Zhang Y et al (2014) Two-photon-excited fluorescence microscopy as a tool to investigate the efficacy of methylprednisolone in a mouse spinal cord injury model. Spine 39(8):E493–E499

40. Skuba A, Himes BT, Son Y-J (2011) Live imaging of dorsal root axons after rhizotomy. J Vis Exp 55:e3126

41. Di Maio A et al (2011) In vivo imaging of dorsal root regeneration: rapid immobilization and presynaptic differentiation at the CNS/PNS border. J Neurosci 31(12):4569–4582

42. Shi Y et al (2011) Longitudinal in vivo coherent anti-Stokes Raman scattering imaging of demyelination and remyelination in injured spinal cord. J Biomed Opt 16(10):106012

43. Fenrich KK, Weber P, Rougon G, Debarbieux F (2013) Implanting glass spinal cord windows in adult mice with experimental autoimmune encephalomyelitis. J Vis Exp 82:e50826

44. Fenrich KK, Weber P, Hocine M, Zalc M, Rougon G, Debarbieux F (2012) Long-term in vivo imaging of normal and pathological mouse spinal cord with subcellular resolution using implanted glass windows. J Physiol 590(16):3665–3675

45. Fenrich KK, Weber P, Rougon G, Debarbieux F (2013) Long- and short-term intravital imaging reveals differential spatiotemporal recruitment and function of myelomonocytic

cells after spinal cord injury. J Physiol 591(19):4895–4902

46. Figley SA et al (2013) A spinal cord window chamber model for in vivo longitudinal multimodal optical and acoustic imaging in a murine model. PLoS One 8(3):e58081

47. Farrar MJ, Schaffer CB (2014) A procedure for implanting a spinal chamber for longitudinal in vivo imaging of the mouse spinal cord. J Vis Exp (94)

48. Farrar MJ, Bernstein IM, Schlafer DH, Cleland TA, Fetcho JR, Schaffer CB (2012) Chronic in vivo imaging in the mouse spinal cord using an implanted chamber. Nat Methods 9(3):297–302

49. Cheng YT, Lett KM, Schaffer CB (2019) Surgical preparations, labeling strategies, and optical techniques for cell-resolved, in vivo imaging in the mouse spinal cord. Exp Neurol 318:192–204

50. Chisholm et al (2021) Encoding of cutaneous stimuli by lamina I projection neurons. https://doi.org/10.1097/j.pain.0000000000002226

51. Romanelli E, Sorbara CD, Nikić I, Dagkalis A, Misgeld T, Kerschensteiner M (2013) Cellular, subcellular and functional in vivo labeling of the spinal cord using vital dyes. Nat Protoc 8(3):481–490

52. Todd AJ (2010) Neuronal circuitry for pain processing in the dorsal horn. Nat Rev Neurosci 11(12):823–836

53. Podor B, Hu Y-L, Ohkura M, Nakai J, Croll R, Fine A (2015) Comparison of genetically encoded calcium indicators for monitoring action potentials in mammalian brain by two-photon excitation fluorescence microscopy. Neurophotonics 2(2):021014

54. Coyle DE (1998) Partial peripheral nerve injury leads to activation of astroglia and microglia which parallels the development of allodynic behavior. Glia 23(1):75–83

55. Colburn RW, Rickman AJ, Deleo JA (1999) The effect of site and type of nerve injury on spinal glial activation and neuropathic pain behavior. Exp Neurol 157(2):289–304

56. Herzberg U, Sagen J (2001) Peripheral nerve exposure to HIV viral envelope protein gp120 induces neuropathic pain and spinal gliosis. J Neuroimmunol 116(1):29–39

57. Tsuda M et al (2011) JAK-STAT3 pathway regulates spinal astrocyte proliferation and neuropathic pain maintenance in rats. Brain 134(4):1127–1139

58. Guo W et al (2007) Glial-cytokine-neuronal interactions underlying the mechanisms of

persistent pain. J Neurosci 27(22): 6006–6018

59. Zarpelon AC et al (2016) Spinal cord oligodendrocyte-derived alarmin IL-33 mediates neuropathic pain. FASEB J 30(1):54–65

60. Shi Y, Shu J, Liang Z, Yuan S, Tang SJ (2016) Oligodendrocytes in HIV-associated pain pathogenesis. Mol Pain 12: 1744806916656845

61. Echeverry S, Shi XQ, Zhang J (2008) Characterization of cell proliferation in rat spinal cord following peripheral nerve injury and the relationship with neuropathic pain. Pain 135(1–2):37–47

62. Beggs S, Salter MW (2007) Stereological and somatotopic analysis of the spinal microglial response to peripheral nerve injury. Brain Behav Immun 21(5):624–633

63. Tsuda M et al (2003) P2X4 receptors induced in spinal microglia gate tactile allodynia after nerve injury. Nature 424(6950):778–783

64. Coull JAM et al (2005) BDNF from microglia causes the shift in neuronal anion gradient underlying neuropathic pain. Nature 438(7070):1017–1021

65. Ledeboer A et al (2005) Minocycline attenuates mechanical allodynia and proinflammatory cytokine expression in rat models of pain facilitation. Pain 115(1–2):71–83

66. Denk F, Crow M, Didangelos A, Lopes DM, McMahon SB (2016) Persistent alterations in microglial enhancers in a model of chronic pain. Cell Rep 15(8):1771–1781

67. Clark AK et al (2007) Inhibition of spinal microglial cathepsin S for the reversal of neuropathic pain. Proc Natl Acad Sci U S A 104(25):10655–10660

68. Guan Z et al (2015) Injured sensory neuron-derived CSF1 induces microglial proliferation and DAP12-dependent pain. Nat Neurosci 19(1):94–101

69. Dibaj P et al (2010) NO mediates microglial response to acute spinal cord injury under ATP control in vivo. Glia 58(9):1133–1144

70. Chrobok NL et al (2017) Monocyte behaviour and tissue transglutaminase expression during experimental autoimmune encephalomyelitis in transgenic CX3CR1 gfp/gfp mice. Amino Acids 49(3):643–658

71. Davalos D et al (2012) Fibrinogen-induced perivascular microglial clustering is required for the development of axonal damage in neuroinflammation. Nat Commun 3:1227

72. Dibaj P et al (2011) In vivo imaging reveals distinct inflammatory activity of CNS microglia versus PNS macrophages in a mouse model for ALS. PLoS One 6(3):e17910

73. Dibaj P et al (2012) Influence of methylene blue on microglia-induced inflammation and motor neuron degeneration in the SOD1G93A model for ALS. PLoS One 7(8):e43963

74. Schain AJ, Hill RA, Grutzendler J (2014) Label-free in vivo imaging of myelinated axons in health and disease with spectral confocal reflectance microscopy. Nat Med 20(4): 443–449

75. Yang Z, Xie W, Ju F, Khan A, Zhang S (2017) In vivo two-photon imaging reveals a role of progesterone in reducing axonal dieback after spinal cord injury in mice. Neuropharmacology 116:30–37

76. Dibaj P, Steffens H, Zschüntzsch J, Kirchhoff F, Schomburg ED, Neusch C (2011) In vivo imaging reveals rapid morphological reactions of astrocytes towards focal lesions in an ALS mouse model. Neurosci Lett 497(2):148–151

77. Yoshihara K, Matsuda T, Kohro Y, Tozaki-Saitoh H, Inoue K, Tsuda M (2018) Astrocytic Ca2+ responses in the spinal dorsal horn by noxious stimuli to the skin. J Pharmacol Sci 137(1):101–104

78. Lutz SE et al (2017) Caveolin1 is required for Th1 cell infiltration, but not tight junction remodeling, at the blood-brain barrier in autoimmune neuroinflammation. Cell Rep 21(8):2104–2117

79. Farrar MJ, Rubin JD, Diago DM, Schaffer CB (2015) Characterization of blood flow in the mouse dorsal spinal venous system before and after dorsal spinal vein occlusion. J Cereb Blood Flow Metab 35:667–675

80. Ishikawa M et al (2002) Platelet adhesion and arteriolar dilation in the photothrombosis: observation with the rat closed cranial and spinal windows. J Neurol Sci 194(1):59–69

81. Ishikawa M, Sekizuka E, Krischek B, Sure U, Becker R, Bertalanffy H (2002) Role of nitric oxide in the regulation of spinal arteriolar tone. Neurosurgery 50(2):371–377

82. Li Y et al (2017) Pericytes impair capillary blood flow and motor function after chronic spinal cord injury. Nat Med 23(6):733–741

83. Miyazaki K et al (2012) Early and progressive impairment of spinal blood flow-glucose metabolism coupling in motor neuron degeneration of ALS model mice. J Cereb Blood Flow Metab 32(3):456–467

84. Barkauskas DS, Evans TA, Myers J, Petrosiute A, Silver J, Huang AY (2013) Extravascular CX3CR1+ cells extend intravascular dendritic processes into intact central

nervous system vessel lumen. Microsc Microanal 19(4):778–790

85. Bartholomäus I et al (2009) Effector T cell interactions with meningeal vascular structures in nascent autoimmune CNS lesions. Nature 462(7269):94–98

86. Sathiyanadan K, Coisne C, Enzmann G, Deutsch U, Engelhardt B (2014) PSGL-1 and E/P-selectins are essential for T-cell rolling in inflamed CNS microvessels but dispensable for initiation of EAE. Eur J Immunol 44(8):2287–2294

87. Coisne C, Lyck R, Engelhardt B (2013) Live cell imaging techniques to study T cell trafficking across the blood-brain barrier in vitro and in vivo. Fluids Barriers CNS 10(1):7

88. Johannssen HC, Helmchen F (2010) In vivo Ca2+ imaging of dorsal horn neuronal populations in mouse spinal cord. J Physiol 588 (Pt 18):3397–3402

89. Tian F et al (2011) In vivo optical imaging of motor neuron autophagy in a mouse model of amyotrophic lateral sclerosis. Autophagy 7(9):985–992

90. Muzumdar MD, Tasic B, Miyamichi K, Li L, Luo L (2007) A global double-fluorescent Cre reporter mouse. Genesis 45(9):593–605

91. Sadeghian M et al (2016) Mitochondrial dysfunction is an important cause of neurological deficits in an inflammatory model of multiple sclerosis. Sci Rep 6:33249

92. Flatters SJL (2015) The contribution of mitochondria to sensory processing and pain. Prog Mol Biol Transl Sci 131:119–146

93. Mantyh PW et al (1997) Inhibition of hyperalgesia by ablation of lamina I spinal neurons expressing the substance P receptor. Science 278(5336):275–279

94. Nichols ML et al (1999) Transmission of chronic nociception by spinal neurons expressing the substance P receptor. Science 286(5444):1558–1561

95. Ikeda H, Heinke B, Ruscheweyh R, Sandkühler J (2003) Synaptic plasticity in spinal lamina I projection neurons that mediate hyperalgesia. Science 299(5610):1237–1240

96. Johannssen HC, Helmchen F (2013) Two-photon imaging of spinal cord cellular networks. Exp Neurol 242:18–26

97. Yang G, Pan F, Parkhurst CN, Grutzendler J, Gan W-B (2010) Thinned-skull cranial window technique for long-term imaging of the cortex in live mice. Nat Protoc 5(2):201–208

98. Xu H-T, Pan F, Yang G, Gan W-B (2007) Choice of cranial window type for in vivo imaging affects dendritic spine turnover in the cortex. Nat Neurosci 10(5):549–551

99. Helmchen F, Imoto K, Sakmann B (1996) Ca2+ buffering and action potential-evoked Ca2+ signaling in dendrites of pyramidal neurons. Biophys J 70(2):1069–1081

100. Rose T, Goltstein PM, Portugues R, Griesbeck O (2014) Putting a finishing touch on GECIs. Front Mol Neurosci 7:88

101. Gobel W, Helmchen F (2007) In Vivo Calcium Imaging of Neural Network Function. Physiology 22(6):358–365

102. Neher E (1995) The use of fura-2 for estimating ca buffers and ca fluxes. Neuropharmacology 34(11):1423–1442

103. McMahon SM, Jackson MB (2018) An inconvenient truth: calcium sensors are calcium buffers. Trends Neurosci 41(12):880–884

104. Yang Y et al (2018) Improved calcium sensor GCaMP-X overcomes the calcium channel perturbations induced by the calmodulin in GCaMP. Nat Commun 9(1):1504

105. Thestrup T et al (2014) Optimized ratiometric calcium sensors for functional in vivo imaging of neurons and T lymphocytes. Nat Methods 11(2):175–182

106. Tian L et al (2009) Imaging neural activity in worms, flies and mice with improved GCaMP calcium indicators. Nat Methods 6(12):875–881

107. Chen T-W et al (2013) Ultrasensitive fluorescent proteins for imaging neuronal activity. Nature 499(7458):295–300

108. Galbraith JA, Terasaki M (2003) Controlled damage in thick specimens by multiphoton excitation. Mol Biol Cell 14(5):1808–1817

109. Yanik MF, Cinar H, Cinar HN, Chisholm AD, Jin Y, Ben-Yakar A (2004) Functional regeneration after laser axotomy. Nature 432(7019):822

110. Jongen JLM et al (2010) Autofluorescent flavoprotein imaging of spinal nociceptive activity. J Neurosci 30(11):4081–4087

111. Farrar MJ, Wise FW, Fetcho JR, Schaffer CB (2011) In vivo imaging of myelin in the vertebrate central nervous system using third harmonic generation microscopy. Biophys J 100(5):1362–1371

112. Williams PR et al (2014) A recoverable state of axon injury persists for hours after spinal cord contusion in vivo. Nat Commun 5:5683

113. Grinvald A, Lieke E, Frostig RD, Gilbert CD, Wiesel TN (1986) Functional architecture of cortex revealed by optical imaging of intrinsic signals. Nature 324(6095):361–364

114. Malonek D, Dirnagl U, Lindauer U, Yamada K, Kanno I, Grinvald A (1997)

Vascular imprints of neuronal activity: relationships between the dynamics of cortical blood flow, oxygenation, and volume changes following sensory stimulation. Proc Natl Acad Sci U S A 94(26):14826–14831

115. Frostig RD, Lieke EE, Ts'o DY, Grinvald A (1990) Cortical functional architecture and local coupling between neuronal activity and the microcirculation revealed by in vivo high-resolution optical imaging of intrinsic signals. Proc Natl Acad Sci U S A 87(16):6082–6086

116. MacVicar BA, Hochman D (1991) Imaging of synaptically evoked intrinsic optical signals in hippocampal slices. J Neurosci 11(5): 1458–1469

117. Lieke EE, Frostig RD, Arieli A, Ts'o DY, Hildesheim R, Grinvald A (1989) Optical imaging of cortical activity: real-time imaging using extrinsic dye-signals and high resolution imaging based on slow intrinsic-signals. Annu Rev Physiol 51:543–559

118. Haglund MM, Ojemann GA, Hochman DW (1992) Optical imaging of epileptiform and functional activity in human cerebral cortex. Nature 358(6388):668–671

119. Cannestra AF et al (1998) Topographical and temporal specificity of human intraoperative optical intrinsic signals. Neuroreport 9(11): 2557–2563

120. Toga AW, Cannestra AF, Black KL (1995) The temporal/spatial evolution of optical signals in human cortex. Cereb Cortex 5(6): 561–565

121. Sato K (2002) Intraoperative intrinsic optical imaging of neuronal activity from subdivisions of the human primary somatosensory cortex. Cereb Cortex 12(3):269–280

122. Pouratian N et al (2000) Optical imaging of bilingual cortical representations. Case report. J Neurosurg 93(4):676–681

123. Lu HD, Chen G, Cai J, Roe AW (2017) Intrinsic signal optical imaging of visual brain activity: tracking of fast cortical dynamics. NeuroImage 148:160–168

124. Jongen JLM et al (2014) Spinal autofluorescent flavoprotein imaging in a rat model of nerve injury-induced pain and the effect of spinal cord stimulation. PLoS One 9(10): e109029

125. Chisholm et al (2016) In vivo imaging of flavoprotein fluorescence during hypoxia reveals the importance of direct arterial oxygen supply to cerebral cortex tissue. https://doi.org/10.1007/978-1-4939-3023-4_29

126. Weigelin B, Bakker G-J, Friedl P (2016) Third harmonic generation microscopy of cells and tissue organization. J Cell Sci 129(2):245–255

127. Stosiek C, Garaschuk O, Holthoff K, Konnerth A (2003) In vivo two-photon calcium imaging of neuronal networks. Proc Natl Acad Sci U S A 100(12):7319–7324

128. Tsien RY (1981) A non-disruptive technique for loading calcium buffers and indicators into cells. Nature 290(5806):527–528

129. Nimmerjahn A, Kirchhoff F, Kerr JND, Helmchen F (2004) Sulforhodamine 101 as a specific marker of astroglia in the neocortex in vivo. Nat Methods 1(1):31–37

130. Sajic M et al (2013) Impulse conduction increases mitochondrial transport in adult mammalian peripheral nerves in vivo. PLoS Biol 11(12):e1001754

131. Kohro Y et al (2015) A new minimally-invasive method for microinjection into the mouse spinal dorsal horn. Sci Rep 5:14306

132. Davidson MW, Campbell RE (2009) Engineered fluorescent proteins: innovations and applications. Nat Methods 6(10):713–717

133. Chisholm KI, Khovanov N, Lopes DM, La Russa F, McMahon SB (2018) Large scale in vivo recording of sensory neuron activity with GCaMP6. eNeuro 5(1): ENEURO.0417-17.2018

134. Peron S, Chen TW, Svoboda K (2015) Comprehensive imaging of cortical networks. Curr Opin Neurobiol 32:115–123

135. Looger LL, Griesbeck O (2012) Genetically encoded neural activity indicators. Curr Opin Neurobiol 22(1):18–23

136. Nakai J, Ohkura M, Imoto K (2001) A high signal-to-noise Ca(2+) probe composed of a single green fluorescent protein. Nat Biotechnol 19(2):137–141

137. Lin MZ, Schnitzer MJ (2016) Genetically encoded indicators of neuronal activity. Nat Neurosci 19(9):1142–1153

138. Belousov VV et al (2006) Genetically encoded fluorescent indicator for intracellular hydrogen peroxide. Nat Methods 3(4):281–286

139. Sankaranarayanan S, De Angelis D, Rothman JE, Ryan TA (2000) The use of pHluorins for optical measurements of presynaptic activity. Biophys J 79(4):2199–2208

140. Okumoto S, Looger LL, Micheva KD, Reimer RJ, Smith SJ, Frommer WB (2005) Detection of glutamate release from neurons by genetically encoded surface-displayed FRET nanosensors. Proc Natl Acad Sci U S A 102(24):8740–8745

141. Marvin JS et al (2013) An optimized fluorescent probe for visualizing glutamate neurotransmission. Nat Methods 10(2):162–170

142. Dimitrov D et al (2007) Engineering and characterization of an enhanced fluorescent protein voltage sensor. PLoS One 2(5):e440

143. Abdelfattah AS et al (2016) A bright and fast red fluorescent protein voltage indicator that reports neuronal activity in organotypic brain slices. J Neurosci 36(8):2458–2472

144. Knöpfel T, Gallero-Salas Y, Song C (2015) Genetically encoded voltage indicators for large scale cortical imaging come of age. Curr Opin Chem Biol 27:75–83

145. Spike RC, Puskar Z, Andrew D, Todd AJ (2003) A quantitative and morphological study of projection neurons in lamina I of the rat lumbar spinal cord. Eur J Neurosci 18(9):2433–2448

146. Horton NG et al (2013) In vivo three-photon microscopy of subcortical structures within an intact mouse brain. Nat Photonics 7(3):205–209

147. Ouzounov DG et al (2017) In vivo three-photon imaging of activity of GcamP6-labeled neurons deep in intact mouse brain. Nat Methods 14(4):388–390

148. Levene MJ, Dombeck DA, Kasischke KA, Molloy RP, Webb WW (2004) In vivo multiphoton microscopy of deep brain tissue. J Neurophysiol 91(4):1908–1912

149. Jung JC, Mehta AD, Aksay E, Stepnoski R, Schnitzer MJ (2004) In vivo mammalian brain imaging using one- and two-photon fluorescence microendoscopy. J Neurophysiol 92(5):3121–3133

150. Xu Y, Zou P, Cohen AE (2017) Voltage imaging with genetically encoded indicators. Curr Opin Chem Biol 39:1–10

151. Gong Y et al (2015) High-speed recording of neural spikes in awake mice and flies with a fluorescent voltage sensor. Science 350(6266):1361–1366

152. Abdelfattah AS et al (2019) Bright and photostable chemigenetic indicators for extended in vivo voltage imaging. Science 365(6454):699–704

153. Bando Y, Sakamoto M, Kim S, Ayzenshtat I, Yuste R (2019) Comparative evaluation of genetically encoded voltage indicators. Cell Rep 26(3):802–813.e4

154. Storace DA, Braubach OR, Jin L, Cohen LB, Sung U (2015) Monitoring brain activity with protein voltage and calcium sensors. Sci Rep 5:1–15

155. Berger T et al (2007) Combined voltage and calcium epifluorescence imaging in vitro and in vivo reveals subthreshold and suprathreshold dynamics of mouse barrel cortex. J Neurophysiol 97(5):3751–3762

156. Ferezou I, Bolea S, Petersen CCH (2006) Visualizing the cortical representation of whisker touch: voltage-sensitive dye imaging in freely moving mice. Neuron 50(4):617–629

157. Scott G et al (2014) Voltage imaging of waking mouse cortex reveals emergence of critical neuronal dynamics. J Neurosci 34(50):16611–16620

158. Fisher JAN, Civillico EF, Contreras D, Yodh AG (2004) In vivo fluorescence microscopy of neuronal activity in three dimensions by use of voltage-sensitive dyes. Opt Lett 29(1):71

159. Carandini M, Shimaoka D, Rossi LF, Sato TK, Benucci A, Knopfel X (2015) Imaging the awake visual cortex with a genetically encoded voltage indicator. J Neurosci 35(1):53–63

160. Akemann W, Mutoh H, Perron A, Park YK, Iwamoto Y, Knöpfel T (2012) Imaging neural circuit dynamics with a voltage-sensitive fluorescent protein. J Neurophysiol 108(8):2323–2337

161. Masters BR, So PT, Gratton E (1997) Multiphoton excitation fluorescence microscopy and spectroscopy of in vivo human skin. Biophys J 72(6):2405–2412

Chapter 15

In Vivo Calcium Imaging of Peripheral Ganglia

Mark Lay and Xinzhong Dong

Abstract

Calcium dynamics are a proxy for neuronal firing, yet studying those dynamics requires dissociation protocols that lead to cellular milieu that may differ significantly to the environments in intact animals. Also, traditional methods like patch clamp electrophysiology are very low throughput. In vivo calcium imaging of peripheral ganglia strives to overcome these limitations. In this chapter, we will discuss the development and procedures associated with this powerful new technique.

Key words Calcium imaging, GCamp, DRG, Neural circuits, Pain, Somatosensation

---

1 Introduction

Somatosensation begins with activation of neurons in the peripheral nervous system. Whether the sensation is from external forces or internal signals, an action potential is propagated centrally toward peripheral ganglia, before reaching the spinal cord and the rest of the central nervous system. Using in vivo calcium imaging techniques, especially with genetically encoded calcium indicators (GECI), we can now monitor these somatosensory signals as mice are physically detecting stimuli [1]. In vivo calcium imaging of peripheral ganglia has provided new interpretations of how different stimuli are encoded, helping to confirm or challenge theories of somatosensation generated in the past few decades. This tool has also been instrumental in validating novel claims made in recent years. In this chapter, we will discuss the development of this technique. We will describe the methodology in detail so more pain researchers can take advantage of this exciting new tool.

The most accepted and established technique to investigate somatosensation in the periphery is via in vitro electrophysiology. Data generated via whole cell patch clamp electrophysiological recording of dissociated peripheral ganglia provides high sensitivity and temporal resolution. Unfortunately, during the experiment preparation, the axons of the ganglia are cut and enzymes are

Rebecca P. Seal (ed.), *Contemporary Approaches to the Study of Pain: From Molecules to Neural Networks*, Neuromethods, vol. 178, https://doi.org/10.1007/978-1-0716-2039-7_15, © Springer Science+Business Media, LLC, part of Springer Nature 2022

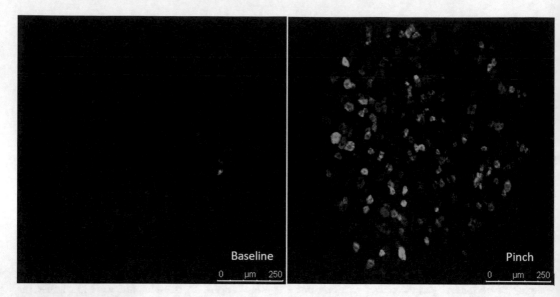

Fig. 1 Representative images from in vivo calcium imaging of DRG from a mouse expressing Pirt-cre; Rosa26-flox-stop-floxGCaMP6s. On the left is the baseline calcium signal from an exposed L4 DRG when no stimulus is presented. On the right is the response observed when the mouse is presented with a 300 g pinch of the hindpaw, which is a receptive field of L4 DRG neurons

used to separate the cell bodies from each other. These dissociated cell bodies represent individual units that compose the overall somatosensation machinery, and electrophysiological recordings can inform us of the firing properties of the different neurons that populate the ganglia. However, the environment of the in vitro preparation is a much different from the milieu within the ganglia, with supporting glia cells and autonomic innervation intact. Furthermore, in vitro electrophysiological experiments are low throughput, capable of only acquiring information from one cell at a time. In vitro calcium imaging of peripheral ganglia is more high throughput, yet suffers from the same caveats due to a need for extraction and dissociation of the ganglia. Meanwhile, in vivo electrophysiological recording of peripheral ganglia provides excellent temporal resolution of neuronal firing properties in an intact system, but suffers as a low-throughput technique. In vivo calcium imaging ultimately provides a window into intact peripheral ganglia, allowing researchers to observe hundreds of neurons at a time (Fig. 1).

To observe calcium dynamics of peripheral ganglia in vivo, we utilized GECIs developed by biotechnologists, who continue to optimize the technology. The GECIs we have used involve genetically engineered green fluorescent proteins (GFPs), which only provide bright green fluorescent signal upon calcium binding. The GECI that accomplishes this, called GCaMP, involves a myosin light chain kinase domain (M13) fused to the N-terminus of GFP. On the C-terminus of the engineered GFP is a fused Calcium binding calmodulin (CaM) [2]. Without calcium binding CaM, the engineered GFP is weakly fluorescent, but a calcium-bound CaM leads to a conformational change in the GFP. The M13 and CaM domains are brought together, allowing for deprotonation of the protein and resulting in high fluorescent activity [3]. The first version of this protein, GCamp1, is not stable under conditions above 30° [2]. Scientists then engineered GCaMP2 to ameliorate the stability issue [4]. Further optimization work resulted in GCamp3, GCaMP5, and GCaMP6 which all have increased signal to noise ratio as well as higher sensitivity compared to the early versions of the protein [4–7]. In our work, we elected to utilize the GECIs GCaMP3 and GCaMP6, by expressing the engineered proteins specifically in peripheral ganglia [8].

We utilized a mouse line with cre expressed under the *Pirt* promoter to express GCaMP3 and GCaMP6 specifically in the neurons of peripheral ganglia, but not in other tissues including skin and spinal cord. In the next section, we will go into more detail about the imaging procedures, but we prepare the mice for imaging by surgically removing the transverse process overlying the Lumbar-4 dorsal root ganglion (DRG). With only the DRG of interest exposed, the anesthetized mice are placed under a confocal microscope for imaging. Under this in vivo calcium imaging setup, we can observe calcium transients that occur in over 1600 individual neurons [8]. We have shown using this technique that DRG neurons can fire in a coupled fashion under inflammatory injury conditions [1]. The in vivo setup provided a novel means to investigate intraganglionic interactions that cannot be accurately interrogated under in vitro conditions. Using this technique, others have shown that most DRG neurons fire specifically to modalities such as heat, cool, noxious mechanical touch [9]. In essence, these new data provide counter evidence against long standing theories of polymodality in C fibers, based on electrophysiological data. More recently, we have used this technique to validate novel mechanistic findings. One such collaboration involves a new technique for quantifying pain behavior using high speed video capture of mouse withdrawal behavior was recently developed. Our in vivo calcium imaging technique provided an important confirmation of the long held views on the effects of certain stimuli on the mouse paw, namely that gentle touch activates mostly large diameter neurons while pinprick stimuli active small diameter neurons preferentially [10]. We have now also used this technique to explore neuroimmune interactions. In a recent paper, we have shown that activation of mast cells in the hind paw using different stimuli can result in dramatic differences in the responses of the DRG neurons [11]. Overall, this in vivo calcium imaging technique provides a high throughput means of investigating calcium dynamics in an intact intraganglionic environment. We believe this technique can help confirm or deny long standing theories of somatosensation, as well as provide a new means of discovery for this field.

2 Materials

In Vivo DRG Calcium Imaging In vivo imaging of whole L4 DRG in live mice was performed for 1–6 h immediately after the surgery. Body temperature was maintained at 37 °C ± 0.5 °C on a heating pad and rectal temperature was monitored. After surgery, mice were laid down in the abdomen-down position on a custom-designed microscope stage (Fig. 2). The spinal column was stabilized using custom-designed clamps to minimize movements caused by breathing and heart beats. In addition, a custom-designed head holder was also used as an anesthesia/gas mask. The animals were maintained under continuous anesthesia for the duration of the imaging experiment with 1–2% isoflurane gas using a gas vaporizer. Pure oxygen air was used to deliver the gas to the animal. The microscope stage was fixed under a laser-scanning confocal microscope (Leica LSI microscope system), which was equipped with macrobased, large-objective, and fast EM-CCD camera. Live images were acquired at typically eight to ten frames with 600 Hz in frame-scan mode per 6–7 s, at depths below the dura ranging from 0 to 70 mm, using a 53 0.5 NA macro dry objective at typically 512 × 512 pixel resolution with solid diode lasers

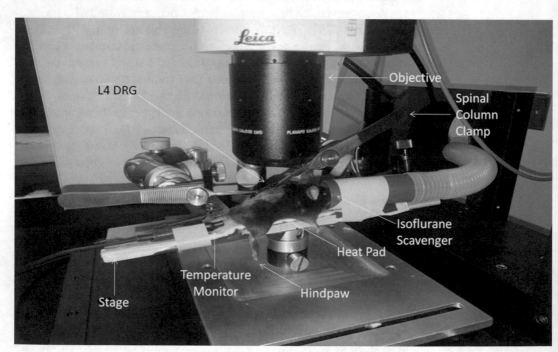

Fig. 2 The in vivo calcium imaging setup with a mouse secured onto a heating pad via gauze tape. A temperature monitor is placed rectally, and an isoflurane scavenger is secured to maintain the anesthetized state. Custom spinal column clamps are used to secure the L3- and L5-associated spinous processes to prevent the effects of breathing from affecting the collected calcium signal. Pictured here is an exposed Lumbar 4 DRG

(Leica) tuned at 488 and at 532 nm wavelength and emission at 500–550 nm for green and 550–650 nm for red fluorescence, respectively. For higher scanning speeds, a z-scan was omitted to acquire images at one frame per 300 ms.

For analysis, raw image stacks (512 × 512 to 1024 × 1024 pixels in the *x–y* plane and 20–30 mm voxel depth; typically 10 optical sections) were imported into ImageJ (NIH) for further analysis. DRG neurons were at the focal plane, and imaging was monitored during the activation of DRG neuron cell bodies by peripheral stimuli. The imaging parameters were chosen to allow repeated imaging of the same cell over many stimuli without causing damage to the imaged cells or to surrounding tissue.

3   Methods

For all imaging experiments, mice 3 months or older were anesthetized by i.p. injection of sodium pentobarbital (40–50 mg/kg). After deep anesthesia was reached, the animal's back was shaved and aseptically prepared, and ophthalmic ointment (Lacrilube; Allergen Pharmaceuticals) was applied to the eyes to prevent drying. During the surgery, mice were kept on a heating pad (DC temperature controller, FHC) to maintain body temperature at 37 °C ± 0.5 °C as monitored by a rectal probe. Dorsal laminectomy in DRG was performed usually at spinal level L6 to S1 below the lumbar enlargement (but occasionally at lower than S1) but without removing the dura (some experimental conditions, such as direct local drug injection into DRG tissue and rhodamine injection into DRG neurons, required the removal of the dura). A 2 cm-long midline incision was made around the lower part of the lumbar enlargement area; next, 0.1 mL of 1% lidocaine was injected into the paravertebral muscles, and these were dissected away to expose the lower lumbar part which surrounds (L3–L5) vertebra bones. The L4 DRG transverse processes were exposed and cleaned. Using small rongeurs, we removed the surface aspect of the L4 DRG transverse process near the vertebra (only the L4 DRG transverse process was removed, but the bone over the spinal cord was intact) to expose the underlying DRG without damaging the DRG and spinal cord. Bleeding from the bone was stopped using styptic cotton.

When using GCaMP for in vivo imaging, there are many technical concerns to consider. Different microscopic avenues offer distinct advantages and disadvantages. Two-photon microscopy may be necessary when imaging targets lie deep within a specimen. Additional adaptations such as adaptive optics, highly sensitive detectors such as GaAsP detectors, or objectives with higher numerical apertures may assist in imaging targets with low fluorescence. When using two-photon imaging, stabilization of the target tissue is paramount. Fortunately, there are ways to circumvent this issue, such as implantable spinal windows, vertebral clamps, a thin layer of agarose over the tissue, or an adaptive focus control unit. One-photon microscopy is less sensitive to movement artifacts in the z-direction due to its higher point spread function; however, tissue-wide delivery of high-intensity laser light introduces the concern of phototoxicity out of the imaging plane. In addition, it can be challenging to use one-photon imaging in highly scattering tissue such as the spinal cord. It is also very important to address variable expression levels of GCaMP between neurons. The level of GCaMP can directly affect the level of fluorescence observed upon activation. Insertion of GCaMP into the mouse genome using a viral vector or gene knock-in approach can result in variable GCaMP expression between neurons. This factor can make it difficult to compare Ca transients between different cell types. In addition, tissue or nerve injury can also affect GCaMP levels in the DRG depending on what promoter drives their expression which potentially make comparison of GCaMP signals between naïve and injury condition more complicated. Importantly, we determined that GCaMP expression under Pirt promoter was not affected by injury. Consequently, when choosing an appropriate strategy for GCaMP insertion into the mouse genome it is imperative that one should take the expression levels into consideration and interpret the results accordingly.

4 Conclusions

GECIs have contributed greatly to the advancement of neuroscience. By expressing GECIs in specific neuron populations of interest, researchers can monitor large numbers of neurons' firing patterns while the mouse is experiencing stimuli. We aimed to accomplish this as well in peripheral ganglia, to better study pain sensation. In our in vivo calcium imaging system, we expressed GECIs in peripheral ganglia via a mouse line with cre recombinase expressed under the promoter *pirt*. We can monitor the firing patterns of thousands of DRG neurons in anesthetized mice while the mice are experiencing somatosensory stimuli. This technique has ultimately led to new hypotheses for somatosensory stimuli encoding as well as the validation of new findings.

Moving forward, we and other scientists aim to utilize new membrane bound GECIs for in vivo calcium imaging of the nerve terminals that innervate specific organs. These proposed studies may provide more precise understanding of the populations of nerve terminals that are responsible for pain. In another direction, some biotechnologists have already begun the optimization of systems for in vivo calcium imaging of freely moving mice. While lower resolution endoscopic methods have been established, higher resolution techniques via fiber optics are currently being developed. These techniques may allow for investigation of firing patterns that may be altered by anesthesia. We are also working with collaborators to complete experiments that begin with in vivo calcium imaging of DRG neurons, followed by in vivo electrophysiological recordings of neurons that fired under the calcium imaging protocol. This combination of techniques will result in high throughput data generated via calcium imaging paired with data from in vivo electrophysiology with high temporal resolution.

Previous studies of pain and itch have heavily utilized in vitro techniques. However, these in vitro preparations poorly simulate the environment present in a live and intact mouse. We believe our technique will help in providing more accurate information about the physiology of pain sensation. We hope more pain researchers will add this powerful method into their toolkits for investigation.

Acknowledgments

The work was supported by grant from the NIH to Xinzhong Dong (NS054791).

References

1. Kim YS, Anderson M, Park K, Zheng Q, Agarwal A et al (2016) Coupled activation of primary sensory neurons contributes to chronic pain. Neuron 91(5):1085–1096

2. Nakai J, Ohkura M, Imoto K (2001) A high signal-to-noise Ca^{2+} probe composed of a single green fluorescent protein. Nat Biotechnol 19:137–141

3. Akerboom J (2009) Crystal structures of the GCaMP calcium sensor reveal the mechanism of fluorescence signal change and aid rational design. J Biol Chem 284(10):6455–6464

4. Ohkura M, Sasaki T, Sadakari J, Gengyo-Ando K, Kagawa-Nagamura Y et al (2012) Genetically encoded green fluorescent Ca^{2+} indicators with improved detectability for neuronal Ca^{2+} signals. PLoS One 7(12):e51286

5. Akerboom J, Chen T, Wardill TJ, Tian L, Marvin JS et al (2012) Optimization of a GCaMP calcium indicator for neural activity imaging. J Neurosci 32(40):13819–13840

6. Muto A, Ohkura M, Kotani T, Higashijima S, Nakai J, Kawakami K (2011) Genetic visualization with an improved GCaMP calcium indicator reveals spatiotemporal activation of the spinal motor neurons in zebrafish. Proc Natl Acad Sci U S A 108(13):5425–5430

7. Zhao Y, Araki S, Wu J, Teramoto T, Chang YF et al (2011) An expanded palette of genetically encoded Ca^{2+} indicators. Science 333(6051): 1888–1891

8. Anderson M, Zheng Q, Dong X (2018) Investigation of pain mechanisms by calcium imaging approaches. Neurosci Bull 34(1):194–199

9. Emery E, Luiz A, Sikander S, Magnusdottir R, Dong X et al (2016) In vivo characterization of distinct modality-specific subsets of somatosensory neurons using GCaMP. Sci Adv 2(11):e1600990

10. Abdus-Saboor I, Fried N, Lay M, Burdge J, Swanson K et al (2019) Development of a mouse pain scale using sub-second behavioral mapping and statistical modeling. Cell Rep 28(6):1623–1634

11. Meixiong J, Anderson M, Limjunyawong N, Sabbagh M, Hu E et al (2019) Activation of mast-cell-expressed mas-related G-protein-coupled receptors drives non-histaminergic itch. Immunity 50(5):1163–1171

Optogenetic Modulation of the Visceromotor Response to Reveal Visceral Pain Mechanisms

Sarah A. Najjar, Emanuel Loeza-Alcocer, Brian M. Davis, and Kristen M. Smith-Edwards

Abstract

Visceral pain is a debilitating condition that is common yet lacks effective treatments. In the case of gastrointestinal disorders (e.g., inflammatory bowel disease and irritable bowel syndrome), pain is thought to be conveyed by primary afferent fibers innervating the colon. These fibers differ in their response properties, anatomical structures, and molecular phenotypes; thus, it is unclear which afferents initiate and maintain changes in visceral sensitivity. Additionally, evidence shows that nonneuronal cell types in the colon (e.g., epithelial cells) can initiate colon afferent activity, indicating that they are an integral part of sensory signaling. Measurement of the visceromotor response (VMR) is an established, reliable method to assess visceral sensitivity to mechanical stimuli. In order to determine the relative contribution of specific cell types to the VMR, our lab has developed techniques utilizing optogenetic mouse models. These techniques will enable researchers to define how subtypes of afferent neurons and nonneuronal cell populations contribute to visceral pain in both normal and pathological conditions. This chapter details how optogenetics can be integrated into VMR analysis, for assessment of colon sensitivity in vivo. These methods can be adapted to other visceral organs and also used in disease models.

Key words Visceral pain, Visceral hypersensitivity, Visceromotor response, Primary afferent neurons, Colon afferents, Optogenetics

1 Introduction

Visceral pain is one of the most common reasons patients seek medical treatment but, until recently, has received far less experimental attention than other areas of pain research. The technical challenge of accessing visceral organs in a noninvasive manner combined with the lack of specificity in responses related to visceral sensation has led to difficulty in developing valid models of visceral pain. Whereas pain arising from the skin is well localized and

Supplementary Information The online version of this chapter (https://doi.org/10.1007/978-1-0716-2039-7_16) contains supplementary material, which is available to authorized users.

Rebecca P. Seal (ed.), *Contemporary Approaches to the Study of Pain: From Molecules to Neural Networks*, Neuromethods, vol. 178, https://doi.org/10.1007/978-1-0716-2039-7_16, © Springer Science+Business Media, LLC, part of Springer Nature 2022

reliably evokes withdrawal reflexes, visceral pain is characterized by poor localization and postural freezing. Stimuli that evoke visceral pain include ischemia, inflammation, traction on the mesentery, and distension of hollow organs [1]. Colorectal distension (CRD) has proven to be a useful model of visceral nociception because the stimulus is easily controlled by the experimenter, it reproduces the intensity and referral pattern of visceral pain in human subjects [1], and it evokes several measurable and reproducible responses (e.g., heart rate, respiration rate) in a number of species (rodents, [2–4]; rabbit, [5]; cat, [6]; dog, [7]; monkey, [8]).

The most reliable and easily measured response to CRD is contraction of the abdominal musculature, known as the viscero-motor response (VMR), which is recorded via EMG electrodes inserted into the abdominal musculature. The VMR resembles a "guarding reflex" and is considered a pseudoaffective response because it requires supraspinal circuitry [9]. The magnitude of the VMR is dependent on stimulus intensity (i.e., larger CRD evokes larger VMR), and the VMR is temporally linked to stimulus onset and termination [10]. Further adding to its utility for visceral pain research, the VMR can be modulated experimentally. Analgesics, such as opioids, inhibit the VMR [11–14], and inflammation-inducing compounds have a sensitizing effect on CRD-evoked VMR. For example, trinitrobenzene sulfonic acid (TNBS; [15]), acetic acid [16], zymosan [17], and dextran sulfate sodium (DSS) [18] lead to increased visceromotor and autonomic responses to CRD and decreased VMR thresholds (the minimum pressure required to evoke a response).

Although the increases in VMR to CRD during inflammation have generally been attributed to the sensitization of spinal primary afferent neurons that innervate the colon, it remains unclear which particular subpopulations initiate, maintain, and/or resolve changes in visceral sensitivity and whether nonneuronal cells (e.g., epithelial cells, immune cells) are involved. The heterogeneity in anatomical structure, response properties, and molecular pheno-type suggests that colon afferents have discrete functional roles in visceral sensation. Anatomically, 13 distinct types of spinal afferent nerve endings have been identified in various layers of the distal colon [19–21] and much effort has been directed toward correlat-ing these anatomical subpopulations with functional properties. In ex vivo preparations, colon afferents have been documented to have either low or high thresholds for response to mechanical distension, but the majority encode distending pressures in the noxious range regardless of their threshold for activation [22, 23]. Interestingly, those with high thresholds exhibit low-frequency firing rates and are correlated to expression of transient receptor potential vanilloid receptor 1 (TRPV1) and the growth factor receptor GFRα3, whereas those with low thresholds fire at high frequencies and lack TRPV1/GFRα3 expression [23]. The use of single cell RNA sequencing from back-labeled colon afferents has supported these findings, but also revealed the full complexity in molecular phenotypes across colon afferent subpopulations with the finding of seven distinct clusters based on gene expression alone [24]. Further adding to this complexity, epithelial cells in the colon have been directly implicated in initiating electrophysiological and behavioral responses to mechanical stimuli [25]. Therefore, in order to unravel the mechanisms underlying visceral pain in pathological conditions and gain insight into the best therapeutic targets, we must know how specific cell populations (neuronal and nonneuronal) contribute to producing or inhibiting CRD-induced VMR.

Optogenetics combines genetic and optical technologies to excite or inhibit genetically defined cells with spatial and temporal precision [26]. The excitatory opsin, channelrhodopsin-2 (ChR2), is a blue light (~470 nm)-activated cation channel that leads to depolarization when open, whereas the inhibitory opsin, archaerhodopsin (Arch), is a proton pump activated by yellow light (~575 nm) and causes hyperpolarization. Several studies have demonstrated that activation of ChR2-expressing sensory neurons can evoke nocifensive behaviors in mice [25, 27–33]. For example, ChR2 activation of TRPV1-expressing colon afferents produces a VMR [25] and selective optical stimulation of bladder afferents has revealed differential regulation of bladder pain and voiding function by sensory afferent subpopulations [34]. Conversely, optical inhibition of bladder afferents has been shown to suppress the VMR [35]. Furthermore, ChR2 activation of colon epithelial cells produces a VMR, demonstrating that nonneuronal cells are capable of activating visceral afferents and evoking a behavioral response normally associated with visceral pain [25]. Thus, the continued use of optogenetic tools combined with VMR recordings will enable researchers to untangle the relative contributions of different molecular subpopulations (visceral afferents and nonneuronal cell populations) to visceral pain in both normal and pathological conditions. This chapter details methods for utilizing optogenetic techniques to assess colon sensitivity in vivo.

2   Materials

2.1   Laser/Balloon Device Materials

1. Fiber-optic.
    (a) Optical fiber (400 µM core; ThorLabs FT400EMT).
    (b) Ceramic ferrules (6.4 mm length, 1.25 mm diameter; ThorLabs CFLC440-10).
    (c) FC/PC connector (440 µM bore; ThorLabs 30440G3).
    (d) Epoxy (e.g., Gorilla clear epoxy).
2. Polyethylene tubing (PE-200).

3. Polyethylene plastic sheet (Saran wrap) 3 cm × 3 cm.

4. 4-0 silk sutures (Ethicon K952H).

5. Stopcock with male Luer lock and 2 female Luers (e.g., Component Supply Co cat. no. SCLP-400C).

6. Modeling clay (e.g., Van Aken Plastalina).

2.2 Experimental Setup

1. Laser and driver (e.g., Laserglow Technologies LRS-0473 and LRS-0589 DPSS series for blue (473 nm) light and yellow (589 nm) light, respectively.

2. Analog-digital converter (e.g., CED Micro 1401).

3. Differential amplifier (A-M Systems Model 1700).

4. BNC cables and splitter.

5. Balloon pressure delivery system: valve interface and valve box (Model B482C, University of Iowa BioEngineering Department).

6. Compressed nitrogen tank and regulator (UN1086, Valley National Gas, Dual-stage general purpose low delivery brass pressure regulator [81-2-580, Matheson Tri-Gas]).

7. Tygon tubing (ID = 1/8 in., OD = 1/4 in., Formula E-3603; Fisher Scientific cat. no. 14-171-103).

8. Y-connectors for the Tygon tubing.

9. Pressure monitor (e.g., World Precision Instruments Pressure Monitor BP-1).

10. Sphygmomanometer.

2.3 Electrode Implantation/Balloon Insertion

1. Urethane (Sigma cat. no. U2500; make up fresh 140 mg/mL in sterile saline solution).

2. Isoflurane.

3. Isoflurane vaporizer (e.g., Parkland Scientific V3000PS).

4. Small compressed oxygen tank (Valley National Gas UN1072).

5. Oxygen E-cylinder regulator (SurgiVet Inc. V7326).

6. Heating pad (preferably water pump operated; e.g. Gaymar TP700 T/Pump).

7. Hair clippers (e.g., Oster Golden A5).

8. Scalpel (blade #10).

9. No. 2 Forceps (Fine Science Tools cat. no. 11223-20).

10. Clip electrodes (Pomona Electronics: Micrograbber® Test Clip model 4233).

11. Cable, 12′ with five-pin connector at one end (A-M Systems cat. no. 692012).

12. Ground electrode.

13. Electrode gel (e.g., SignaGel).

14. Water-soluble lubricant (e.g. Johnson & Johnson).

2.4 EMG Recording of VMR

1. PC Computer.

2. Software to collect/analyze EMG signals (e.g., Spike 2).

3   Methods

3.1 Constructing Balloon/Laser Device

This device was customized for colon distension, thus a 400-µM fiber optic is used. The final product is pictured in Fig. 1. Please note that a thinner fiber optic setup can be used for bladder applications.

1. Assemble fiber optic: cut a 100-cm piece of optical fiber and, using epoxy, glue a ceramic ferrule to one end of the optical fiber and the FC/PC connector to the other end.

2. Cut a 15-cm piece of PE-200 tubing and attach it to the top female end of the stopcock using a Luer connector.

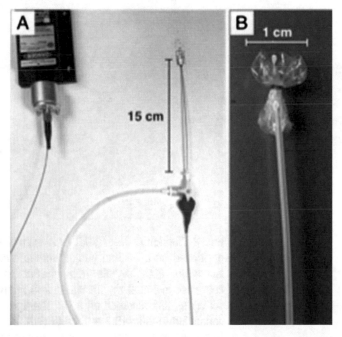

Fig. 1 Laser-balloon device. (**a**) Laser is shown with fiber optic threaded through stopcock and PE-200 tubing. Tygon tubing is connected to the stopcock, to deliver air through the PE-200 tubing. The open side of the stopcock is plugged with clay. (**b**) Laser-balloon device during balloon inflation. Polyethylene balloon is secured around the PE-200 tubing that encloses the fiber optic

3. Thread the fiber optic through the stopcock and tubing so that the end of the fiber is fully encased in the tubing. Use clay to plug the end of the stopcock that is opposite the tubing.

4. Use the perpendicular female end of the stopcock to connect the Tygon tubing that will carry the pressurized air (to inflate balloon).

5. To add the balloon, cut a 3 cm × 3 cm piece of polyethylene film and shape it around the PE-200 tubing containing the fiber optic. Secure the film by wrapping 4-0 silk sutures around the bottom of the film 5–8 times, to create an airtight seal. Ensure that the length of the balloon is 1.5 cm and, when inflated, the diameter is 0.9 cm (*see* **Note 1**).

3.2 Equipment Setup   The following experimental setup is for one mouse (Fig. 2). Testing more than one mouse at a time is possible but will require one laser per mouse. *See* Christianson & Gebhart [10] for a detailed diagram

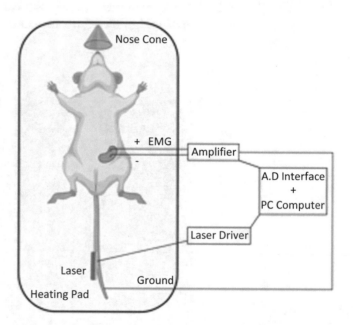

Fig. 2 Diagram of laser-VMR experimental setup. The anesthetized mouse is positioned on a heating pad, abdomen facing up, with nose cone delivering isoflurane. EMG signals from the abdominal musculature are amplified, converted to digital signals via an A–D interface (e.g., CED1401 run by Spike2 software), and recorded on a PC. The laser is connected to a driver, which is connected to the A–D interface to allow computer control of the laser. Please note that this schematic includes only laser stimulation (not balloon distension). *See* Subheading 3 for how to integrate the laser stimulation with colorectal distension

of the all equipment connections other than the laser (this is easily added to the established setup). The connections described here enable balloon and laser stimuli to be triggered via computer control and also enable the balloon pressure and laser stimuli to be demarcated in the Spike2 software.

1. Connect the FC/PC end of the fiber optic to the fiber coupler on the laser. Ensure that driver settings are on TTL+ (if applicable) to enable computer control of the laser.

2. Connect the laser to both DAC output and ADC input on the A-D converter, using 3 BNC cables and a splitter.

3. Plug the cable, 12′ with five-pin connector at one end and another end containing the clip electrodes to Channel 1 on the differential amplifier. Connect a BNC cable from the Channel 1 output to the ADC input on the A–D converter.

4. Attach a ground electrode to the differential amplifier ground.

5. Connect the pressure monitor to ADC input on the A-D converter via BNC cable.

6. Connect the air valve interface to DAC output on the A–D converter via BNC cable.

7. Using Tygon tubing, connect the compressed nitrogen tank to the air valve box.

8. Use Tygon tubing and Y-connectors to attach the air valve box to the stopcock connected to the laser/balloon device, as well as to the pressure monitor and sphygmomanometer.

9. Calibrate the pressure monitor's feedback in Spike2, using the sphygmomanometer readings.

3.3 Electrode Implantation/Balloon Insertion

The following methods describe VMR assessment of a lightly anesthetized mouse. The same balloon/laser device can be used in awake mice, but EMG electrodes must be implanted days before experimental analysis. Detailed methods on analyzing VMR in awake mice can be found in previous publications [10, 36].

1. Anesthetize an adult mouse using intraperitoneal injection of urethane (1.2 g/kg). It is best for the mouse to have been fasted overnight to minimize contents in the digestive tract (*see* **Note 2**).

2. Place mouse on heating pad and into isoflurane nose cone (1.5–2% isoflurane) (*see* **Note 3**).

3. Shave the lower left quadrant of the mouse's abdomen and use a scalpel to make a small (1.5 cm) incision in the skin, revealing abdominal musculature.

4. Using #2 forceps, attach clip electrodes to the abdominal muscle, approximately 3 mm apart (*see* **Note 4**).

5. Apply the water-based lubricant to the balloon and insert the balloon/laser device into the colorectum until the end of the balloon is about 3 mm beyond the anal sphincter. Use tape to secure the tube containing the fiber optic to the mouse's tail.

6. Use electrode gel to secure the ground electrode to the end of the mouse's tail.

3.4 EMG Recording of VMR

1. Turn the isoflurane level down to ~0.25%. Monitor the mouse's level of anesthesia by observing respiration and toe-pinch response (*see* **Note 5**).

2. Set balloon distension pressure to 60 mmHg using the regulator on the nitrogen tank. Start a new Spike2 file and use keyboard commands in the software to control balloon inflation (PC should be connected to the A–D converter to enable this).

3. Begin recording EMG signals in a new Spike2 file. Use controls in Spike2 to apply 60 mmHg of balloon pressure for 10 s every 5–10 min until mouse responds to the distension (this usually takes about 1 h) (*see* **Note 6**).

4. After anesthesia level is stable and mouse gives consistent responses, record baseline EMG activity for at least 2 min. Then apply at least three 10-s distensions, with 4-min intervals between presentations.

5. Adjust the balloon pressure to test other distension parameters as required, again testing at least three times with 4 min between distensions (*see* **Note 7**).

6. Apply laser stimulation protocol. Fig. 3 shows examples of laser parameters that can be applied, and the resulting laser-evoked VMRs (*see* **Note 8**).

7. Analyze EMG data in Spike 2 (*see* **Note 9**).

4 Notes

1. The tip of the balloon should be about 3 mm from the top of the PE-200 tubing, to avoid poking the balloon as it inflates. To ensure that the sutures are secured tightly, inflate the balloon in a small beaker of water and check for air leakage.

2. Testing VMR to laser stimulation requires that the mouse express an optogenetic actuator such as ChR2, Halo, and Arch. These can be expressed in specific cell populations using transgenic or viral techniques. When using these techniques, it is critical to confirm that opsin expression is specific to the target cell population.

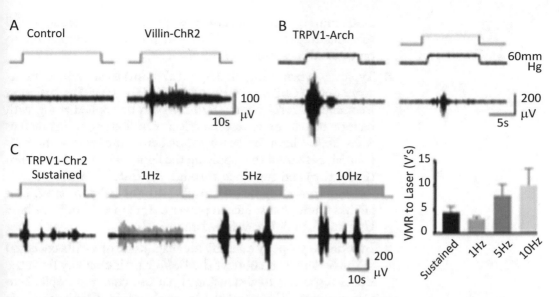

Fig. 3 Examples of visceromotor responses (VMR) during laser stimulation and/or balloon colorectal distension. (**a**) In Control mice (expressing ChR2 but no Cre driver), blue light does not evoke a VMR (left trace). In Villin-ChR2 mice, blue light activates ChR2 expressed in colon epithelial cells and evokes a VMR (right trace). (**b**) In TRPV1-Arch mice, yellow light initiates the inhibitory Arch expressed in colon afferent fibers and suppresses the VMR (right trace) evoked by 60 mmHg balloon distension (left trace). (**c**) In TRPV1-ChR2 mice, blue light delivered at varying frequencies activates ChR2 expressed in colon afferent fibers, in each case eliciting a VMR. Graph shows quantification of the VMR using four different laser stimulation parameters

3. Do not leave the mouse on this high level of isoflurane for too long. The skin incision, electrode attachment, and balloon insertion in the following steps should be done in under 3 min.

4. Securing the electrodes too far apart from each other will create excess noise in the recording signal. After securing the ground electrode to the mouse's tail, turn on the differential amplifier and start recording in Spike2 to optimize signal-to-noise ratio.

5. The isoflurane level may need to be slightly higher or lower than 0.25%. At this low level of isoflurane, the mouse should respond to toe pinch but should not be ambulating. The goal is for the anesthesia to be light enough to enable stable responses to distension (detailed in **Note 6**).

6. If the mouse does not respond to distension after 1 h, turn the isoflurane down by 0.125% and continue to apply distension every 5–10 min until the mouse responds. Repeat as necessary.

7. It is common to apply different distension pressures to test colon sensitivity. Previous studies have used pressures ranging from 15 mmHg to 80 mmHg. For example, some studies have applied pressure at 20, 40, 60, and 80 mmHg [37, 38]. Others have used 15, 30, 45, 60, and 75 mmHg [10], cautioning that pressures above 60 or 75 mmHg can result in colon damage. VMR data will ideally be collected at noxious (typically

>30 mmHg) and nonnoxious pressures to create a full stimulus-response curve. This will provide a comparison to optogenetically induced responses.

8. Typically, when using blue light (473 nm) to activate an excitatory opsin (ChR2), the laser is applied for 10–30 s, at varying frequencies. Yellow light (590 nm) can be applied along with balloon distension to test the effect of inhibitory opsins such as Arch. Yellow light can be turned on before the balloon stimulus (our lab has found that applying the light 1 s before the balloon is effective) and sustained through the distension. Or the yellow light onset can be concurrent with the balloon onset. Our lab has found that sufficient power outputs are 20 mW for blue laser and 12 mW for yellow laser.

9. In order to properly quantify the magnitude of VMRs collected in the Spike2 files, subtract the EMG baseline activity from the EMG responses to distension, then use cursors in Spike2 to measure the EMG signals, as outlined in Christianson & Gebhart [10].

Acknowledgments

This work was supported by NIH OT2 OD023859 (SAN, KMSE, BMD), F32 DK120115 (KMSE), and R01 DK107966 (ELA).

References

1. Ness TJ, Gebhart GF (1990) Visceral pain: a review of experimental studies. Pain 41(2): 167–234. https://doi.org/10.1016/0304-3959(90)90021-5

2. Coutinho SV et al (2002) Neonatal maternal separation alters stress-induced responses to viscerosomatic nociceptive stimuli in rat. Am J Physiol Gastrointest Liver Physiol 282(2): G307–G316. https://doi.org/10.1152/ajpgi.00240.2001

3. Kamp EH et al (2003) Quantitative assessment and characterization of visceral nociception and hyperalgesia in mice. Am J Physiol Gastrointest Liver Physiol 284(3):G434–G444. https://doi.org/10.1152/ajpgi.00324.2002

4. Ji Y, Traub RJ (2002) Differential effects of spinal CNQX on two populations of dorsal horn neurons responding to colorectal distension in the rat. Pain 99(1–2):217–222. https://doi.org/10.1016/s0304-3959(02)00106-9

5. Crawford ME et al (1993) Direct spinal effect of intrathecal and extradural midazolam on visceral noxious stimulation in rabbits. Br J Anaesth 70(6):642–646. https://doi.org/10.1093/bja/70.6.642

6. Janig W, Koltzenburg M (1991) Receptive properties of sacral primary afferent neurons supplying the colon. J Neurophysiol 65(5): 1067–1077. https://doi.org/10.1152/jn.1991.65.5.1067

7. Cevese A et al (1992) Haemodynamic effects of distension of the descending colon in anaesthetized dogs. J Physiol 447:409–423. https://doi.org/10.1113/jphysiol.1992.sp019009

8. Al-Chaer ED, Feng Y, Willis WD (1998) A role for the dorsal column in nociceptive visceral input into the thalamus of primates. J Neurophysiol 79(6):3143–3150. https://doi.org/10.1152/jn.1998.79.6.3143

9. Ness TJ, Gebhart GF (1987) Characterization of neuronal responses to noxious visceral and somatic stimuli in the medial lumbosacral spinal cord of the rat. J Neurophysiol 57(6): 1867–1892. https://doi.org/10.1152/jn.1987.57.6.1867

10. Christianson JA, Gebhart GF (2007) Assessment of colon sensitivity by luminal distension in mice. Nat Protoc 2(10):2624–2631. https://doi.org/10.1038/nprot.2007.392

11. Briggs SL et al (1995) Oxymorphone-induced analgesia and colonic motility measured in colorectal distension. Pharmacol Biochem Behav 52(3):561–563. https://doi.org/10.1016/0091-3057(95)00140-r

12. Ness TJ, Follett KA (1998) The development of tolerance to intrathecal morphine in rat models of visceral and cutaneous pain. Neurosci Lett 248(1):33–36. https://doi.org/10.1016/s0304-3940(98)00327-9

13. Traub RJ, Stitt S, Gebhart GF (1995) Attenuation of c-Fos expression in the rat lumbosacral spinal cord by morphine or tramadol following noxious colorectal distention. Brain Res 701(1–2):175–182. https://doi.org/10.1016/0006-8993(95)00990-5

14. Bjorkman R et al (1990) Central, naloxone-reversible antinociception by diclofenac in the rat. Naunyn Schmiedeberg's Arch Pharmacol 342(2):171–176. https://doi.org/10.1007/bf00166960

15. Julia V, Mezzasalma T, Bueno L (1995) Influence of bradykinin in gastrointestinal disorders and visceral pain induced by acute or chronic inflammation in rats. Dig Dis Sci 40(9):1913–1921. https://doi.org/10.1007/bf02208656

16. Burton MB, Gebhart GF (1995) Effects of intracolonic acetic acid on responses to colorectal distension in the rat. Brain Res 672(1 2):77–82. https://doi.org/10.1016/0006-8993(94)01382-r

17. Coutinho SV, Meller ST, Gebhart GF (1996) Intracolonic zymosan produces visceral hyperalgesia in the rat that is mediated by spinal NMDA and non-NMDA receptors. Brain Res 736(1–2):7–15. https://doi.org/10.1016/0006-8993(96)00661-0

18. Verma-Gandhu M et al (2007) Visceral pain perception is determined by the duration of colitis and associated neuropeptide expression in the mouse. Gut 56(3):358–364. https://doi.org/10.1136/gut.2006.100016

19. Spencer NJ, Kyloh M, Duffield M (2014) Identification of different types of spinal afferent nerve endings that encode noxious and innocuous stimuli in the large intestine using a novel anterograde tracing technique. PLoS One 9(11):e112466. https://doi.org/10.1371/journal.pone.0112466

20. Spencer NJ et al (2016) Spinal afferent nerve endings in visceral organs: recent advances. Am J Physiol Gastrointest Liver Physiol 311(6):G1056–G1063. https://doi.org/10.1152/ajpgi.00319.2016

21. Brierley SM, Hibberd TJ, Spencer NJ (2018) Spinal afferent innervation of the colon and Rectum. Front Cell Neurosci 12:467. https://doi.org/10.3389/fncel.2018.00467

22. Hibberd TJ et al (2016) Identification of different functional types of spinal afferent neurons innervating the mouse large intestine using a novel CGRPalpha transgenic reporter mouse. Am J Physiol Gastrointest Liver Physiol 310(8):G561–G573. https://doi.org/10.1152/ajpgi.00462.2015

23. Malin SA et al (2009) TPRV1 expression defines functionally distinct pelvic colon afferents. J Neurosci 29(3):743–752. https://doi.org/10.1523/JNEUROSCI.3791-08.2009

24. Hockley JRF et al (2019) Single-cell RNAseq reveals seven classes of colonic sensory neuron. Gut 68(4):633–644. https://doi.org/10.1136/gutjnl-2017-315631

25. Makadia PA et al (2018) Optogenetic activation of colon epithelium of the mouse produces high-frequency bursting in extrinsic colon afferents and engages visceromotor responses. J Neurosci 38(25):5788–5798. https://doi.org/10.1523/JNEUROSCI.0837-18.2018

26. Boyden ES et al (2005) Millisecond-timescale, genetically targeted optical control of neural activity. Nat Neurosci 8(9):1263–1268. https://doi.org/10.1038/nn1525

27. Baumbauer KM et al (2015) Keratinocytes can modulate and directly initiate nociceptive responses. Elife 4:e09674. https://doi.org/10.7554/eLife.09674

28. Boada MD et al (2014) Fast-conducting mechanoreceptors contribute to withdrawal behavior in normal and nerve injured rats. Pain 155(12):2646–2655. https://doi.org/10.1016/j.pain.2014.09.030

29. Copits BA, Pullen MY, Gereau RW (2016) Spotlight on pain: optogenetic approaches for interrogating somatosensory circuits. Pain 157(11):2424–2433. https://doi.org/10.1097/j.pain.0000000000000620

30. Daou I et al (2013) Remote optogenetic activation and sensitization of pain pathways in freely moving mice. J Neurosci 33(47):18631–18640. https://doi.org/10.1523/JNEUROSCI.2424-13.2013

31. Iyer SM et al (2014) Virally mediated optogenetic excitation and inhibition of pain in freely moving nontransgenic mice. Nat Biotechnol 32(3):274–278. https://doi.org/10.1038/nbt.2834

32. Montgomery KL et al (2015) Wirelessly powered, fully internal optogenetics for brain,

spinal and peripheral circuits in mice. Nat Methods 12(10):969–974. https://doi.org/10.1038/nmeth.3536

33. Park SI et al (2015) Soft, stretchable, fully implantable miniaturized optoelectronic systems for wireless optogenetics. Nat Biotechnol 33(12):1280–1286. https://doi.org/10.1038/nbt.3415

34. DeBerry JJ et al (2018) Differential regulation of bladder pain and voiding function by sensory afferent populations revealed by selective Optogenetic activation. Front Integr Neurosci 12:5. https://doi.org/10.3389/fnint.2018.00005

35. Samineni VK et al (2017) Optogenetic silencing of nociceptive primary afferents reduces evoked and ongoing bladder pain. Sci Rep 7(1):15865. https://doi.org/10.1038/s41598-017-16129-3

36. O'Mahony SM et al (2012) Rodent models of colorectal distension. Curr Protoc Neurosci. Chapter 9:Unit 9 40. https://doi.org/10.1002/0471142301.ns0940s61

37. Salvatierra J et al (2018) NaV1.1 inhibition can reduce visceral hypersensitivity. JCI Insight 3(11):e121000. https://doi.org/10.1172/jci.insight.121000

38. Grundy L et al (2018) Chronic linaclotide treatment reduces colitis-induced neuroplasticity and reverses persistent bladder dysfunction. JCI Insight 3(19):e121841. https://doi.org/10.1172/jci.insight.121841

Chapter 17

Use of Optogenetics for the Study of Skin–Nerve Communication

Ariel Epouhe, Marsha Ritter Jones, Sarah A. Najjar, Jonathan A. Cohen, Daniel H. Kaplan, H. Richard Koerber, and Kathryn M. Albers

Abstract

This chapter describes the use of light-activated opsin proteins for study of keratinocyte and immune cell communication with sensory neurons that innervate the skin. We describe the optogenetic mouse models and approaches used in two of our recent studies. The first study analyzed keratinocyte–sensory afferent signaling in mice that express ChR2 in skin keratinoctyes. Blue light activation of keratinocytes was shown to be sufficient to generate action potential firing in many types of cutaneous sensory afferents and cause behavioral hypersensitivity (Baumbauer et al. eLife 4:e09674, 2015). In another study we examined communication of nerve fibers with immune cells and showed that light activation of cutaneous afferents was sufficient to generate a Type17 immune response and that this response was protective against the spread of injury (Cohen et al. Cell 178(4):919–32 e14, 2019). Here, we describe the mouse models, stimulation protocols and equipment used in these studies.

Key words Sensory neuron, Keratinocyte, Laser stimulation, Immune response

---

1 Introduction

The skin is a complex tissue that serves as a first line of defense against injury, water loss and pathogen invasion. It is composed of an overlying epidermis and a connective tissue rich dermis. The epidermis is a cornified, stratified epithelium composed of keratinocyte epithelial cells, melanocytes and immune cells, for example, Langerhan's cells and T cells. Glabrous skin on the palms and soles is hairless and thicker than hairy skin. Hairy skin has epidermal appendages (hair follicles, sebaceous glands) that project into dermal tissue that is rich in collagen and elastin fibers, blood and lymph vasculature and immune cells. Coursing through the dermis and penetrating the epidermis is a network of sensory nerve fibers. Signaling between sensory fibers and skin cells facilitates transduction of external sensory stimuli. Disruption in this communication contributes to chronic inflammatory pathologies of the skin,

Rebecca P. Seal (ed.), *Contemporary Approaches to the Study of Pain: From Molecules to Neural Networks*, Neuromethods, vol. 178, https://doi.org/10.1007/978-1-0716-2039-7_17, © Springer Science+Business Media, LLC, part of Springer Nature 2022

particularly those associated with paresthesia, pain, and itch (e.g., psoriasis, atopic dermatitis). Neuronal sprouting and neuronal hyperexcitability frequently accompany these pathologies [1–5].

Cutaneous sensory nerve fibers are diverse in their cellular and functional properties. Larger, myelinated fibers have terminations that form specialized endings such as Merkel cell complexes, Meissner corpuscles or lanceolate endings that encircle hair follicles. Other fibers have "free" endings with reduced myelination. These fibers can be classified as peptidergic or nonpeptidergic based on expression of known biomarkers and many project into the epidermis. In mouse, peptidergic fibers are rich in calcitonin gene-related peptide (CGRP), the neurotrophin receptor tropomyosin receptor kinase A (trkA), and the thermal and capsaicin receptor channel transient receptor potential vanilloid 1 (TRPV1). Nonpeptidergic fibers bind the isolectin B4 and express mas-related genes (Mrgs) and the glial cell line–derived neurotrophic factor (GDNF) growth factor receptor, Ret. The majority of fibers that penetrate the epidermis are nonpeptidergic whereas peptidergic, TRPV1-positive fibers primarily terminate in lower regions of the epidermis and within the dermal layer.

In vivo and ex vivo studies have shown that peptidergic and nonpeptidergic fibers that innervate skin have different yet overlapping electrophysiologic response properties to applied stimuli such as heat, cold, mechanical, and chemical stimuli [6]. These fibers also transmit noxious stimuli and act as pain sensors. How neural endings are activated in response to noxious and threatening stimuli, and approaches to block this activation, are important issues for understanding mechanisms that underlie acute and chronic pain conditions, response to infectious agents and inflammatory stimuli. Evidence supports the idea that activation of nerves in response to noxious stimuli involves not only activation of neuronal terminals, but also activation of surrounding keratinocytes and immune cells of the skin. These complex interactions are difficult to study because of the intimate relationship between nerve terminals with skin and immune cells, that is, stimuli that activates nerves also have the potential to activate surrounding nonneuronal cells. Understanding of neural–epithelial communication is also complicated by the diversity of neuronal afferents and skin cell heterogeneity. In addition, establishing neuronal–epidermal cocultures that mimic in vivo neural–epithelial interactions is challenging.

To examine neural–epithelial communication in the in vivo system we and others have developed optogenetic mouse models that target light-activated opsin molecules to sensory neurons or epithelial keratinocytes [7–14]. Blue light–activated channelrhodopsin (ChR2) targeted to sensory neurons (e.g., under control of the Nav1.8, peripherin, or TrpV1 gene promoters), causes strong, instantaneous nocifensive behaviors in response to blue light illumination of the skin. In an ex vivo skin–nerve–ganglia and spinal cord preparation, light stimulation elicits activity in a variety of sensory afferents, including C-fiber nociceptors [15]. Although light activation of ChR2 in nerve fibers can elicit robust axon firing, neural responses were often at lower frequencies than those obtained by direct mechanical or thermal stimulation of the skin. Given the range of ChR2-induced responses and the high level of ChR2 expression in afferents, we hypothesized that light activation of the afferents was not sufficiently reproducing the firing pattern elicited by natural stimulation and that other cell types in the skin contribute to neural activation. We therefore isolated an optogenetic mouse model in which ChR2 expression was driven in keratin 14 (Krt14)-expressing epidermal keratinocytes. We found that specific stimulation of skin keratinocytes with blue light was alone sufficient to elicit action potential firing in sensory afferents and behavioral paw withdrawal responses [15].

In this review, we describe the approach used to analyze keratinocyte–sensory afferent signaling in Krt14-ChR2 mice. Light mediated activation of keratinocytes was shown to be sufficient to generate action potential firing in nearly all types of cutaneous sensory afferents [15]. Nerve activation correlated with parallel studies that showed blue light stimulation of keratinocytes was also sufficient to elicit behavioral hypersensitivity. We also describe the approach used in a recent study where we examined communication of nerve fibers with immune cells. We found that light activation of cutaneous afferents could generate a Type 17 immune response and that this response was protective against the spread of infection, an effect referred to as anticipatory immunity [16]. Here we describe the models and stimulation protocols used in these studies.

2 Materials

2.1 Animals

2.1.1 Keratinocyte–Sensory Afferent Communication Studies

Mice that express excitatory (channelrhodopsin, ChR2) or inhibitory (halorhodopsin (NpHR3) or archaerhodopsin (ArchT)) opsins in keratinocytes were generated by crossing mouse lines obtained from Jackson Laboratory (Bar Harbor, ME). Male and female mice ages 6 weeks to 3 months were used. To express ChR2 –EYFP in basal keratinocytes, Krt14cre mice (Jax #004782) were crossed with Ai32 mice (Jax #012569). Mice with two copies of the opsin allele provide a stronger response than the heterozygous mice. The Cre allele is also kept as one copy to avoid potential cre-mediated off-target effects.

2.1.2 Sensory Afferent–Immune Communication

For studies of sensory afferent–immune signaling, mice that express Cre under control of the Trpv1 promoter were used (Jackson Laboratory, # 017769). Male mice 7–12 weeks of age that express

ChR2-eYFP in TRPV1 sensory neurons (TRPV1-Ai32) were generated by crossing homozygous Ai32 mice with homozygous TrpV1-cre mice (*see* **Note 1**). Mice were genotyped by PCR amplification of tail DNA using primer sequences and amplification protocols provided on each Jax mouse page (under Technical Support tab). Mice were housed in microisolator cages and fed irradiated food and acidified water.

2.2 Preparation of Hairy Skin Prior to Laser Stimulation

In these studies light stimulation of keratinocytes was done on the hairy skin of the dorsum and on ear skin. Dorsal skin requires hair removal prior to laser stimulation. At 12–18 h prior to laser stimulation skin is first shaved using a professional hair clipper (Arco clipper, Wahl Outlet Store, Syosset, NY). To lessen chemical damage (from calcium hydroxide and sodium hydroxide), remaining fur is removed using the Sally Hansen hair remover wax strip kit. During laser exposure, the treatment area is outlined using a Mini XL Surgical Marker (Viscot Medical, East Hanover, NJ). Following laser treatment skin biopsies can be collected using a Miltex 4 mm disposable biopsy punch (Integra LifeSciences, Plainsboro, NJ) or using sharp scissors and forceps that are RNase free.

2.3 Lasers

Lasers, power supplies and cable connectors were purchased from Laserglow Technologies, Toronto, Canada. We used a diode-pumped solid state (DPSS) 473 nm blue laser that provides 5–100 mW output power (average output power of 95.18 mW). For halorhodopsin and archaerhodopsin we used a similar configuration with a 589 nm yellow laser used at average power of 34 mW. Lasers are connected using a FC/PC fiber coupler/collimator attached to an armored fiber-optic cable with SMA 905 connectors (Laserglow Technology) to provide flexibility in positioning and delivery of light. A carrying case (included with purchase) and safety goggles are also required.

Laser power was controlled using a power supply (LRS-0473-GFM-00050-05). Laser stimulus was controlled using Spike2 software (Cambridge Electronic Design Limited). The power of the laser should be recorded before the start of each experiment. We used a Laser Check power meter (Model #2168219, Coherent, Inc., Santa Clara, CA).

2.4 Construction of Place Preference Boxes for Behavioral Assessment

Behavior boxes were first designed and constructed in the Koerber laboratory by Drs. Kuan Hsien (Milly) Lee and Peter Adelman. Plexiglass boxes were constructed with two chambers and a passageway between them to allow mice to choose a chamber with either floor mounted blue light-emitting diode (LED) strips (for activating ChR2) or yellow strips (for activation of halorhodopsin and archaerhodopsin).

2.4.1 Components of LED Array—Can Be Ordered from Environmental Lights, San Diego, California, Unless Stated Otherwise

1. Amber LED Strip Light (dominant wavelength is 595 nm), 120/m, 10 mm wide, by the 5 m Reel, 12 VDC, 8.9 Watts/meter, 742 mA/meter (cat. no. amber3528–120-10-reel).

2. Blue LED Strip Light (dominant wavelength is 460 nm), 120/m, 10 mm wide, by the 5 m Reel, 12 VDC, 8.9 Watts/meter, 742 mA/meter (cat. no. blue3528–120-10-reel).

3. Two 60 W 12 VDC Dimming Power Supply (cat. no. 60W12VDim).

4. 44 LED Strip Light Connector-2 Conductor 10 mm Ribbon to Ribbon Jumper (cat. no. rf2-jumper-10).

5. Two Power Cords with North American Plug (cat. no. PowerCord-NA).

6. Four Male Mini Power Supply Plugs and 20″ Cord, 18AWG (cat. no. MaleMiniPlug).

7. Two BNC cables.

8. Two 11 in. × 17 in. shallow baking trays (not Teflon-coated).

9. Two Solid State Relays (cat. no. AD-SSR6M12-DC-200D; AutomationDirect, Cumming, Georgia).

10. Analog–Digital Interface Micro1401 (CED, Milton, Cambridge, England).

11. Personal computer.

12. Spike 2 software (Cambridge Electronic Design Limited, Cambridge, England).

13. Temperature controlled glass surface (e.g., thermal plantar testing unit, IITC Life Science, Woodland Hills, CA).

14. Camera to record behavior (e.g., Digital HD Video Camera Recorder, Sony Corporation, Tokyo, Japan).

2.4.2 Assembly of LED Platform

1. From blue LED light reel cut twelve 8 in strips. One strip should be taken from the end of the reel and therefore have a female connector attached to it. Repeat for the amber LED light strip.

2. Align the 12 blue LED strips side-by-side on one half of bottom of the baking tray (strips have an adhesive backing, which enables them to be attached directly to the tray) and 12 of the amber LED strips side-by-side on the other half of the tray (Fig. 1a). Place a 2¼ in. gap between each group of 12 strips.

3. Connect each of the 12 blue LED strips to its neighbor with LED Strip Light Connectors so that the set of 12 strips carries a continuous current. Repeat for the amber LED strips.

4. The Dimming Power Supply is attached with a power cord and connected to the LED array using the protocol described in the

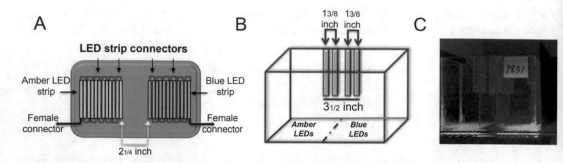

Fig. 1 Assembly of LED platform used to construct place preference behavior box. (**a**) Diagram shows arrangement of LED arrays assembled on non–Teflon coated baking tray. See text for details on assembly. (**b**) Arrangement of acrylic pieces that serve as holders for vertically arranged compartment dividers. Shown is one set of 5 in. × ½-in. spacers adhered to the internal surface of the 10 in. × 12 in. box. Space pieces such that the first is 4¼ in. from one edge, the next is ⅜ in. away, the third is ¾ in. away and the next is ⅜ in. away. (**c**) An assembled place preference box with both arrays on

Dimming Supplies Manual (*see* www.environmentalights.com). Briefly, the power cord should be connected to the input (primary) connectors of the dimming power supply. Connect the output (secondary) connectors to two male mini power supply plugs and to the solid-state relay.

5. Using a BNC cable, connect the solid-state relay to a DAC output on the Micro 1401. Then connect the male mini power supply plugs to the LED array.

6. The solid-state relay enables the power supply to be connected, through the Micro 1401, to the PC computer. The power supply/LED arrays can therefore be controlled using commands in the Spike 2 software.

2.4.3 Components and Assembly of Place Preference Behavior Box

Components

1. 48-in × 96-in. × ⅛-in. plexiglass sheet (purchased from Lowe's building supply).

2. SCIGRIP 1610315 Acrylic cement, low VOC, medium bodied, 5 fl oz. tube, clear (Model #10315; SCIGRIP).

3. 1⅛-in. × 1⅛-in. × 8 feet plastic corner guard—Commercial (Model #01188G; Trimaco).

Box Assembly

1. Have a machine shop cut plexiglass into two 12-in. × 10-in. pieces, four 5-in. × 10-in. pieces and eight 5-in. × ½-in. pieces.

2. Use acrylic cement to attach the 12-in. × 10-in. pieces to two of the 5-in. × 10-in. pieces to construct a 12-in. × 5-in. rectangular box. The other two 5-in. × 10-in. pieces will be used as chamber dividers.

3. To allow for compartment separation, assemble stationary guides to hold the 5-in. × 10-in. chamber dividers vertically in the box. Cement pairs of the 5-in. × ½-in. pieces together to form four 5-in. × ½-in. × ¼-in. pieces.

4. Glue the four 5-in. × ½-in. × ¼-in. pieces to the inside of the box so that they form guides to hold the other 5-in. × 10-in. plexiglass sheets (Fig. 1b).

5. Cut plastic corner guard into 4 × 10-in. pieces. Cement to each box corner to reinforce structure. The box should be 12-in. × 5-in. in dimension; there is no top or bottom.

2.5 Data Analysis

Version 7.16 of Spike2 can be used to control the voltage of the LED array and for offline analysis of time spent in each LED chamber.

3 Methods

To examine the role of keratinocytes in the activation of cutaneous afferents, electrophysiological measures were done using an ex vivo preparation of the skin, saphenous nerve and dorsal root ganglia. Teased fiber or intracellular recordings were made in response to laser stimulation of the skin of the hind paw. Details of this preparation are provided in [6, 15, 17, 18]; in addition, see Stucky chapter, this book). Here we describe procedures used to laser stimulate the skin in this ex vivo preparation.

3.1 Keratinocytes and Sensory Afferent Communication—Ex Vivo Analysis

3.1.1 Photostimulation of ChR2 Expressed by Keratinocytes to Evoke Sensory Afferent Firing

1. Ex vivo skin–nerve–DRG preparations were used as described in Baumbauer et al. [15]. Light stimulation was done using a blue (473 nm) laser positioned so as to illuminate a 1-2 mm diameter circle of skin for recordings. Laser power at 80 mW was optimal for stimulation of fiber activity, although with time wattage could vary down to 30 mW, which we found was still sufficient to evoke a physiological response. For reference, measures of the laser power delivered over the skin area can be estimated for these conditions where for a 1 mm diameter circle, the power per area (W/cm^2) using an 80 mW laser is 25.5 mW/mm^2 (80 mW divided by 3.14 mm^2, with $A = \pi r^2$) and, using 39.7 mW (as in [15]), is 12.6 mW/mm^2. No significant heating of the skin was measured under these conditions.

2. Several types of sensory afferents can be activated and recorded from using *ex vivo* preparations from Krt14-ChR2 mice: SA1, slowly adapting type 1; RA (Aβ), rapidly adapting A beta low-threshold mechanoreceptor; RA (Aδ), rapidly adapting A delta low-threshold mechanoreceptor, A-HTMR, high-threshold mechanoreceptor; CM, C mechanoreceptor; CC, C cold receptor; CH, C heat receptor; CMC, C mechano-cold receptor; CMH, C mechano-heat receptor; CMHC, C mechano-heat and cold receptor.

3. Inhibitory responses can also be elicited by Krt14-Cre mediated expression of archaerhodopsin or halorhodopsin. Yellow light (589 nm) is applied in the same manner as blue light. When coupled with mechanical or thermal stimuli, it should precede the natural stimulus by 1 s.

4. Summation of effect can also be examined by presenting multiple stimuli to the skin–nerve preparation; for example, the tip of a small diameter (1 mm) mechanical stimulator is applied concurrently with the laser. Because the stimulator tip is 1 mm in diameter it does not block the entire receptive field available for laser stimulation. In addition, the light is delivered on an angle of approximately 45°, so will also penetrate the skin beneath the probe.

3.2 Keratinocytes and Sensory Afferent Communication—In Vivo Behavioral Analysis

Mice that express excitatory or inhibitory opsins in keratinocytes were generated by crossing lines obtained from the Jackson Laboratory (described in Subheading 2.1). Behavioral responses were measured by illumination of the footpad skin of hind paws using a 473 nm laser that produced up to 80 mW power. However, for best responses, stimulation should be made with a laser output of 60–80 mW.

3.2.1 Paw Withdrawal Latency as a Measure of Keratinocyte-Induced Behavioral Response

Measures of paw withdrawal latency requires two people and should be done in a blinded manner. One person transfers mice to and from a chamber and one applies the laser. Animals are confined to a small space (e.g., a 500 mL glass beaker can be used) and once the animal is settled, laser stimulation is applied to the ventral paw skin from a distance of 8–10 mm. The number of nocifensive responses (paw lifting, biting, licking) out of 10 stimulations is recorded. Under these conditions mice with ChR2 expressed in keratinocytes should exhibit a robust withdrawal response at least once out of 10 trials. Laser stimulation was restricted to a 30 s maximum (as done in plantar thermal testing). Measures using an infrared thermometer are done to confirm no change in skin temperature occurs under these conditions. Plots of the distribution of the latencies, for example, percent of the responders in 0–5 s, is one way to express the measures.

3.2.2 Place Preference Assay to Measure Behavioral Response to Light Stimulation of Keratinocytes

Place preference assays are an unbiased measure of behavioral responses. Avoidance behaviors in response to either blue or yellow light can be measured using place preference boxes (see Subheading 2.4). Plexiglass boxes (12-in. × 5-in.) contain two chambers, one lined with blue LED strips and the other with yellow LED strips. Prior to acclimation and testing, place the box on top of a temperature-controlled glass top held at 30 °C. Prop the LED array on top of a holder (e.g., mouse transfer buckets work well) below the heated glass plate. This arrangement allowed a consistent surface temperature of 30 °C to be held during testing without significantly diminishing laser intensity.

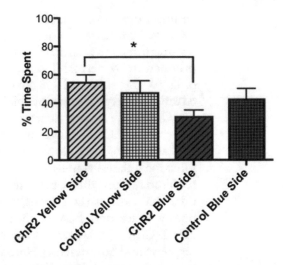

Fig. 2 Place preference assay of Krt14-ChR2 mice. Krt14-ChR2 mice ($n = 11$) spend less time in the chamber containing blue light whereas wildtype mice (Krt14-Cre mice, $n = 7$) spend equal time in the blue or yellow chamber. Data were analyzed using a one-way analysis of variance with Tukey's post hoc test. Asterisk indicates significance with $p < 0.05$

Mice are acclimated to the box for 15 min at least 1 day prior to testing. On the day of testing, individual mice are placed in the box and allowed to freely roam for 15 min under low light room conditions and with the LED array off. On the day of testing, mice are place in the center of the box (a 2 in. space devoid of LED lights, that is, the home chamber). The 5-in. × 10-in. plexiglass dividers can be used to constrain movement between chambers. One or both LED arrays (with blue or yellow light) are switched on and mice roam freely between each side for up to 15 min. During this time movements are recorded using a camcorder facing the box so that each side of the box is in view. Following testing, movements are analyzed manually or using software to quantify the time spent in each chamber over the 15 min interval. Total time spent on each side is calculated (in 5 min bins) and measures compared using analysis software (e.g., Prism). Place preference behavioral testing illustrating sensitivity to blue light in Krt14-ChR2 mice is shown in Fig. 2. Also, *see* **Notes 2** and **3**.

3.3 Immune System and Sensory Afferent Communication

The skin is a major barrier organ skin exposed to bacterial, viral, and fungal pathogens. Resident immune cells detect pathogens and mediate immunity to these challenges by recruitment of immune effector cells and antimicrobial responses. Recent studies have shown Trpv1$^+$ cutaneous sensory afferents can be activated by inflammatory mediators and pathogen-derived molecules [19, 20]. Afferent activation was shown to initiate an immune response through release of neuropeptides, for example, CGRP.

In mice, infection of skin with *Candida albicans* evokes release of CGRPα from Trpv1$^+$ afferents [19]. CGRP is required for IL-23 production by the dermal dendritic cells and IL-23 promotes IL-17 production by dermal T cells, resulting in recruitment of neutrophils to the skin and effective control of *C. albicans* infection. Ablation of Trpv1$^+$ afferents blocks this immune response.

Cultured sensory neurons exposed to heat killed *C. albicans* also release CGRP, suggesting pathogens alone can activate afferents and in so doing, activate an innate immune responses, that is, infection-related signals from other cells in the skin may not be required. To test this, we studied neural immune communication in mice that express the blue light-activated ChR2 channel in Trpv1-lineage afferents. Results show that specific light activation of Trpv1-ChR2$^+$ afferents is sufficient to initiate an innate immune response and host defense. Here we describe the methodology used to optogenetically stimulate the skin to examine neural–immune communication.

3.3.1 Mouse Model

Male Trpv1-ChR2 mice and littermate controls 7 to 12 weeks of age were used. Mice were generated as described in Subheading 2.1. Control mice in these studies were Ai32 mice (mice that have floxed ChR2-YFP but lack Cre recombinase).

3.3.2 Mouse Dorsal Skin—Optogenetic Stimulation

Cutaneous afferents of Trpv1-ChR2 mice of either ear skin or dorsal skin were stimulated using 473 nm blue light. Studies on dorsal skin sites require hair depilation (*see* Subheading 2.2). This should be done 12–18 h before laser stimulation.

Light stimulation of TrpV1-lineage afferents is painful, producing a rapid and intense nocifensive response. Thus, prior to exposure to blue light mice must be anesthetized. Anesthesia induction is done using a nose cone delivering 5% isoflurane followed by a light plane of anesthesia maintained using 2% isoflurane. During this time mice are placed prone on a heating pad to maintain body temperature. If a laser protocol is longer than 2 h, inject 0.5 mL of sterile saline subcutaneously to maintain hydration.

We tested several protocols of light stimulation to identify a time, frequency and power dosage that produced consistent readouts without detectable off-target effects. For photostimulation, the optical fiber is positioned at 1.5 in. perpendicular to the skin surface (treated areas can be demarcated using a skin marking pen). This produces a circle of ~4 mm diameter at the site. The parameters of exposure (frequency, pulse width, exposure time) are controlled through the power source (*see* Subheading 2.3), which is externally controlled by Spike2 software. For exposure of dorsal skin we used a frequency of 10 Hz, pulse width of 50 ms and total time of exposure of 30 min. Over the 30 min interval the laser was set in a 1 s on, 1 s off sequence. Up to 4 different areas on the mouse skin can be light treated during one session.

Following light treatment 4 mm biopsy punches are collected using a disposable punch biopsy tool. As control for animal and treatment variability, an area of identical size is collected from the contralateral untreated side.

3.3.3 Mouse Ear Skin— Optogenetic Stimulation

Similar to treatment of dorsal skin, mice are maintained on a light anesthetic plane using isoflurane, placed on their stomachs and maintained on a heating pad. For light stimulation of ear skin, the laser was placed in a holder 1 cm from the surface of the skin. As for back skin, different light stimulation protocols can be used. In all cases stimulation is done in 30 min sessions. For 6 h photostimulation, skin is stimulated in two 30 min sessions.

Repeated 30 min treatment sessions, four times per day (with at least 2 h between each session) can also be used over a 3–5 day period. This paradigm evokes robust changes at histological and gene expression levels. In these experiments stimulation was at a frequency of 10 Hz, pulse width of 10 ms, and a 10 s on 5 s off sequence for 30 min. This produces a power density of 8 mW/mm^2.

In Trpv1-ChR2 mice, repeated activation of TRPV1-lineage afferents over a 5 day period led to increased ear thickness, erythema and scaling dermatitis [16]. Importantly, laser stimulated Ai32 control mice did not develop this inflammatory reaction. Light-evoked changes can be assessed by measure of ear thickness using a micrometer (Mitutoyo), as described in [21]. Inflammation scores can also be obtained using a modified PASI scoring system as in [22].

Afferent-evoked changes can also be assessed using sections of fresh-frozen or paraffin-embedded ear skin. Hematoxylin and eosin–stained skin sections of Trpv1-Ai32 mice stimulated for 4 days should exhibit a dense neutrophilic and lymphocytic infiltrate along with acanthosis and parakeratosis.

Because the ChR2 used in these studies is tagged with eYFP, the distribution of Trpv1 eYFP fluorescent fibers in skin can be visualized. Colabeling with other neural markers, for example, rabbit anti-PGP9.5 (various sources; Ultraclone Ltd., Isle of Wight, Sigma, AbCam), CGRP (Fisher), and DAPI (to demarcate nuclei of skin cells), can also be utilized.

Flow-cytometric analysis can also be done on infiltrate collected from the ear as described in [16, 19]. In this assay, repeated light stimulation of ChR2 expressing TrpV1-lineage afferents produces an increase in the number of dermal TCRγδ T cells, CD4$^+$ TCRαβ T cells and neutrophils without change in DETC, CD8$^+$ T cells, or ILC [16].

Sensory afferent–immune cell communication can also be assessed at the transcriptional level using RNA isolated from the whole ear. Following light stimulation whole ears are homogenized in Trizol (Sigma-Aldrich, St. Louis, MO). Purified RNA was reverse transcribed using a High Capacity cDNA Reverse

Transcription Kit (Applied Biosystems, Carlsbad, CA) following manufacturer's protocols. Relative levels of cytokine mRNAs were measured using Power SYBR Green PCR Master Mix (Applied Biosystems), supplemented with gene specific primers [16]. Ct values were normalized to Hprt expression. Changes in cytokine expression can be detected as early as 6 h after initiation of photostimulation [16].

Light activation of cutaneous Trpv1-lineage afferents also increases cytokine proteins. Whole ears harvested at 48 h post light stimulation are homogenized in 500 mL of Cell Extraction Buffer (Invitrogen) supplemented with phenylmethylsulfonyl fluoride (PMSF) and protease inhibitor (Sigma-Aldrich), used according to the manufacturer's guidelines. Cytokine levels in tissue supernatants can be quantified using a bead-based LEGENDplex immunoassay (BioLegend, San Diego, CA) according to the manufacturer's protocol.

3.3.4 Cultured Dorsal Root Sensory Neurons— Optogenetic Stimulation

The effects of ChR2-mediated stimulation can also be assessed in dissociated and cultured dorsal root ganglia (DRG) neurons. We found a 30 min light stimulation of DRG neurons isolated from TRPV1-Ai32 mice (but not control mice) is sufficient to release CGRPα protein into the media (measured by EIA using a kit obtained from Cayman Chemicals, #589001) [16]. Laser stimulation was done on cells cultured on laminin/polyornithine coated 48-well culture plates placed on a humidified platform maintained at 37 °C. The parameters for laser stimulation were the same as used for dorsal skin (frequency of 4 Hz, pulse width of 50 ms, total time of exposure of 30 min with the laser set in a 1 s on, 1 s off sequence).

With targeted activation of specific cell types, optogenetic mouse models have provided new insights into how epithelial and immune cell types of the skin communicate with sensory afferents. This new knowledge provides additional insights into how communication occurs under homeostatic and pathogen challenged conditions in the skin, and how this communication might regulate pain and immune responses [23]. These approaches can be easily adapted and expanded to include other cell types in the skin as well as to study communication in other organ systems, for example, the colon [24].

4 Notes

1. The Trpv1 gene is activated in precursors of peptidergic and nonpeptidergic neurons during development [25]. Thus, in the adult, the Trpv1-Cre driver line targets TrpV1[+], peptidergic neurons and some nonpeptidergic neurons, which are referred to as "TRPV1-lineage" neurons in this chapter.

2. Using LED arrays in place preference boxes works well when the opsin is expressed in the primary afferent. Behavioral measures of mice with opsin expression in skin keratinocytes were less robust but measurable by preference. However, stimulation using a hand held laser gave more consistent, repeatable outcomes. This may be related to thickness of the keratinized epithelium and the lesser power generated by the LED arrays.

3. We also generated and tested Krt14-Cre mice crossed with Ai39 mice (Jax #014539) or Ai40D mice (Jax #021188) to target halorhodopsin (eNpHR3.0-EYFP) or archaerhodopsin (aop3/EYFP), respectively, to basal keratinocytes. Analysis using the ex vivo preparation showed activation of these inhibitory opsins blocked neural activity [15]. However, behavioral testing (post CFA or formalin challenge) using the place preference boxes produced results that showed high variability. Thus, solid conclusions on effectiveness could not be made.

References

1. Groneberg DA, Serowka F, Peckenschneider N, Artuc M, Grutzkau A, Fischer A et al (2005) Gene expression and regulation of nerve growth factor in atopic dermatitis mast cells and the human mast cell line-1. J Neuroimmunol 161(1–2):87–92

2. Johansson O, Liang Y, Emtestam L (2002) Increased nerve growth factor- and tyrosine kinase A-like immunoreactivities in prurigo nodularis skin—an exploration of the cause of neurohyperplasia. Arch Dermatol Res 293(12): 614–619

3. Kinkelin I, Motzing S, Koltenzenburg M, Brocker EB (2000) Increase in NGF content and nerve fiber sprouting in human allergic contact eczema. Cell Tissue Res 302(1):31–37

4. Truzzi F, Marconi A, Pincelli C (2011) Neurotrophins in healthy and diseased skin. Dermatoendocrinol 3(1):32–36

5. Urashima R, Mihara M (1998) Cutaneous nerves in atopic dermatitis. A histological, immunohistochemical and electron microscopic study. Virchows Arch 432(4):363–370

6. Koerber HR, Woodbury CJ (2002) Comprehensive phenotyping of sensory neurons using an ex vivo somatosensory system. Physiol Behav 77(4–5):589–594

7. Daou I, Tuttle AH, Longo G, Wieskopf JS, Bonin RP, Ase AR et al (2013) Remote optogenetic activation and sensitization of pain pathways in freely moving mice. J Neurosci 33(47):18631–18640

8. Ji ZG, Ito S, Honjoh T, Ohta H, Ishizuka T, Fukazawa Y et al (2012) Light-evoked somatosensory perception of transgenic rats that express channelrhodopsin-2 in dorsal root ganglion cells. PLoS One 7(3):e32699

9. Maksimovic S, Nakatani M, Baba Y, Nelson AM, Marshall KL, Wellnitz SA et al (2014) Epidermal Merkel cells are mechanosensory cells that tune mammalian touch receptors. Nature 509(7502):617–621

10. Igarashi H, Ikeda K, Onimaru H, Kaneko R, Koizumi K, Beppu K et al (2018) Targeted expression of step-function opsins in transgenic rats for optogenetic studies. Sci Rep 8(1):5435

11. Uhelski ML, Bruce DJ, Seguela P, Wilcox GL, Simone DA (2017) In vivo optogenetic activation of Nav1.8(+) cutaneous nociceptors and their responses to natural stimuli. J Neurophysiol 117(6):2218–2223

12. Tashima R, Koga K, Sekine M, Kanehisa K, Kohro Y, Tominaga K et al (2018) Optogenetic activation of non-nociceptive Abeta fibers induces neuropathic pain-like sensory and emotional behaviors after nerve injury in rats. eNeuro 5(1):ENEURO.0450-17.2018

13. Moehring F, Cowie AM, Menzel AD, Weyer AD, Grzybowski M, Arzua T et al (2018) Keratinocytes mediate innocuous and noxious touch via ATP-P2X4 signaling. eLife 7:e31684

14. Ritter-Jones M, Najjar S, Albers KM (2016) Keratinocytes as modulators of sensory afferent firing. Pain 157(4):786–787

15. Baumbauer KM, DeBerry JJ, Adelman PC, Miller RH, Hachisuka J, Lee KH et al (2015) Keratinocytes can modulate and directly initiate nociceptive responses. eLife 4:e09674

16. Cohen JA, Edwards TN, Liu AW, Hirai T, Jones MR, Wu J et al (2019) Cutaneous TRPV1(+) neurons trigger protective innate type 17 anticipatory immunity. Cell 178(4): 919–32 e14

17. Lawson JJ, McIlwrath SL, Woodbury CJ, Davis BM, Koerber HR (2008) TRPV1 unlike TRPV2 is restricted to a subset of mechanically insensitive cutaneous nociceptors responding to heat. J Pain 9(4):298–308

18. McIlwrath SL, Lawson JJ, Anderson CE, Albers KM, Koerber HR (2007) Overexpression of neurotrophin-3 enhances the mechanical response properties of slowly adapting type 1 afferents and myelinated nociceptors. Eur J Neurosci 26(7):1801–1812

19. Kashem SW, Riedl MS, Yao C, Honda CN, Vulchanova L, Kaplan DH (2015) Nociceptive sensory fibers drive Interleukin-23 production from CD301b+ dermal dendritic cells and drive protective cutaneous immunity. Immunity 43(3):515–526

20. Riol-Blanco L, Ordovas-Montanes J, Perro M, Naval E, Thiriot A, Alvarez D et al (2014) Nociceptive sensory neurons drive interleukin-23-mediated psoriasiform skin inflammation. Nature 510(7503):157–161

21. Kaplan DH, Jenison MC, Saeland S, Shlomchik WD, Shlomchik MJ (2005) Epidermal Langerhans cell-deficient mice develop enhanced contact hypersensitivity. Immunity 23(6):611–620

22. van der Fits L, Mourits S, Voerman JSA, Kant M, Boon L, Laman JD et al (2009) Imiquimod-induced psoriasis-like skin inflammation in mice is mediated via the IL-23/IL-17 Axis. J Immunol 182(9):5836–5845

23. Talagas M, Lebonvallet N, Berthod F, Misery L (2019) Cutaneous nociception: role of keratinocytes. Exp Dermatol 28:1466–1469

24. Makadia PA, Najjar SA, Saloman JL, Adelman P, Feng B, Margiotta JF et al (2018) Optogenetic activation of colon epithelium of the mouse produces high-frequency bursting in extrinsic colon afferents and engages visceromotor responses. J Neurosci 38(25): 5788–5798

25. Cavanaugh DJ, Chesler AT, Braz JM, Shah NM, Julius D, Basbaum AI (2011) Restriction of transient receptor potential vanilloid-1 to the peptidergic subset of primary afferent neurons follows its developmental downregulation in nonpeptidergic neurons. J Neurosci 31(28): 10119–10127

Chapter 18

Channelrhodopsin-2 Assisted Circuit Mapping in the Spinal Cord Dorsal Horn

Kelly M. Smith and Brett A. Graham

Abstract

Identifying how neural networks communicate through the organization of microcircuits in the central nervous system is a long-standing challenge in neuroscience. Collecting this information in areas that lack clear cellular organization such as the dorsal horn of the spinal cord, where heterogeneous populations of cells are intermingled, has been especially difficult. Improvements in optical technologies in combination with advanced genetic techniques, collectively termed optogenetics, have greatly improved our ability to address this issue. Several studies have now employed optogenetics to study the connectivity of various dorsal horn interneuron populations, as well as modality-specific input provided by primary afferent populations. This work allows for a circuit-based understanding of spinal sensory processing mechanisms to be assembled, something that has been sought since the publication of the gate control theory in 1965. This chapter seeks to provide a practical, experimental-based description of the various optogenetic approaches available to characterize dorsal horn circuits at a level of resolution not possible using more classical approaches.

Key words Channelrhodopsin-2, Electrophysiology, Patch clamp, Circuit mapping, Optogenetics, Photoexcitation, In vitro, Dorsal horn

1 Introduction

The dorsal horn of the spinal cord is a primary site to receive multisensory information from the body related to pain, light touch, thermal, and itch sensations. Importantly, this region does not passively transmit sensory signals, instead a key characteristic of the dorsal horn is that sensory information can undergo substantial processing before refined signals are relayed to the brain where perception occurs [1–3]. Reinforcing this point, the vast majority of neurons in the superficial dorsal horn (laminae I and II) are interneurons, constituting approximately 99% of the population

Supplementary Information The online version of this chapter (https://doi.org/10.1007/978-1-0716-2039-7_18) contains supplementary material, which is available to authorized users.

Rebecca P. Seal (ed.), *Contemporary Approaches to the Study of Pain: From Molecules to Neural Networks*, Neuromethods, vol. 178, https://doi.org/10.1007/978-1-0716-2039-7_18, © Springer Science+Business Media, LLC, part of Springer Nature 2022

in this region [4]. Thus, microcircuits within the dorsal horn are considered crucial for normal somatosensory experience as well as representing a key locus for dysfunction under pathological conditions such as neuropathic pain and itch. Yet despite these broadly accepted functions and implications for disease, defining functionally discrete populations of dorsal horn interneurons and their connectivity in microcircuits has remained a significant challenge.

The intermingled nature of populations of interneurons in the dorsal horn dramatically complicates circuit mapping efforts in this region, unlike other CNS regions like the cortex, hippocampus, and cerebellum, where various cell populations and circuits are well defined and regionally discrete. In addition, the substantial heterogeneity of cellular features such as morphology, electrophysiology, and neurochemistry has made the task of even identifying functional discrete populations of interneurons a long-standing challenge [5–7]. Thus, the main approach has been to assess several of these features together in genetically labelled subpopulations using fluorescent proteins such as green fluorescent protein or TdTomato. Together, these approaches have begun to define and dissect dorsal horn heterogeneity, providing insights into the role of various populations in sensory processing [7, 8]. These insights have been further aided with new transgenic technologies such as genetic ablation or silencing of specific dorsal horn subpopulations confirming functional roles for various neurons in sensory experience [9, 10]. While this heterogeneity continues to be challenging, recent molecular taxonomy techniques are providing an alternative approach to further segregate dorsal horn subpopulations [11, 12]. As this progress continues, however, the task of assembling these neurons into circuits is also vital.

For some time, the mainstay for collecting dorsal horn connectivity information has been to either use neuroanatomical techniques or make paired patch clamp recordings in spinal cord slices. In fact, paired recordings remain the gold standard for identifying synaptically connected neurons but also represents a significant challenge when applied within the dorsal horn. Specifically, paired recordings have shown relatively low levels of connectivity in this region (~15% of pairs connected), making the task of collecting an adequate sample of any particular type of connection difficult when coupled with the substantial heterogeneity of dorsal horn neurons [13, 14]. Nevertheless, heroic work employing this approach has identified some recurring motifs that support the existence of specific inhibitory pathways within these sensory processing circuits. Likewise, a level of modularity has been proposed among excitatory circuits in same region. Further, technical modifications to the paired recording approach have enabled greater sampling, suggesting that overall, excitatory connections outweigh inhibitory synapses by an order of 2:1 [15]. Complementing these datasets, glutamate uncaging slice experiments have added spatial information to the arrangement of dorsal horn microcircuits, though this approach does not allow connected neurons to both be identified [16–18]. Thus, paired recordings and glutamate uncaging approaches have advanced our understanding of sensory processing microcircuits within the dorsal horn; however, the unique features of this region have hampered progress.

Fortunately, the advent of optogenetics has addressed many of the difficulties in studying dorsal horn connectivity, allowing for optical activation of defined neuronal subpopulations. Specifically, the expression of light gated ion channels (opsins) in neurons was first described in 2005 by Boyden et al. and has subsequently become the technique of choice to study connectivity throughout the nervous system [19]. As the use of optogenetics has increased in circuit mapping studies, so too has the number of optogenetic probes, providing an ever-expanding toolkit for these experiments [20]. Channelrhodosin-2 (ChR2) remains the most popular optogenetic probe for excitation studies, a phototaxis-related protein from *Chlamydomonas reinhardtii*. When exposed to blue light (~470 nm) ChR2 channels open allowing a nonspecific cation conductance to flow until light exposure is terminated, closing the channel. Thus, expression of these channels in neurons provides a means to depolarize cells with light (photostimulation) and induce action potential discharge (Fig. 1). Further, by experimentally directing ChR2-expression to a specific neuron population the connectivity of these cells can be rapidly assessed making patch clamp recordings from spinal cord slices while applying photostimulation to activate the target population. Together, this represents the foundation of optogenetic circuit mapping, with changes to various experimental conditions allowing different types of connections to be studied. In addition, there is an ever-expanding range of optogenetic proteins. This includes red-shifted opsins that are compatible with neuronal activity monitoring proteins such as genetically encoded calcium indicators that have blue/green excitation/emission profiles [21, 22]. Alternatively, a number of inhibitory opsins such as halorhodopsin and archaerhodopsin are available to suppress neuronal activation [23]. A number of reviews provide excellent overviews of the current available options and their suitability for different experimental designs, below we briefly discuss the most commonly used approaches [24–26].

All optogenetic experiments must start by establishing optogenetic protein expression in a target neuron population. Two main approaches achieve this objective, either breeding transgenic mice that constitutively express opsin proteins in a defined neuronal subset or injecting viral constructs into the brain or spinal cord to achieve local opsin protein expression. In both cases, a variety of two component strategies can be employed to restrict opsin expression. For example, the commonly used Cre-Lox recombination approach either uses transgenic mice expressing Cre recombinase

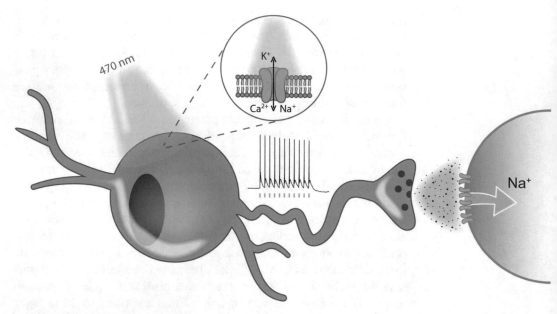

Fig. 1 *Channelrhodopsin-2 function*. ChR2 expression on a neuron (purple, left) is activated by exposure to blue light (~470 nm). These light-gated channels open allowing cation flux (inset), depolarization, and ultimately resulting in action potential discharge. Action potentials in the ChR2 neuron (spiking trace) then cause neurotransmitter release, synaptic transmission (yellow) and depolarization of postsynaptic neurons (green, right)

in a genetically distinct neuron population or injects virus to drive Cre recombinase expression. Both approaches establish a population of neurons that can be referred to as responders, in that subsequent protein (opsin) expression is limited to these "genetically marked" neurons. A driver gene is then introduced, either by breeding or a second virus, that codes for the desired opsin protein; however, incorporation and expression is limited to the Cre-expressing "responder" population. Experimental objectives will determine the most appropriate strategy to establish opsin expression with the literature containing many good reviews of the advantages and disadvantages of each [23, 27–29]. Some important considerations include developmental expression patterns, the need for localized or widespread expression, and the level of expression required to obtain sufficient optogenetic control. Regarding the dorsal horn interneuron populations, transgenic breeding and viral-based approaches are both amenable to study the connectivity of these neurons. Both breeding and viral strategies are also adequate to study sensory input conveyed by primary afferents, with the viral approach allowing for organ or tissue selective expression and breeding allowing restricted expression in different classes of genetically defined afferents. Finally, connections originating from descending pathways, and spinal projection neurons that relay information to the brain are typically targeted using viral approaches where anatomically restricted viral injections limit opsin expression to populations within specific nuclei. Regardless of the approach taken, a critical study control before experiments proceed is validation of targeted opsin expression. For example, it is integral to compare opsin expression with endogenous protein expression patterns using neuroanatomical and immunolabeling techniques to validate restricted opsin expression in the subpopulation of interest [10]. Alternatively, when connections projecting from distant brain or spinal regions are studied, viral injection sites should be assessed to verify restricted and appropriately targeted injections and to determine the time required for adequate opsin expression levels. Failure to undertake these controls renders any subsequent results potentially uninterpretable.

The specific experimental conditions and response features analyzed when connections are found will depend on the type of connection to be studied. This is because sensory circuits within the dorsal horn contain a number of different connection types including primary afferent synapses onto dorsal horn neurons; excitatory synaptic connections intrinsic to this region among interneurons and projection neurons; inhibitory synaptic connections intrinsic to this region between interneurons and projection neurons; inhibitory synaptic connections between local interneurons and primary afferent terminals; and descending inputs from higher order brain structures, which may be excitatory or inhibitory (Fig. 2). Importantly, the literature contains examples of experiments characterizing each of these configurations [30]. For example, studies of the primary afferent synapse using optogenetics are typically undertaken with ChR2 selectively expressed in a specific afferent modality [31, 32], or afferents originating from a specific body region or organ [33, 34]. In this way, a neuron exhibiting a connection can, by extension, be implicated in the central processing of that sensory modality or organ structure. As the DRG may not be retained in acutely prepared spinal slices, these experiments often rely on photostimulation evoked neurotransmitter release evoked by ChR2-expression in the afferents central terminal. In contrast, when the intrinsic connectivity within the dorsal horn is studied, connections arising from interneurons can be studied by photostimulation evoked somatic APs or terminal release, highlighting the importance of the photostimulation–recruitment relationship. Finally, as is the case with afferent synapses, studies targeting connections arising from descending brain structures rely on terminal release as the cell bodies located in the brain are excised during spinal cord dissection and slice preparation. Much of the following methods are common for all connectivity approaches, with some additional considerations noted when required for specific connection types.

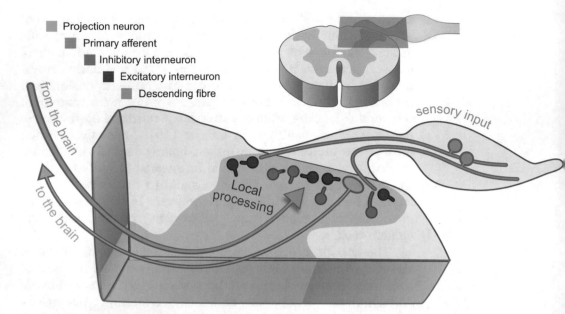

Fig. 2 *Dorsal Horn Organization.* Schematic summarizes the major neuronal categories and connections observed in the spinal cord dorsal horn including; projection neurons (green), excitatory interneurons (red), and inhibitory interneurons (blue). Inset highlights dorsal horn in a transverse spinal cord slice. Sensory information is relayed to the dorsal horn via central terminals of primary afferents (orange), with cell bodies located in the dorsal root ganglia. Descending fibers (teal) from various brain regions also provide inhibitory or excitatory input to this region. The interplay of primary afferent inputs, intrinsic excitatory and inhibitory connections, and descending inputs process sensory signals and determine the overall output of the dorsal horn that is relayed to the brain via projection neurons

2   Materials

2.1   *Animals*

Consideration must be given to the strategy for inducing opsin expression in the neurons of interest. For simplicity we will refer to channelrhodopsin-2 (ChR2) throughout the following text, but the same principles and methods apply for most other opsins. Current technology has several options with the approach chosen influenced by each specific experimental goal. A number of transgenic mice are available through commercial providers (e.g. Jackson Laboratory, Bar Harbor, ME, USA) or laboratory groups where ChR2 is stably expressed in specific cell types. Often the literature contains publications using these animals and thus the fidelity and reliability of ChR2 expression in the target population may already be assessed. A potentially valuable addition is the availability of inducible transgenic animals where administration of a factor such as tetracycline or tamoxifen (for the tet-operon/repressor and estrogen receptor ligand-binding domain systems) drives gene expression. These allow for gene expression timing to be controlled, a property that can be used to avoid expressing ChR2 in neuronal populations that only transiently express a target protein during development. Cre-expressing transgenic mice are also commercially available to target specific cell types and can be used to direct ChR2 expression into specific populations through either a breeding strategy when crossing with floxed (cre-dependent) ChR2 mice or following viral injection of cre-dependent virus encoding ChR2 into the region of interest. A purely viral strategy can also be used in wildtype mice, with several commercial repositories stocking a large range of viral tools (e.g. Addgene, Watertown, MA, USA). Considerations for using a viral approach include virus type (e.g., adenoassocated viruses and lentiviruses), serotype, promoter, and whether pan neuronal or selective expression is desired (e.g., cre/lox or flp/frt). More recently, combinations of transgenic mouse and conditionally dependent viruses are being used for intersectional approaches that combine two genetic factors to further restrict ChR2 expression to more selective populations. Table 1 summarizes some of the applications, advantages, and disadvantages for each approach, or *see* [24] for an excellent review discussing current technologies available for gene delivery to neural tissue.

2.2 Equipment

The equipment required for *I* ChR2-assisted circuit mapping (CRACM) experiments is broadly identical to that required for a standard slice-based electrophysiology experiment (Fig. 3). This includes the laboratory instruments needed for spinal cord dissection and isolation from the vertebral column, a vibratome for tissue slicing, and some form of incubator (submersion or interface) for slice storage until recording [35]. The experiments themselves require an electrophysiology rig capable of whole-cell patch clamp recordings including, a fluorescence-capable upright microscope equipped with long working distance water dipping objective and superfusion tissue bath setup. Experiments also require a temperature control unit to manipulate the temperature of bath superfusate, for example, when assessing presynaptic inhibition in spinal cord slices. Data collection requires a patch clamp amplifier, digitizer, computer interface, and data acquisition software. A critical consideration in these experiments is the illumination setup to deliver precisely timed photostimulation for optogenetic control of ChR2-expressing neurons. This is typically done using either a high intensity LED system or laser that delivers illumination through the microscope fluorescence light path, or via a micromanipulator-mounted probe positioned in the recording bath focused directly on the tissue slice (Fig. 3). Higher resolution mapping can also be achieved using a laser-scanning microscope [36], however, this is not necessary for basic circuit mapping protocols. Regardless of the illumination strategy, illumination wavelength is critical and must match the excitation profile of the opsin chosen for experiments (~470 nm for ChR2), with a second wavelength also typically necessary to identify any genetic label used to target recordings (e.g., mCherry, TdTomato) during connectivity searches. Finally,

Table 1
expression pattern, advantages and disadvantages of methods to cause opsin expression in central and peripheral nervous system. *WT* wild type ICV-intracerebroventricular, *CM* cisterna magna, *IT* intrathecal, *DRG* dorsal root ganglia

Administration route	Expression Location	Advantages	Disadvantages
Transgenic breeding (e.g., cre/lox)	Where cre is expressed in the peripheral or central nervous system throughout development	• Many commercially available animal lines • No invasive procedures required • Technically easy	• Opsin expression in lineage cells
Conditional transgenic breeding (e.g., CreER)	Opsin is expressed in cells that are creER positive at the time of tamoxifen delivery	• Many commercially available animal lines • No invasive procedures required • Technically easy	• Timing and dose of tamoxifen delivery must be troubleshooted
Injection into CNS region of interest (e.g., spinal cord dorsal horn)	Opsin expression in all cells around in injection site *can also be used in cre dependent animals for marker specific expression	• Serotype and promotor selection can restrict expression to specific cell types (e.g., synapsin can be used to restrict expression in neurons) • Expression can be localized to a specific area • Small volume required (nL) • Damage to injected area	• Not able to restrict expression to specific populations if using WT animals • Requires surgery • Cells with projections to the injections site may also be infected
CSF delivery	ICV CM and IT delivery can cause transduction in brain (ICV and CM) and spinal cord and DRGs (IT)	• Widespread delivery • ICV and CM can be performed at P0 without full surgery • IT can be performed without anesthesia	• ICV and CM in adults require surgery • ICV and CM in young pups can be difficult • In adults expression is not uniform throughout CNS. • Not always confined to CNS.
IP injection at P0		• Expression restricted to periphery • Can use with cre driver lines for expression in specific populations	• Large volume required

control of illumination must be linked to data acquisition to ensure the latency from photostimulation onset can be determined for responses. This is commonly achieved by sending an external trigger signal from the acquisition software, with most illumination systems accepting an external control signal (trigger).

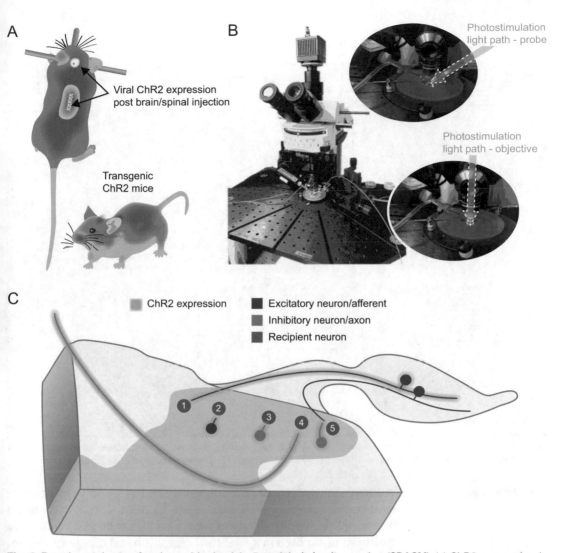

Fig. 3 *Experimental setup for channelrhodopdsin-2 assisted circuit mapping (CRACM)*. (**a**) ChR2 expression in dorsal horn neurons can be achieved by viral injection either directly into the spinal cord to transduce dorsal horn neurons, or into the brain to transduce projection neurons via retrograde transport. Importantly this can be done using wild type animals and a pan neuronal virus to drive expression in all neurons at the injection site, or cre dependent virus to direct expression to a specific neuron population. Alternatively, breeding techniques using (cre/lox or flp/frt) can also be used for selective ChR2 expression. (**b**) A standard patch-clamp electrophysiology rig can be used in CRACM experiments, equipped with an LED or laser system to deliver precisely timed photostimulation either through the microscope light path, or via an optic fiber probe positioned over the preparation. (**c**) Schematic summarizes neuronal populations that have been transduced with ChR2 to study dorsal horn connectivity using the CRACM approach. This includes ChR2 expression to study (1) primary afferent input, (2) postsynaptic excitation from local interneurons, (3) postsynaptic inhibition from local interneuron, (4) descending inputs, and (5) presynaptic inhibition from local interneurons

Table 2
Recipe for dissection and recording solutions

Chemical	Molecular weight	Concentration (mM)
Sucrose Substituted ACSF for dissection and cutting		
Sucrose	342.3	250
$NaHCO_3$	84.01	25
D-Glucose	180.16	10
KCl	74.55	2.5
$NaH_2PO_4 \cdot H_2O$	137.99	1
$MgCl_2$	95.211	6
$CaCl_2$	147.01	1
ACSF for recording		
NaCl	118	118
$NaHCO_3$	84.01	25
D-Glucose	180.16	10
KCl	74.55	2.5
$NaH_2PO_4 \cdot H_2O$	137.99	1
$MgCl_2$	95.211	1
$CaCl_2$	147.01	2.5

Both dissection and recording solutions should be pH 7.3. Last four ingredients in each solution can be made as 1 M stocks and stored at 4 °C for ~2 weeks. All solutions should be made up to volume in dH_2O (*see* **Note 1**)

2.3 Solutions

As above for equipment, the solutions needed for in vitro optogenetic circuit mapping experiments are identical to those required for normal electrophysiology experiments. This generally includes a cutting solution used in spinal cord dissection and slice preparation, bath/recording solution for use during experiments, recording (pipette internal) solution, and a variety of antagonists/agonists for pharmacological dissection of photostimulation responses (*see* **Notes 1** and **2**). The exact composition of these solutions varies in the literature and between experimental objectives. Example compositions for cutting, recording, and internal solutions and are listed in Tables 2 and 3. In addition, the literature contains many references on other recipes, including a paper identifying alternative solutions for best preserving cortical neurons [37], which can also be used for spinal cord and an excellent systematic assessment on cutting and recording solution ingredients for preserving tissue viability in adult spinal cord slices prepared to record from motoneurons [38, 39]. Importantly, motoneurons are widely considered among the populations most sensitive to tissue slicing approaches and thus solutions optimized to retain motoneuron health are likely to benefit the general viability of all neurons in spinal cord slices.

Table 3
Recipe for internal (pipette) solutions

Chemical	Molecular Weight	Concentration (mM)
Potassium Gluconate-based internal solution		
Potassium gluconate	150.2	135
NaCl	58.4	6
HEPES	238.3	10
EGTA	380.4	0.1
Mg_2ATP	507.18	2
Na_3GTP	523	0.3
Cesium-based internal solution		
CsCl		130
HEPES		10
EGTA		10
$MgCl_2$		1
Mg_2ATP		2
Na_3GTP		0.3

EGTA and Na_3GTP can be made as stock solutions (in dH_2O) and stored in a desiccator at $-20\ °C$ for several months. We recommend 10 mM stock for EGTA (use 200 μL in 20 mL of internal) and 25 mg/mL stock for Na3GTP (use 125 μL in 20 mL of internal)

Internal should be pH 7.3 (with KOH for potassium-based solution and CsOH for Cs-based solution) and osmolarity 290 mOsm (with sucrose). Internal solution should be made with dH_2O (*see* **Note 2**)

Regarding internal solutions, composition can vary more widely depending on the data to be acquired, but generally potassium-based internal solutions are used when membrane potentials, action potential discharge, or excitatory currents are to be studied. Alternatively, cesium-based internal solutions are often employed when recording inhibitory currents, with chloride concentration also elevated by some groups (including ours) to better resolve these currents at hyperpolarized holding potentials (-70 mV).

Finally, many agonists and antagonists may be used in dorsal horn connectivity studies, but some are typically necessary. For example, the fast voltage-gated fast sodium channel antagonist tetrodotoxin (1 μM) is necessary to test the action potential dependence of photostimulation-evoked synaptic responses, help identify terminal release evoked responses, as well as differentiating direct (monosynaptic) inputs from indirect (polysynaptic) responses. Also related to these distinctions, 4-aminopyridine (4-AP, 0.1–1 mM) is often used to enhance terminal release by inhibiting voltage gated potassium channels and consequently enhancing the excitability and recruitment of ChR2-expressing synaptic terminals during photostimulation.   Pharmacological   dissection   of photostimulation-evoked synaptic responses as excitatory or inhibitory, as well as the type of receptor mediating these responses, necessitates a range of drugs. AMPA-receptor antagonists such as CNQX (10 μM), DNQX (10 μM) and NBQX (10 μM) are typically used along with the NMDA-receptor antagonist D-AP5 (100 μM) to verify and study excitatory connections. Alternatively, GABA$_A$-receptor antagonists such as bicuculline (10 μM), picrotoxin (50 μM), or gabazine (10 μM), and the glycine receptor antagonist strychnine (1 μM) are required to verify and study inhibitory connections (*see* **Note 3**). Depending on the identity of ChR2-expressing neurons in each preparation, other neurotransmitters systems and signaling pathways may be engaged requiring additional drugs to verify and dissect photostimulation responses.

3   Methods

As outlined above, the general workflow for a series of optogenetic mapping experiments requires two fundamental procedures that have been described extensively elsewhere in the literature: spinal slice preparation; and patch clamp recordings [35, 40]. The following sections now step through the specific techniques and protocols that can be used to optogenetically interrogate connectivity in the dorsal horn.

For in vitro connectivity studies, preparation of acute spinal cord slices is required. During this process, consideration should be given to the orientation of slicing, with transverse and sagittal sectioning the two most common cutting planes. Transverse sectioning yields many slices and allows easy orientation of recordings across the mediolateral extent of the dorsal horn, whereas sagittal slicing better preserves much of the neuropil, with the dendritic processes of many dorsal horn neurons oriented in this plane. This yields more complete morphological reconstruction, and subsequent classification, of filled neurons. Another consideration may be to intentionally preserve attached dorsal roots during slicing, especially if primary afferent activation is desired to study these inputs, or presynaptic inhibition of afferent terminals.

3.1   Characterization of ChR2 Recruitment

Following tissue preparation an important task to allow confident interpretation of photostimulation mapping experiments is to characterize the recruitment of ChR2-expressing neurons during photostimulation (Fig. 4). The necessity of this step lies in the interpretation of subsequent photostimulation responses in connected neurons. Specifically, multi component response could arise if photostimulation evokes a burst of action potentials (APs) in ChR2-expressing neurons, or if a single AP in this population activates both direct monosynaptic input as well as delayed input through polysynaptic connections. Additionally, unless the

A Cell Attached

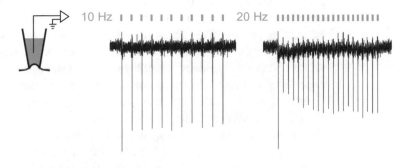

B Whole Cell

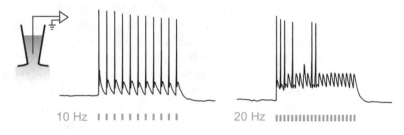

C Sources of delay

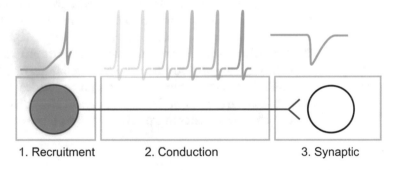

1. Recruitment   2. Conduction   3. Synaptic

D Post synaptic response

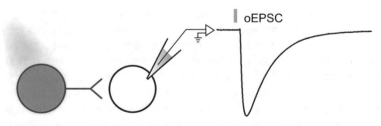

Fig. 4 *Characterization of channelrhodopsin-2 function and photostimulation recruitment.* (**a** and **b**) Summarize two approaches to assess photostimulation parameters necessary to evoke ChR2-mediated action potential discharge. (**a**) In cell-attached recordings a tight seal is established between recording pipette and cell membrane, but the membrane remains intact (left schematic). Photostimulation is then applied at different intensities, durations and frequencies while extracellular spiking is monitored (right traces). (**b**) In the whole

capability of ChR2 to evoke AP discharge is tested, absence of a photostimulation response could suggest no connection is present, or that photostimulation has failed to recruited the ChR2-expressing population. Thus, the ability of a stimulus to confidently evoke a single AP is critical for results to be interpreted. This relationship is typically established by visualizing the ChR2-expressing neurons (in most cases ChR2 is expressed along with a fluorophore like YFP, mCherry, or TdTom) and then recording electrical activity evoked during photostimulation (*see* **Note 4**). Recordings in voltage clamp allow photocurrent properties such as amplitude to be assessed, providing a functional measure of ChR2-expression level (*see* **Note 5**). Recording in current clamp mode allows photostimulation intensity (typically 1–20 mW) and duration (typically 1–50 ms) to be adjusted to reliably induce a single action potential with each stimulus. In addition, trains of repetitive photostimulation at different frequencies (e.g., 1–50 Hz) can be applied to characterize the reliability of recruitment, which is routinely reported to remain robust up to 10–20 Hz, but then often falls to include AP failures at higher stimulation intensities.

For AP spiking parameters, the above assessments can be undertaken in loose (cell attached) patch or whole cell mode (*see* Fig. 4, **Note 6**). One reason for determining the photostimulation/spiking relationship in ChR2-expressing cells is that photostimulation responses in mapping experiments can arise from two sources. Specifically, photostimulation can induce ChR2-mediated spiking of neurons in slices that synapse on a recorded cell, or ChR2-mediated terminal release of neurotransmitter from synapses on a recoded cell, without the need for a cell body or AP spiking. As terminal release is typically more difficult to induce, relating to the necessity of high levels of ChR2 because of the small conductance of these channels, reducing photostimulation power and duration

---

Fig. 4 (continued) cell configuration a tight seal is established before the membrane overlying the pipette tip is ruptured (left schematic) to establish whole cell recording. Photostimulation is then applied at different intensities, durations and frequencies with action potential spiking responses recorded in current clamp mode (right traces). Example traces in (**a** and **b**) show responses to 10 and 20 Hz photostimulation. Note spiking responses are reliable at 10 Hz photostimulation in cell-attached and whole cell modes. Cell-attached recordings remain reliable at 20 Hz, but spiking responses begin to fail when recorded in whole-cell. (**c**) Schematic summarizes sources of delay between photostimulation onset and a recorded postsynaptic response. (1) Recruitment delay is the time between photostimulation onset and action potential generation in the ChR2-expressing population. (2) Conduction delay is the time taken for an action potential to propagate along the ChR2-expressing neurons axon and reach the synaptic terminal. (3) Synaptic delay is time taken for neurotransmitter release and activation of ligand-gated ion channels on postsynaptic neuron (*see* **Note 12**). (**d**) Schematic summarizing the principle of CRACM. A known population of neurons expresses ChR2 and is photostimulated while patch clamp recordings are monitored in search of neurons that exhibit optically evoked postsynaptic reposes (synaptic currents or potentials) that are time locked to photostimulation onset. (Traces in (**a**) are unpublished data kindly provided by Mark Gradwell)

to the minimum required to activate AP spiking in the targeted population can be useful, limiting responses to those when synaptically connected neurons are retained in the slice preparation. An extension of these mapping control experiments is to determine the optical footprint of ChR2-expressing neurons using photostimulation delivered by a laser-scanning microscope [36]. This further allows the photostimulation parameters (intensity and duration) to be optimized so AP discharge is only evoked when photostimulation is applied to a ChR2-expressing neurons soma, but not dendrites or axon. Subsequent experiments applying focal photostimulation in a grid array across a slice preparation can then be used to functionally identify the cell bodies of putative ChR2-expressing neurons connected to the recorded cell. In summary, initial characterization of how photostimulation recruits ChR2-expressing neurons represents best practice for this work and enables a more detailed and confident interpretation of subsequent results.

While the above characterization is an important step for optogenetic circuit mapping, once completed subsequent connectivity experiments are typically completed using these established photostimulation parameters, rather that assessing the photostimulation recruitment relationship in every experiment. Recordings then proceed, typically targeting non-ChR2 expressing neurons and applying photostimulation to search for connections. This can either be undertaken by random selection of ChR2-negative profiles, or in tissue that includes a second target population genetically labelled by expression of a fluorophore with a different excitation/emission profile to the fluorophore tagged to ChR2 (e.g., mCherry or TdTom; see Note 7).

3.2 Excitatory Connections

Optogenetics can (and has) been used to study and map excitatory synaptic connections arising from primary afferents, and local excitatory interneurons in the dorsal horn. For example, in the case of afferents, transgenic and viral-based strategies have been used to direct ChR2-expression in nonpeptidergic nociceptive C-fibers [41], low-threshold mechanoreceptive C-fibers [42], low threshold mechanoreceptive Aβ fiber afferents [43], and MrgprA3-expressing itch primary afferents [32, 44]. This work was undertaken in spinal cord slices and either photostimulated primary afferent terminals within the slice (terminal release), or the dorsal root ganglion (DRG) with its connection to the spinal slice preserved via the dorsal root (action potential dependent activation, see Note 8). The results of these studies have mapped the connectivity of specific afferent types within dorsal horn circuits, studied the pharmacology of these specialized connections, assessed injury induced remodeling of the related circuits, and confirmed the existence of modality-specific dorsal horn populations. Together, these findings highlight the strength of an optogenetic approach to primary afferent stimulation, which allows selective activation of specific afferent modalities in dorsal horn circuits not possible using classical electrical stimulation of dorsal roots.

The connectivity and function of a number of dorsal horn excitatory interneuron populations have also been studied with optogenetic approaches, rapidly increasing our understanding of these cells. For example, transgenic expression of ChR2 in neurotensin- [45] and calretinin-expressing dorsal horn neurons [46] has highlighted the widespread connectivity of each population, including extensive interconnectivity within the populations. The later observation can be made when brief photostimulation pulses (1–5 ms) are applied while recording from a ChR2-expressing cell in voltage clamp. When the targeted cell type (ie, the population expressing ChR2) is part of an interconnected network the resulting photostimulation response includes an immediate inward photocurrent mediated by ChR2 as well as a second longer latency inward current arising because neighboring ChR2-expressing neurons are also activated and synapse on the recorded cell. Importantly, in the case of an excitatory connection, bath application of an AMPA/kainite receptor antagonist such as CNQX will abolish synaptic responses without altering photocurrent responses. These pharmacologically isolated photocurrents can then be subtracted from the original photostimulation responses to better resolve and characterize the synaptic properties of these connections (Fig. 5). Defining and studying connections with other cell populations (ChR2 negative) is not complicated by the presence of photocurrents, with photostimulation resulting in optically evoked excitatory postsynaptic currents (oEPSCs) that occur at stable and short latency relative to photostimulation onset and exhibit CNQX sensitivity. A number of properties that will be discussed below can also be assessed to distinguish between direct (monosynaptic) and indirect (polysynaptic) responses (see monosynaptic vs. polysynaptic section). Finally, in addition to single photostimulation pulses to identify connections, longer duration pulses or trains of photostimulation can be applied to further interrogate excitatory connectivity. Using extended photostimulation protocols has demonstrated the ability of interconnected excitatory interneuron populations to amplify dorsal horn signaling and contribute to reverberating activity patterns and wind-up [45, 46]. Repetitive photostimulation has also shown how excitatory interneurons that express gastrin releasing peptide (GRP) gate spinal itch signaling [32]. This work demonstrated that high frequency stimulation of GRP cells was required to evoke corelease of glutamate and GRP and recruit the postsynaptic networks that convey itch signals to the brain. Alternatively, glutamate alone was released when stimulation frequency was low, and was not sufficient to activate postsynaptic itch signaling. These methods can also be used to assess direct connections between interneurons and projection neurons

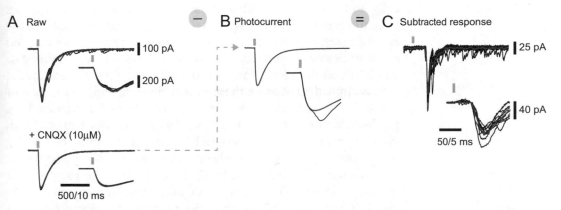

Fig. 5 Photocurrent subtraction method to resolve connectivity in ChR2-expressing neurons. (**a**) Overlaid traces (unsubtracted, left) show voltage clamp recordings of photostimulation responses recorded over 10 trials (average in red) under baseline conditions (upper) and after bath addition of CNQX (10 μM). Insets show same traces on expanded timescale (blue line indicates timing of 1 ms photostimulation period). Note baseline response has two components at onset, the direct photocurrent and a synaptic current, but the synaptic component is blocked by CNQX, leaving an isolated photocurrent. (**b**) The averaged photocurrent trace is then rescaled to match the amplitude of individual baseline traces before it is subtracted to isolate the synaptic responses component. (**c**) Overlaid traces show the outcome of photocurrent subtraction allowing analysis of synaptic responses in ChR2-expressing neurons. Insets are the same traces on expanded timescale (blue line indicates timing of 1 ms photostimulation period). (Figure adapted from [46]). Note that this example is for excitatory postsynaptic currents, but the same principle can be used to dissect inhibitory connections using GABA and glycine receptor antagonists (e.g. bicuculline (10 μM) and strychnine (1 μM)) in place of CNQX

[46, 47], furthering our understanding of how sensory information is relayed to the brain. These new insights on excitatory interneuron connectivity and the functional implications for these connections within dorsal horn sensory processing circuits have been made possible with the advent of optogenetic stimulation approaches. Future work continuing this strategy, targeting smaller more discrete excitatory subpopulations in health and disease will continue building our understanding of these cells that constitute the largest population within the dorsal horn.

3.3 Inhibitory Connections

The identity and function of inhibitory interneurons has been a major focus in pain research since the publication of the Gate Control Theory in 1965 [48]. It follows that much like the excitatory populations above, studies are increasingly using optogenetic approaches to characterize the connectivity of inhibitory dorsal horn populations. This work is complicated by the combination of two neurotransmitters, GABA and glycine, mediating fast synaptic inhibition within dorsal horn circuits. In addition, two forms of inhibition are known to play critical roles in sensory processing: postsynaptic inhibition acting through axondendritic and axosomatic synapses from inhibitory populations onto other dorsal horn cells; and presynaptic inhibition acting through axoaxonic synapses from inhibitory populations onto primary afferent terminals. In the case of postsynaptic inhibition, optogenetic activation can be used as for excitatory populations. ChR2 is expressed (transgenically or virally) in an inhibitory interneuron type and then photostimulation is applied to activate these cells while recording from various other dorsal horn neurons. This approach has been reported for glycinergic interneurons [49], enkephalinergic interneurons [50], NPY-expressing interneurons [51], and parvalbumin expressing interneurons [52, 53], with photostimulation evoked postsynaptic currents recorded that could be abolished by GABA$_A$-receptor and glycine receptor antagonists (bicuculline, picrotoxin, and strychnine). In the case of NPY-expressing neurons, a photostimulation-activation approach (using red shifted channelrhodopsin) was combined with genetic labelling to confirm functional inhibitory synaptic connections onto NPY Y1-receptor expressing neurons labeled with GFP [51]. In the enkephalinergic interneuron study photostimulation responses were more complex, containing fast excitatory and inhibitory components, consistent with enkephalin-containing dorsal horn neurons including a mixed GABAergic/glutamatergic population. These differences were discriminated as inward and outward currents in voltage clamp but were not pharmacologically dissected (see **Note 9**). In addition, a slower outward current with kinetics matching GIRK channel activation followed photostimulation in some recordings. Although pharmacology was not used to verify GIRK channel identity, this response supports the likelihood photostimulation evoked enkephalin release and postsynaptic opioid receptor activation. The postsynaptic targets receiving this input were also assessed, showing a specific action potential discharge pattern (delayed firing) in these cells that was used to imply that excitatory interneurons preferentially received these connections. Using similar approaches, our group has defined a specific inhibitory connection between inhibitory parvalbumin positive interneurons and a morphologically classified dorsal horn population termed vertical cells, identified by a prominent bias to ventrally projected dendrites and delayed firing AP discharge. Importantly, the vertical population is known to be excitatory and possess axons projecting into lamina I and receive myelinated low threshold mechanoreceptor input. Thus, inhibition from parvalbumin positive neurons appears to suppress touch signals from activating nociceptive circuits [52]. These conclusions were made possible by neurobiotin filling recorded cells, allowing recovery and verification of morphology (vertical-like) and neuroanatomical confirmation of ChR2 inputs (arising from the parvalbumin population) onto these cells (see **Note 10**).

Another known source of postsynaptic inhibition in dorsal horn circuits is that coming from the brainstem and in particular the rostral ventromedial medulla (RVM) [54]. The synaptic substrate of these pathways has been investigated using in vivo electrical and chemical stimulation of RVM while patch clamp recording in the dorsal horn confirming these connections are mediated by GABA and glycine [55]. Despite this, the type of dorsal horn neurons receiving these inputs, and therefore the precise impact on spinal sensory processing circuits, was not identified. More recently, however, optogenetics has allowed a reexamination of this issue using virally mediated ChR2 expression in RVM neurons [50]. Spinal cord slices were prepared from mice several weeks following viral injections in RVM, allowing sufficient time for ChR2 transduction in RVM neurons as well as in the associated axons and synaptic terminals projecting to the spinal cord. In addition, these experiments were undertaken in transgenic mice expressing TdTomato in enkephalin-expressing dorsal horn neurons to allow targeted recording from this population. Photostimulation reliably evoked outward currents that were sensitive to bicuculline and strychnine confirming these inputs originated from RVM, released GABA and glycine, and terminated on the enkephalinergic population. Assessment of action potential discharge in recordings showed that inhibitory RVM inputs selectively targeted tonic firing enkephalin positive neurons but not those exhibiting delayed discharge. Using the literature that associates action potential discharge with neurotransmitter phenotype in the dorsal horn [6, 7], this observation was used as evidence the inhibitory RVM projection only targeted inhibitory enkephalinergic neurons. Together with a range of other experiments, this work defined a descending disynaptic circuit that facilitated mechanical pain via disinhibition within the dorsal horn. It also demonstrated a straightforward approach to study connectivity between the brain and spinal cord, relevant to sensory processing circuits.

Unlike the above examples that directly assess postsynaptic inhibition with optogenetics, presynaptic inhibition of primary afferent terminals in the dorsal horn cannot be directly recorded due to the small nature of afferent terminals. Nevertheless, our work in the parvalbumin population as well as the analysis published on enkephalin positive neurons has provide examples of a number of indirect approaches that can be used [50, 52]. The first and perhaps most obvious is to prepare spinal cord slices with a dorsal root (DR) attached. This allows recordings to be obtained from dorsal horn neurons while electrically stimulating the DR, thereby activating primary afferents to search for cells that receive afferent input. Once a cell with primary afferent input is identified, exhibiting a reliable excitatory postsynaptic current time locked to DR stimulation, ChR2-expressing cell photostimulation can be applied prior to DR stimulation. If the photostimulated target cells do mediate presynaptic inhibition, DR-evoked excitatory postsynaptic current amplitude will either be reduced or abolished by prior photostimulation, as shown in our parvalbumin study and the enkephalin neuron work [50, 52]. Our work also took advantage of an alternative means to study parvalbumin cell mediated presynaptic inhibition, building off observations made in classical studies in the ventral spinal cord under in vivo conditions [56]. Specifically, presynaptic inhibition differs from its postsynaptic counterpart in that activation of these synapses results in chloride efflux (rather than influx) because primary afferents maintain a higher intracellular chloride concentration. Thus, primary afferents depolarize during presynaptic inhibition, preventing subsequent signaling. Original work on this mechanism suggested that as the primary afferent is depolarized by an inhibitory input, transient neurotransmitter release can be evoked, observed as an excitatory postsynaptic response in the target of these afferent synapses (motoneurons in the original experiments). Furthermore, local cooling of the spinal cord accentuated this phenomenon [57]. These characteristics were more recently combined with optogenetics to study presynaptic inhibition under *in vitro* conditions in the ventral spinal cord [58]. Photostimulation of inhibitory interneurons produced delayed latency excitatory responses in motoneurons that could be abolished by elevated recording bath temperature or blockade of the inhibitory inputs between these interneurons and primary afferents using a $GABA_A$-receptor antagonist. This approach not only represents an alternative strategy for studying presynaptic inhibition, but also identifies the postsynaptic neurons targeted by these specialized afferent inputs that are under inhibitory regulation. Given these outcomes, we have used the same approach to identify parvalbumin positive interneurons as the first defined population to mediate presynaptic inhibition of myelinated afferents in the dorsal horn [52]. These experiments confirmed that optogenetic stimulation of parvalbumin neurons evoked long latency excitatory inputs in other dorsal horn cells. Furthermore, these responses were temperature sensitive, bicuculline sensitive; could be reduced or abolished by prefatiguing dorsal root stimulation; and exhibited longer latencies and higher onset jitter than monosynaptic connections. This assay also allowed us to identify a specific morphological type of dorsal horn neuron, the vertical cells, as receiving input from myelinated afferents under parvalbumin neuron mediated presynaptic inhibition, as well as postsynaptic inhibition (also arising from the parvalbumin population). In summary, optogenetic mapping has provided new opportunities to study presynaptic inhibition at a level of detail that was previously impossible. This, when coupled with the advantages optogenetics brings to study postsynaptic inhibition, ideally positions the field of spinal pain processing to complete a detailed understanding of inhibitory microcircuits within the dorsal horn.

3.4 Monosynaptic Versus Polysynaptic

A key distinction when mapping connectivity in any CNS region is the ability to distinguish direct (monosynaptic) connections from those that are indirect (polysynaptic). A number of approaches have been used to differentiate mono- from polysynaptic connections in past optogenetic studies, with techniques generally either analyzing the temporal properties of responses, or using a pharmacological strategy (Fig. 6). In the response latency approach, several photostimulation trials are repeated while recording from a test neuron in voltage clamp and ensuring a long interphotostimulation interval that avoids rundown of the response (*see* **Note 11**). Each record is then analyzed to determine the latency between photostimulation onset and the response. The mean response latency can then be calculated, with monosynaptic connections exhibiting short latency compared to polysynaptic responses. Important considerations for this analysis are discussed in **Note 12**. In addition, a measure of the stability of the latency, often termed jitter, can also be assessed by calculating the standard deviation of latency values across trials (*see* **Note 13**). Typically, response latencies <10 ms and jitter <1 ms are indicative of monosynaptic connections, whereas larger values suggest polysynaptic connections. Calculating failure rate (the percentage of trails that do not elicit a photostimulation-evoked response) provides an additional index of connectivity, with monosynaptic connections exhibiting lower failure rates that polysynaptic connections.

A pharmacological approach to determine connectivity characteristics relies on the ability of photostimulation to induce action potential independent release from synaptic terminals, often referred to as terminal release (Fig. 6). This feature means that photostimulation evoked synaptic reposes that persist in the presence of the fast-activating sodium channel blocker tetrodotoxin (TTX, 1 µM) can only occur if a ChR2 terminal is directly apposed to the recorded neuron, and is hence monosynaptic. Alternatively, photostimulation responses that are blocked by TTX require fast activating sodium channels and action potential discharge and are therefore interpreted as indicating photostimulation requires the spiking of an interposed neuron to produce photostimulation responses (ie, a polysynaptic connection). One caveat to this approach is that ChR2 expression at synaptic terminals may not be adequate to evoked neurotransmitter release, raising the potential for monosynaptic inputs to be misclassified as polysynaptic. This can be addressed by bath application of 4-aminopyrimadine (4-AP, 0.1–1 mM), which blocks potassium channels and heightens the excitability of synaptic terminals. Only monosynaptic photostimulation-evoked connections can persist under these conditions with the presence of TTX and 4-AP. Interestingly, presynaptic inhibition represents a potential exception when it evokes transmitter release from afferent terminals at room temperature. Such responses can be evoked by terminal release (i.e., from a

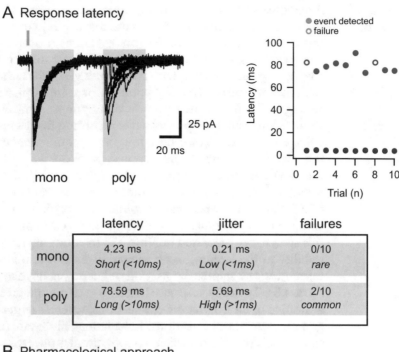

	latency	jitter	failures
mono	4.23 ms *Short (<10ms)*	0.21 ms *Low (<1ms)*	0/10 *rare*
poly	78.59 ms *Long (>10ms)*	5.69 ms *High (>1ms)*	2/10 *common*

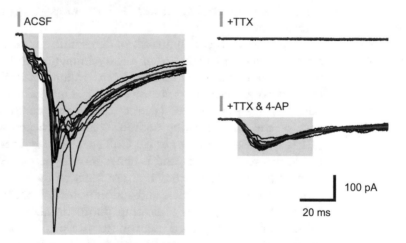

Fig. 6 *Differentiating monosynaptic and poly synaptic connections.* (**a**) Overlaid traces (top left) show 10 photostimulation trials (blue line indicates timing of 1 ms photostimulation period). Note responses contains short (monosynaptic) and long (polysynaptic) latency components. A plot of response latency versus trial number (top right) shows the low jitter observed as constant latency, and high reliability with no synaptic failures in the monosynaptic response (red markers); versus the higher jitter and reduced reliability in the polysynaptic response (blue markers). Synaptic failures denoted by open markers. Table (below) presents values derived from the above data (upper) and general criteria (lower) summarizing latency, jitter, and failure rate for monosynaptic and polysynaptic connections. (**b**) Overlaid traces (left) show 10 photostimulation trials (blue line indicates timing of 1 ms photostimulation period). Note the complex waveform containing a combination of putative mono and polysynaptic components to the photostimulation response. No photo-stimulation responses can be resolved from the same neuron following tetrodotoxin (TTX, 1 μM) exposure to block voltage-gated Na$^+$ channels and action potential generation (upper right). Bath addition of 4-aminopyradine (4-AP, 0.2 mM), still in the presence of TTX, recovers a more uniform photostimulation

ChR2-expressing inhibitory terminal), depolarizing a primary afferent terminal and producing a postsynaptic response through a disynaptic connection. These connections may therefore continue to signal in the presence of TTX and 4-AP, though the extra connection (and associated synaptic delay) will produce a longer latency and higher jitter than truly monosynaptic connections.

In conclusion, the above sections detail how optogenetics can be used to study the full range of connectivity in spinal sensory processing circuits including primary afferent, excitatory and inhibitory interneuronal, and connections to and from the brain. A systematic approach that fully characterizes optogenetic control of ChR2-expressing neurons represents best practice for these experiments and yields the greatest insight on connectivity. Continued refinement of these techniques and the discrete populations studied will accelerate the field toward completing our view of the dorsal horn connectome and how this region achieves its remarkable multisensory processing role. Finally, addition of other variations such as optogenetic inhibition, chemogenetic control, genetically encoded activity imaging markers with the integration of in vitro and in vivo approaches will allow an understanding of spinal sensory processing in health and disease. Information that is necessary to improve how we treat debilitating conditions such as chronic pain, itch, and spinal cord injury.

4   Notes

1. Cutting (sucrose ACSF) and recording (normal ACSF) solutions should be made fresh every day. Often the pH of these solutions can be adjusted to 7.3 by simply bubbling with carbonox (95% CO_2, 5% O_2) for several minutes.

2. Internal solutions can be made and stored at $-80\ °C$ in parafilm sealed eppendorfs for ~3 months.

3. Antagonists, and other drugs that will be bath applied can be prepared at 1000x final concentration and stored at $-80\ °C$ for several months in small aliquots (100–200 μL). Aliquots can then be thawed and bath applied to reach final concentration during experiments, we typically use 50 μL of stock to 50 mL ACSF in a smaller drug pot.

Fig. 6 (continued) response. This confirms that the recovered component arises from a monosynaptic connection. The optically evoked response occurs independent of action potential discharge, instead arising from ChR2-mediated depolarization of a synaptic terminal directly apposed to the recorded neuron. This process is commonly referred to as terminal release and can be used to confirm monosynaptic nature of photostimulated inputs. (Traces in (**b**) are unpublished data kindly provided by Mark Gradwell)

4. If the opsin is tagged to a fluorophore with a similar excitation wavelength as the opsin (e.g., ChR2 and YFP) care must be taken to not keep the illumination on for extended periods when searching for target neurons as this will cause depolarization and action potential discharge in opsin expressing neurons.

5. When measuring the amplitude of the channelrhodopsin-2 current in voltage clamp a longer stimulation (1 s) is recommended to allow measurement of the steady state current.

6. As photostimulation of ChR2 expressing cells during mapping experiments occurs without the electrical load of a recording pipette or intracellular dialysis of pipette contents, we suggest that loose patch (on-cell) mode is more appropriate to better reflect photostimulation evoked spiking.

7. Unintentional selection of a ChR2-expression neuron for recording is easily identified if an immediate photocurrent is observed in the recorded neuron at the onset of photostimulation. Due to this we suggest the first step either in cell attached or whole cell mode is to run a brief photostimulation to verify cell identity.

8. If not preserving DRG and relying on terminal activation of central afferent terminals it is important to ensure high terminal expression of the opsin. Alternatively, if preserving the DRG cell bodies and dorsal root it should be secured within the recording bath to ensure consistent stimulation each time. For this method we recommend not using a light source through the microscope and instead positioning the laser or LED over the DRG cell bodies.

9. Using a potassium gluconate-based internal solution, recording in voltage clamp mode at a holding potential of -70 mV, inward currents (downward deflections) are optically evoked excitatory postsynaptic currents (oEPSCs) while outward currents (upward deflections) are optically evoked inhibitory postsynaptic currents (oIPSCs). To better distinguish the outward currents, the holding potential of the cell can be brought closer to 0 mV (usually approximately −40 mV).

10. Recovery of neuron morphology can be achieved by including neurobiotin (0.2%, Vector Laboratories, product #SP-1120) in the pipette (internal) solution. When finished recording in the slice, immediately place the tissue in 4% paraformaldehyde (this can be made in large batches at stored at −20 °C until required) for 2 h. Slices should then be washed in 0.1 M PBS and stored until ready for tissue processing. Slices can be incubated in a streptavidin secondary antibody for 2 h, washed again in 0.1 M PBS, mounted and imaged. As every recorded cell is unlikely to be recovered we recommend careful

documentation of cell locations during recording to ensure correct matching of morphology with electrophysiological profile.

11. Ten sweeps of photostimulation with a 10 s interevent interval should be sufficient to characterize these responses including failure rate, response latency, and jitter.

12. Several careful measurements must be taken to ensure correct analysis of EPSC (or IPSC) latency: recruitment delay, conduction delay, and synaptic delay (Fig. 4c).

  (a) Recruitment delay—time between photostimulation and action potential initiation in ChR2 expressing neurons. Recruitment delay can be determined by recording directly from ChR2-expresing neurons during photostimulation and measuring the time between the onset of photostimulation and action potential initiation. This can be either done in the on-cell recording configuration (preferable) or in whole-cell mode. Multiple trials can be recorded to calculate an average recruitment delay, used to help determine a window for connections to be considered monosynaptic. For example, two standard deviations above and below average recruitment delay, with conduction and synaptic delays then added.

  (b) Conduction and synaptic delay—time for action potential to reach synaptic terminals (conduction delay) and time for neurotransmitter to reach postsynaptic neuron and cause an observable response (synaptic delay). In order to set this window, we have assigned conduction and synaptic delay in dorsal horn circuits using past paired recording studies. Across multiple studies and laboratories, this work suggests the combined contribution of conduction and synaptic delay is on average 2.5 ms [13, 15].

13. Jitter is defined as the standard deviation of oEPSC (or oIPSC) latency onset assessed across several trials.

References

1. Koch SC, Acton D, Goulding M (2018) Spinal circuits for touch, pain, and itch. Annu Rev Physiol 80:189–217

2. Moehring F et al (2018) Uncovering the cells and circuits of touch in Normal and pathological settings. Neuron 100(2):349–360

3. Peirs C, Seal RP (2016) Neural circuits for pain: recent advances and current views. Science 354(6312):578–584

4. Todd AJ (2010) Neuronal circuitry for pain processing in the dorsal horn. Nat Rev Neurosci 11(12):823–836

5. Graham BA, Brichta AM, Callister RJ (2007) Moving from an averaged to specific view of spinal cord pain processing circuits. J Neurophysiol 98(3):1057–1063

6. Yasaka T et al (2010) Populations of inhibitory and excitatory interneurons in lamina II of the

adult rat spinal dorsal horn revealed by a combined electrophysiological and anatomical approach. Pain 151(2):475–488

7. Browne TJ et al (2020) Transgenic cross-referencing of inhibitory and excitatory interneuron populations to dissect neuronal heterogeneity in the dorsal horn. Front Mol Neurosci 13:32

8. Gatto G et al (2019) Neuronal diversity in the somatosensory system: bridging the gap between cell type and function. Curr Opin Neurobiol 56:167–174

9. Smith KM et al (2014) The search for novel analgesics: re-examining spinal cord circuits with new tools. Front Pharmacol 5:22

10. Graham BA, Hughes DI (2019) Rewards, perils and pitfalls of untangling spinal pain circuits. Curr Opin Physio 11:35–41

11. Häring M et al (2018) Neuronal atlas of the dorsal horn defines its architecture and links sensory input to transcriptional cell types. Nat Neurosci 21(6):869–880

12. Sathyamurthy A et al (2018) Massively parallel single nucleus transcriptional profiling defines spinal cord neurons and their activity during behavior. Cell Rep 22(8):2216–2225

13. Lu Y, Perl ER (2003) A specific inhibitory pathway between substantia gelatinosa neurons receiving direct C-fiber input. J Neurosci 23(25):8752–8758

14. Lu Y, Perl ER (2005) Modular organization of excitatory circuits between neurons of the spinal superficial dorsal horn (laminae I and II). J Neurosci 25(15):3900–3907

15. Santos SFA et al (2007) Excitatory interneurons dominate sensory processing in the spinal substantia gelatinosa of rat. J Physiol 581 (Pt 1):241–254

16. Kato G et al (2007) Differential wiring of local excitatory and inhibitory synaptic inputs to islet cells in rat spinal lamina II demonstrated by laser scanning photostimulation. J Physiol 580(Pt.3):815–833

17. Kosugi M et al (2013) Subpopulation-specific patterns of intrinsic connectivity in mouse superficial dorsal horn as revealed by laser scanning photostimulation. J Physiol 591(7): 1935–1949

18. Callaway EM, Katz LC (1993) Photostimulation using caged glutamate reveals functional circuitry in living brain slices. Proc Natl Acad Sci U S A 90(16):7661–7665

19. Boyden ES et al (2005) Millisecond-timescale, genetically targeted optical control of neural activity. Nat Neurosci 8(9):1263–1268

20. Brown J et al (2018) Expanding the Optogenetics toolkit by topological inversion of Rhodopsins. Cell 175(4):1131–1140.e11

21. Mattis J et al (2011) Principles for applying optogenetic tools derived from direct comparative analysis of microbial opsins. Nat Methods 9(2):159–172

22. Chen IW et al (2019) In vivo submillisecond two-photon Optogenetics with temporally focused patterned light. J Neurosci 39(18): 3484–3497

23. Wiegert JS et al (2017) Silencing neurons: tools, applications, and experimental constraints. Neuron 95(3):504–529

24. Haery L et al (2019) Adeno-associated virus technologies and methods for targeted neuronal manipulation. Front Neuroanat 13:93–93

25. Ting JT, Feng G (2013) Development of transgenic animals for optogenetic manipulation of mammalian nervous system function: progress and prospects for behavioral neuroscience. Behav Brain Res 255:3–18

26. Madisen L et al (2015) Transgenic mice for intersectional targeting of neural sensors and effectors with high specificity and performance. Neuron 85(5):942–958

27. Navabpour S, Kwapis JL, Jarome TJ (2019) A neuroscientist's guide to transgenic mice and other genetic tools. Neurosci Biobehav Rev 108:732–748

28. Kim H et al (2018) Mouse Cre-LoxP system: general principles to determine tissue-specific roles of target genes. Lab Anim Res 34(4): 147–159

29. Guru A et al (2015) Making sense of Optogenetics. Int J Neuropsychopharmacol 18(11): pyv079-pyv079

30. Cordero-Erausquin M et al (2016) Neuronal networks and nociceptive processing in the dorsal horn of the spinal cord. Neuroscience 338:230–247

31. Snyder LM et al (2018) Kappa opioid receptor distribution and function in primary afferents. Neuron 99(6):1274–1288.e6

32. Pagani M et al (2019) How gastrin-releasing peptide opens the spinal gate for itch. Neuron 103(1):102–117.e5

33. Iyer SM et al (2014) Virally mediated optogenetic excitation and inhibition of pain in freely moving nontransgenic mice. Nat Biotechnol 32(3):274–278

34. Spencer NJ et al (2018) Visceral pain - novel approaches for optogenetic control of spinal afferents. Brain Res 1693(Pt B):159–164

35. Zhu M et al (2019) Preparation of acute spinal cord slices for whole-cell patch-clamp recording in substantia Gelatinosa neurons. J Vis Exp 143

36. Kim J et al (2014) Optogenetic mapping of cerebellar inhibitory circuitry reveals spatially biased coordination of interneurons via electrical synapses. Cell Rep 7(5):1601–1613

37. Ting JT et al (2014) Acute brain slice methods for adult and aging animals: application of targeted patch clamp analysis and optogenetics. Methods Mol Biol 1183:221–242

38. Carp JS et al (2008) An in vitro protocol for recording from spinal motoneurons of adult rats. J Neurophysiol 100(1):474–481

39. Leroy F, Lamotte d'Incamps B (2016) The preparation of oblique spinal cord slices for ventral root stimulation. J Vis Exp 116:54525

40. Segev A, Garcia-Oscos F, Kourrich S (2016) Whole-cell patch-clamp recordings in brain slices. J Vis Exp 112:54024

41. Wang H, Zylka MJ (2009) Mrgprd-expressing polymodal nociceptive neurons innervate most known classes of substantia gelatinosa neurons. J Neurosci 29(42):13202–13209

42. Honsek SD, Seal RP, Sandkuhler J (2015) Presynaptic inhibition of optogenetically identified VGluT3+ sensory fibres by opioids and baclofen. Pain 156(2):243–251

43. Tashima R et al (2018) Optogenetic activation of non-nociceptive Abeta fibers induces neuropathic pain-like sensory and emotional behaviors after nerve injury in rats. eNeuro 5(1): ENEURO.0450-17.2018

44. Albisetti GW et al (2019) Dorsal horn gastrin-releasing peptide expressing neurons transmit spinal itch but not pain signals. J Neurosci 39(12):2238–2250

45. Hachisuka J et al (2018) Wind-up in lamina I spinoparabrachial neurons: a role for reverberatory circuits. Pain 159(8):1484–1493

46. Smith KM et al (2019) Calretinin positive neurons form an excitatory amplifier network in the spinal cord dorsal horn. Elife 8:e49190

47. Petitjean H et al (2019) Recruitment of Spino-parabrachial neurons by dorsal horn Calretinin neurons. Cell Rep 28(6):1429–1438.e4

48. Melzack R, Wall PD (1965) Pain mechanisms: a new theory. Science 150(3699):971–979

49. Foster E et al (2015) Targeted ablation, silencing, and activation establish glycinergic dorsal horn neurons as key components of a spinal gate for pain and itch. Neuron 85(6): 1289–1304

50. Francois A et al (2017) A brainstem-spinal cord inhibitory circuit for mechanical pain modulation by GABA and Enkephalins. Neuron 93(4): 822–839.e6

51. Acton D et al (2019) Spinal neuropeptide Y1 receptor-expressing neurons form an essential excitatory pathway for mechanical itch. Cell Rep 28(3):625–639.e6

52. Boyle KA et al (2019) Defining a spinal microcircuit that gates myelinated afferent input: implications for tactile allodynia. Cell Rep 28(2):526–540.e6

53. Yang K et al (2015) Optoactivation of parvalbumin neurons in the spinal dorsal horn evokes GABA release that is regulated by presynaptic GABAB receptors. Neurosci Lett 594:55–59

54. Fields HL, Basbaum AI (1978) Brainstem control of spinal pain-transmission neurons. Annu Rev Physiol 40:217–248

55. Kato G et al (2006) Direct GABAergic and glycinergic inhibition of the substantia gelatinosa from the rostral ventromedial medulla revealed by in vivo patch-clamp analysis in rats. J Neurosci 26(6):1787–1794

56. Eccles JC, Schmidt R, Willis WD (1963) Pharmacological studies on presynaptic inhibition. J Physiol 168(3):500–530

57. Brooks CM, Koizumi K, Malcolm JL (1955) Effects of changes in temperature on reactions of spinal cord. J Neurophysiol 18(3):205–216

58. Fink AJP et al (2014) Presynaptic inhibition of spinal sensory feedback ensures smooth movement. Nature 509(7498):43–48

Chapter 19

Production of AAVs and Injection into the Spinal Cord

Hendrik Wildner, Jean-Charles Paterna, and Karen Haenraets

Abstract

Adeno-associated virus (AAV) vector–mediated gene transfer has become a widely applied tool in neurobiology. More recently, it has also become crucial in the analysis of somatosensory spinal circuits as it allows for the local restriction of transgene expression and determination of the onset of expression. The combination of recombinase-dependent AAV vectors with transgenic mice enables cell type-specific expression of any protein, within the limits of the genome size of wild-type (wt) AAVs (~4700 nucleotides). Recombinase-dependent AAV vectors encoding for marker, sensor or effector proteins such as fluorescent proteins, calcium sensors or pharmacogenetic receptors and toxins can be used to label, monitor, or functionally manipulate recombinase-expressing spinal neurons. This allows interrogating the function of any genetically identifiable spinal neuron in specific somatosensory modalities such as itch or in different modalities of pain. Here, we first discuss the capabilities of AAV vectors and describe how to modify and use their properties. In the subsequent protocol we outline how to generate AAV vectors and how to deliver them into the spinal cord of mice.

Key words intraspinal injection, Adeno-associated virus, AAV production, Spinal cord, Serotype

1 Introduction

The advance of transgenic and virus-based technologies has recently enabled researchers to interrogate at an unprecedented resolution, the spinal circuits that process itch and pain [1–3]. In many of these studies, adeno-associated virus (AAV) vectors (i.e., an AAV containing an artificially produced genome) have become a crucial tool [4–7]. AAV belongs to the family of Parvoviridae, genus Dependoparvovirus. It is composed of a nonenveloped protein shell (capsid) coating its approximately 4.7 kb single-stranded DNA genome, which contains two ORFs encoding only six nonstructural (regulatory) and three structural (capsid) proteins translated from unspliced and spliced transcripts produced from three different promoters. The viral genome is flanked by inverted

Supplementary Information The online version of this chapter (https://doi.org/10.1007/978-1-0716-2039-7_19) contains supplementary material, which is available to authorized users.

Rebecca P. Seal (ed.), *Contemporary Approaches to the Study of Pain: From Molecules to Neural Networks*, Neuromethods, vol. 178, https://doi.org/10.1007/978-1-0716-2039-7_19, © Springer Science+Business Media, LLC, part of Springer Nature 2022

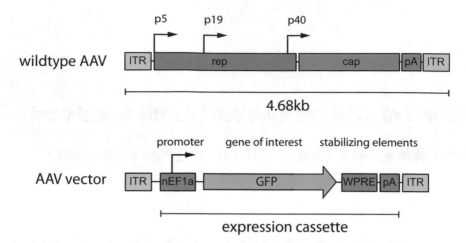

Fig. 1 Schematic representation of a wildtype AAV genome and an exemplary AAV vector genome. Wildtype AAV genomes are composed of two inverted terminal repeats (ITR) at the respective ends of the genome and two open reading frames encoding the proteins required for replication (rep) and the viral capsid (cap). Viral promoters are indicated by arrows. In an AAV vector rep and cap genes are replaced by a heterologous promoter (e.g., nEF1a) a gene of interest (e.g., green fluorescent protein (GFP)) and elements stabilizing the transcribed RNA (e.g., woodchuck hepatitis virus posttranscriptional regulatory element (WPRE) and a poly adenylation site (pA)) thus increasing transgene expression

terminal repeats (ITRs), which are essential for rescue, replication, and packaging of the viral genome into newly assembled capsids. In contrast to the wild-type (wt) AAV genome only the ITRs are retained in AAV vector genomes, while the two ORFs are replaced by an expression cassette of choice (Fig. 1a), which renders AAV vectors replication-defective (even in the putative presence of a helper virus). Wt AAVs can infect dividing and nondividing cells. Upon infection virions translocate to the nucleus where they are uncoated in order to release their genome. While wt AAV genomes are efficiently integrated into the host genome (provirus) at specific integration sites (IS; for example AAVS1 on human chromosome 19q13.3-qter) the genomes of AAV vectors stay mostly episomal as circular concatemers because nonstructural proteins that mediate site-specific integration are no longer encoded. This low frequency of integration means a low risk of insertional mutagenesis by AAV vectors compared to integrating viral vectors (such as retroviral vectors). The variety of naturally occurring and artificially engineered (for example by directed evolution) capsids offers the creation of tailored AAV vectors with specific tropism, unprecedented transduction efficiency and favorable immunological properties. In principle any kind of RNAs and (polycistronic) proteins can be encoded by AAV vectors. However, a major restriction exerted on the construction of expression cassettes is imposed by the small ("parvus") wt AAV genome size, which limits the choice of ORFs and regulatory elements.

In the context of analyzing neural networks required for processing aversive stimuli such as those evoking pain and itch, AAV vectors have been designed to mainly encode three types of molecules.

1. *Marker proteins—to label the cells of interest (e.g., fluorescent proteins such as eGFP or tdTomato).*

2. *Sensor proteins—to monitor the activity of the cells of interest (e.g., calcium sensors such as GCaMP).*

3. *Effector proteins—to activate, inhibit, silence, or ablate the cells of interest (e.g., light-inducible ion channels such as ChR2 or NpHR, pharmacogenetic receptors such as hM3Dq, hM4Di, PSAM4-GlyR or toxins such as TeLC or DTA).*

A main advantage of AAV vector-mediated gene transfer is that transduction (i.e., the expression of RNAs or proteins by cells that transcribe an expression cassette delivered by viral vectors; not followed by the production of progeny wt viruses, which defines an infection), is mostly restricted to the injection site, which is, in the example outlined in Subheading 3, the spinal cord. In addition, transduction and therefore expression of the encoded transgene can only occur after local delivery of the AAV vector. As a consequence, unwanted expression at other CNS sites or at time points before the injection, which may lead to developmental adaptations, can be ruled out. However, delivery of AAV vectors will lead to transduction of the majority of cells at the injection site. In the analysis of neural circuits this is often unwanted as experimental designs usually aim to address the function of a distinct neuronal subpopulation. To achieve selective expression of a marker, sensor or effector protein in a neuronal subpopulation, a combination of transgenic mice expressing a recombinase in the cell population of interest and injection of recombinase dependent AAV vectors has been used. In this scenario, still the majority of cells at the injection site receive a particular expression cassette, but only those neurons expressing the recombinase are able to switch on the expression of the encoded gene (Fig. 2a). Therefore, only those neurons expressing the recombinase become labeled or can be activated or silenced, depending on which marker or effector is encoded by the injected AAV vector. To be even more specific or in cases where a single marker is not sufficient to label the neuronal subpopulation of interest, intersectional strategies have been developed (Fig. 2b [8]). Here, the activities of two recombinases are required to switch on the expression of a marker or effector gene. If expression of the two recombinases is under the control of different driver genes, only those cells can be manipulated in which both driver genes are active.

Two other strategies have been used either alternatively or in combination with recombinase-dependent gene expression to limit the expression of the AAV vector-encoded transgene. Several

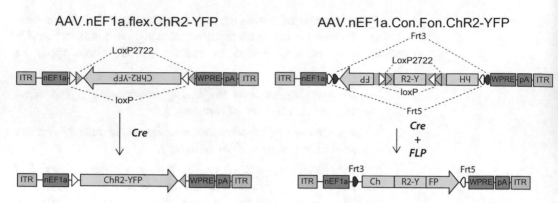

Fig. 2 Recombinase-dependent transgene expression. (**a**) Cre recombinase dependent expression of the transgene of interest (e.g., a channelrhodopsin-yellow fluorescent fusion protein (ChR2-YFP)) can be achieved by flanking the inverted transgene with two pairs of incompatible LoxP sites (LoxP and LoxP2722). Cre dependent recombination will lead to the irreversible inversion and thus expression of the LoxP flanked transgene. (**b**) Intersectional expression of the transgene can be achieved if the transgene of interest will only be expressed after two different recombination events (e.g., recombination by the Cre and the FLP recombinases). In the depicted example the transgene (ChR2-YFP) is split into three exons (yellow) which are separated by two LoxP sites containing introns. For transgene expression to occur, Cre mediated inversion of the LoxP flanked middle exon and FLP mediated inversion of the FRT flanked entire open reading frame is required

groups have studied the effect of serotypes (capsids) on transduction efficiencies to identify serotype (capsid) transduction profiles. In addition, molecular evolution has been used to develop serotypes that show a specific transduction pattern. Those efforts lead to characterization of differences in viral spread depending on the serotype (capsid) used. Most significantly, it was found that AAV vectors composed of the wt AAV serotype 2-derived capsid variant "retro" (AAV-retro) are an order of magnitude more efficient in retrogradely transducing neurons at axon terminals and thus allow efficient labeling of projection neurons whose cell bodies are not located at the injection site [9, 10]. Second, while many AAV vectors that are used for the analysis of neural circuits utilize ubiquitous promoters, others have made use of more specific promoter fragments such as the human synapsin promoter fragment 1 (hSyn1). The hSyn1 promoter fragment has been found to be specific for neurons and therefore avoids expression in spinal glia [11]. Furthermore, it has been shown that by utilizing the hSyn1 promoter fragment especially in combination with an AAV serotype (capsid) that displays reduced spread at the spinal injection site (e.g., AAV-8), unwanted transduction of projections of primary afferent neurons can be limited [5, 9]. Yet, in the context of spinal circuits, the use of specific promoter fragments to drive expression in distinct cell populations has so far been poorly exploited, potentially due to the lack of identified fragments which would be able to drive neuronal subtype-specific gene expression in the context of

AAV vectors. However, the power of using specific promoter fragments has been demonstrated in other CNS regions such as the retina [12]. Recently developed techniques such as ATAC-seq [13, 14] hold the promise of leading to the identification of numerous new cell type-specific promoter/enhancer fragments, which might be utilized in the future to drive effector gene expression in a desired spinal subpopulation. Using transgenic mice could thus be avoided and AAV vector-mediated circuit manipulation could be more easily extended to other mammals [12].

Finally, a combination of AAV vector delivery and CRISPR/Cas9 technology has recently been used to achieve somatic mutations in postmitotic neurons enabling the knock-in of an AAV vector-delivered transgene into a specific endogenous gene of interest [15, 16]. This technique may be used in the future to directly manipulate an endogenous gene of interest that is important for pain transmission in the spinal cord. Alternatively, an effector gene may be integrated into the gene of interest, thereby achieving expression in the population of interest and thus enabling manipulation of the activity of the given population without the need for transgenic mice or the identification of cell type-specific promoters.

Regardless of how subtype-specific expression of AAV vector-delivered transgenes into spinal neurons is achieved, the basis for all described experimental scenarios is the production of high-quality, high-titer AAV vectors and the subsequent injection into spinal tissue. Here we describe in detail how to produce AAV vectors and how to inject them into spinal tissue in order to mediate successful transduction of marker, sensor or effector proteins into spinal neurons (The outlined protocol for spinal injections is adapted from [17]).

1.1 AAV Vector Production

Helper virus-free production of Adenovirus-associated virus (AAV) vectors for preclinical research by discontinuous iodixanol density gradient ultracentrifugation.

2 Materials

2.1 HEK 293T Cells

1. AAVpro® 293T cell line (TaKaRa, 632,273, stored in the vapor phase of liquid nitrogen).

2. Mycoplasmacheck (Eurofins Genomics).

3. Tissue culture plates for adherent cells, growth area 152 cm^2 (Sarstedt, 83.3903, or equivalent).

4. Dulbecco's Modified Eagle's Medium, high glucose with GlutaMAX (DMEM) (Gibco™, 61965026, stored at 4 °C).

5. Fetal bovine serum (FBS), South America, for standard applications (Gibco™, 10270106, or equivalent, stored at −20 to −80 °C).

6. TrypLE™ Express Enzyme solution (1×), no phenol red (Gibco™, 12604021, stored at 4 °C).

7. Cryotubes (VWR, 479-1254, 479-1256, or equivalent).

8. Sodium chloride (NaCl) (Sigma-Aldrich, 71380, or equivalent).

9. Potassium chloride (KCl) (VWR, 437023F, or equivalent).

10. Disodium hydrogen phosphate (Na_2HPO_4) (VWR, 102495D, or equivalent).

11. Potassium dihydrogen phosphate (KH_2PO_4) (VWR, 26936, or equivalent).

12. 20× phosphate buffered solution (PBS), pH 7.4.

   (a) Weigh 2737.85 mmol/l NaCl, 53.66 mmol/l KCl, 202.87 mmol/l Na_2HPO_4, 35.27 mmol/l KH_2PO_4 into a beaker.

   (b) Add ultrapure water to about 80% of the final volume.

   (c) Add a magnetic stir bar and stir at room temperature (RT) until all components dissolved.

   (d) Adjust to final volume with ultrapure water.

   (e) Aliquot if required.

   (f) Heat-sterilize (20 min at 121 °C, 1 bar).

   (g) Store at RT.

13. 1× phosphate buffered solution (PBS), pH 7.4.

   (a) Dilute 20× PBS, pH 7.4 20fold with ultrapure water to 1× PBS, pH 7.4.

   (b) Aliquot if required.

   (c) Heat-sterilize (20 min at 121 °C, 1 bar).

   (d) Store at RT.

14. CO_2 cell culture incubator (Binder, CB 220, or equivalent).

2.2 Plasmid DNA

1. MDS™42 LowMut ΔrecA (Scarab Genomics LLC, C-6262-20K or any other suitable.

2. Bacterial strain to amplify AAV vector plasmids, stored at −80 °C).

3. Anion exchange columns (Macherey & Nagel, NucleoBond® Xtra, 740414, or equivalent).

4. Tryptone (Sigma-Aldrich, T9410).

5. Yeast extract (PanReac AppliChem, A1552).

6. Glycerol (VWR, 24388, or equivalent).

7. Dipotassium hydrogen phosphate (K_2HPO_4) (VWR, 26930, or equivalent).

8. Adenoviral (AV) helper plasmid (Addgene or any other provider)

9. AAV helper plasmid (Addgene or any other provider).

10. AAV vector plasmid (Addgene or any other provider).

11. 50% glycerol solution.

    (a) Add ultrapure water to 50% of the final volume into a flask.

    (b) Add 100% glycerol to 50% of the final volume.

    (c) Gently mix by inverting the flask several times.

    (d) Heat-sterilize (20 min at 121 °C, 1 bar).

    (e) Store at RT.

12. Terrific Broth (TB) medium.

    (a) Weigh 1.2% tryptone, 2.4% yeast extract, 54 mmol/l K_2HPO_4, 1.6 mmol/l KH_2PO_4 and 0.8% of a 50% glycerol solution into a beaker.

    (b) Add ultrapure water to about 90% of the final volume.

    (c) Add a magnetic stir bar and stir at RT until all components dissolved.

    (d) Adjust to final volume with ultrapure water.

    (e) Aliquot if required.

    (f) Heat-sterilize (20 min at 121 °C, 1 bar).

    (g) Store at 4 °C protected from light.

2.3 PEI-Mediated Split Transfection of HEK 293T Cells

1. 15 ml reaction tubes (VWR, 525-0150, or equivalent).

2. 50 ml reaction tubes (VWR, 525-0304, or equivalent).

3. Syringe-driven filtration unit [Millipore, Millex GF, 0.22 μm polyethersulfone (PES), SLGP033RS].

4. Bottle top vacuum-driven filtration unit [Millipore, Stericup-GP Express Plus®, 0.22 μm polyethersulfone (PES), SCGPU02RE].

5. Polyethylenimine (PEI) "Max" [Polysciences, Inc., transfection grade linear polyethylenimine hydrochloride, molecular weight (MW) 40,000, 24765-2].

6. [1 mg/ml] PEI "Max" solution.

    (a) Depending on the final volume, weigh the required amount of PEI "Max" into a vessel.

    (b) Add ultrapure water to about 80% of the final volume.

    (c) Close the vessel and manually and gently inverse the vessel until the PEI "max" has dissolved.

    (d) Adjust to final volume with ultrapure water.

(e) Sterile filter (depending on the final volume, either use a syringe-driven filter unit, such as Millex GF or a bottle top vacuum-driven filtration unit such as Stericup-GP express plus®).

(f) Aliquot if required.

(g) Store at −20 °C (*see* **Note 1**).

2.4 Harvesting and Processing HEK 293T Cells and Supernatant Posttransfection

1. 500 ml polypropylene (PP) centrifuge tubes with plug seal cap (Corning, 431123).

2. Support cushions for 500 ml PP centrifuge tubes with plug seal cap (Corning, 431124).

3. Protective sleeves (VWR, 113-0718, or equivalent).

4. Polyethylene glycol 8000 (PEG 8000) (PanReac AppliChem, A2204.0500).

5. Tris(hydroxymethyl)aminomethane (Tris base) (Biosolve, 200923, or equivalent).

6. 5 N NaOH solution (VWR, 1.09913.0001, or equivalent).

7. 5 N HCl solution (VWR, 85400.320, or equivalent).

8. DENARASE (c-LEcta GmbH, 20804, or equivalent).

9. 1 M Tris, pH 8.5 solution.

   (a) Weigh 1 mol/l Tris base into a beaker.

   (b) Add ultrapure water to about 80% of the final volume.

   (c) Add 5 N HCl (a final volume of 1 l requires about 84 ml 5 N HCl).

   (d) Add ultrapure water to about 95% of the final volume.

   (e) Remove 0.5 ml of the Tris base solution and add to 4.5 ml ultrapure water (1:10 dilution).

   (f) Measure pH of the 1:10 dilution.

   (g) If necessary, add more 5 N HCl to the stock solution and remeasure the pH of a new 1:10 dilution until pH reaches 8.5.

   (h) Adjust to final volume with ultrapure water.

   (i) Aliquot if required.

   (j) Heat-sterilize (20 min at 121 °C, 1 bar).

   (k) Store at RT.

10. 5 M NaCl solution

    (a) Weigh 5 mol/l NaCl into a beaker.

    (b) Add ultrapure water to about 80% of the final volume.

    (c) Add a magnetic stir bar and stir at RT until the NaCl has dissolved.

    (d) Adjust to final volume with ultrapure water.

(e) Aliquot if required.

(f) Heat-sterilize (20 min at 121 °C, 1 bar).

(g) Store at RT.

11. AAV resuspension buffer, pH 8.5.

(a) Add ultrapure water to about 80% of the final volume into a beaker.

(b) Add 150 mmol/l of a 5 M NaCl solution.

(c) Add 50 mmol/l of a 1 M Tris, pH 8.5 solution.

(d) Add a magnetic stir bar and stir at RT until the solutions are mixed.

(e) Measure pH and adjust to 8.5 by adding dropwise a 5 N NaOH solution.

(f) Adjust to final volume with ultrapure water.

(g) Aliquot if required.

(h) Heat-sterilize (20 min at 121 °C, 1 bar).

(i) Store at RT.

12. 40% PEG 8000 solution.

(a) Depending on the final volume, weigh the required amount of PEG 8000 into a beaker.

(b) Add ultrapure water to about 90% of the final volume.

(c) Use a rod and manually stir the solution until the PEG 8000 partially dissolved.

(d) Add a magnetic stir bar, cover the beaker with a plastic foil and stir at RT until the PEG 8000 completely dissolved (several hours to overnight).

(e) When all PEG 8000 dissolved adjust to final volume with ultrapure water and mix.

(f) Aliquot if required.

(g) Heat-sterilize (20 min at 121 °C, 1 bar).

(h) Store at RT.

13. Precellys® Evolution homogenizer (Bertin Instruments, P000062-PEVO0-A) or Minilys Personal homogenizer (Bertin Instruments, P000673-MLYS0-A).

14. 7 ml soft tissue homogenizing CK14 tubes (Bertin Instruments, P000940-LYSK0-A).

2.5 Discontinuous Iodixanol Density Gradient Ultracentrifugation

1. OptiPrep (60% iodixanol in water; Axis-Shield Density Gradient Media, Alere.

2. Technologies AS, 1114542; store at RT protected from light).

3. Magnesium chloride hexahydrate ($MgCl_2 \cdot 6H_2O$; VWR, 87060, or equivalent).

4. Potassium chloride (KCl, VWR, 437023F, or equivalent).

5. Heat-sterilized needle (gauge 14/length 150 mm/point style 3) (Hamilton Company, 7749-05).

6. 35 ml PA Ultracrimp tubes (including plugs and caps) (Thermo Fisher Scientific, 03989).

7. Ultracrimp/Clearcrimp sealing tool (Thermo Fisher Scientific, 03920).

8. Tube rack (for 35 ml PA Ultracrimp tubes in combination with Ultracrimp/Clearcrimp) (Thermo Fisher Scientific, 03935).

9. Rotor caps (for 35 ml PA Ultracrimp tubes in combination with T-865 fixed angle titanium rotor) (Thermo Fisher Scientific, 03996).

10. T-865 fixed angle titanium rotor (Thermo Fisher Scientific, 51411).

11. Sorvall WX 80 ultracentrifuge (Thermo Fisher Scientific, 46900, or equivalent).

12. Precision balance ($d = 0.001$ g).

13. 18-G, 1.2×40 mm injection needle (Sterican, B. Braun Medical AG, 4550400-01).

14. 10 ml syringes (Injekt® Luer Solo, B. Braun Medical AG, 4606108V).

15. Tube holder (Carl Friedrich Usbeck, VWR, 241-7510).

16. 1 M $MgCl_2$ solution.

    (a) Weigh 1 mol/l $MgCl_2 \cdot 6H_2O$ into a beaker.

    (b) Add ultrapure water to about 80% of the final volume.

    (c) Add a magnetic stir bar and stir at RT until the $MgCl_2 \cdot 6H_2O$ dissolved.

    (d) Adjust to final volume with ultrapure water.

    (e) Sterile filter (do not autoclave).

    (f) Aliquot under sterile conditions if required.

    (g) Store at RT.

17. 1 M KCl solution.

    (a) Weigh 1 mol/l KCl into a beaker.

    (b) Add ultrapure water to about 80% of the final volume.

    (c) Add a magnetic stir bar and stir at RT until the KCl dissolved.

    (d) Adjust to final volume with ultrapure water.

    (e) Sterile filter (do not autoclave).

    (f) Aliquot under sterile conditions if required.

    (g) Store at RT.

18. 15% OptiPrep (Iodixanol), 1 M NaCl solution.

 (a) Work under sterile conditions.

 (b) Depending on the final volume, add about 90% of the required sterile ultrapure water into a measuring cylinder (a final volume of 200 ml contains 104.8 ml sterile ultrapure water; 500 ml: 262 ml; 1000 ml: 523.9 ml; 2000 ml: 1047.8 ml).

 (c) Add 20× PBS, pH 7.4 to reach a final concentration of 1× PBS, pH 7.4 (a final volume of 200 ml contains 10 ml 20× PBS, pH 7.4; 500 ml: 25 ml; 1000 ml: 50 ml; 2000 ml: 100 ml).

 (d) Add 1 M MgCl$_2$ solution to reach a final concentration of 1 mmol/l (a final volume of 200 ml contains 0.2 ml 1 M MgCl$_2$ solution; 500 ml: 0.5 ml; 1000 ml: 1 ml; 2000 ml: 2 ml).

 (e) Add 1 M KCl solution to reach a final concentration of 2.5 mmol/l (a final volume of 200 ml contains 0.5 ml 1 M MgCl$_2$ solution; 500 ml: 1.25 ml; 1000 ml: 2.5 ml; 2000 ml: 5.0 ml).

 (f) Add 5 M NaCl solution to reach a final concentration of 1 M (a final volume of 200 ml contains 34.5 ml 5 M NaCl solution; 500 ml: 86.3 ml; 1000 ml: 172.6 ml; 2000 ml: 345.2 ml; note: The NaCl provided with the 1× PBS, pH 7.4 is taken into account).

 (g) Add 60% OptiPrep (Iodixanol) to reach a final concentration of 15% OptiPrep (a final volume of 200 ml contains 50 ml 60% OptiPrep; 500 ml: 125 ml; 1000 ml: 250 ml; 2000 ml: 500 ml).

 (h) Adjust to final volume with ultrapure water.

 (i) Seal measuring cylinder with Parafilm.

 (j) Mix the 15% OptiPrep, 1 M NaCl solution by gently rocking.

 (k) Transfer the 15% OptiPrep, 1 M NaCl solution into a sterile and labeled glass bottle.

 (l) Prepare 9 ml aliquots using 15 ml reaction tubes.

 (m) Store protected from light at RT.

19. 25% OptiPrep (Iodixanol) solution.

 (a) Work under sterile conditions.

 (b) Depending on the final volume, add about 90% of the required sterile ultrapure water into a measuring cylinder (a final volume of 200 ml contains 106 ml sterile ultrapure water; 500 ml: 264.9 ml; 1000 ml: 529.9 ml; 2000 ml: 1059.7 ml).

(c) Add 20× PBS, pH 7.4 to reach a final concentration of 1× PBS, pH 7.4 (a final volume of 200 ml contains 10 ml 20× PBS, pH 7.4; 500 ml: 25 ml; 1000 ml: 50 ml; 2000 ml: 100 ml).

(d) Add 1 M $MgCl_2$ solution to reach a final concentration of 1 mmol/l (a final volume of 200 ml contains 0.2 ml 1 M $MgCl_2$ solution; 500 ml: 0.5 ml; 1000 ml: 1 ml; 2000 ml: 2 ml).

(e) Add 1 M KCl solution to reach a final concentration of 2.5 mmol/l (a final volume of 200 ml contains 0.5 ml 1 M $MgCl_2$ solution; 500 ml: 1.25 ml; 1000 ml: 2.5 ml; 2000 ml: 5.0 ml).

(f) Add 60% OptiPrep (Iodixanol) to reach a final concentration of 25% OptiPrep (a final volume of 200 ml contains 83.3 ml 60% OptiPrep; 500 ml: 208.3 ml; 1000 ml: 416.6 ml; 2000 ml: 833.3 ml).

(g) Adjust to final volume with ultrapure water.

(h) Seal measuring cylinder with Parafilm.

(i) Mix the 25% OptiPrep solution by gently rocking.

(j) Transfer the 25% OptiPrep solution into a sterile and labeled glass bottle.

(k) Prepare 8 ml aliquots using 15 ml reaction tubes.

(l) Store protected from light at RT.

20. 40% OptiPrep (Iodixanol) solution.

(a) Work under sterile conditions.

(b) Depending on the final volume, add about 90% of the required sterile ultrapure water into a measuring cylinder (a final volume of 200 ml contains 56 ml sterile ultrapure water; 500 ml: 139.9 ml; 1000 ml: 279.9 ml; 2000 ml: 559.7 ml).

(c) Add 20× PBS, pH 7.4 to reach a final concentration of 1× PBS, pH 7.4 (a final volume of 200 ml contains 10 ml 20× PBS, pH 7.4; 500 ml: 25 ml; 1000 ml: 50 ml; 2000 ml: 100 ml).

(d) Add 1 M $MgCl_2$ solution to reach a final concentration of 1 mmol/l (a final volume of 200 ml contains 0.2 ml 1 M $MgCl_2$ solution; 500 ml: 0.5 ml; 1000 ml: 1 ml; 2000 ml: 2 ml).

(e) Add 1 M KCl solution to reach a final concentration of 2.5 mmol/l (a final volume of 200 ml contains 0.5 ml 1 M $MgCl_2$ solution; 500 ml: 1.25 ml; 1000 ml: 2.5 ml; 2000 ml: 5.0 ml).

(f) Add 60% OptiPrep (Iodixanol) to reach a final concentration of 40% OptiPrep (a final volume of 200 ml contains 133.3 ml 60% OptiPrep; 500 ml: 333.3 ml; 1000 ml: 666.6 ml; 2000 ml: 1333.3 ml).

(g) Adjust to final volume with ultrapure water.

(h) Seal measuring cylinder with Parafilm.

(i) Mix the 40% OptiPrep solution by gently rocking.

(j) Transfer the 40% OptiPrep solution into a sterile and labeled glass bottle.

(k) Prepare 8 ml aliquots using 15 ml reaction tubes.

(l) Store protected from light at RT.

2.6 Diafilstration

1. Vivaspin® 20 ultrafiltration device 100,000 MWCO PES [Sartorius, VS2042; note: it is essential to use polyethersulfone (PES) membranes].

2. 1× PBS-MK, pH 7.4 solution (prepare fresh as required).

(a) Add 1 M $MgCl_2$ solution to reach a final concentration of 1 mmol/l and 1 M KCl solution to reach a final concentration of 2.5 mmol/l to 1× PBS, pH 7.4 solution.

(b) Aliquot under sterile conditions if required.

(c) Store at RT.

2.7 Physical Titer Quantification

1. Qubit™ 3.0 (or higher) fluorometer (Thermo Fisher Scientific, Q33226).

2. Qubit assay tubes (Thermo Fisher Scientific, Q32856, or equivalent)

3. Qubit® dsDNA (double-stranded DNA) HS assay kit (Thermo Fisher Scientific, Q32854)

4. 1× PBS-MK, pH 7.4 solution (*see* Subheading 2.6).

2.8 Identity Check

1. Deoxynucleotide set, 100 mM (Sigma-Aldrich, DNTP100-1KT, or equivalent).

2. 25 nmol desalted sequencing-primers (Thermo Fisher Scientific, or equivalent).

3. Taq DNA polymerase (including ThermoPol® Buffer, New England Biolabs, M0267L, or equivalent).

3 Methods

3.1 HEK 293T Cells

We use the human embryonic kidney (HEK) cell line AAVpro® 293T (HEK 293T) [18, 19] maintained in antibiotic-free cell culture medium to produce AAV vectors by polyethylenimine (PEI)-mediated split transfection (*see* **Note 2**). If HEK 293T cells are properly handled, their growth rate and morphology will remain stable over time. We did not notice any drop in AAV vector yield when using HEK 293T cells passaged more than 100 times.

3.1.1 Starting Adherent HEK 293T Cells from Frozen Stocks

Starting adherent HEK 293T cells from frozen stocks requires about 30 min.

1. Add RT 9 ml DMEM/GlutaMAX cell culture medium supplemented with 10% non–heat-inactivated FBS ($DMEM_c$) into a 15 ml reaction tube (use a biosafety cabinet class II).

2. Heat a water bath to 37 °C.

3. Transfer a cryovial of HEK 293T cells from liquid nitrogen storage to a −20 °C freezer immediately before thawing.

4. Partially thaw the frozen HEK 293T cells by dipping the lower half of the cryovial for about 1 to 1.5 min into the 37 °C water bath.

5. Remove the cryovial, wipe the outside with a 70% EtOH soaked single-use paper-wipe and dry.

6. Transfer the HEK 293T cell suspension using a 1000 μl micropipette (*see* **Note 3**) into the 15 ml reaction tube containing the RT 9 ml $DMEM_c$.

7. Close the 15 ml reaction tube and gently inverse a couple of times to mix the HEK 293T cells.

8. Centrifuge for 3 min at $600 \times g$/RT in a swing-out rotor.

9. Aspirate the supernatant using a sterile Pasteur glass pipette connected to a vacuum aspiration system.

10. Add 10 ml RT $DMEM_c$ using a 10 ml serological pipette and resuspend the HEK 293T cell pellet by either pipetting up-and-down several times or by a brief cycle of vortexing.

11. Transfer the resuspended HEK 293T cells into a 15-cm cell culture plate.

12. Complement with 10 ml RT $DMEM_c$.

13. Move the 15-cm cell culture plate horizontally forward and backward as well as sideward several times in order to distribute the HEK 293T cells as evenly as possible.

14. Incubate at 37 °C/5% pCO_2 until the HEK 293T cells have reached almost confluency (*see* **Note 4**).

15. Continue passaging HEK 293T cells.

3.1.2 Passaging HEK 293T Cells

The preparation of eighty 15-cm cell culture plates from ten 15-cm cell culture plates (1:10 split ratio, about 72 h of continuous growth) requires about 1 h. Not more than 30 min are required if only a couple of 15-cm cell culture plates are prepared.

1. When HEK 293T cells have reached almost confluency, aspirate DMEM$_c$ using a sterile Pasteur glass pipette connected to a vacuum aspiration system (use a biosafety cabinet class II) (*see* **Note 5**).

2. Add about 11 ml RT 1× PBS, pH 7.4 solution to each 15-cm cell culture plate (*see* **Note 6**).

3. Gently tilt the 15-cm cell culture plate(s) several times to rinse the HEK 293T cells.

4. Aspirate the 1× PBS, pH 7.4 solution.

5. Add about 4 ml TrypLE™ express Enzyme solution per 15-cm cell culture plate directly onto the HEK 293T cells.

6. Gently tilt the 15-cm cell culture plate(s) several times to evenly distribute the TrypLE™ express Enzyme solution.

7. Incubate for about 5 min at 37 °C/5% pCO$_2$.

8. Remove the 15-cm cell culture plate(s) from the CO$_2$ incubator and gently tap until the HEK 293T cells detach completely.

9. Add about 8 ml RT DMEM$_c$ per 15-cm cell culture plate to inactivate the TrypLE™ express Enzyme solution.

10. Add another 15 to 17 ml RT DMEM$_c$ per 15-cm cell culture plate immediately before detaching the HEK 293T cells.

11. Aspirate and expel the HEK 293T cells suspension using a 25 ml serological pipette until all HEK 293T cells have detached (*see* **Note 7**).

12. In a final step, aspirate as much as possible of the HEK 293T cells suspension and force the HEK 293T cells through the 25 ml serological pipette while holding the tip almost perpendicular onto the bottom of the 15-cm cell culture plate.

13. Depending on the intended passaging ratio, transfer the required volume of the HEK 293T cell suspension into RT DMEM$_c$ to reach the final volume (*see* **Note 8**).

14. Mix the entire HEK 293T cell suspension by pipetting several times up-and-down and evenly distribute about 20 to 25 ml of the HEK 293T cell suspension into the required number of 15-cm cell culture plates (*see* **Note 9**).

15. Move the 15-cm cell culture plates horizontally forward and backward as well as sideward several times in order to distribute the HEK 293T cells as evenly as possible.

16. Incubate at 37 °C/5% pCO$_2$.

17. Continue with PEI-mediated split transfection of HEK 293T cells.

3.2 Plasmid DNA   The preparation of plasmid DNA after retransformation usually requires 6 working days.

1. Amplify and characterize plasmid DNA to be transfected (*see* **Note 10**).

2. Continue with PEI-mediated split transfection of HEK 293T cells.

3.3 PEI-Mediated Split Transfection of HEK 293T Cells

PEI-mediated transfection of cell lines has been described previously as a cost-effective and efficient method to produced AAV vectors [20]. Furthermore, in the context of lentiviral vector production, it was suggested to split HEK 293T cells immediately followed by PEI-mediated transfection [21]. We think that the PEI-mediated split transfection of HEK 293T cells to produce AAV vectors is more flexible with respect to the preparation of the HEK 293T cells compared to the more frequently applied transfection of HEK 293T cells prepared the day before transfection. Furthermore, the split transfection method allows more HEK 293T cells to be prepared per surface area (confluency of 90% to 95% at the time of transfection) compared to the 70% to 80% for HEK 293T cells prepared the day before transfection.

3.3.1 Plasmid DNA and PEI Solutions

The preparation of plasmid DNA and PEI solutions for a single AAV vector preparation (based on five 15 cm cell culture plates) requires about 60 min. If multiple AAV vector preparations are prepared simultaneously, significantly less time is required per AAV vector preparation (ten AAV vector preparations require about 3–4 h).

1. For five 15-cm cell culture plates prepare two 50 ml reaction tubes and label both with the name of the AAV vector to be produced (*see* **Note 11**).

2. Additionally, label one of the two 50 ml reaction tubes with "PEI."

3. Calculate the volumes of sterile ultrapure water, 5 M NaCl solution, plasmid DNA, and PEI solution to be added to the 50 ml reaction tubes (*see* **Note 12**).

4. Add the calculated volume of sterile ultrapure water to each 50 ml reaction tube (use a biosafety cabinet class II).

5. Add 450 µl 5 M NaCl solution to each of the 50 ml reaction tubes.

6. Add the plasmid DNA to the tube labeled with the name of the AAV vector.

7. Add the PEI solution to the tube labeled with "PEI."

8. Keep both 50 ml reaction tubes at RT and proceed to the split transfection of HEK 293T cells.

<table>
<tr><td>

3.3.2 Split Transfection of HEK 293T Cells

</td><td>

To complete four individual AAV vector preparations (each composed of five 15-cm cell culture plates) about 1 h and 15 min (standard protocol) or 1 h (advanced protocol) are required (*see* **Note 13**).

</td></tr>
</table>

1. Prepare the number of 15-cm cell culture plates required on the day of transfection containing HEK 293T cells grown almost to confluency (use a biosafety cabinet class II).

2. Split (passage) the HEK 293T cells 1:1 (*see* **Note 9**) and add the PEI solution to the plasmid DNA solution, vigorously vortex for a few seconds and incubate for 20 min at RT (*see* **Note 13**).

3. Split-transfect the HEK 293T cells.

4. Incubate at 37 °C/5% pCO_2.

5. The next day, take images of the HEK 293T cells (*see* **Note 14**).

6. Incubate for a total of 5–7 days without changing the $DMEM_c$.

7. Continue with harvesting and processing of HEK 293T cells and their supernatants posttransfection.

3.4 Harvesting and Processing of HEK 293T Cells and Their Supernatants Posttransfection

About 10 to 15 min are required to collect an individual AAV vector preparation (based on five 15-cm cell culture plates). Processing of the HEK 293T cells requires another 20 min, while the resuspension of an individual PEG 8000 pellet needs about 5 min.

1. For each AAV vector preparation prepare two 500 ml centrifuge tubes with plug seal cap and place each on a support cushion.

2. Label each 500 ml centrifuge tubes with the AAV vector name, number of 15-cm cell culture plates and one with "supernatant."

3. Take images of the HEK 293T cells immediately before collecting (*see* **Note 15**).

4. Collect the HEK 293T cells and their supernatant by flushing them within their own $DMEM_c$ using a 25 ml serological pipette and pool the HEK 293T cell suspension of an entire AAV vector preparation (maximal ten 15 cm cell culture plates) into the 500 ml centrifuge tube not labeled with "supernatant" (*see* **Note 16**).

5. Note the average volume of each 15-cm culture plate and calculate the combined total volume.

6. Centrifuge the 500 ml centrifuge tube placed in its support cushion 15 min at $1500 \times g$/4 °C in a swing-out rotor.

7. Meanwhile, prepare the PEG 8000 precipitation of the supernatant by adding in this order the calculated volumes of 40% PEG 8000 solution (20% of the combined total volume, final:

8%) and 5 M NaCl solution (10% of the combined total volume, final: 0.5 M) to the 500 ml centrifuge tube labeled "supernatant."

8. After completion of the centrifugation, pour the supernatant into the 500 ml centrifuge tube labeled "supernatant," close the centrifuge tube, mix thoroughly by orbital circular moving and incubate for 1 to 2 days at 4 °C.

9. Store the HEK 293T cell pellet at −20 °C until further processing (*see* **Note 17**).

10. Complete the PEG 8000 precipitation of the supernatant by a 1 h centrifugation at $3488 \times g/4$ °C in a swing-out rotor (*see* **Note 18**).

11. Pour as much as possible of the PEG 8000 precipitated supernatant into a waste bin avoiding spills (*see* **Note 19**).

12. Add 1.6 ml AAV resuspension buffer, pH 8.5 to the PEG 8000 pellet and resuspend it using a 1000 µl micropipette until completely dissolved (rinse the wall of the 500 ml centrifuge bottle as well) (*see* **Note 20**).

13. Store the resuspended PEG 8000 pellet at −20 °C until further processing (*see* **Note 17**).

14. Continue with discontinuous iodixanol density gradient ultracentrifugation.

3.5 Discontinuous Iodixanol Density Gradient Ultracentrifugation

We follow the recommendations of the pioneer work described by Zolotukhin [22] to separate AAV vectors from most of the HEK 293T-derived proteins, genomic DNAs and RNAs, plasmid DNA, and "empty" AAV vectors (lacking the recombinant genome) by discontinuous iodixanol density gradient ultracentrifugation (isopycnic ultracentrifugation, buoyant or equilibrium separation).

3.5.1 Preparing the Crude and Cleared Cell Lysates

About 30 min are required to prepare the crude cell lysates and to initiate the enzymatic treatment of eight individual AAV vector preparations (the maximal capacity of the ultracentrifugation rotor). The enzymatic treatment of the crude cell lysate requires between 30 min to 3 h. Another 40 min are needed to prepare the cleared cell lysates of eight individual AAV vector preparations.

1. Remove a pair of 500 ml centrifuge tubes of an AAV vector preparation (a total of not more than eight, the maximal capacity of the ultracentrifugation rotor) containing the frozen HEK 293T cell pellet and the resuspended PEG 8000 pellet from −20 °C and thaw at 4 °C.

2. Prepare a labeled 35 ml PA Ultracrimp tube for each AAV vector preparation.

3. Place plugs and caps of the labeled 35 ml PA Ultracrimp tubes into a 10-cm cell culture plate in order to prevent losing them in the biosafety cabinet class II.

4. For each AAV vector preparation connect a sterile gauge 14/length 150 mm/point style 3 needle to an unlabeled 10 ml syringe and insert the needle into a labeled 35 ml PA Ultracrimp tube.

5. Label four 2.0 ml reaction tubes per labeled 35 ml PA Ultracrimp tube (representing an AAV vector preparation originating either from five or ten 15-cm cell culture plates).

6. Transfer the 500 ml centrifuge tubes containing the defrosted HEK 293T cell pellets from 4 °C to a biosafety cabinet class II.

7. Add 4 ml AAV resuspension buffer, pH 8.5 using a 10 ml serological pipette to each HEK 293T cell pellet originating from five 15-cm cell culture plates (*see* **Note 21**).

8. Resuspend the HEK 293T cell pellet.

9. Transfer the HEK 293T cell suspension into a labeled 7 ml soft tissue homogenizing CK14 tube and close firmly.

10. Place the 7 ml soft tissue homogenizing CK14 tube into a Precellys® evolution homogenizer (*see* **Note 22**).

11. Homogenize two times 45 s at 5000 rpm/RT with an intermediary pause of 15 s (*see* **Note 23**).

12. Add 750 U DENARASE (3 µl) per 7 ml soft tissue homogenizing CK14 tube containing the crude cell lysate and incubate for 30 min to 3 h at 37 °C in a water bath.

13. Remove the 7 ml soft tissue homogenizing CK14 tubes from the water bath, add 1 ml 5 M NaCl solution to each and mix well by pipetting up-and-down several times.

14. Evenly distribute the DENARASE-treated crude cell lysate of a 7 ml soft tissue homogenizing CK14 tube into the corresponding four labeled 2.0 ml reaction tubes (*see* **Note 24**).

15. Centrifuge for 10 min at 17,000 × g/4 °C in a fixed angle rotor.

16. During the 10 min centrifugation, continue with preparing the resuspended pellet of the PEG 8000 precipitated supernatant for discontinuous iodixanol density gradient ultracentrifugation (*see* Subheading 3.5.2).

17. After completion of the 10 min centrifugation of the DENARASE-treated crude cell lysate, transfer the cleared cell lysate (supernatant) of the four 2.0 ml reaction tubes into the correspondingly labeled 35 ml PA Ultracrimp tube (*see* **Note 25**).

18. Continue with preparing the discontinuous iodixanol density gradients.

Only a couple of minutes are required for this step.

1. Transfer the 500 ml centrifuge tubes containing the defrosted resuspended PEG 8000 pellet from 4 °C to a biosafety cabinet class II.

2. Add 500 µl 5 M NaCl solution to each 500 ml centrifuge tube.

3. Aspirate the resuspended pellet of the PEG 8000 precipitated supernatant with the sterile gauge 14/length 150 mm/point style 3 needle connected to an unlabeled 10 ml syringe and transfer the resuspended pellet of the PEG 8000 precipitated supernatant into the correspondingly labeled 35 ml PA Ultracrimp tube.

4. Continue with preparing the discontinuous iodixanol density gradients.

The preparation of each discontinuous iodixanol density gradient requires about 5 min (experienced user) to 10 min (standard user). We process eight individual AAV vector preparations for each ultracentrifugation run. The discontinuous iodixanol density gradients remain stable during this period and are kept at RT.

1. Underly the AAV vector containing combined cleared lysate–supernatant solution with (in this order).

    (a) 9 ml 15% OptiPrep (Iodixanol), 1 M NaCl solution.

    (b) about 5 ml 25% OptiPrep (Iodixanol) solution.

    (c) 8 ml 40% OptiPrep (Iodixanol) solution.

    (d) 3 ml 60% OptiPrep (Iodixanol) (see **Note 26**).

2. Aspirate most (but not all) of the remaining 25% OptiPrep (Iodixanol) solution without aspirating air.

3. Insert the needle a few millimeters away from the boundary between the 40% OptiPrep (Iodixanol) and 25% OptiPrep (Iodixanol) solution (see **Note 27**).

4. Slowly release the 25% OptiPrep (Iodixanol) solution without disturbing the 40% OptiPrep (Iodixanol) layer until the AAV vector containing combined cleared lysate–supernatant solution is lifted to the nozzle of the labeled 35 ml PA Ultracrimp tube (see **Note 28**).

5. Slowly retract the needle while simultaneously and gently squeezing the labeled 35 ml PA Ultracrimp tube so that any air bubbles stay at the top of the nozzle.

6. Aspirate any air bubbles present at the top of the nozzle with the needle and slowly release the labeled 35 ml PA Ultracrimp tube (see **Note 29**).

7. Insert a plug into the nozzle and cover with a cap.

8. Place the labeled 35 ml PA Ultracrimp tube into a tube rack and seal using the Ultracrimp/Clearcrimp sealing tool.

9. Counterbalance the sealed and labeled 35 ml PA Ultracrimp tube (*see* **Note 30**).

10. Continue with the ultracentrifugation.

3.5.4 Ultracentrifugation   The ultracentrifugation is completed within 2 h and 30 min.

1. Precool a Sorvall WX 80 ultracentrifuge to 15 °C and set acceleration to "7" and deceleration to "4" [this can be done while preparing the discontinuous iodixanol density gradient (s)].

2. Insert the counterbalanced, sealed and labeled 35 ml PA Ultra-crimp tubes into a T-865 fixed angle titanium rotor and centri-fuge for 2 h and 15 min at 365,929 × g/15 °C (*see* **Note 31**).

3. Continue with collecting the AAV vectors.

3.5.5 Collecting the AAV Vectors   The collection of eight individual AAV vector preparations requires between 20 and 30 min.

1. Connect an 18-G, 1.2 × 40 mm injection needle to a labeled 10 ml syringe.

2. Remove the centrifuged 35 ml PA Ultracrimp tube from the ultracentrifugation rotor and fix it into a tube holder.

3. Place a waste bin below the 35 ml PA Ultracrimp tube.

4. Wipe at the injection site the 35 ml PA Ultracrimp tube with a 70% EtOH soaked single-use paper-wipe and dry.

5. Horizontally insert the 18-G, 1.2 × 40 mm injection needle connected to the labeled 10 ml syringe just above the 60% OptiPrep (Iodixanol) solution (*see* **Note 32**).

6. Place the bevel tip of the 18-G, 1.2 × 40 mm injection needle upward and retract the plunger slightly in order to reduce any overpressure inside the 35 ml PA Ultracrimp tube generated during the puncture.

7. Insert a second 18-G, 1.2 × 40 mm injection needle (without a syringe connected) at a 90° angle at the dome of the 35 ml PA Ultracrimp tube.

8. Retract the plunger and slowly collect part of the 40% OptiPrep (Iodixanol) solution (*see* **Note 33**).

9. Stop collecting when the 25% OptiPrep (Iodixanol) solution (containing most of the cellular proteins and empty AAV vector capsids) is still a few millimeters away from the tip of the 18-G, 1.2 × 40 mm injection needle (*see* **Note 34**).

10. Retract the 18-G, 1.2 × 40 mm injection needle and release the 35 ml PA Ultracrimp tube from the tube holder into the waste bin.

11. Place the 18-G, 1.2 × 40 mm injection needle and its connected syringe upright (needle to the top) into a holder and let the sheath of the needle passively drop down along the 18-G, 1.2 × 40 mm injection needle, keeping your fingers all time away from the 18-G, 1.2 × 40 mm injection needle (*see* **Note 35**).

12. Store the AAV vector containing 40% OptiPrep (Iodixanol) solution until further processing at 4 °C (*see* **Note 36**).

13. Continue with diafiltration.

3.6 Diafiltration   Diafiltration requires about 2 to 3 h for eight individual AAV vector preparations.

1. Add 20 ml 1× PBS-MK, pH 7.4 solution to a labeled Vivaspin® 20 ultrafiltration device (*see* **Note 37**).

2. Centrifuge for 2 min at 3000 × g/20 °C in a swing-out rotor.

3. Aspirate the filtrate using a sterile Pasteur glass pipette connected to a vacuum aspiration system.

4. Add about 13 ml 1× PBS-MK, pH 7.4 solution to the rinsed Vivaspin® 20 ultrafiltration device.

5. Disconnect the needle from the 10 ml syringe containing the collected AAV vector/40% OptiPrep (Iodixanol) solution and dispose into a disposable sharps waste bin.

6. Release the collected AAV vector/40% OptiPrep (Iodixanol) solution into the Vivaspin® 20 ultrafiltration device (*see* **Note 38**).

7. Dispose of the 10 ml syringe and close the Vivaspin® 20 ultrafiltration device.

8. Centrifuge for 10 min at 3000 × g/20 °C in a swing-out rotor.

9. Verify the progress of the diafiltration and continue the centrifugation until about 0.5 to 1 ml of the retentate remain (*see* **Note 39**).

10. Replenish with 1× PBS-MK, pH 7.4 solution to 20 ml, close the Vivaspin® 20 ultrafiltration device and mark the completion of the first diafiltration step by a dash on the cap.

11. Seal the hole in the center of the cap of the Vivaspin® 20 ultrafiltration device with an adhesive tape and invert the Vivaspin® 20 ultrafiltration device four times (*see* **Note 40**).

12. Remove the adhesive tape and continue the centrifugation until the retentate reaches again a volume of 0.5 to 1 ml.

13. Repeat the procedure as described after the first diafiltration for another two rounds (a total of three diafiltrations).

14. After the third (final) diafiltration step, adjust the volume of the retentate with 1× PBS-MK, pH 7.4 solution to about 1.2 ml.

15. Pipet the retentate up-and-down about 10 times using a 1000 μl micropipette without introducing air bubbles.

16. Transfer the retentate into a labeled 50 ml reaction tube and store at 4 °C until further processing (*see* **Note 41**).

17. Continue with physical titer quantification.

3.7 Physical Titer Quantification

Physical titer quantification requires about 1.5 h for eight individual AAV vector preparations (including calculations). Additional time may be required in case AAV vector preparations need to be remeasured (in case physical titers are outside the range of the assay).

3.7.1 Measuring Genomic AAV Vector DNA Concentration

1. For each AAV vector to be quantified, prepare a pair of 0.5 ml Qubit assay tubes and label them on the lid (not the side) with the name of the AAV vector.

2. Additionally label one of the two 0.5 qubit assay tubes with a "− "(minus, untreated, nondenatured) and the other with a "+" (plus, heat-denatured).

3. Prepare another pair of 0.5 ml qubit assay tubes to be used for the DNA standards and label one on the lid with "1 "and the other with "2."

4. Prepare a 1:10 dilution of the AAV vector preparation in a 1.5 ml reaction tube (45 μl 1× PBS-MK, pH 7.4 solution + 5 μl undiluted AAV vector preparation), mix by vortexing and briefly spin down (*see* **Note 42**).

5. Add 5 μl of a 1:10 diluted AAV vector preparation to each 0.5 ml Qubit assay tube labeled "−" and "+".

6. Incubate the 0.5 ml Qubit assay tube labeled "+" 5 to 6 min at 95 °C to heat-denature the capsid an liberate the AAV vector genomes.

7. In the meantime prepare for each 0.5 ml Qubit assay tube 200 μl of the Qubit™ working solution according to the instructions provided with the Qubit™ dsDNA HS assay kit and vigorously vortex (dilute the Qubit™ dsDNA HS reagent 1:200 in Qubit™ dsDNA HS buffer; prepare a master mix for all 0.5 ml Qubit assay tubes, including standards) (*see* **Note 43**).

8. After the incubation at 95 °C let the "+" 0.5 ml Qubit assay tubes 2 to 3 min equilibrate to RT and briefly spin down.

9. Add 190 μl Qubit™ working solution to each of the two 0.5 ml Qubit assay tubes prepared for the standards.

10. Add 195 μl Qubit™ working solution to each of the 0.5 ml Qubit assay tubes containing the untreated ("−", nondenatured) and heat-denatured ("+") AAV vector samples (final volume per 0.5 ml Qubit assay tube: 200 μl) (*see* **Note 44**).

11. Add 10 μl of the Qubit™ dsDNA HS Standard #1 [0 ng dsDNA/μl] to the 0.5 ml Qubit assay tube labeled "1" (final volume: 200 μl).

12. Add 10 μl of the Qubit™ dsDNA HS Standard #2 [10 ng dsDNA/μl] to the 0.5 ml Qubit assay tube labeled "2" (final volume: 200 μl).

13. Thoroughly mix each 0.5 ml Qubit assay tube by vortexing followed by a brief spin down.

14. Incubate for between 2 and 5 min at RT before proceeding to the measurements.

15. Switch on the Qubit™ 3.0 (or higher) fluorometer.

16. Select "dsDNA" assay, then "High sensitivity."

17. Select "Read standards."

18. Insert the 0.5 ml Qubit assay tube "1" into the sample chamber, close the lid and select "Read standard" (*see* **Note 45**).

19. Insert the 0.5 ml Qubit assay tube "2" into the sample chamber, close the lid and select "Read standard" (*see* **Note 45**).

20. Select "Run samples," enter the original sample volume (5 μl) and select the desired output sample units (ng/ml).

21. For each AAV vector preparation, insert first the 0.5 ml Qubit assay tube labeled "−" into the sample chamber, close the lid, select "Read tube" and note the Qubit tube concentration value in ng/ml (*see* **Note 46**).

22. Continue with the 0.5 ml Qubit assay tube labeled "+".

23. Repeat measurements for each AAV vector preparation (*see* **Note 47**).

24. Continue with the calculations.

3.7.2 Calculations

Calculations are completed within a few minutes when using a template.

1. Calculate the nonencapsidated dsDNA stock concentration (extraviral, background) in ng/ml according to the following Eq. (A) (*see* **Note 48**).

$$\left.\begin{array}{c} \text{measured} \\ \text{non}-\text{denatured} \\ \text{sample} \\ \text{in\,ng/ml} \end{array}\right\} \times \left(\frac{200\,\mu\text{l\,total\,volume}}{\text{sample\,volume\,in}\,\mu\text{l}}\right) = \text{extraviral\,dsDNA\,stock\,concentration\,in\,ng/ml}$$

2. Calculate the total dsDNA stock concentration [nonencapsidated (extraviral, background) + encapsidated (intraviral)] in ng/ml according to the following Eq. (B).

$$\left(\begin{array}{c} \text{measured} \\ \text{heat\,denatured} \\ \text{sample} \\ \text{in\,ng/ml} \end{array}\right) \times \left(\frac{200\;\mu\text{l\,total\,volume}}{\text{sample\,volume\,in}\;\mu\text{l}}\right)$$

$$= \text{extraviral} + \text{intraviral\,dsDNA\,stock\,concentration\,in\,ng/ml} \qquad (B)$$

3. Calculate the encapsidated (intraviral) dsDNA concentration in ng/ml according to the following Eq. (C).

$$(B) - (A) = \text{intraviral\,dsDNA\,stock\,concentration\,in\,ng/ml}$$

4. Calculate the number of vector genomes (vg)/ng according to the following Eq. (D) (*see* **Note 49**).

$$\frac{6.023 \times 10^{23}/\text{mol}}{[(\text{bp\,dsDNA\,genome} \times 607.4\;\text{g/mol}) + 159.9\;\text{g/mol}]} \times 10^{9}$$

$$= \text{number\,of\,vector\,genomes/ng} \qquad (D)$$

5. Calculate the number of vg/ml (physical titer) according to the following Eq. (E).

$$(C) \times (D) = \text{number\,of\,vector\,genomes/ml} \qquad (E)$$

6. Multiply the calculated physical titer of Eq. (E) by the dilution factor applied ($10\times = 1{:}10$ dilution) and in case of ssAAV vectors by a factor of 2 (*see* **Note 48**).

3.8   Quality Controls

Each lot of an AAV vector preparation may be analyzed by a variety of assays (for example purity at the level of proteins and nucleic acids). Currently, there is no methodology available to routinely verify and confirm the identity of the capsid of an AAV vector preparation without using large sample volumes.

We have established a quality control that we named "Identity check." While the main purpose of the identity check is to verify and confirm the identity of the packaged AAV vector genomes (and the presence of different genomes in case of AAV vector libraries), we have discovered that certain AAV vector genomes (or parts thereof) contained within their AAV vector plasmids will not be rescued, replicated and packaged properly into AAV capsids, regardless of the particular AAV capsid. For example, if a specific nucleotide sequence of tandem dimer (td) Tomato is present on the AAV vector plasmid, one of the two dTomato copies of tdTomato will inevitably be deleted from the packaged AAV vector genome, leaving behind an AAV vector genome that encodes dTomato instead of tdTomato. While the deletion of one dTomato copy from tdTomato will not be notice for many applications, the possible loss of functional elements (for example of sensors) may be problematic.

3.8.1 *Identity Check*

1. add 1 µl of an AAV vector preparation adjusted to a physical titer of 1.0×10^{12}.

2. vg/ml to a Taq DNA polymerase-based PCR containing two primers generating an amplicon that encompasses most of the expression cassette (*see* **Note 50**).

3. Run the PCR (*see* **Note 51**).

4. Analyze the amplicon on an agarose gel (*see* **Note 52**).

5. Submit the remaining of the PCR to a commercial company for purification and subsequent Sanger DNA-sequencing.

6. Analyze the chromatograms (*see* **Note 53**).

4   Notes

1. The PEI "Max" solution remains stable for years at −20 °C. Thaw and store aliquots in use at 4 °C (PEI "Max" solution may be stored for several weeks at 4 °C).

2. The quality of the HEK 293T cells—besides the quality of the plasmid DNA to be transfected—determines the quality (integrity of the AAV vector genome, transduction efficiency, absence of contaminations) and quantity (physical titers) of AAV vector preparations. It is therefore fundamental to maintain HEK 293T in excellent condition. We prepare working cell banks of low passage HEK 293T grown to subconfluency and stored in the vapor phase of liquid nitrogen. Each vial contains 1 ml HEK 293T cells to reach confluency the next day on a 15-cm cell culture plate. We monitor HEK 293T periodically during passaging for potential mycoplasma contaminations using Mycoplasmacheck. CO_2 incubators are periodically heat-sterilized and serviced.

3. We use for all micropipetting (0.5 to 1000 µl) aerosol tight sterile pipette filter tips.

4. Check morphology and number of detached HEK 293T cells the next day. A rounded morphology and an elevated number

of detached HEK 293T cells (a few detached HEK 293T cells are always present) are indicative of a negative situation. HEK 293T cells should be discarded and replaced. Passage HEK 293T cells at least three times before PEI-mediated split transfection.

5. Depending on the growth rate and the density of the HEK 293T cells immediately before passaging, we generally passage HEK 293T at a ratio of 1:6 to 1:8 to allow about 48 h of continuous growth or 1:8 to 1:10 for about 72 h of continuous growth. This will result in almost confluent HEK 293T cells at the end of the incubation.

6. Place the tip of the 25 ml serological pipette at the interior rim of the 15-cm cell culture plate a few millimeters away from the bottom and release the 1× PBS, pH 7.4 with the pipetting aid at such as speed, that the HEK 293T cells do not detach.

7. Never expel the HEK 293T cell suspension entirely out of the 25 ml serological pipette in order to prevent the generation of excess aerosols and spills. Keep the 15-cm culture plate slightly tilted and move the tip of the 25 ml serological pipette stepwise either clockwise or counter-clockwise 360° along the wall of the 15-cm cell culture plate while simultaneously expelling and aspirating the HEK 293T cell suspension. Proceed to the center of the 15-cm cell culture plate. Repeat until all HEK 293T are detached.

8. We use empty DMEM bottles to add the required volume of $DMEM_c$ kept at RT.

9. We reuse the 15-cm cell culture plates from the previously passaged HEK 293T cells.

10. The quality of the plasmid DNA—besides that of the HEK 293T cells (see **Note 2**) —determines transfection efficiency and hence the quality (integrity of the AAV vector genome, transduction efficiency, absence of contaminations) and quantity (physical titers) of AAV vector preparations. It is therefore important to construct, amplify, and characterize high-quality plasmid DNA for subsequent AAV vector production. While the human adenovirus (AV) helper plasmid (providing the AV serotype 5 helper functions E2A, VA, E4 in trans in combination with the AV serotype 5 E1A and E1B helper functions stably integrated into the genome of HEK 293T cells) and the various AAV helper plasmids (generally providing the AAV serotype 2 Rep proteins and the Cap proteins of the desired capsid/serotype) are not critical with respect to stability during amplification, the AAV vector plasmids (providing the to-be packaged AAV vector genome) are prone to genetic instability (deletions) when bacteria are (re)transformed and subsequently amplified. The instability is attributed to the inverted

terminal repeats [ITRs, in most cases originating from wild-type (wt) AAV serotype 2], which are the only noncoding cis elements retained from the wt AAV genome and flanking the expression cassette at its 5′- and 3′-ends. Furthermore, repetitive sequences within the expression cassette are as well prone to genetic instability (deletions). Therefore, suitable bacterial strains and incubation conditions need to be determined empirically for each AAV vector plasmid if genetic instability occurs during amplification. We generally use MDS™42 Low-Mut ΔrecA bacteria in combination with Terrific Broth (TB) medium and vary incubation time and temperature to determine the most suitable amplification conditions for each AAV vector plasmid. If MDS™42 LowMut ΔrecA bacteria are not suitable to propagate a specific AAV vector plasmid, we test other commercially available strains (NEB® Stable Competent E. coli, Thermo Fisher Scientific Stbl3™, Agilent Technologies SURE 2). We use anion exchange columns followed by isopropanol precipitation to prepare plasmid DNA dissolved over night at 4 °C in sterile ultrapure water ready for PEI-mediated transfection (certain buffers—such as TE buffer—lower PEI-mediated transfection efficiency). Integrity of ITRs and expression cassettes are verified by SmaI (or XmaI) endonuclease restriction analysis in combination with Sanger DNA-sequencing and plasmid DNA is stored at −80 °C in cryovials (working aliquots are stored at −20 °C).

11. We generally use five 15-cm cell culture plates per AAV vector preparation. Certain AAV serotypes/capsids usually result in lower physical titers than others (for example AAV serotypes 1, 2, 6). In those cases, we prepare ten 15-cm cell culture plates per AAV vector preparation. If higher physical titers are needed, prepare accordingly more 15-cm cell culture plates.

12. We use a molar ratio of 1:1:1 for the AAV vector plasmid, AAV helper plasmid and AV helper plasmid. Our AV helper plasmid [23] has a size of 14,084 bp and we have determined the optimal amount for efficient AAV vector production per 15-cm cell culture plate to be 24 µg (120 µg for five 15-cm cell culture plates). In the event a hybrid AAV/AV helper plasmid is used, we multiply 24 µg by the size of the hybrid AAV/AV helper plasmid (in bp) divided by 14,084 bp in order to determine the amount of the hybrid AAV/AV helper plasmid required (the molar ratio of the AAV/AV helper plasmid and the AAV vector plasmid remains at 1:1). Each 15-cm cell culture plate to be transfected requires a final volume of 3 ml containing the plasmid DNAs and 150 mM NaCl solution (90 µl of a 5 M NaCl solution, 450 µl for five 15-cm cell culture plates) dissolved in sterile ultrapure water and another 3 ml containing the PEI solution and 150 mM NaCl solution (90 µl

of a 5 M NaCl solution, 450 µl for five 15-cm cell culture plates) dissolved in sterile ultrapure water. The volume of the PEI solution depends on the total amount of plasmid DNA. We use a weight ratio of PEI to total plasmid DNA of 3:1 (which results in a nitrogen to phosphate ratio of 23:1).

13. To efficiently process the passaging of HEK 293T cells during the PEI–plasmid DNA solution incubation, we have established a standard protocol suitable for most users and an advanced protocol for experienced users with respect to passaging HEK 293T cells in a timely fashion. The difference between the two protocols is that the standard protocol requires about 1 h 15 min to process four different AAV vector preparations (each composed of five 15-cm cell culture plates), while the advanced protocol requires about 1 h for the same throughput. The standard protocol is based on the following order and timing: (a) passage ten 15-cm cell culture plates up to (and including) the point when the TrypLE™ Express Enzyme solution is added, (b) pour the PEI solution of the first AAV vector preparation into the corresponding plasmid DNA solution, vigorously vortex for a few seconds, incubate at RT and start a timer ($t = 0$ min), (c) continue incubating the ten 15-cm cell culture plates supplemented with the TrypLE™ Express Enzyme solution at 37 °C/5% pCO_2, (d) during the 5 min TrypLE™ Express Enzyme incubation at 37 °C/5% pCO_2 prepare about 220 ml RT $DMEM_c$, (e) at $t = 3$, pour the PEI solution of the second AAV vector preparation into the corresponding plasmid DNA solution, vigorously vortex for a few seconds, and incubate at RT, (f) at $t = 5$, finalize the passaging of the HEK 293T cells (reuse the ten 15-cm cell culture plates; we use a final volume per 15-cm cell culture plate of about 25 to 27 ml, depending on the overall volume; the passaging of the HEK 293T is usually completed at $t = 17$ to 18 and HEK 293T cells are kept in the biosafety cabinet class II at RT for the remaining few minutes until the PEI–plasmid DNA solution is added), (g) vigorously vortex for a few seconds the first PEI–plasmid DNA solution at $t = 19$ and prepare a 10 ml serological pipet, (h) at $t = 20$ add about 6 ml of the first PEI–plasmid DNA solution dropwise to the first 15-cm cell culture plate and process the remaining four identically, (i) move all five 15-cm cell culture plates horizontally forward and backward as well as sideward several times in order to distribute the HEK 293T cells as evenly as possible, (j) incubate at 37 °C/5% pCO_2, (k) at $t = 22$ vigorously vortex for a few seconds the second PEI–plasmid DNA solution and prepare a new 10 ml serological pipet, (l) at $t = 23$ transfect the remaining five 15-cm cell culture plates, (m) incubate at 37 °C/5% pCO_2. The advanced protocol is based on the

following order and timing: (a) passage twenty 15-cm cell culture plates up to (and including the first 4 min) of the TrypLE™ Express Enzyme incubation at 37 °C/5% pCO_2 (start a timer when TrypLE™ Express Enzyme solution was added to the last 15-cm cell culture plate, $t = 0$), (b) during the 4 min TrypLE™ Express Enzyme incubation at 37 °C/5% pCO_2 prepare about 450 ml RT $DMEM_c$, (c) at about $t = 4$ transfer the twenty 15-cm cell culture plates back to the bio-safety cabinet class II and gently tap the 15-cm cell culture plates until most of the HEK 293T cells detach, (d) at $t = 5$ pour the PEI solution of the first AAV vector preparation into the corresponding plasmid DNA solution, vigorously vortex for a few seconds, and incubate at RT, (e) add about 8 ml RT $DMEM_c$ to each of the twenty 15-cm cell culture plates, (f) at $t = 8$ pour the PEI solution of the second AAV vector preparation into the corresponding plasmid DNA solution, vigorously vortex for a few seconds, and incubate at RT, (g) continue with passaging HEK 293T cells (about eight to nine 15-cm cell culture plates can be processed at this stage), (h) at $t = 14$ pour the PEI solution of the third AAV vector preparation into the corresponding plasmid DNA solution, vigorously vortex for a few seconds, and incubate at RT, (i) continue passaging HEK 293T cells (about three to four 15-cm cell culture plates can be processed at this stage), (j) at $t = 17$ pour the PEI solution of the fourth (last) AAV vector preparation into the corresponding plasmid DNA solution, vigorously vortex for a few seconds, and incubate at RT, (k) transfer the remaining HEK 293T cells into the 450 ml RT DMEMc, (l) mix the HEK 293T cell suspension by pipetting several times up-and-down and evenly distribute about 25 to 27 ml (depending on the overall volume) of the HEK 293T cell suspension into each of the first ten reused 15-cm cell culture plates, (m) cap the bottle containing the remaining HEK 293T cell suspension, (n) continue with the transfection at $t = 25$ and $t = 28$ for the first two AAV vectors as described for the standard proto-col, (o) mix the HEK 293T cells suspension by pipetting several times up-and-down and evenly distribute about 25 to 27 ml (depending on the overall volume) of the HEK 293T cell suspension into each of the remaining ten reused 15-cm cell culture plates, (p) continue with the transfection at $t = 34$ and $t = 37$ for the last two AAV vectors as described for the standard protocol.

14. In case a fluorescent protein is encoded and expressed by the AAV vector plasmid, a transfection efficiency of 80% to 90% (and higher, depending on the transcriptional activity of the promoter in HEK 293T cells) should be detectable, while very few HEK 293T cells are detached at this early stage of AAV vector production.

15. About 50% to 70% of the HEK 293T cells are detached at this stage and in case a fluorescent protein is encoded and expressed by the AAV vector plasmid virtually all HEK 293T will be positive for the fluorescent protein.

16. HEK 293T cells should detach readily. Avoid spills (do not expel the HEK 293T cell suspension completely). Wear protective sleeves. When processing multiple AAV vector preparations in parallel store HEK 293T cell suspension of each AAV vector preparation at 4 °C until further processing.

17. Collected HEK 293T cells and resuspended PEG 8000 pellet can be stored at least for several months at −20 °C.

18. Higher g-forces may be applied.

19. The PEG 8000 precipitated supernatant may be aspirated instead.

20. A brownish PEG 8000 pellet at the conical bottom of the 500 ml centrifuge bottle should be easily detectable. Furthermore, precipitate usually forms along the conical wall of the 500 ml centrifuge bottle (usually on opposite sides, with intermediate precipitation-free areas). It requires multiple pipetting steps to detach and resuspend the PEG 8000 pellet.

21. Add only 3.5 ml AAV resuspension buffer, pH 8.5 if the HEK 293T cell pellet originates from then 15-cm cell culture plates. It is important not to exceed a final volume of about 5 to 5.5 ml.

22. The Precellys® Evolution homogenizer allows the simultaneous processing of twelve 7 ml soft tissue homogenizing CK14 tubes. If only a few 7 ml soft tissue homogenizing CK14 tubes need to be processed, a Minilys Personal homogenizer may be used instead.

23. The homogenization does not negatively affect the transduction efficiency of the AAV vectors when compared to the more frequently applied multiple freeze–thaw cycles.

24. Instead of evenly distributing the DENARASE-treated crude cell lysate of a 7 ml soft tissue homogenizing CK14 tube into the corresponding four labeled 2.0 ml reaction tubes, the lysate may be processed directly in its 7 ml soft tissue homogenizing CK14 tube: Centrifuge for 10 min at >7100 × g/4 °C in a swing-out rotor and transfer the cleared lysate (supernatant) into the correspondingly labeled 35 ml PA Ultracrimp tube as described. Add the resuspended pellet of the PEG 8000 precipitated supernatant and 1.5 ml 1 M NaCl and continue with the preparation of the discontinuous iodixanol density gradients.

25. Do not aspirate the pellet of the crude cell lysate. Use the needle connected to its syringe for all subsequent transfers

and place the needle connected to its syringe into the 35 ml PA Ultracrimp tube when not in use.

26. Insert the needle into the labeled 35 ml PA Ultracrimp tube (but still outside of the reddish AAV vector containing combined cleared lysate–supernatant solution). Gently press the plunger of the syringe until no more air and bubbles are released at the tip of the needle. Then carefully insert the needle close to (but without touching) the bottom of the labeled 35 ml PA Ultracrimp tube before manually and regularly (about 0.3 to 0.5 ml per s) releasing the different OptiPrep (Iodixanol) solution sequentially. The reddish AAV vector containing combined cleared lysate–supernatant solution will be lifted by the 9 ml 15% OptiPrep (Iodixanol), 1 M NaCl solution, forming a clear cushion. Of the 8 ml 25% OptiPrep (Iodixanol) solution aspirate about 7.5 ml without aspirating air and underly the 9 ml 15% OptiPrep (Iodixanol), 1 M NaCl solution with about 5 ml. Transfer the remaining 25% OptiPrep (Iodixanol) solution back to its 15 ml reaction tube. Proceed with the remaining 40% and 60% OptiPrep (Iodixanol).

27. The boundary between the 40% OptiPrep (Iodixanol) and 25% OptiPrep (Iodixanol) solution is visible as an opaque rim. We do not use coloring agents that help to visually discriminate between the individual OptiPrep (Iodixanol) fractions.

28. Any air bubbles present within the AAV vector containing combined cleared lysate–supernatant solution need to reach the top of the nozzle. Take care not to spill over.

29. The top of the AAV vector containing combined cleared lysate–supernatant solution will reach its final position just below the nozzle.

30. If only a single sealed and labeled 35 ml PA Ultracrimp tube was prepared (or an uneven number), identically prepare a 35 ml PA Ultracrimp tube using AAV resuspension buffer, pH 8.5 instead of the combined cleared lysate–supernatant solution to counterbalance. If an even number of multiple sealed and labeled 35 ml PA Ultracrimp tubes were prepared, identify a combination of 35 ml PA Ultracrimp tubes and rotor caps that will result in a difference in weight below 15 mg.

31. Other ultracentrifuges and rotors may be used [calculate g-forces and compare clearing (k)-factors].

32. It is possible to collect the AAV vectors from the bottom of the 35 ml PA Ultracrimp tube.

33. In case the plunger cannot be retracted with ease (because no or very little air enters the 35 ml PA Ultracrimp tube through the second 18-G, 1.2 × 40 mm injection needle inserted at the

dome of the 35 ml PA Ultracrimp tube), insert a third 18-G, 1.2 × 40 mm injection needle at the dome of the 35 ml PA Ultracrimp tube.

34. The bevel tip of the 18-G, 1.2 × 40 mm injection needle may be gently turned toward the bottom of the 35 ml PA Ultracrimp tube when approaching the 25% OptiPrep (Iodixanol) solution. The collected volume of the initial 8 ml 40% OptiPrep (Iodixanol) solution is between 5 and 6 ml.

35. Do not recap the needle by firmly covering the 18-G, 1.2 × 40 mm injection needle with its sheath.

36. The AAV vector containing 40% OptiPrep (Iodixanol) solution may be kept several days at 4 °C without affecting the transduction efficiency of the AAV vectors.

37. The Vivaspin® 20 ultrafiltration device may be sterilized by a rinse with 20 ml 70% EtOH solution followed by a 1× PBS-MK, pH 7.4 solution wash. Label the cap and the side of the filtration unit.

38. Insert the tip of the 10 ml syringe into the 13 ml 1× PBS-MK, pH 7.4 solution (without touching the wall of the Vivaspin® 20 ultrafiltration device with the body of the syringe) and when almost all AAV vector containing 40% OptiPrep (Iodixanol) solution is released, suck back about 8 ml of the solution and release again to rinse the 10 ml syringe.

39. After 10 min of centrifugation the retentate reaches a volume of about 10 ml.

40. The residual 40% OptiPrep (Iodixanol) solution will be visible as viscous solution that gradually disappears when inverting the Vivaspin® 20 ultrafiltration device four times.

41. The AAV vector containing retentate may be kept for several days at 4 °C without affecting the transduction efficiency of the AAV vectors.

42. Depending on the expected physical titer, 5 µl of the undiluted AAV vector preparation may be used directly instead of a 1:10 dilution. Furthermore, the sample volume of the Qubit™ dsDNA HS assay kit can be adjusted between 1 and 20 µl, providing additional flexibility.

43. It is important to use reactions tubes made of plastic (not glass). We use 2.0 ml reaction tubes to prepare the Qubit™ working solution.

44. Use a new filter tip for each 0.5 ml Qubit assay tube.

45. Record the RFU value displayed to compare the variability between different measurements over time.

46. Do not use the original calculated sample concentration displayed for further calculations. In case the concentration falls

outside of the range of the Qubit™ dsDNA HS assay, either further dilute the AAV vector sample (if too high) or add more AAV vector sample (if too low) and remeasure.

47. We remeasure the same samples again about 30 min after their first measurement to confirm the initial values. Measured values tend to slightly increase over time, whereas a decrease sometimes occurs as well. Nevertheless, calculated physical titers do not change significantly over time.

48. Use the Qubit tube concentration value (in ng/ml) as the measured nondenatured sample value (not the original calculated sample concentration). Single-stranded (ss) and self-complementary (sc) AAV vectors contain linear genomes composed of a ssDNA molecule. The ssDNA molecules of ssAAV and scAAV vector preparations are present at either polarity (+ and − ssDNA) and at equal frequency (50:50). Therefore, heat-denatured ssAAV vector preparations will liberate ssDNA genomes of either polarity at equal frequency that preferentially hybridize intermolecularly to form dsDNA molecules (two ssAAV vector genomes are required to form one dsDNA molecule). On the other hand, heat-denatured scAAV vector preparations will liberate ssDNA genomes of either polarity at equal frequency that predominantly hybridize intramolecularly, because the ssDNA genomes flanked at both ends by inverted-terminal repeats (ITRs) are composed of two halves that are complementary to each other and that are separated in the middle of the genome by a noncleavable ITR (a single scAAV vector genome is required to form one dsDNA molecule).

49. For ssAAV vectors, the dsDNA genome size in bp equals the total genome length of the ssAAV vector (including the ITRs). For scAAV vectors, the dsDNA genome size in bp is calculated by doubling the total genome length between (and including) the ITRs, subtracting the length of the noncleavable ITR present in the middle of the genome (around 106 nucleotides) and finally dividing by a factor of 2.

50. It may be necessary to test other DNA-dependent DNA polymerases, to add additives to the PCR or to even subdivide an amplicon into two or more shorter amplicons (using additional primers) in case the PCR amplification is not successful. For example, PCR encompassing internal ribosomal entry sites (IRES) usually do not work.

51. Use a 50 μl final volume including 5 μl 10× Taq DNA polymerase buffer, 2.5 μl 4 mM dNTPs, 1 μl 10 μM forward primer, 1 μl 10 μM reverse primer, 1 μl 1.0×10^{12} vg/ml AAV vector as the template DNA, 0.25 U Taq DNA polymerase. Run an initial denaturation for 1 min at 95 °C, followed by

35 cycles (denaturation 20 s at 95 °C, annealing 20 s between 50 and 60 °C, depending on the primer binding properties, and extension 1 min at 68 °C per kb) and a final extension for 5 min at 68 °C.

52. Use a 0.8–1.2% TBE agarose gel, depending on the size of the calculated amplicon. Besides the calculated amplicon, it may be that additional bands are detectable that are either shorter or longer compared to the calculated amplicon.

53. Align in silico the Sanger DNA-sequencing results (chromatograms) and the amplicon against the entire AAV vector genome. The chromatograms usually reach a length between 500 and 700 nucleotides, which is slightly shorter when using plasmid DNA as the template DNA.

5 Intraspinal Injections

5.1 Materials

5.1.1 Equipment

1. Micropipette puller (Zeitz, DMZ-Universal-Electrode Puller).
2. Surgical microscope (Zeiss, OPMI Pico).
3. Anesthesia unit (Weinmann, Oxymat3 oxygen concentrator).
4. Heat mat.
5. Electric shaver.
6. Small animal stereotaxic frame (Kopf).
7. Neurostar StereoDrive.
8. Syringe pump (Harvard Apparatus, PHD Ultra syringe pump with nanomite).
9. Removable needle compression fitting 1 mm (Hamilton).
10. Dentistry drilling apparatus.

5.1.2 Surgical Tools

1. Glass microliter syringe (Hamilton 701 RN 10 μl).
2. Spherical cutter, 0.5 mm.
3. Cunningham mouse spinal adaptor for stereotaxic frame (Harvard Apparatus, Model 51690).
4. Dumont #2 laminectomy forceps.
5. Scalpel handle.
6. Extra fine Bonn scissors.
7. Adson forceps.
8. Friedman-Pearson rongeurs.
9. Olsen-Hegar needle holders, serrated.

1. Ethanol, 70%.
2. Iodine solution.
3. Anesthetics (e.g., isoflurane).
4. Activated carbon scavenger for anesthetic gases (Cardiff, Aldasorber).
5. Ophthalmic ointment.
6. Analgesics (e.g., buprenorphine).
7. Saline, sterile, 0.9%.
8. Deionized water, sterile.
9. Thin-wall glass capillaries, 1 mm outside diameter (World Precision Instruments).
10. Needles, 26G beveled.
11. Syringes, 1 and 5 ml.
12. Scalpel blades, sterile.
13. Surgery drapes.
14. Facial tissues.
15. Cotton wool.
16. Cotton swabs.
17. Collagen compress.
18. Surgical sutures, absorbable and nonabsorbable.
19. Parafilm.

6 Methods

6.1 Intraspinal Injections

The protocol for intraspinal injections presented here is based on Haenraets et al. In the first part we describe the necessary preparations for the surgery, in the second the exposure of the spinal cord and finally, the injections and completion of the surgery.

6.1.1 Preparation for the Surgery

Before starting the surgery, the virus solution needs to be diluted in sterile saline to the desired physical titer. The optimal physical titer should be determined for each experiment (see **Note 1**). A good physical titer to first try is about 3×10^{12} vg/ml with three times 300 nl injected. The diluted virus should be kept on ice until injected and repeated freeze–thaw cycles should be avoided, as they lead to reduced functional titers.

The working area and all equipment should be cleaned and disinfected. Surgical tools should be autoclaved and placed ready for the procedure. A 1 ml syringe is filled with appropriately diluted analgesics and 5 ml syringes are filled with sterile water and saline.

The micropipettes are prepared from the glass capillaries with a micropipette puller. They should have a long (about 5.5 mm in our hands), shallow shank. The tip is then clipped to create an opening with an inner diameter of about 25–35 μm (Fig. 1). The micropipettes should be stored in a closed container to minimize the risk of clogging due to dust particles.

6.1.2 *Surgical Exposure of the Spinal Cord*

1. Anesthesia is induced (at about 5% isoflurane). Throughout the surgery, the breathing rate (should be steady) and the absence of reflexes are monitored and the anesthesia is adapted accordingly (usually between 1.5% and 2% isoflurane). In addition, the heat mat on which the mouse is placed within the frame is adjusted to keep the animal warm. The anesthetic is injected subcutaneously and ophthalmic ointment is applied to the eyes to keep the corneas moist. The back of the animal is shaved, remaining hair is removed using ethanol-soaked tissues and the skin is disinfected using iodine-soaked cotton wool.

2. When the iodine has dried, a cut of about 1.5 cm length is made into the skin along the middle of the back above the vertebrae to be exposed. For an injection into the lumbar spinal cord, this is at the point where the back arches highest. The most caudal ribs can be palpated at this level. The skin is lifted with forceps and detached from the underlying tissue by inserting the closed scissors into the cut and opening them horizontally below the skin. The exposed tissue is kept moist throughout the surgery by application of saline.

3. The target vertebra is identified (Fig. 2). The skin is pulled back toward the tail so that the iliac crests and the most caudal pair of visible intertransverse ligaments can be seen. This pair of ligaments is attached to the spinous process of the L6 vertebra. Counting backward in the rostral direction, the T13 vertebra, below which the spinal cord level L4 is located, can be identified (*see* **Note 2**). The following can serve as additional anatomical landmarks. The tendons that run along the sides of the vertebral column are most white just caudal to the T13 vertebra. The most caudal rib pair is located just rostral to the T13 vertebra.

4. The animal is placed onto a cushion made from rolled up tissues to elevate the spinal column to the height at which the spinal adaptors are fixed within the frame. The spinal column is held in place using the Adson forceps and the target vertebra is clamped between the spinal adaptors.

5. A cut is made on both sides just medial to the tendons visible along the side of the spinal column and additional cuts are made perpendicular to these rostral and caudal to the target

vertebra. The paraspinous muscle in between the cuts is removed using rongeurs. The dorsal blood vessel will now be visible in the intervertebral space marking the midline of the spinal cord.

6. Holding the drill at a 45° angle, a hole is made into one side of the vertebra (for unilateral injections) at an equal distance from the intervertebral spaces. No pressure should be exerted to avoid damage of the underlying spinal cord tissue. Remaining bone fragments can be removed using a 26G needle and forceps. For bilateral injections, an additional hole can be made into the contralateral side.

7. Using a 26G needle the dura is perforated in the intervertebral spaces and below the hole in the vertebra. If the dura has remained intact so far, some cerebrospinal fluid will escape, and the spinal cord will slightly bulge out.

6.1.3 Intraspinal Injection

1. The injection syringe is assembled by attaching the micropipette to the Hamilton syringe with the compression fitting kit. It is filled with water using the previously prepared 5 ml water syringe. The plunger is inserted to about the 3 μl level on the scale. If there is considerable resistance to insertion of the plunger or if there are bubbles within the syringe, the syringe should be disassembled and then reassembled with a new micropipette.

2. The assembled syringe is inserted into the pump and a small air bubble is drawn into the micropipette to separate the virus solution from the water. A 2.5 μl droplet is placed onto a piece of Parafilm just below the tip of the micropipette. The syringe is lowered into the droplet and the virus solution is sucked up. A little droplet of virus needs to be dispensed to ensure the syringe is ready for injection. Using a pen, a scale is drawn onto the micropipette from the bubble down.

3. The syringe is lowered until the tip of the micropipette touches the spinal cord in one of the holes, then it is further lowered in 100 μm steps to the desired depth. The cord will first indent and then the micropipette will insert into the cord. If it fails to penetrate the spinal tissue, it has likely caught on the dura. Retraction and a repositioning of the tip may help. Alternatively, the needle can be used again to ensure the hole in the dura is in the desired place. For targeting the dorsal horn, we lower the tip to a depth of 500 μm and then retract to a depth of 300 μm (*see* **Note 3**).

4. The pump is programmed to the desired injection volume and speed (e.g., 300 nl, 50 nl/min) and the injection is started. After the termination of the injection, the syringe is kept in place for an additional 3 min to allow for pressure equilibration of the solution in the tissue and to reduce backflow along the tract. The scale on the micropipette is inspected to confirm that the viral solution was delivered successfully. If the level of the virus has not dropped, the scale is monitored while retracting the syringe. If the level suddenly drops, the tip is inserted again and the injection is repeated. If the level does not drop, the syringe is reassembled with a new micropipette and the injection is repeated.

5. Repeat **steps 3** and **4** for the other two injection sites.

6.1.4 Completion of the Surgery

The spinal clamps and the cushion below the animal are removed. The superficial tissue layers are sutured with interrupted stitches using absorbable sutures. The skin is detached again from the underlying tissues using scissors and sutured with nonabsorbable sutures. The wound is disinfected using iodine-soaked cotton wool. After termination of the anesthesia, the animal is placed into a warmed recovery cage or left to recover on the heat mat before placing it back into its home cage. The health of the mouse is monitored during the following hours and days by inspection of its overall appearance (weight, gait, fur, eyes, and behavior) and the healing of the wound. Analgesic treatment is continued as necessary.

7   Notes

1. The optimal physical titer will vary depending on the mouse line, the viral serotype (capsid) and promoter, the transgene to be expressed and the readout. For instance, a minimal expression level of diphtheria toxin results in ablation of the respective cell, whereas activating pharmaco- and optogenetic receptors need to be expressed at suprathreshold levels to evoke action potentials and thus induce behavioral changes in the animals. Similarly, fluorophores need to be expressed in sufficient quantity to be detectable with standard immunofluorescence methods. Very high physical titers on the other hand may induce toxicity.

2. The corresponding levels of the spinal cord and the vertebral column are not aligned. The different development leads to an increasing shift between the two. In the adult animal, for instance the spinal cord level L4 is located within the thoracic vertebra T13.

3. The number of injections as well as injection volume and depth can be adapted to the aim of the respective experiment. The three injections described here result in a targeting of dorsal spinal neurons from L3 to L5, the area that receives input from the sensory neurons innervating the hindlimb. Unilateral injections allow for using the contralateral paw as an internal control. A larger injection volume will lead to a larger rostrocaudal spread, but also increases the likelihood of spillover to the contralateral side and may result in the transduction of neurons in the ventral horn. Additional injections can also be used to transduce neurons in a larger area, but will increase the time required for the surgery.

References

1. Gatto G et al (2019) Neuronal diversity in the somatosensory system: bridging the gap between cell type and function. Curr Opin Neurobiol 56:167–174

2. Moehring F et al (2018) Uncovering the cells and circuits of touch in Normal and pathological settings. Neuron 100(2):349–360

3. Peirs C, Seal RP (2016) Neural circuits for pain: recent advances and current views. Science 354(6312):578–584

4. Foster E et al (2015) Targeted ablation, silencing, and activation establish glycinergic dorsal horn neurons as key components of a spinal gate for pain and itch. Neuron 85(6): 1289–1304

5. Peirs C et al (2015) Dorsal horn circuits for persistent mechanical pain. Neuron 87(4): 797–812

6. Petitjean H et al (2019) Recruitment of Spinoparabrachial neurons by dorsal horn Calretinin neurons. Cell Rep 28(6):1429–1438 e4

7. Petitjean H et al (2015) Dorsal horn Parvalbumin neurons are gate-keepers of touch-evoked pain after nerve injury. Cell Rep 13(6): 1246–1257

8. Fenno LE et al (2014) Targeting cells with single vectors using multiple-feature Boolean logic. Nat Methods 11(7):763–772

9. Haenraets K et al (2017) Spinal nociceptive circuit analysis with recombinant adeno-associated viruses: the impact of serotypes and promoters. J Neurochem 142(5):721–733

10. Tervo DG et al (2016) A designer AAV variant permits efficient retrograde access to projection neurons. Neuron 92(2):372–382

11. Kugler S, Kilic E, Bahr M (2003) Human synapsin 1 gene promoter confers highly neuron-specific long-term transgene expression from an adenoviral vector in the adult rat

brain depending on the transduced area. Gene Ther 10(4):337–347

12. Juttner J et al (2019) Targeting neuronal and glial cell types with synthetic promoter AAVs in mice, non-human primates and humans. Nat Neurosci 22(8):1345–1356

13. Buenrostro JD et al (2013) Transposition of native chromatin for fast and sensitive epigenomic profiling of open chromatin, DNA-binding proteins and nucleosome position. Nat Methods 10(12):1213–1218

14. Buenrostro JD et al (2015) ATAC-seq: a method for assaying chromatin accessibility genome-wide. Curr Protoc Mol Biol 109: 21.29.1–21.29.9

15. Mikuni T et al (2016) High-throughput, high-resolution mapping of protein localization in mammalian brain by in vivo genome editing. Cell 165(7):1803–1817

16. Suzuki K et al (2016) In vivo genome editing via CRISPR/Cas9 mediated homology-independent targeted integration. Nature 540(7631):144–149

17. Haenraets K et al (2018) Adeno-associated virus-mediated transgene expression in genetically defined neurons of the spinal cord. J Vis Exp 135:57382

18. Graham FL, Smiley J, Russell WC, Nairn R (1977) Characteristics of a human cell line transformed by DNA from human adenovirus type 5. J Gen Virol 36:59–74. https://doi.org/10.1099/0022-1317-36-1-59

19. DuBridge RB, Tang P, Hsia HC et al (1987) Analysis of mutation in human cells by using an Epstein-Barr virus shuttle system. Mol Cell Biol 7(1):379–387. https://doi.org/10.1128/mcb.7.1.379

20. Reed SE, Staley EM, Mayginnes JP et al (2006) Transfection of mammalian cells using linear

polyethylenimine is a simple and effective means of producing recombinant adeno-associated virus vectors. J Virol Methods 138(1-2):85–98. https://doi.org/10.1016/j.jviromet.2006.07.024

21. Kuroda H, Kutner RH, Bazan NG, Reiser J (2009) Simplified lentivirus vector production in protein-free media using polyethylenimine-mediated transfection. J Virol Methods 157(2):113–121. https://doi.org/10.1016/j.jviromet.2008.11.021

22. Zolotukhin S, Byrne BJ, Mason E et al (1999) Recombinant adeno-associated virus purification using novel methods improves infectious titer and yield. Gene Ther 6(6):973–985. https://doi.org/10.1038/sj.gt.3300938

23. Paterna JC, Moccetti T, Mura A et al (2000) Influence of promoter and WHV post-transcriptional regulatory element on AAV-mediated transgene expression in the rat brain. Gene Ther 7(15):1304–1311. https://doi.org/10.1038/sj.gt.3301221

Use of Intraspinally Delivered Chemogenetic Receptor, PSAM-GlyR, to Probe the Behavioral Role of Spinal Dorsal Horn Neurons

Cynthia M. Arokiaraj, Myung-chul Noh, and Rebecca P. Seal

Abstract

Chemogenetic receptors are revolutionizing our understanding of the neural circuitry that underlies somatosensation at all levels: primary sensory ganglia, spinal cord, and brain. The receptors, also known as designer receptors exclusively activated by designer drugs (DREADDs), control the excitability of cells when activated by otherwise inert chemical ligands, thus permitting delineation of the behavioral role of select cell populations. Here we describe in detail, experimental paradigms to probe the role of specific dorsal horn excitatory interneuron populations in acute sensation and persistent pain using an ion channel-based inhibitory designer receptor, $PSAM^{L141F}$-GlyR. We also describe methods for the intraspinal injection of the designer receptor packaged in adeno-associated virus as well as for the post-hoc examination of its expression.

Key words Chemogenetics, Spinal cord, Intraspinal, DREADD, PSAM-GlyR, Behavior

---

1 Introduction

Our understanding of how the nervous system encodes behavior has accelerated with the advent of genetically encoded optogenetic and chemogenetic receptors that allow for the acute and reversible in vivo manipulation of the excitability of select cell populations while measuring behavior [1, 2]. The optogenetic receptors are light-gated excitatory and inhibitory ion channels that act on a millisecond timescale and can be applied acutely or chronically to circuits in vivo or in vitro in a spatially restricted manner, as is described in detail in other chapters within this volume. Chemogenetic receptors on the other hand are engineered to be selectively activated by otherwise inert chemical ligands while unresponsive to

Supplementary Information The online version of this chapter (https://doi.org/10.1007/978-1-0716-2039-7_20) contains supplementary material, which is available to authorized users.

Rebecca P. Seal (ed.), *Contemporary Approaches to the Study of Pain: From Molecules to Neural Networks*, Neuromethods, vol. 178, https://doi.org/10.1007/978-1-0716-2039-7_20, © Springer Science+Business Media, LLC, part of Springer Nature 2022

physiological concentrations of endogenous ligands [3]. With the use of chemogenetic receptors, cell excitability can be controlled on the order of minutes to hours depending on the pharmacokinetics of the ligand. The first chemogenetic receptors to be commonly used are based on the human muscarinic receptor subtypes, hM3Dq and hM4Di [4, 5]. Both are G-protein coupled receptors that regulate ion channel activity through second messengers, with the former producing membrane depolarization (excitatory) and the latter producing membrane hyperpolarization (inhibitory). The ligand, clozapine-n-oxide (CNO), is designed to have high selectivity for these receptors, but is also metabolized to clozapine [6, 7]. The latter compound, a known psychotropic drug, acts on a range of endogenous receptors, but importantly shows significantly greater potency for the hM3Dq and hM4Di designer receptors. Selective activation of these DREADDs thus requires using CNO at <10 mg/kg or a subthreshold dose of clozapine at 0.1 mg/kg [8]. Other synthetic agonists for these DREADDs like the JHU3712, JHU37160 and deschloroclozapine (DCZ) appear to be promising alternatives due to their high brain penetrability and potency with minimal off-target effects [9, 10]. A second Gi coupled DREADD (inhibitory) is based on the kappa opioid receptor, KORD. This designer receptor is engineered to have a high affinity for the otherwise inert ligand, salvinorin B, and thus, when combined with the muscarinic based DREADD, allows for the differential control of multiple cell populations within the same experiment [11]. Lastly, the proper control to interpret the outcome of such studies requires testing in parallel mice that lack the receptor (e.g., injected with a reporter only).

Ionotropic chemogenetic receptors, such as the PSAM series, comprise another designer receptor system [12]. The PSAM receptors incorporate a mutated version of the ligand binding domain of the human alpha-7 nicotinic receptor fused to either the ion channel domain of the human serotonin (5-HT) subtype 3 receptor to generate an excitatory receptor, PSAM-5HT3, or to the human glycine receptor to generate an inhibitory receptor, PSAM-GlyR. Further refinements have been made to the ligand binding domain of these receptors to increase their potency for highly selective FDA approved ligands, such as varenicline as well as other newly designed ligands called μPSEMs, such as $PSEM^{792}$, $PSEM^{793}$, and $PSEM^{819}$ [13].

Chemogenetic studies of the neural circuitry underlying somatosensation in mice have been applied to populations throughout the nervous system including peripheral sensory neurons, spinal cord, and brain [14–25]. Studies have also utilized these receptors to study the contribution of glial cells to the transmission of pain [26–28]. To gain access to select cell populations, many laboratories have taken advantage of the Cre-lox system in mice, though efforts to identify regulatory elements that can drive cell-type specific expression is allowing researchers to perform these types of experiments independent of the Cre-lox system as well as in other species [26, 29]. Expression of the chemogenetic receptor is typically accomplished through injection of an AAV encoding the Cre-dependent receptor either intrathecally or intraperitoneally in the case of primary sensory ganglia, intraspinally for spinal cord cells or intracranially to target brain cells, in mice expressing Cre recombinase in the target cell population. Alternatively, mouse lines that have incorporated the second messenger coupled receptors in the genome are also available [23]. Because the ligands are typically delivered systemically, precise cellular targeting of the receptor is key. Focal delivery of the ligand is, however, another strategy that can be used to further refine which cells are manipulated.

Here we describe the use of the chemogenetic receptor, PSAM^{L141F}-GlyR targeted to specific excitatory neurons in the spinal cord to determine their role in acute somatosensory behavior and in persistent pain after injury [18]. The paradigm involves intraspinal injection of Cre-dependent PSAM^{L141F}-GlyR packaged into AAV8 and then behavioral testing conducted 3 weeks later. Use of the AAV8 serotype injected intraspinally confers selective expression in the spinal cord with little to none in primary afferents or brain [16, 18, 30]. Mice are typically injected at P21, but can be injected at younger or older ages as needed. Behaviors are typically assessed when the mice reach 6 weeks of age.

In initial experiments, we have demonstrated that activation of PSAM^{L141F}-GlyR inhibits dorsal horn excitatory neurons in spinal cord slices using patch clamp electrophysiology [18]. For this purpose, we injected AAV8 hSyn-Flex-PSAM^{L141F}-GlyR-IRES-GFP in mice expressing Cre in a large population of excitatory neurons (Tlx3^{Cre}). After 2 weeks, we recorded from the neurons. Bath application of the PSAM^{L141F}-GlyR agonist, PSEM^{89S} (30 µM) hyperpolarized the membrane and blocked action potentials in response to current injection in GFP$^+$ neurons. The ligand had no effect on GFP$^-$ neurons. A recent report using the newer PSAM4-GlyR inhibitory DREADD suggests that it may alter the chloride gradient of striatal dopamine receptor 1 containing medium spiny neurons, resulting in excitation rather than inhibition with receptor activation [31]. This may extend to chloride channel DREADDs in general. It is therefore recommended that inhibition by the DREADDs of all cell types are tested in slices prior to performing behavioral experiments. This may extend to cell types that have altered chloride gradients due to injury. Interestingly, although the behavioral outcome was as expected for activation of hM4Di in primary afferents that express TRPV1, it was reported that the primary sensory neurons showed altered sodium and calcium currents and that the activity of morphine receptors in these cells was altered, presumably due to changes in the availability in second messengers [17]. These findings thus provide notes of caution for those using designer receptors.

To assess the role of PSAM-GlyR expressing neurons in somatosensory behaviors, we measure punctate mechanical thresholds with von Frey filaments, dynamic mechanical responses with a cotton swab, the sensitivity to a piece of sticky tape, the noxious mechanical pain response with a pinprick as well as thermal response with the Hargreaves assay or acetone test. All are measured on the plantar surface of the hind paw. We also test the mice with the rotarod to check for any deficits in forelimb and hind limb motor coordination. Measurements are taken 30–90 min following injection of saline or ligand (30–50 mg/kg).

Next, we determine whether the targeted excitatory neuronal population has a role in the transmission of persistent pain using both neuropathic and inflammatory pain models since our findings indicate that the dorsal horn network for mechanical allodynia differs depending on the type of injury [18]. We also typically assay for both dynamic and punctate allodynia and we also assay for heat or cold hypersensitivity depending on the pain model used. There are a number of validated chronic pain models. The spared nerve injury (SNI) is a commonly used neuropathic pain model because it is highly reproducible and produces robust mechanical and cold allodynia [32]. In the sural version of this model, which produces both punctate and dynamic allodynia, the tibial and common peroneal nerves are cut and ligated and the sural nerve remains intact. The region of the hind paw innervated by the sural nerve becomes hypersensitive (the lateral edge of the paw at the border between the plantar surface and the hairy skin). In the spared tibial version of the model, the common peroneal and sural nerves are cut and ligated and the tibial nerve is spared [33]. This model only shows punctate mechanical allodynia and cold allodynia and the hypersensitivity is located in the middle of the plantar surface of the hind paw, which corresponds to where the tibial nerve innervates. We typically perform the surgery 2 weeks after the virus injection and measure mechanical thresholds at 1, 2, and 6 weeks after the surgery. Cold hypersensitivity using the acetone assay can also be measured at these time points. A statistically significant decrease in the mechanical threshold to von Frey filaments or a statistically significant increase in the response to the cotton swab indicates the expression of punctate and dynamic mechanical allodynia, respectively. A significant increase in the time spent responding to the acetone indicates cold allodynia. A significant reversal in the sensitivity to punctate or dynamic mechanical stimuli or to acetone in mice expressing PSAM-GlyR after receiving the ligand, compared to saline injection, indicates a role for the neurons in the transmission of allodynia caused by neuropathic injury.

To test the role of the neurons in an inflammatory pain model, we use carrageenan (acute) or complete Freund's adjuvant (CFA), which model acute and more chronic inflammatory injuries, respectively. For the carrageenan model, 10 µL of lambda-carrageenan is injected in the plantar hind paw and then punctate and dynamic allodynia as well as heat hypersensitivity are measured 1–3 days later. For the CFA model, the plantar hind paw is injected with 10 μL of a solution that is equal parts CFA and sterile 0.9% saline. The time courses for the expression of punctate allodynia, dynamic allodynia, and heat hypersensitivity differ in this model, with heat hypersensitivity peaking typically 3 days following injection, dynamic allodynia peaking 4 to 5 days following injection, and punctate allodynia lasting at least 1 week. Behavioral measurements are therefore taken at these time points.

Following behavioral testing, we confirm expression of the PSAM^{L141F}-GlyR receptor in the dorsal horn of the mice by staining fixed sections with alpha-bungarotoxin conjugated to a fluorophore (e.g., Alexa 647). The α-bungarotoxin binds to the ligand binding domain of the α7-nicotinic receptor of PSAM^{L141F}-GlyR. Fortuitously, we do not observe α-bungarotoxin staining in the dorsal horn of mice lacking the PSAM^{L141F}-GlyR, thus simplifying the interpretation [18]. If α-bungarotoxin staining is absent at the expected site of injection, behavioral data from the mouse is excluded from the study.

1.1 "Minimally Invasive" Intraspinal Injection

Herein, a modified intraspinal injection method based on Piers et al. (2015), which is similar to Kohro et al. (2015), is described [16, 34]. This "minimally invasive" microinjection method allows for delivery of AAV vectors into the spinal dorsal horn without laminectomy or drilling. Protocols described in this chapter were approved by the University of Pittsburgh's Institutional Animal Care and Use Committee.

2 Materials

2.1 Reagents

1. Ethanol, 70%.
2. Iodine solution.
3. Saline, sterile, 0.9%.
4. Lactated Ringer's Solution with 5% dextrose.
5. Ketoprofen, sterile (5 mg/kg).
6. Isoflurane.
7. Ocular lubricant, sterile.
8. Mineral Oil.

2.2 Surgical Tools

1. 2 Dumont #5 fine forceps
2. Spring scissors 8 mm Cutting Edge.
3. 2 Schwartz Micro Serrefines.
4. Fine Scissors.

5. Noyes Spring Scissors.

6. Needle holder.

7. Dumont #7 Forceps.

2.3 Equipment

1. Sterilization pack.

2. 6-0 Silk sutures

3. 6-0 Nylon sutures.

4. Sterile cotton swabs.

5. Sterile eye spears.

6. Rechargeable clippers.

7. Sterile gauze.

8. Parafilm.

9. Kimwipes.

10. sHeating pad.

11. Absorbent pad.

12. Isoflurane vaporizer.

13. Anesthesia induction chamber.

14. Isoflurane scavenger.

15. Micropipette puller (P-97, Sutter).

16. Fire polished glass capillaries (World Precision Instruments, #504949).

17. CV-5-1GU headstage and compatible glass capillary holder (Axon Instruments).

18. 5 µL glass microliter syringe (Hamilton, Model 7105 KH).

19. Dental cement.

20. Polyethylene tubing.

21. Stereotaxic injector (Stoelting).

22. Small animal stereotaxic frame with small animal spinal unit (Kopf).

    (a) Spinal base plate (Kopf, model 912).

    (b) 2 Adjustable base mounts with post and clamp (Kopf, model 982).

    (c) "V" notch spikes (Kopf, model 987).

23. Surgical microscope.

24. Mini microcentrifuge.

25. 30-gauge needle.

26. 10 mL syringe.

27. 3 way Luer lock.

28. 1 mL syringes.

3   Methods

3.1   Custom Stereotaxic Injection Setup (Optional)

In our laboratory, we created a custom stereotaxic injection setup which uses patch clamp amplifier headstage as a glass capillary holder (Fig. 1). We attached a CV-5-1GU headstage from Axon Instruments to the stereotaxic frame. Then, we connected a 5 µL Hamilton model 7105 syringe to the glass pipette holder attached to the CV-5-1GU headstage by polyethylene tubing. Finally, to ensure a tight seal, we applied dental cement to each ends of the polyethylene tube. While this custom setup works consistently in our hands, any stereotaxic injection setup will be fully compatible with the methods described in this chapter.

3.2   Preparation for the Surgery

1. Autoclave surgical tools in sterilization pack prior to surgery.
2. Pull glass microinjection needle using micropipette puller and store it in a container.
   (a) P-97 protocol: Heat = 580, Pull = 200, Vel. = 20 Time = 250 (You may decrease or increase the heat setting to control the length of tapered tip).
3. Obtain necessary AAV aliquots containing vector of your choice and store it in an ice bucket.
   (a) Titers between 10^{12} and 10^{13} has worked consistently in our hands.
   (b) Multiple AAV vectors can be injected at the same time.
4. Load 1 mL syringe with sterile saline.
   (a) This will be used to keep the surgical area moist throughout the procedure.

Fig. 1 Intraspinal setup (Image on the left): In this image can be seen the small animal stereotaxic instrument with digital display console, heat pad, isoflurane anesthesia unit, and stereotaxic injector. **On the right** can be seen, glass capillary for injection as well as the spinal clamp area

5. Load 1 mL syringes with 50% Lactated Ringer's Solution (LRS) with 5% dextrose and 50% sterile saline.

   (a) This will be injected subcutaneously after the procedure to prevent severe dehydration.

6. Load 1 mL syringe with necessary amount of ketoprofen solution (5 mg/kg final dose).

7. All solutions prepared in **steps 4–6** should be kept warm on heating pad before injection.

3.3 Preparation for the Stereotaxic Injector

Prior to surgery, glass microinjection needle should be mounted, trimmed, and loaded with mineral oil. Methods described below is specific to our setup and this may vary depending on type of stereotaxic injector unit (*see* Fig. 1)

1. Mount the headstage to the stereotaxic frame.

2. Mount the 5 μL Hamilton syringe to stereotaxic injector.

3. Prepare a 10 mL syringe with mineral oil and three-way Luer lock connected to a 30-guage needle. This is used to flush the injection loop and remove any air bubbles.

   (a) Ensure there is no air bubbles in the 10 mL syringes.

4. Carefully pull out the needle from Hamilton syringe and insert 10 mL syringe needle into the opening of Hamilton syringe needle.

5. Slowly flush out any bubbles in the injection loop. It is critical to completely fill the loop with mineral oil to ensure a smooth injection procedure without clog.

   (a) Excess mineral oil on the electrode holder should be cleaned with Kimwipes.

6. Once the stereotaxic injector loop is filled with mineral oil, close the three-way Luer lock, and glass microinjection needle may be inserted into the electrode holder.

   (a) Tightly close the electrode holder by rotating the fitting counter-clockwise.

7. Using the surgical microscope trim the tip of the glass microinjection needle.

   (a) Having a colored paper underneath the needle helps with contrast.

   (b) Tip diameter should be small enough to minimize the trauma to the spinal cord, and large enough to allow sufficient flow rate of AAV solution.

   (c) Gently touch the tip of the needle with spring scissors and trim where the tip no longer sharply bends with little pressure from the spring scissors. The tip must be rigid and sharp enough to break through the pia membrane.

8. Opening the three-way Luer lock on the 10 mL syringe will push mineral oil through glass needle.

   (a) Pull the 10 mL syringe out when mineral oil fills approximately 30% of the glass microinjection needle. This is done to account for Hamilton syringe's plunger volume.

9. Reassemble the Hamilton syringe by inserting the needle through the plunger.

   (a) Mineral oil will have been pushed out of the needle. Using the stereotaxic injector, draw back the Hamilton syringe to equilibrate the pressure. No mineral oil should be coming out of the needle.

10. Set the injection volume to 1 µL and rate of injection to 200 to 300 nL/min. Ensure the stereotaxic injector is set to dispense.

3.4 Surgical Procedures to Expose the Spinal Cord Segment

1. Weigh the animal before inducing anesthesia.

2. Deeply anesthetize the mice in the anesthesia induction chamber with 5% isoflurane and oxygen.

3. Transfer the mice from induction chamber to heating pad into a nose cone with 1.5–2% isoflurane and oxygen flow.

   (a) We prefer to orient the mice horizontally during the surgical procedure; however, any orientation may be used.

   (b) It is critical to maintain and monitor the respiration rate of the mice throughout the procedure and adjust the concentration of isoflurane accordingly.

4. Apply ocular lubricant with sterile cotton swab.

5. Subcutaneously inject 5 mg/kg Ketoprofen.

6. Shave the thoracolumbar area with electric clippers.

7. Apply 70% ethanol to exposed skin area with sterile gauze to remove excess hair and let it dry.

8. Subsequently, apply iodine solution to the shaved area with sterile cotton swab and let it dry.

9. Lift the skin up with Dumont #7 forceps and make a small cut with Noyes spring scissors. Then, cut along the midline to expose the thoracolumbar area underneath the skin. Carefully tease away the facia by opening the spring scissors from closed position. No damage to the muscle tissue should be done during this step. Finally, clip the skin away from the exposed muscle tissue with Schwartz Micro Serrefines.

10. Make sure to keep the surgical area moist by applying small volume of sterile saline throughout the procedure.

11. Intervertebral space between T12 and T13 vertebral spine allows injection to L3-L4 region of the spinal cord. Alternatively, you may choose to inject into intervertebral space

between T13 and L1 vertebral spine to target L4-L5 region of the spinal cord. There are landmarks that may help with identification of these regions (*See* Harrison et al., 2013 for detailed MRI images of mice spinal cord segments) [35]. The T13 vertebra is connected to the last rib of the mice, and this can be identified by gently pushing along the ribs of the animal with forceps (below T13 there will be no resistance when you push with forceps). Alternatively, one may choose to count the vertebral column from the iliac crest which is attached to the L6 vertebra (*see* Haenraets et al., 2018 for an excellent video guide to identifying vertebral landmarks) [36]. Once the correct vertebral spines have been identified, you may gently push between the vertebral spines with forceps to identify the intervertebral space (intervertebral space will have no resistance when pushed with forceps). This step is critical for this procedure as correct identification of the intervertebral space minimizes the damage to the muscle tissue.

12. Gently tease away the muscle tissue by blunt dissection using 2 Dumont #5 fine forceps under the surgical microscope.

    (a) If bleeding occurs during this step, use the sterile eye spears to apply gentle pressure to the area until the bleeding stops.

    (b) Slightly lifting the muscle tissue with one forceps, and teasing away the muscle tissue with the other forceps may minimize the potential damage to the spinal cord. It is absolutely critical that you do not damage the spinal cord during this step.

    (c) Spinal cord underneath the muscle tissue will be clearly visible when blunt dissection is done properly.

13. Once spinal cord is exposed, dura must be removed before the injection procedure. Gently scrape the dura with sterile 30 gauge needle to create an opening. Once dura is opened you will notice cerebrospinal fluid (CSF) leaks out and spinal cord bulges out slightly. Alternatively, you may use the Dumont #5 fine forceps to scrape the dura repeatedly. Great care must be taken not to damage the spinal cord.

14. Transfer the animal to the isoflurane adapter on the stereotaxic frame. Depending on the frame, you will need to lift the mice up to proper height (we use a Styrofoam block layered with absorbent pad to lift up the mice).

15. Clamp the vertebral column with "V" notch spikes. Other spinal adaptor kits and vertebrae clamp compatible with stereotaxic frame may be used. Vertebrae clamp (Kopf) may provide the best stability during the injection; however, this cannot be used for younger mice.

16. Using the surgical microscope, ensure exposed spinal cord and dorsal blood vessel is clearly visible. If bleeding occurs, it is critical to stop the bleeding with eye spear before proceeding with intraspinal injection. Bleeding during the injection will cause clog.

3.5 Intraspinal Injection

1. Spin down the virus solution with mini microcentrifuge.

2. Dispense viral solution to parafilm. For single AAV solution, dispense 1.2 μL (for 1 μL injection), and for two AAV solutions dispense 0.6 μL of each virus solution and mix thoroughly by pipetting the droplet up and down.

3. Under the surgical microscope, lower the microinjection needle to the droplet and draw up the virus solution gradually with stereotaxic injector. Once enough virus solution is drawn up, apply small positive pressure with the stereotaxic injector to ensure virus solution flows out.

4. Align the microinjection needle to the dorsal blood vessel of the exposed spinal cord segment. Reset the x-axis of stereotaxic coordinates.

5. From the dorsal blood vessel adjust the x coordinate 300 to 350 μm away from the dorsal blood vessel.

    (a) For targeting L4 segment, target the most caudal spinal cord segment between T12 and T13 intervertebral space or most rostral segment between T13 and L1 intervertebral space.

    (b) Align the needle to an area where the dura is open (Critical).

6. Slowly lower the microinjection needle to touch the spinal cord. Small dimple will form when you have touched the spinal cord (*see* **Note 1**).

7. Reset the z-axis of the stereotaxic coordinates. Then, slowly lower the z-axis to penetrate the spinal cord. Depending on the age and location of the injection, you may need to lower the needle as deep as 500 μm to break through (If severe deformation of spinal cord occurs, get a new microinjection needle and trim. Then repeat **steps 2** to **7**). Once needle is inside the spinal cord, slowly retract the needle back up to 250 μm to target the spinal dorsal horn and wait at least 2–3 min at this position before starting the injection to reach equilibrium (*see* **Note 2**).

8. Start the injector to dispense virus solution. Use the surgical microscope's grid to monitor the meniscus of the virus solution to monitor the flow.

(a) We are injecting relatively high volume (1 μL) of virus solution to ensure good mediolateral and rostrocaudal spread. However, this may cause overflow of the virus solution.

(b) Occasionally, a pool of CSF may form immediately after needle breaks through the spinal cord. Use the eye spears to carefully absorb any CSF overflow.

9. Once the injection is complete, wait at least 2–3 min before retracting the microinjection needle.

(a) You must ensure no mineral oil leaks out of the microinjection needle.

10. Slowly retract the microinjection needle out of the spinal cord. Clean the area with eye spear (*See* **Note 3**).

11. Remove the "V" notch clamps and suture the *latissimus dorsi* muscle with 6-0 nylon sutures.

12. Suture the skin with 6-0 silk sutures.

13. Subcutaneously inject appropriate volume of 50/50 LRS with 5% dextrose and saline solution. Maximum volume will depend on weight and age of the animal (we have been giving 300–400 μL to p14-p30 mice).

14. Stop the isoflurane and allow the animal to fully recover on heating pad before returning the animal to its cage.

3.6 Post-operative Care of Animals

Once animals have been returned to the cage, health of the animal must be monitored closely over the next 48 h. We always provide mushed pellets during this period, and we monitor the weight, sutures, activity levels, gait, and signs of dehydration. Furthermore, post-operative analgesic (5 mg/kg ketoprofen) is given by subcutaneous injection with 50/50 LRS and saline mixture once a day over 48 h.

4 Notes

1. This may be difficult to confirm if the spinal cord opening is filled with CSF or blood. It is a good idea to remove as much CSF as possible with an eye spear before this step.

2. It is critical to monitor the tip of the needle for any blood. Occasionally you may hit small blood vessels inside the spinal cord. This will clog the microinjection needle and ruin the injection. If you see any blood in the microinjection needle, immediately apply small positive pressure to push it out.

3. Wait period and slow retraction minimizes backflow of injected virus solution.

5   Chemogenetics to Assess the Behavioral Role of Dorsal Horn Neurons

5.1   Materials

1. PSEM^{89s} (Tocris Cat. No. 6426).

2. 0.9% sterile NaCl.

3. Von Frey/Semmes-Weinstein Monofilaments.

4. IITC Mesh Stand and Animal Enclosures for mice (Part #410 and #435).

5. Cotton buds.

6. Plantar Analgesia Meter (IITC).

7. CaviCide® disinfectant.

8. Paper towels.

9. Acetone.

10. Stopwatch.

11. 1 mL syringes.

5.2   Methods

5.2.1   Acclimation Prior to Behavior

It is important to habituate the mice in their animal enclosures prior to beginning behavioral testing. The measurements obtained will be more reliable when the mice are not stressed from being away from their home cages and in a new environment. It is best to acclimate the mice a day before von Frey or cotton swab tests. The Hargreaves assay requires more time for acclimation, so it is recommended to acclimate the mice at least 2–3 times to the Hargreaves apparatus before the day of experimentation.

5.2.2   Acclimation Period for Cotton Swab and von Frey

1. Choose a quiet room for your behavioral testing so that the mice do not get agitated by other people or noise.

2. Spray the animal enclosures and mesh stand with disinfectant.

3. Place the animal enclosures on the mesh stand and keep paper towels below the wire mesh to prevent mice droppings from making a mess.

4. Remove the mice from their home cages and place each of them in their own compartment in the Plexiglass chambers (animal enclosures).

5. Let them remain in their animal enclosures for an hour. It is ideal to remain in the room for half or more of that time so that the mice get acclimated to your scent as well.

6. After an hour, remove the mice from the Plexiglass chamber and return them to their home cages.

7. Spray down the animal enclosures and mesh stand with disinfectant.

5.2.3 Acclimation Period for Hargreaves

1. Choose a quiet room for your behavioral testing so that the mice do not get agitated by other people or noise.

2. Spray the temperature-regulated glass panel and Plexiglass chambers with disinfectant. Wipe it down.

3. Place the animal enclosures on the glass panel.

4. Make sure the glass panel is set at 30 °C.

5. Remove mice from their home cage and set them each in a compartment in the Plexiglass chamber.

6. Allow them to explore their compartments for an hour.

7. Return the mice to their home cages.

8. Spray the glass panel and Plexiglass chambers with disinfectant and wipe down.

5.3 Control Experiments for Off-target Behavioral Effects of the Ligand

Prior to testing the mice intraspinally injected with the PSAM-GlyR virus, it is critical to rule out any effects the ligand PSEM[89s] has on somatosensation. To determine this, we intraperitoneally inject mice that have no virus with PSEM[89s] and test behavior. We measure cotton swab before von Frey because the former assay requires the mice not be at all agitated. After von Frey, mice are sometimes agitated (*see* **Note 1**) is also important to avoid taking behavioral measurements when the mice are grooming as this results in false negative responses.

5.3.1 Cotton Swab and von Frey Assay

1. Weigh the mice to be tested and note down their measurements.

2. Before behavioral testing, habituate the mice in their Plexiglass chambers for 30 min. Make sure the surfaces of the wire mesh and Plexiglass chambers are clean.

3. Prepare PSEM[89s] (30–50 mg/kg) in 0.9% sterile saline 15 min before testing. Keep on ice. Control animals will receive 0.9% saline alone.

4. Measure the dynamic response using the cotton swab test. Puff out the cotton bud until it is three times its original size. The cotton swab is gently brushed against the plantar hind paw in the heel-to-toe direction. This is done six times for each paw, leaving 3 min between each stimulation. A response is when the mice withdraw their paw, flick their paw or start licking the paw. The percentage of times the mice respond to the cotton swab can then be calculated for both the left and right hind paws separately.

5. Next, the mechanical threshold of the mice is determined using the von Frey assay by the Up-Down (Chaplan) or simplified Up-Down (SUDO) method, starting with the 0.4 g filament [37, 38]. Each filament is applied to the plantar hind paw until

the filament starts to bend and held in place for 3 s or until a sharp response is observed. A response includes paw withdrawal, shaking or licking. An interval of 5 min is given between each stimulation.

6. Mice are then injected (30–50 mg/kg, i.p) with PSEM^{89s} or saline. **The tester should be blinded to the treatments**.

7. The mice are placed back in the Plexiglass chambers in the same compartment as they were in before.

8. Twenty minutes after drug/saline treatment, the dynamic response is measured again with the cotton swab.

9. After the cotton swab test, we measure the von Frey threshold in the mice.

10. After testing, return the mice back to their home cages.

11. Clean the Plexiglass chambers, wire mesh and surrounding area.

5.3.2 Hargreaves Thermal Assay [39]

It is recommended that the thermal assay be conducted on a different day (*see* **Note 2**).

1. Acclimate the mice in their Plexiglass chambers for 30 min. Make sure the surfaces of the glass panel and Plexiglass chambers are clean. The temperature-regulated glass panel should be set at 30 °C.

2. During the acclimation time, you can prepare the PSEM^{89s} (30–50 mg/kg) in 0.9% sterile saline. Control animals will receive 0.9% saline alone.

3. The Plantar Analgesia Meter is set at 20% intensity with a cut off time of 20 s.

4. Keep paper towels nearby and wipe off any droppings on the glass panel. Liquid on the glass panel can reduce the temperature of the glass panel, hindering accurate measurement.

5. Focus the heat source on the plantar hind paw. As soon as the paw is withdrawn, shut off the heat source. The latency to thermal response is noted. If the mice don't withdraw their paw even after 20 s, the heat source will automatically turn off.

6. Take three measurements per paw with an interval of 5 min between each stimulation. The measurements can then be averaged for each paw.

7. Mice are then injected (30–50 mg/kg, i.p) with PSEM^{89s} or saline. **The tester should be blinded to the treatments**.

8. Twenty minutes after drug/saline treatment, the paw withdrawal latency is measured as before. Three measurements are taken for each paw with an interval of 5 min between each stimulation. Make sure to wipe any droppings on the glass panel before testing.

9. Return the mice to their home cage after testing.

10. Spray the Plexiglass chambers and glass panel with disinfectant.

5.4 Baseline Measurements

To test the functional role of spinal cord dorsal horn neurons in pain transmission, transgenic Cre-driver lines are utilized. This allows for acute inhibition of specific excitatory neuron populations expressing the PSAM-GlyR virus by delivering the PSEM[89S] ligand (i.p.). In mice injected with the PSAM-GlyR virus, the baseline somatosensory responses are measured 2 weeks after the intraspinal injection. After ligand administration, we measure their mechanical threshold with von Frey, dynamic response using the cotton swab test, heat sensitivity with the Hargreaves assay and acetone evaporation test to measure sensitivity to cold. You may choose to include other behavioral pain assays such as the pinprick test for noxious pain detection.

5.4.1 Cotton Swab and von Frey Assay

1. Two weeks after intraspinal injection of the PSAM-GlyR virus, acclimate the mice in the Plexiglass chambers a day before testing as instructed above.

2. On the day of testing, acclimate the mice in the Plexiglass chambers for 30 min.

3. Prepare PSEM[89s] (30–50 mg/kg) in 0.9% sterile saline 15 min before testing. Keep on ice. Control animals will receive 0.9% saline alone.

4. Measure the dynamic response and paw withdrawal threshold with the cotton swab and von Frey assays respectively.

5. Inject (i.p.) mice with PSEM[89s] or sterile saline. **The tester should be blinded to the treatments.**

6. Twenty minutes after injection, perform the cotton swab and von Frey assay.

7. Return the mice to their home cages and clean up the surfaces with disinfectant.

5.4.2 Hargreaves Thermal Assay

Interweave the days on which you measure the static and dynamic thresholds with latency to thermal response.

1. Two weeks after intraspinal injection of the PSAM-GlyR virus, acclimate the mice as instructed above.

2. On the day of testing, acclimate the mice in the Plexiglass chambers for at least 30 min. Make sure the glass panel is at 30 °C.

3. Prepare PSEM[89s] (30–50 mg/kg) in 0.9% sterile saline 15 min before testing. Keep on ice. Control animals will receive 0.9% saline alone.

4. Measure the paw withdrawal latency to heat. Wipe any droppings from the mice found on the glass floor with paper towels.

5. Inject (i.p.) mice with PSEM^{89s} or sterile saline. **The tester should be blinded to the treatments**.

6. Twenty minutes after injection, perform the Hargreaves assay again.

7. Return the mice to their home cages and clean up the surfaces with disinfectant.

5.4.3 Acetone Evaporation Test

Here, we measure the time taken by the mouse in responding (which includes quick paw withdrawal, flicking or licking of the paw) to acetone application [40].

1. The von Frey threshold setup with the plexiglass chambers on the wire mesh can be used for the acetone evaporation test.

2. On the day of testing, acclimate the mice injected with PSAM-GlyR in the Plexiglass chambers for at least 30 min.

3. Prepare PSEM^{89s} (30–50 mg/kg) in 0.9% sterile saline 15 min before testing. Keep on ice. Control animals will receive 0.9% saline alone.

4. Once the mice are acclimated, draw up acetone into a 1 mL syringe. Form uniform drops of acetone and apply one drop of acetone to the plantar surface of the mouse hind paw (*see* **Note 3**).

5. The amount of time spent in flicking or licking the paw after the acetone application is measured using a stopwatch up to 30 s.

6. Perform three trials on each hind paw, with an interval of 5 min between each trial.

7. Inject (i.p.) mice with PSEM^{89s} or sterile saline. **The tester should be blinded to the treatments**.

8. Twenty minutes after injection, perform the acetone test again.

9. Return the mice to their home cages and clean up the surfaces with disinfectant.

5.5 Notes

1. The cotton swab assay requires the mice to not be at all agitated. After von Frey, mice are sometimes agitated and become sensitized to the mechanical stimulation. It is recommended therefore to start with the least stressful assay before proceeding with the more stressful assay.

2. Conducting too many behavioral pain assays on the same day can stress out the mice, resulting in variable results.

3. One limitation of the acetone evaporation assay is the difficulty in creating consistent droplets of acetone. We insert a sharply cut piece of a P20 pipette tip on to the 1 mL syringe in order to draw up approximately 10 µL of acetone for each application.

6 Persistent Pain Models

6.1 Materials

6.1.1 Spared Nerve Injury Neuropathic Pain Model

1. 2% Isoflurane
2. Electric shaver.
3. Silk, nonabsorbable suture, 6-0.
4. Nylon suture, 8-0.
5. Forceps.
6. Fine scissors.
7. Dissection scissors.
8. Spring scissors.
9. Blue drapes.
10. Eye ointment.
11. Betadine.

6.1.2 Complete Freund's Adjuvant Inflammatory Pain Model

1. 2% Isoflurane.
2. Complete Freund's adjuvant (1 mg/mL heat killed and dried mycobacterium tuberculosis).
3. 0.9% sterile NaCl.
4. 70% Ethanol.
5. 3 mL glass syringe.
6. 23-G, ¾ in. needle.
7. Blue drapes.

6.2 Methods

6.2.1 Spared Nerve Injury Neuropathic Pain Model

1. After carrying out baseline measurements in the mice injected with PSAM-GlyR virus, perform SNI surgery ipsilateral to the side of injection. Here, we spare the sural nerve and cut and ligate the tibial and common peroneal nerves.

2. One week after the SNI surgery, acclimate the mice for 30 min in the Plexiglass chambers.

3. Measure the dynamic and static mechanical thresholds using the cotton swab and von Frey assays respectively by applying the stimulations to lateral region of the injured hind paw.

4. Return the mice to their home cages and clean up the surrounding area and equipment.

5. The next day, acclimate the mice in the Plexiglass chambers for 30 min.

6. Prepare PSEM[89s] (30–50 mg/kg) in 0.9% sterile saline 15 min before testing. Control animals will receive 0.9% saline alone.

7. Inject (i.p.) mice with PSEM[89s] or sterile saline. **The tester should be blinded to the treatments**.

8. Twenty minutes after injection, measure the dynamic and static mechanical thresholds as before.

9. To measure cold allodynia, perform the acetone evaporation assay on the injured mice, preferably on a separate day from the cotton swab and von Frey assays. Acclimate the mice for 30 min in the Plexiglass chambers.

10. Prepare PSEM^{89s} (30–50 mg/kg) in 0.9% sterile saline 15 min before testing. Control animals will receive 0.9% saline alone.

11. Measure the time spent in responding to the acetone application with a stopwatch. Again, disregard the initial 10 s where the mice show reaction to the acetone.

12. Inject (i.p.) mice with PSEM^{89s} or sterile saline. **The tester should be blinded to the treatments**.

13. Twenty minutes after injection, perform the acetone assay again and measure the time response.

6.2.2 Complete Freund's Adjuvant Inflammatory Pain Model

1. After baseline measurements carried out as stated above, prepare an equal parts emulsion of Complete Freund's adjuvant and 0.9% sterile saline. This should be prepared fresh prior to injection into the footpad. Draw the solution into the glass syringe with a 23-G, ¾ in. needle attached and mix by drawing the solution up and down.

2. Inject 20 μL of CFA in the glabrous skin of the hind paw ipsilateral to the side of injection.

3. Twenty four hours later, check the ipsilateral paws for signs of inflammation. If there is no inflammation, inject an extra 5 μL.

4. Five days after CFA injection, acclimate the mice in the Plexiglass chambers for 30 min.

5. Prepare PSEM^{89s} (30–50 mg/kg) in 0.9% sterile saline 15 min before testing. Control animals will receive 0.9% saline alone.

6. After the acclimation period, measure the dynamic and static mechanical thresholds or heat latency as instructed before.

7. Inject (i.p.) mice with PSEM^{89s} or sterile saline. **The tester should be blinded to the treatments**.

8. Twenty minutes after injection, measure the dynamic and static mechanical thresholds or heat latency as before.

7 Validation of Viral Injection with Immunohistochemistry

7.1 Materials

1. 4% paraformaldehyde (Prepare fresh).

2. 30% sucrose in 1× PBS.

3. α-bungarotoxin, Alexa Fluor™ 647 conjugate (Catalog number B35450, Thermofisher).

4. O.C.T Compound (Tissue-Tek).

5. 1 × PBS.

6. 1% Triton in 1× PBS.

7. 1.5 mL microcentrifuge tubes.

8. Tube Revolver/Rotator (Thermo Scientific).

9. Fluoromount-G® (SouthernBiotech).

10. Coverslips.

11. Superfrost Plus Microscope Slides (Fisherbrand).

7.2 Methods

1. Perfuse the mice with 1× PBS, followed by cold 4% paraformaldehyde.

2. Dissect out the spinal cords, taking care not to damage the lumbar region.

3. Place the spinal cords in 4% PFA for 2 h at 4 °C.

4. Transfer the cords to 30% sucrose in 1× PBS and leave at 4 °C for 1–2 days.

5. Embed the lumbar spinal cord in OCT and freeze down in the −80 °C.

6. When you are ready to section the tissue, transfer the embedded tissue to the cryostat. Let it sit at −20 °C for 30–45 min.

7. Section the tissue using the cryostat at 30 μm into wells containing 1× PBS. You can later transfer the tissues into 1.5 mL microcentrifuge tubes for staining.

8. Wash the sections with 1× PBS again.

9. Prepare a staining solution of 1% Triton, 2% NDS (optional) and α-bungarotoxin, Alexa Fluor™ 647 (1:1000).

10. Remove the 1× PBS and add the staining solution to the tissues.

11. Rotate using the Tube Rotator in the dark for 2 h at room temperature.

12. Remove the staining solution and wash with 1× PBS.

13. Rotate for 2 min and then repeat the wash with fresh 1× PBS.

14. Mount the sections on to Superfrost Slides and cover with one or two drops of mounting medium (e.g., Fluoromount-G®).

15. Coverslip and view under the microscope to confirm viral expression in the dorsal horn (*see* Fig. 2 and [18]).

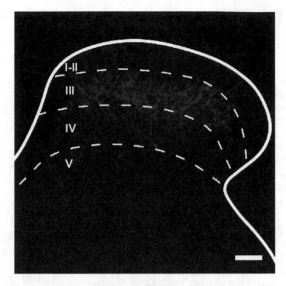

Fig. 2 Dorsal horn expression of PSAM^{L141F}:GlyR-IRES-GFP. Lumbar dorsal horn of a *Cck*$^{Cre/+}$ mouse injected with AAVDJhSyn-Flex-PSAM^{L141F}:GlyR stained with α-bungarotoxin (red). Scale bar is 100 μm

Acknowledgments

Authors would like to acknowledge grant support from NINDS-NS107364 (RPS) and NS111791 (CMA).

References

1. Sternson SM, Roth BL (2014) Chemogenetic tools to interrogate brain functions. Annu Rev Neurosci 37:387–407. https://doi.org/10.1146/annurev-neuro-071013-014048

2. Rajasethupathy P, Ferenczi E, Deisseroth K (2016) Targeting neural circuits. Cell 165(3):524–534. https://doi.org/10.1016/j.cell.2016.03.047

3. Ying Pei, Shuyun Dong, Bryan L Roth Generation of designer receptors exclusively activated by designer drugs (DREADDs) using directed molecular evolution. 2010. Chapter 4:Unit 4.33

4. Armbruster B, Roth B (2005) Creation of designer biogenic amine receptors via directed molecular evolution. In: Neuropsychopharmacology. Nature Publishing Group Macmillan Building, 4 Crinan St, London N1 9xw, London, England, p S265

5. Armbruster BN, Li X, Pausch MH, Herlitze S, Roth BL (2007) Evolving the lock to fit the key to create a family of G protein-coupled receptors potently activated by an inert ligand. Proc Natl Acad Sci U S A 104(12):5163–5168

6. Jann MW, Lam YW, Chang WH (1994) Rapid formation of clozapine in guinea-pigs and man following clozapine-N-oxide administration. Arch Int Pharmacodyn Ther 328(2):243–250

7. Manvich DF, Webster KA, Foster SL, Farrell MS, Ritchie JC, Porter JH, Weinshenker D (2018) The DREADD agonist clozapine N-oxide (CNO) is reverse-metabolized to clozapine and produces clozapine-like interoceptive stimulus effects in rats and mice. Sci Rep 8(1):3840. https://doi.org/10.1038/s41598-018-22116-z

8. Gomez JL, Bonaventura J, Lesniak W, Mathews WB, Sysa-Shah P, Rodriguez LA, Ellis RJ, Richie CT, Harvey BK, Dannals RF, Pomper MG, Bonci A, Michaelides M (2017) Chemogenetics revealed: DREADD occupancy and activation via converted clozapine. Science 357(6350):503–507. https://doi.org/10.1126/science.aan2475

9. Nagai Y, Miyakawa N, Takuwa H, Hori Y, Oyama K, Ji B, Takahashi M, Huang XP, Slocum ST, DiBerto JF, Xiong Y, Urushihata T, Hirabayashi T, Fujimoto A, Mimura K, English JG, Liu J, Inoue KI, Kumata K, Seki C, Ono M, Shimojo M, Zhang MR, Tomita Y, Nakahara J, Suhara T, Takada M, Higuchi M, Jin J, Roth BL, Minamimoto T (2020) Deschloroclozapine, a potent and selective chemogenetic actuator enables rapid neuronal and behavioral modulations in mice and monkeys. Nat Neurosci 23(9):1157–1167. https://doi.org/10.1038/s41593-020-0661-3

10. Bonaventura J, Eldridge MAG, Hu F, Gomez JL, Sanchez-Soto M, Abramyan AM, Lam S, Boehm MA, Ruiz C, Farrell MR, Moreno A, Galal Faress IM, Andersen N, Lin JY, Moaddel R, Morris PJ, Shi L, Sibley DR, Mahler SV, Nabavi S, Pomper MG, Bonci A, Horti AG, Richmond BJ, Michaelides M (2019) High-potency ligands for DREADD imaging and activation in rodents and monkeys. Nat Commun 10(1):4627. https://doi.org/10.1038/s41467-019-12236-z

11. Vardy E, Robinson JE, Li C, Olsen RHJ, DiBerto JF, Giguere PM, Sassano FM, Huang XP, Zhu H, Urban DJ, White KL, Rittiner JE, Crowley NA, Pleil KE, Mazzone CM, Mosier PD, Song J, Kash TL, Malanga CJ, Krashes MJ, Roth BL (2015) A new DREADD facilitates the multiplexed Chemogenetic interrogation of behavior. Neuron 86(4):936–946. https://doi.org/10.1016/j.neuron.2015.03.065

12. Magnus CJ, Lee PH, Atasoy D, Su HH, Looger LL, Sternson SM (2011) Chemical and genetic engineering of selective ion channel-ligand interactions. Science 333(6047):1292–1296

13. Magnus CJ, Lee PH, Bonaventura J, Zemla R, Gomez JL, Ramirez MH, Hu X, Galvan A, Basu J, Michaelides M, Sternson SM (2019) Ultrapotent chemogenetics for research and potential clinical applications. Science 364(6436):eaav5282. https://doi.org/10.1126/science.aav5282

14. Francois A, Low SA, Sypek EI, Christensen AJ, Sotoudeh C, Beier KT, Ramakrishnan C, Ritola KD, Sharif-Naeini R, Deisseroth K, Delp SL, Malenka RC, Luo L, Hantman AW, Scherrer G (2017) A brainstem-spinal cord inhibitory circuit for mechanical pain modulation by GABA and Enkephalins. Neuron 93(4):822–839. e826. https://doi.org/10.1016/j.neuron.2017.01.008

15. Montgomery KL, Iyer SM, Christensen AJ, Deisseroth K, Delp SL (2016) Beyond the brain: Optogenetic control in the spinal cord and peripheral nervous system. Sci Transl Med 8(337):337rv335. https://doi.org/10.1126/scitranslmed.aad7577

16. Peirs C, Williams SP, Zhao X, Walsh CE, Gedeon JY, Cagle NE, Goldring AC, Hioki H, Liu Z, Marell PS, Seal RP (2015) Dorsal horn circuits for persistent mechanical pain. Neuron 87(4):797–812. https://doi.org/10.1016/j.neuron.2015.07.029

17. Saloman JL, Scheff NN, Snyder LM, Ross SE, Davis BM, Gold MS (2016) Gi-DREADD expression in peripheral nerves produces ligand-dependent analgesia, as well as ligand-independent functional changes in sensory neurons. J Neurosci 36(42):10769–10781. https://doi.org/10.1523/JNEUROSCI.3480-15.2016

18. Peirs C, Williams SG, Zhao X, Arokiaraj CM, Ferreira DW, Noh MC, Smith KM, Halder P, Corrigan KA, Gedeon JY, Lee SJ, Gatto G, Chi D, Ross SE, Goulding M, Seal RP (2021) Mechanical Allodynia Circuitry in the Dorsal Horn Is Defined by the Nature of the Injury. Neuron 109(1):73–90.e7

19. Petitjean H, Pawlowski SA, Fraine SL, Sharif B, Hamad D, Fatima T, Berg J, Brown CM, Jan LY, Ribeiro-da-Silva A, Braz JM, Basbaum AI, Sharif-Naeini R (2015) Dorsal horn Parvalbumin neurons are gate-keepers of touch-evoked pain after nerve injury. Cell Rep 13(6):1246–1257. https://doi.org/10.1016/j.celrep.2015.09.080

20. Sharif B, Ase AR, Ribeiro-da-Silva A, Seguela P (2020) Differential coding of itch and pain by a subpopulation of primary afferent neurons. Neuron 106(6):940–951. e944. https://doi.org/10.1016/j.neuron.2020.03.021

21. Huang J, Polgar E, Solinski HJ, Mishra SK, Tseng PY, Iwagaki N, Boyle KA, Dickie AC, Kriegbaum MC, Wildner H, Zeilhofer HU, Watanabe M, Riddell JS, Todd AJ, Hoon MA (2018) Circuit dissection of the role of somatostatin in itch and pain. Nat Neurosci 21(5):707–716. https://doi.org/10.1038/s41593-018-0119-z

22. Taylor NE, Pei J, Zhang J, Vlasov KY, Davis T, Taylor E, Weng FJ, Van Dort CJ, Solt K, Brown EN (2019) The role of glutamatergic and dopaminergic neurons in the periaqueductal gray/dorsal raphe: separating analgesia and anxiety. eNeuro 6(1):ENEURO.0018-18.2019. https://doi.org/10.1523/ENEURO.0018-18.2019

23. Zhu H, Aryal DK, Olsen RH, Urban DJ, Swearingen A, Forbes S, Roth BL, Hochgeschwender U (2016) Cre-dependent DREADD (Designer Receptors Exclusively Activated by Designer Drugs) mice. Genesis 54(8):439–446

24. Barik A, Thompson JH, Seltzer M, Ghitani N, Chesler AT (2018) A brainstem-spinal circuit controlling Nocifensive behavior. Neuron 100(6):1491–1503 e1493. https://doi.org/10.1016/j.neuron.2018.10.037

25. Paretkar T, Dimitrov E (2019) Activation of enkephalinergic (Enk) interneurons in the central amygdala (CeA) buffers the behavioral effects of persistent pain. Neurobiol Dis 124: 364–372. https://doi.org/10.1016/j.nbd.2018.12.005

26. Grace PM, Wang X, Strand KA, Baratta MV, Zhang Y, Galer EL, Yin H, Maier SF, Watkins LR (2018) DREADDed microglia in pain: implications for spinal inflammatory signaling in male rats. Exp Neurol 304:125–131. https://doi.org/10.1016/j.expneurol.2018.03.005

27. Kohro Y, Matsuda T, Yoshihara K, Kohno K, Koga K, Katsuragi R, Oka T, Tashima R, Muneta S, Yamane T, Okada S, Momokino K, Furusho A, Hamase K, Oti T, Sakamoto H, Hayashida K, Kobayashi R, Horii T, Hatada I, Tozaki-Saitoh H, Mikoshiba K, Taylor V, Inoue K, Tsuda M (2020) Spinal astrocytes in superficial laminae gate brainstem descending control of mechanosensory hypersensitivity. Nat Neurosci 23(11):1376–1387. https://doi.org/10.1038/s41593-020-00713-4

28. Jayaraj ND, Bhattacharyya BJ, Belmadani AA, Ren D, Rathwell CA, Hackelberg S, Hopkins BE, Gupta HR, Miller RJ, Menichella DM (2018) Reducing CXCR4-mediated nociceptor hyperexcitability reverses painful diabetic neuropathy. J Clin Invest 128(6):2205–2225. https://doi.org/10.1172/JCI92117

29. Tashima R, Koga K, Yoshikawa Y, Sekine M, Watanabe M, Tozaki-Saitoh H, Furue H, Yasaka T, Tsuda M (2021) A subset of spinal dorsal horn interneurons crucial for gating touch-evoked pain-like behavior. Proc Natl Acad Sci U S A 118(3):e2021220118. https://doi.org/10.1073/pnas.2021220118

30. Cui L, Miao X, Liang L, Abdus-Saboor I, Olson W, Fleming MS, Ma M, Tao YX, Luo W (2016) Identification of early RET+ deep dorsal spinal cord interneurons in gating pain. Neuron 91(6):1413. https://doi.org/10.1016/j.neuron.2016.09.010

31. Gantz SC, Ortiz MM, Belilos AJ, Moussawi K (2021) Excitation of medium spiny neurons by "inhibitory" ultrapotent chemogenetics via shifts in chloride reversal potential. Elife 10: e64241. https://doi.org/10.7554/eLife.64241

32. Decosterd I, Woolf CJ (2000) Spared nerve injury: an animal model of persistent peripheral neuropathic pain. Pain 87(2):149–158

33. Shields SD, Eckert WA 3rd, Basbaum AI (2003) Spared nerve injury model of neuropathic pain in the mouse: a behavioral and anatomic analysis. J Pain 4(8):465–470. https://doi.org/10.1067/s1526-5900(03)00781-8

34. Kohro Y, Sakaguchi E, Tashima R, Tozaki-Saitoh H, Okano H, Inoue K, Tsuda M (2015) A new minimally-invasive method for microinjection into the mouse spinal dorsal horn. Sci Rep 5(1):14306. https://doi.org/10.1038/srep14306

35. Harrison M, O'Brien A, Adams L, Cowin G, Ruitenberg MJ, Sengul G, Watson C (2013) Vertebral landmarks for the identification of spinal cord segments in the mouse. NeuroImage 68:22–29. https://doi.org/10.1016/j.neuroimage.2012.11.048

36. Haenraets K, Albisetti GW, Foster E, Wildner H (2018) Adeno-associated Virus-mediated Transgene Expression in Genetically Defined Neurons of the Spinal Cord. J Vis Exp 135: 57382

37. Chaplan SR, Bach FW, Pogrel JW, Chung JM, Yaksh TL (1994) Quantitative assessment of tactile allodynia in the rat paw. J Neurosci Methods 53(1):55–63. https://doi.org/10.1016/0165-0270(94)90144-9

38. Bonin RP, Bories C, De Koninck Y (2014) A simplified up-down method (SUDO) for measuring mechanical nociception in rodents using von Frey filaments. Mol Pain 10(1):26. https://doi.org/10.1186/1744-8069-10-26

39. Hargreaves K, Dubner R, Brown F, Flores C, Joris J (1988) A new and sensitive method for measuring thermal nociception in cutaneous hyperalgesia. Pain 32(1):77–88. https://doi.org/10.1016/0304-3959(88)90026-7

40. Golden JP, Hoshi M, Nassar MA, Enomoto H, Wood JN, Milbrandt J, RWT G, Johnson EM Jr, Jain S (2010) RET signaling is required for survival and normal function of nonpeptidergic nociceptors. J Neurosci 30(11):3983–3994. https://doi.org/10.1523/JNEUROSCI.5930-09.2010

Chapter 21

Measuring Mouse Somatosensory Reflexive Behaviors with High-Speed Videography, Statistical Modeling, and Machine Learning

Ishmail Abdus-Saboor and Wenqin Luo

Abstract

Objectively measuring and interpreting an animal's sensory experience remains a challenging task. This is particularly true when using preclinical rodent models to study pain mechanisms and screen for potential new pain treatment reagents. How to determine their pain states in a precise and unbiased manner is a hurdle that the field will need to overcome. Here, we describe our efforts to measure mouse somatosensory reflexive behaviors with greatly improved precision by high-speed video imaging. We describe how coupling subsecond ethograms of reflexive behaviors with a statistical reduction method and supervised machine learning can be used to create a more objective quantitative mouse "pain scale." Our goal is to provide the readers with a protocol of how to integrate some of the new tools described here with currently used mechanical somatosensory assays, while discussing the advantages and limitations of this new approach.

Key words Pain, Touch, Somatosensation, Reflexive behaviors, High-speed videography, Principle component analysis, Machine learning

---

1 Introduction

Pain is a complicated and subjective experience. In humans, pain is usually assessed from the patient's own description of experience, which is roughly quantified by self-assignment along a single-score pain scale [1–5]. To investigate pain mechanisms and screen for new pain treatment reagents, animal models, such as rodents, are commonly used [6–9]. However, more and more researchers are realizing that objectively measuring and interpreting pain sensation in rodents is a big challenge [10–14].

Since rodents are nonverbal, researchers have relied on their behaviors to infer their pain state. The reflexive withdrawal assays are the most widely used measurements of pain sensation in rodents, in which a stimulus is applied to a region of the rodent, such as the paw or tail, and the withdrawal frequency is quantified as a readout for the animal's pain state [15–19]. In this chapter, we

Rebecca P. Seal (ed.), *Contemporary Approaches to the Study of Pain: From Molecules to Neural Networks*, Neuromethods, vol. 178, https://doi.org/10.1007/978-1-0716-2039-7_21, © Springer Science+Business Media, LLC, part of Springer Nature 2022

will focus exclusively on the paw withdrawal assay and our recent efforts to improve it.

The reason why pain researchers like to use the paw withdrawal test to measure pain sensation in rodents is because it is easy to perform this test as well as train novice experimenters to collect data [20–23]. Additionally, the results of the paw withdrawal test are instantaneous (i.e., presence or absence of paw lift). Moreover, the behavioral setup to conduct this test is inexpensive and relatively easy to set up, as opposed to more expensive platforms [24, 25]. Other pain behavioral tests, such as spontaneous pain behaviors or the operant assays, may require hours or even days for animals to perform or learn to perform tasks [26–33]. Lastly, the striking similarity between the paw withdrawal reflexes of rodents and human withdrawal responses evoked by noxious stimuli (i.e., humans also display withdrawal reflexes to calibrated von Frey hair filaments or noxious thermal stimuli) adds face validity for using this test as a readout of pain [34–36]. As a field, we have learned much mechanisms and physiology underlying pain processing by using this simple behavioral assay [37–39]. Notably, however, for bench-to-bedside translation of new pain treatment reagents, the successful rate from discoveries made in rodents has not been impressive [40–43]. Though many reasons could contribute to this low success rate, this has caused many pain researchers, clinicians, and pharmaceutical companies to question the specificity, robustness, and validity of the behavior assays used to assess pain sensation in rodents [44–46]. In this chapter, we describe innovation pieces added to the commonly used paw withdrawal assay that might help to mitigate some of the traditional limitations.

A major limitation from focusing solely on the incidence rate of paw withdrawal in response to application of sensory stimuli, as the field traditionally does, is that behaving rodents often display a similar responsive rate to both noxious and innocuous mechanical stimuli. For example, the dynamic brush test uses a soft innocuous brush applied to the mouse paw, and many researchers report a paw withdrawal frequency in approximately 80% of trials [47–50]. This incidence rate is similar to what is commonly reported for potentially noxious stimuli like pinpricking needles or high force von Frey hair filaments [47, 48, 51]. Since the withdrawal frequency itself is not sufficient to distinguish a touch or mechanical pain response, the experimenters have to assign what a given stimulus might mean to the animal, that is, "noxious" or "innocuous." The problem with this "assignment" process is that rodents and humans, or even different human individuals, may perceive some stimuli quality differently. There are massive debates in the field on what some of the most commonly used sensory stimuli are actually measuring in rodents. For example, some reports suggest that 1.4 g of force with von Frey hair filaments is a noxious stimulus to mice, while other reports suggest that this stimulus is perceived as gentle touch

[52, 53]. This kind of discrepancy makes it difficult to translate findings across labs. How can we avoid this subjective "stimulus quality assignment" process?

Mechanical somatosensory stimuli turn on sensorimotor neuronal circuits of an animal, which oftentimes trigger very rapid movements, within 100 ms from the stimulus onset [54–57]. Therefore, when researchers tried to measure rodent reflexes to mechanical sensory stimuli with the unaided eyes, extracting movement features besides incidence of paw lifting has been challenging if not impossible. To gather more detailed information related to the animal's behavioral response, others and we have turned toward high-speed videography to capture subsecond paw withdrawal features. For example, the Woolf and Ginty labs recorded state dependent subsecond pain behaviors at 1000 frames per second (fps) with an optogenetic approach to activate nociceptive neurons [58]. The Lechner group also recorded subsecond pain behaviors with optogenetic activate of nociceptors at 240 fps using an iphone6 smartphone [59]. The Iadarola group likewise recorded unrestrained rats with a high-speed camera at 500 fps, using noxious mechanical and thermal stimuli and observed multisegmental body movements in response to the stimuli [60]. In our methodology, as detailed below in Subheading 3, we also recorded subsecond behaviors at 500 fps or 1000 fps in freely behaving mice [61]. A major distinction in our work compared to some of the other studies is the application of both innocuous and noxious stimuli and the quantification and a composite statistical analysis of behaviors, transforming the multifactorial datasets into a single dimension with principal component analyses. This analysis of behaviors to diverse stimuli allowed us to generate a threshold that distinguishes mouse paw withdrawal reflexes indicating touch from those indicating pain. Furthermore, with a supervised machine learning method, we could determine the "pain-like" probability from the paw withdrawal reflex on a trial-by-trial basis. Taken together, our results reveal that capturing rapid behaviors using high speed imaging and merging multiple behavior features into one numeric parameter can greatly improve precision and rigor of rodent somatosensory behavioral assays.

2 Materials

2.1 Animals

We performed our high-speed assessment analysis with wild type mice purchased from the Jackson Laboratories and Charles River first. Specifically we used the C57BL/6J inbred strain (stock no. 000664) and the CD-1 (stock no. 022) outbred strain. For optogenetic activation of nociceptors we used the following three mouse lines, which are available from the Jackson Laboratories: ChR2 f/f Ai32 (stock no. 012569) [62], Mrgprd^{CreERT2} (stock no. 031286) [63], and TrpV1^{Cre} (stock no. 017769) [64].We conducted behavior assays with both male and female mice to control for the potential sex differences.

2.2 Somatosensory Stimuli

1. Natural mechanical stimuli with predefined quality: We first applied four mechanical stimuli with predefined quality for our studies: cotton swab, dynamic brush (makeup brush from CVS, e.l.f.), and light or heavy application of a pinprick needle (Austerlitz). Details on how these stimuli are applied to the mouse hind paw are described in Subheading 3. Cotton swab and dynamic brush are commonly used by the field as "innocuous touch" stimuli whereas pinprick is commonly used as a "noxious" stimulus. In collaboration with Dr. Xinzhong Dong's group at the Johns Hopkins University using whole animal dorsal root ganglion calcium imaging [65–67], we demonstrated that the cotton swab and dynamic brush mainly activate large diameter touch neurons, while the pinprick stimuli mainly activate small diameter nociceptors [61], further validating the stimulus quality. These stimuli were applied on different days in limited number of trials (usually no more than 3 times/day) to avoid any sensitization to stimuli. Further testing should be conducted if labs want to give the animals multiple stimuli in a single session or day.

2. Von Frey hair filaments (VFHs): These are the most commonly used mechanical stimuli [20, 68–70], which deliver a known quantity of mechanical force to a surface. We used three different VFHs (Stoelting Company, 58011) that are frequently used in the field: 0.6 g, 1.4 g, and 4 g.

3. Peripheral optical stimuli: To activate the blue light sensitive ion channel channelrhodopsin (ChR2) transdermally through the paw skin, we delivered 473 nm blue laser light (Shanghai Laser and Optics Century, BL473T8-150FC/ADR-800A) pulsed at 10hz sine frequency controlled by a pattern generator (Agilent 10 MHZ Function Waveform Generator (33210A). The laser light source was connected to an FC/PC optogenetic patch cable with a 200 mm core opening (Thorlabs, M72L01), in order to get the light from the source to the animal's hind paw. Light power intensity was held between 5 and 20 mW for each experiment and was measured with a digital power meter with a 9.5 mmaperture (Thorlabs, PM100A). The technique of using optogenetics to activate somatosensory afferents through the skin of mice has also been performed by many other research groups [71–73].

2.3 Testing Platform and Holding Chambers

For a platform to place mice on top of during testing (part a in Fig. 1), we used a custom designed mesh floor with perforated openings from below that allows experimenters to reach the mouse

High-Speed Videography Behavioral Setup

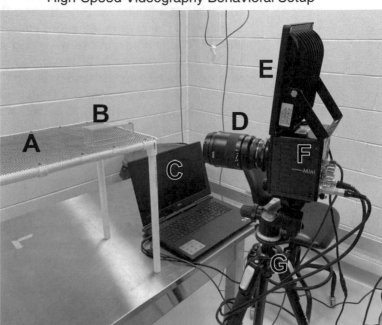

Fig. 1 Setup for the High-speed imaging method. This picture shows all the equipment and parts necessary for our high-speed image recordings. (**a**) Raised mesh platform where mice stand while behavioral studies are performed, (**b**) A mouse holding chamber that allows free but restricted movement, which also facilitates side imaging of behavior. (**c**) Computer attached to the camera for real-time viewing of the behavioral data that is being captured and stored. (**d**) Lens attached to the high speed camera for focusing on the mouse. (**e**) Infrared lighting attached to the camera, which illuminates the mouse to the camera. (**f**) High-speed camera. (**g**) Tripod outfitted to hold to the high-speed camera

hind paw to directly deliver a mechanical stimulus. For holding the mice in chambers that allow free yet restricted movement, small plexiglass rectangular holding chambers with dimensions (11.5 × 11.5 × 16 cm) were used that did not allow mouse standing up or rearing (Part b in Fig. 1). We have found that this design helps with behavioral characterizations, as mice do not stand up with front paws in the air. Although we built these items in house, the similar testing platform and mouse holding chambers can be purchased commercially from many vendors.

2.4 High-Speed Camera

To capture quick paw withdrawal and nocidefensive behaviors at subsecond speeds we used a high-speed camera, Photron FastCAM UX100 (800K-M-4GB—Monochrome 800K with 4 GB memory) (Part f in Fig. 1). This particular camera allows capturing of behaviors at much fast speeds around 1000 fps or higher. Therefore, it is possible that some less expensive cameras, whose maximum frame rate is between 500 and 1000 fps, may also be sufficient to use our method. In addition, a Zeiss lens (Milvus 2/100M ZF.2) was attached to the camera for zoom and focus capabilities (Part d in Fig. 1). The camera and lens were held with a fitted tripod (Part G in Fig. 1) approximately 1–2 feet away from the mice and directed at a 45° angle from the testing setup. To maximally activate the camera, we included a far-red shifted LED light (Tech Imaging) (Part e in Fig. 1), which mice cannot see. This is an important consideration for having better imaging effects without affecting the mouse behaviors.

3   Methods

3.1   Performing Somatosensory Behaviors

We ordered testing mice directly from commercial vendors and allowed the animals to habituate to our animal facility for at least 2 weeks prior to testing. Next, we habituated animals for 1 h per day for approximately 1 week in our behavior room with animals in the holding chambers on top of the testing platform. Additionally, the experimenter performing the behavior remained in the room with the animals during the entire habituation period and the experimenter waved a gloved hand under the animal platform to mimic stimulus delivery. These habituating measures were taken to avoid startle to the animals when tests were commenced.

On the actual testing day, mice were allowed to habituate for 15–30 min before we applied somatosensory stimuli one hind paw on either side of the animal, depending upon which paw was in view of the imaging setup. To obtain high quality data that is easy to interpret, it is imperative that the animals are calm and still when a sensory stimulus is delivered. Our testing platform allowed five animals to be mounted on the platform at once, and tested in an assembly line fashion one-by-one. We put dividers between the animals so that they could not see one another and thus affect each other's behavioral responses. We only delivered one kind of stimuli per day to each mouse tested.

For application of the cotton swab stimulus, we used a q-tip for gentle application to the mouse hind paw. For dynamic brush, we used a soft make-up brush to stroke the hind paw gently from back to front. For light pinprick, we used an insect-pinning needle to apply a weak application of the stimulus to the hind paw and the stimulus was withdrawn when the mouse began paw lift. With heavy pinprick, we used the same needle as above, but applied more intense force upward and withdrawing the needle when approximately 1/3 of the pin's length had passed through the mesh platform. Lastly, for application of von Frey hair filaments, the stimulus was delivered upward to the mouse hind paw until the filament bent. As the majority of positive responses occurred within 100 ms of stimulus delivery, if an animal failed to respond within 2 s, we considered it as a negative response. We chose this criterion to limit the false positive rate of scoring a response to normal animal movement that was unrelated to the actual stimulus.

For all of these stimuli, if mice did not respond in the first trial, we would repeat the stimulus until we could record a response, with trials separated by at least 5 min. We curated and saved each video prior to moving to the next animal to ensure a high quality video and correct delivery of stimuli. On average, it took 2–3 min for one video to save to the internal memory within the high-speed camera.

3.2 Scoring Subsecond Withdrawal and Escape Behaviors

After completing the behavioral testing, the next step is to work at a computer to quantify the behaviors at subsecond resolution. Since we used the Photron high-speed camera, we did behavioral analysis in the companion Photron Fastcam Viewer software. This software is freely downloadable and should be compatible with videos taken from other cameras. Although we initially scored many movement features, we settled upon scoring six that contributed to the majority of variance in the data (Fig. 2). We completed this task using an exploratory factor analysis with orthogonal Varimax rotation conducted with SPSS software. Basically, we chose features that explained the majority of the observed results and excluded other features that were likely redundant or not significant for distinguishing touch from pain. The six features can be divided into sensory-reflexive behaviors (maximum paw height and velocity) and defensive/coping/affective behaviors (paw guarding, paw shaking, jumping, orbital tightening) (Fig. 2a–d). For measuring paw height, we simply drew a vertical line in the software to measure the distance between the paw on the floor to its maximal height (Fig. 2e). For measuring paw velocity we tracked the paw during its ascent away from the stimulus and captured a snapshot of velocity during its maximal peak, recording the distance between two points and the time it took the paw to get from one point to the next (Fig. 2e). For the "affective" features [74, 75], we used a yes/no (i.e., 0 or 1) binary system for presence or absence of the corresponding behaviors (Fig. 2a–d). We combined these 4 measures into one number that we called a "pain score." For example, if an animal displayed orbital tightening and paw shaking it received a score of 2. With these measures, we observed clear statistical separation between the touch stimuli (cotton swab and dynamic brush) and the pain stimuli (light and heavy pinprick) [61]. Lastly, we tested both male and female mice, and did not observe statistically significant sex differences in the behaviors [61].

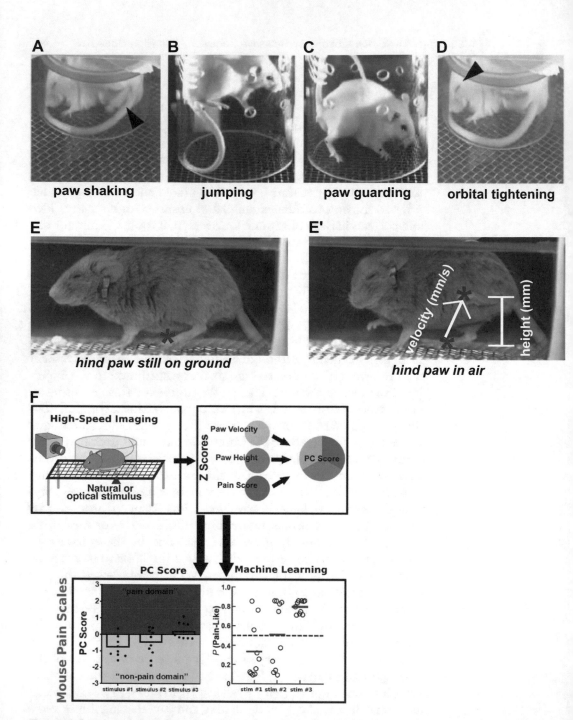

Fig. 2 Subsecond behavioral movements that distinguish touch from pain. Pain sensation associated defensive and coping behaviors: paw shaking (**a**), jumping (**b**), paw guarding (**c**), and orbital tightening (**d**). (**e, e'**) Reflexive behavioral measurements were made by determining the maximal height of the stimulated paw and the speed taken to reach that height. Panel **e** shows the animal close to the time of stimulation just before the paw ascends in the air. The red asterisk labels the "starting rest position" of the stimulated paw. Panel **e'** shows where the paw at rest (red asterisk) to its maximum withdrawal height (blue asterisk). Panels **e** and **e'** are offset by approximately 15 ms for this example. (**f**) Graphical abstract summarizing how we move from behavioral assessment to creation of a "pain scale" by collapsing the data into a single dimension via PCA analysis or predicting pain-like probabilities via machine learning on a trial-by-trial basis. (Images in panels **a–d**, **f** were adapted from [61], with permission from the publisher Elsevier)

3.3 Transforming Data into a Single Dimension with Principal Component Analysis

To produce a simple one-dimensional score that encompassed the six subsecond movement features we scored, we used principal component analysis. We first converted the raw data measured in different units (i.e., mm, mm/s, 0.1) into z-scores to normalize numbers and units and to visualize how far individual datapoints diverged from the mean of the entire group. Next we used principal component analysis [76] for dimension reduction using SAS software, which generated contributing weights, or eigenvalues, for each measurement. After plotting the principal component analysis scores (or PC scores) the data aligned along a continuous scale from −3 to 3. We noticed that the PC scores to innocuous stimuli were typically negative, while the scores to painful stimuli were typically positive. Therefore, "0" appeared to serve as a nice cutoff with this tool to statistically separate touch from pain (Fig. 2f). Therefore, this linear distribution of PC scores seems to make up a mouse "acute pain scale."

3.4 Supervised Machine Learning to Make Predictions about Pain-like Probabilities

To provide a more "user friendly" output for interpreting mouse paw withdrawal reflex, we could use part of the collected data (cotton swab and heavy pinprick) to train a SVM algorithm [77], which was then able to predict "pain-like" probability of any given mouse paw withdrawal trial. For details of our SVM machine-learning pipeline, we direct you to our prior published study [61]. This "pain-like" probability basically mirrored our findings using the PC scores (Fig. 2f).

3.5 Application of our Method to Study VFH and Peripheral Optical Stimuli Triggered Paw Withdrawal Reflexes

After creating this new mouse "pain scale" method to measure and interpret mouse paw withdrawal reflexes, we asked whether we could use this new method and collected database to map the mouse sensory experiences in response to VFHs and optogenetic activation of different types of nociceptors.

For VFHs, we used three stimuli that are frequently used to measure pain in mice (0.6 g, 1.4 g, and 4 g, Fig. 3a–c) and tested them on CD1 male mice. Consistent with other publications, we found ~50% withdrawal rate with 0.6 g VFH and close to 100% withdrawal rate for both 1.4 g and 4 g VFHs [61, 78]. As described above, we quantified the paw height and velocity, orbital tightening, jumping, and paw shaking and guarding. From the raw data, we calculated Z score and principal component (PC) scores using the CD-1 baseline data with cotton swab, dynamic brush, and pinpricks as the reference. With this approach, we observed that responses to 0.6 g and 1.4 g had on average negative PC scores, while the responses to 4 g had positive PC scores (Fig. 3a). In other words, 0.6 g and 1.4 g map in the nonpain domain with behavioral responses reminiscent of cotton swab and dynamic brush—despite 1.4 g having close to 100% paw withdrawal rates. With a SWM trained by CD1 male cotton swab and heavy pinprick data, it predicted that 0.6 g had a low probability of being pain-like

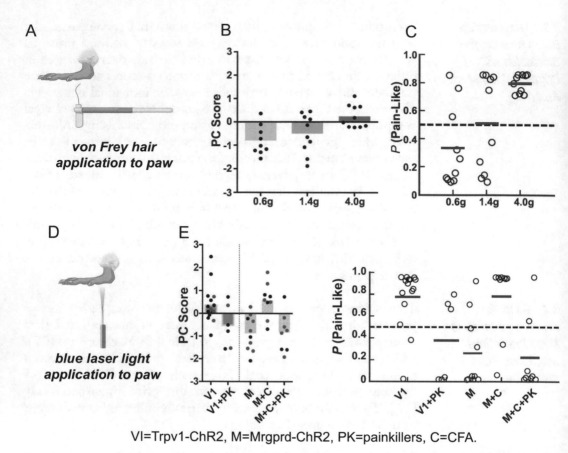

VI=Trpv1-ChR2, M=Mrgprd-ChR2, PK=painkillers, C=CFA.

Fig. 3 Measuring mouse sensation in response to VFHs and optical stimuli. **(a–c)** PC scores and SVM machine learning reveal different sensations and the pain-like probabilities triggered by three different forces of von Frey hair filaments. **(d–f)** PC scores and SVM machine learning reveal different sensations and the pain-like probabilities triggered by optogenetic activation of two distinct nociceptor populations. (These images were adapted from [61], with permission from the publisher Elsevier)

(~30%), 1.4 g was near the threshold that separates touch from pain (~50%), and 4 g had a high probability of being pain-like (~80%, Fig. 3b). It is worth noting that the field usually used 50% paw withdrawal, 0.6 g mechanical force here, as the mechanical pain threshold [49, 78]. However, our analysis argued that this 50% withdrawal threshold indicates mechanical touch rather than mechanical pain.

To measure the sensory experience evoked by optogenetic activation, we used two transgenic mouse lines to optically activate two different populations of nociceptors. We used Mrgprd-ChR2 to activate C-fiber nonpeptidergic nociceptors [63] and Trpv1-ChR2 which targets Trpv1-lineage neurons including the majority of nociceptors [58, 74, 79] (Fig. 3d–f). We activated the sensory afferent terminals in the hind paw skin using 473 nm blue laser light. For negative control experiments, we applied the same blue light to littermate animals that did not express the light-sensitive ChR2 and we did not observe nonspecific responses. In the experimental animals, we observed high paw withdrawal rates in the entire group of mice. We next measured the behavioral parameters described in Fig. 2, followed by calculating PC scores and pain-like probabilities with a SVM trained by all C57 and CD1 cotton swab and heavy pinprick data. Optically activating Trpv1-ChR2 line triggered pain sensation, indicated by both the PC scores and pain-like probability (Fig. 3e, f), which is further validated by a reversion of responses after painkiller administration (Fig. 3e, f). Despite high paw withdrawal rates (Fig. 3e, f), optical stimulation of MrgD-ChR2 mice triggered nonpainful paw withdrawals. Only after local inflammation by paw injection of the heat-killed bacteria complete Freund's adjuvant (CFA), optical stimulation triggered pain-like paw withdrawal reflexes (Fig. 3e, f), which was also validated by painkillers (Fig. 3e, f). Our data strongly suggests that activation of Mrgprd+ nociceptors at baseline states does not cause a pain response. Other research groups have arrived at similar conclusions about these neurons using chemogenetic approaches and/or operant assays [80–82].

Together, these data revealed the power of our new "pain scale" method to measure mouse pain sensation from paw withdrawal reflexes.

4 Discussion

In summary, the relatively simple improvements we described here seem to facilitate greater accuracy in assessing the mouse somatosensory experience from their paw withdrawal reflexes. Since this method is congruent with the behavioral setups that most labs currently use, we do not anticipate major barriers for adopting this approach as long as a high-speed camera is available. Although we have delineated some of the advantages of using this technology, there are still some limitations to be considered. What we have presented is a platform to measure responses to mechanical stimuli under baseline and inflammatory states, but not during chronic neuropathic pain conditions. Neuropathic pain is of the most common, yet hard to treat, chronic pain in the clinic, and we need more robust ways to measure it in preclinical animals. Since many neuropathic pain models involve injury of a nerve, usually the sciatic nerve [6, 83–91], two movement features we emphasized here, paw withdrawal height and velocity, will be affected. Whether our approach could be used for studying neuropathic pain or not will need to be tested in different neuropathic pain models. In addition, this method measured evoked but not spontaneous responses. Spontaneous pain is an important dimension of the pain experience in both the clinic and in animal models [92–101]. Therefore, it would be the best to combine this platform with technologies measuring mouse spontaneous pain behaviors, such as the grimace scale for example or other spontaneous behavior detection platforms [27, 102–111], for achieving a more complete picture of the rodent pain state. Moreover, we manually delivered the stimuli and quantified the behaviors for the previously published study [61]. Automation for some or all of these steps would help to reduce human errors as well as subjectivity. Another issue that will be mitigated with time and when more researchers adopt this technology is that it is currently unclear whether the subsecond behavioral features we defined are similar for different mouse strains or other rodent species such as rats. Nonetheless, these limitations do not preclude adopting this approach now to move beyond the simple yes-no incidence rate quantification of paw withdrawal reflexes. The methodology described here, alongside others with the capacity to detect spontaneous bouts of pain, should greatly increase our confidence in measuring the mouse sensory experience and potentially speed up the rate of translating basic science findings into the clinic.

References

1. Melzack R (1975) The McGill pain questionnaire: major properties and scoring methods. Pain 1(3):277–299

2. Hawker GA et al (2011) Measures of adult pain: visual analog scale for pain (VAS pain), numeric rating scale for pain (NRS pain), McGill pain questionnaire (MPQ), short-form McGill pain questionnaire (SF-MPQ), chronic pain grade scale (CPGS), short Form-36 bodily pain scale (SF-36 BPS), and measure of intermittent and constant osteoarthritis pain (ICOAP). Arthritis Care Res (Hoboken) 63(Suppl 11):S240–S252

3. Garra G et al (2010) Validation of the Wong-baker FACES pain rating scale in pediatric emergency department patients. Acad Emerg Med 17(1):50–54

4. Daut RL, Cleeland CS, Flanery RC (1983) Development of the Wisconsin brief pain questionnaire to assess pain in cancer and other diseases. Pain 17(2):197–210

5. Bennett M (2001) The LANSS pain scale: the Leeds assessment of neuropathic symptoms and signs. Pain 92(1–2):147–157

6. Kim KJ, Yoon YW, Chung JM (1997) Comparison of three rodent neuropathic pain models. Exp Brain Res 113(2):200–206

7. Gregory NS et al (2013) An overview of animal models of pain: disease models and outcome measures. J Pain 14(11):1255–1269

8. Burma NE et al (2017) Animal models of chronic pain: advances and challenges for clinical translation. J Neurosci Res 95(6):1242–1256

9. Mogil JS (2009) Animal models of pain: progress and challenges. Nat Rev Neurosci 10(4):283–294

10. Fried NT et al (2020) Improving pain assessment in mice and rats with advanced videography and computational approaches. Pain 161(7):1420–1424

11. Vardeh D, Mannion RJ, Woolf CJ (2016) Toward a mechanism-based approach to pain diagnosis. J Pain 17(9 Suppl):T50–T69

12. Berge OG (2011) Predictive validity of behavioural animal models for chronic pain. Br J Pharmacol 164(4):1195–1206

13. Kissin I (2010) The development of new analgesics over the past 50 years: a lack of real breakthrough drugs. Anesth Analg 110(3):780–789

14. Woolf CJ (2010) Overcoming obstacles to developing new analgesics. Nat Med 16(11):1241–1247

15. Tyers MB (1980) A classification of opiate receptors that mediate Antinociception in animals. Br J Pharmacol 69(3):503–512

16. Basbaum AI (1973) Conduction of the effects of noxious stimulation by short-fiber

multisynaptic systems of the spinal cord in the rat. Exp Neurol 40(3):699–716

17. Hardy JD (1956) The nature of pain. J Chronic Dis 4(1):22–51

18. Le Bars D, Gozariu M, Cadden SW (2001) Animal models of nociception. Pharmacol Rev 53(4):597–652

19. Barrot M (2012) Tests and models of nociception and pain in rodents. Neuroscience 211:39–50

20. Deuis JR, Dvorakova LS, Vetter I (2017) Methods used to evaluate pain behaviors in rodents. Front Mol Neurosci 10:284

21. Taiwo YO, Coderre TJ, Levine JD (1989) The contribution of training to sensitivity in the nociceptive paw-withdrawal test. Brain Res 487(1):148–151

22. LaBuda CJ, Fuchs PN (2000) A behavioral test paradigm to measure the aversive quality of inflammatory and neuropathic pain in rats. Exp Neurol 163(2):490–494

23. Pitcher GM, Ritchie J, Henry JL (1999) Paw withdrawal threshold in the von Frey hair test is influenced by the surface on which the rat stands. J Neurosci Methods 87(2):185–193

24. Hargreaves K et al (1988) A new and sensitive method for measuring thermal nociception in cutaneous hyperalgesia. Pain 32(1):77–88

25. Cheah M, Fawcett JW, Andrews MR (2017) Assessment of thermal pain sensation in rats and mice using the Hargreaves test. Bio Protoc 7(16):e2506

26. Santos ARS, Calixto JB (1997) Further evidence for the involvement of tachykinin receptor subtypes in formalin and capsaicin models of pain in mice. Neuropeptides 31(4):381–389

27. Langford DJ et al (2010) Coding of facial expressions of pain in the laboratory mouse. Nat Methods 7(6):447–449

28. Neubert JK et al (2008) Characterization of mouse orofacial pain and the effects of lesioning TRPV1-expressing neurons on operant behavior. Mol Pain 4:43

29. Nolan TA et al (2012) Placebo-induced analgesia in an operant pain model in rats. Pain 153(10):2009–2016

30. Mauderli AP, Acosta-Rua A, Vierck CJ (2000) An operant assay of thermal pain in conscious, unrestrained rats. J Neurosci Methods 97(1):19–29

31. Dolan JC et al (2010) The dolognawmeter: a novel instrument and assay to quantify nociception in rodent models of orofacial pain. J Neurosci Methods 187(2):207–215

32. Rohrs EL et al (2015) A novel operant-based behavioral assay of mechanical allodynia in the orofacial region of rats. J Neurosci Methods 248:1–6

33. Neubert JK et al (2005) Use of a novel thermal operant behavioral assay for characterization of orofacial pain sensitivity. Pain 116(3):386–395

34. Andrews K, Fitzgerald M (1994) The cutaneous withdrawal reflex in human neonates: sensitization, receptive fields, and the effects of contralateral stimulation. Pain 56(1):95–101

35. Andersen OK et al (2005) Gradual enlargement of human withdrawal reflex receptive fields following repetitive painful stimulation. Brain Res 1042(2):194–204

36. Morch CD et al (2007) Nociceptive withdrawal reflexes evoked by uniform-temperature laser heat stimulation of large skin areas in humans. J Neurosci Methods 160(1):85–92

37. Basbaum AI et al (2009) Cellular and molecular mechanisms of pain. Cell 139(2):267–284

38. Besson JM (1999) The neurobiology of pain. Lancet 353(9164):1610–1615

39. Dubner R, Gold M (1999) The neurobiology of pain. Proc Natl Acad Sci U S A 96(14):7627–7630

40. Huggins JP et al (2012) An efficient randomised, placebo-controlled clinical trial with the irreversible fatty acid amide hydrolase-1 inhibitor PF-04457845, which modulates endocannabinoids but fails to induce effective analgesia in patients with pain due to osteoarthritis of the knee. Pain 153(9):1837–1846

41. Hill R (2000) NK1 (substance P) receptor antagonists--why are they not analgesic in humans? Trends Pharmacol Sci 21(7):244–246

42. Negus SS et al (2006) Preclinical assessment of candidate analgesic drugs: recent advances and future challenges. J Pharmacol Exp Ther 319(2):507–514

43. Borsook D et al (2014) Lost but making progress--Where will new analgesic drugs come from? Sci Transl Med 6(249):249sr3

44. Yekkirala AS et al (2017) Breaking barriers to novel analgesic drug development. Nat Rev Drug Discov 16(8):545–564

45. Clark JD (2016) Preclinical pain research: can we do better? Anesthesiology 125(5):846–849

46. Yekkirala AS et al (2017) Breaking barriers to novel analgesic drug development. Nat Rev Drug Discov 16(11):810

47. Murthy SE et al (2018) The mechanosensitive ion channel Piezo2 mediates sensitivity to mechanical pain in mice. Sci Transl Med 10(462):eaat9897

48. Bourane S et al (2015) Identification of a spinal circuit for light touch and fine motor control. Cell 160(3):503–515

49. Cheng L et al (2017) Identification of spinal circuits involved in touch-evoked dynamic mechanical pain. Nat Neurosci 20(6): 804–814

50. Liu Y et al (2018) Touch and tactile neuropathic pain sensitivity are set by corticospinal projections. Nature 561(7724):547–550

51. Dhandapani R et al (2018) Control of mechanical pain hypersensitivity in mice through ligand-targeted photoablation of TrkB-positive sensory neurons. Nat Commun 9(1):1640

52. Francois A et al (2015) The low-threshold Calcium Channel Cav3.2 determines low-threshold mechanoreceptor function. Cell Rep 10(3):370–382

53. Woo SH et al (2014) Piezo2 is required for Merkel-cell mechanotransduction. Nature 509(7502):622–626

54. Severson KS et al (2017) Active touch and self-motion encoding by Merkel cell-associated afferents. Neuron 94(3): 666–676 e9

55. Douglass AD et al (2008) Escape behavior elicited by single, channelrhodopsin-2-evoked spikes in zebrafish somatosensory neurons. Curr Biol 18(15):1133–1137

56. Krupa DJ et al (2001) Behavioral properties of the trigeminal somatosensory system in rats performing whisker-dependent tactile discriminations. J Neurosci 21(15):5752–5763

57. May ES et al (2017) Behavioral responses to noxious stimuli shape the perception of pain. Sci Rep 7:44083

58. Browne LE et al (2017) Time-resolved fast mammalian behavior reveals the complexity of protective pain responses. Cell Rep 20(1): 89–98

59. Arcourt A et al (2017) Touch receptor-derived sensory information alleviates acute pain signaling and fine-tunes nociceptive reflex coordination. Neuron 93(1):179–193

60. Blivis D et al (2017) Identification of a novel spinal nociceptive-motor gate control for Adelta pain stimuli in rats. Elife 6:e23584

61. Abdus-Saboor I et al (2019) Development of a mouse pain scale using sub-second behavioral mapping and statistical modeling. Cell Rep 28(6):1623–1634 e4

62. Madisen L et al (2012) A toolbox of Cre-dependent optogenetic transgenic mice for light-induced activation and silencing. Nat Neurosci 15(5):793–802

63. Olson W et al (2017) Sparse genetic tracing reveals regionally specific functional organization of mammalian nociceptors. Elife 6: e29507

64. Cavanaugh DJ et al (2011) Trpv1 reporter mice reveal highly restricted brain distribution and functional expression in arteriolar smooth muscle cells. J Neurosci 31(13):5067–5077

65. Kim YS et al (2016) Coupled activation of primary sensory neurons contributes to chronic pain. Neuron 91(5):1085–1096

66. Anderson M, Zheng Q, Dong X (2018) Investigation of pain mechanisms by calcium imaging approaches. Neurosci Bull 34(1): 194–199

67. Han L et al (2018) Mrgprs on vagal sensory neurons contribute to bronchoconstriction and airway hyper-responsiveness. Nat Neurosci 21(3):324–328

68. Dixon WJ (1980) Efficient analysis of experimental observations. Annu Rev Pharmacol Toxicol 20:441–462

69. Chaplan SR et al (1994) Quantitative assessment of tactile allodynia in the rat paw. J Neurosci Methods 53(1):55–63

70. Woolf CJ (1983) Evidence for a central component of post-injury pain hypersensitivity. Nature 306(5944):686–688

71. Daou I et al (2013) Remote optogenetic activation and sensitization of pain pathways in freely moving mice. J Neurosci 33(47): 18631–18640

72. Husson SJ et al (2012) Optogenetic analysis of a nociceptor neuron and network reveals ion channels acting downstream of primary sensors. Curr Biol 22(9):743–752

73. Carr FB, Zachariou V (2014) Nociception and pain: lessons from optogenetics. Front Behav Neurosci 8:69

74. Corder G et al (2017) Loss of mu opioid receptor signaling in nociceptors, but not microglia, abrogates morphine tolerance without disrupting analgesia. Nat Med 23(2):164–173

75. Corder G et al (2019) An amygdalar neural ensemble that encodes the unpleasantness of pain. Science 363(6424):276–281

76. Ringner M (2008) What is principal component analysis? Nat Biotechnol 26(3):303–304

77. Tarca AL et al (2007) Machine learning and its applications to biology. PLoS Comput Biol 3(6):e116

78. Duan B et al (2014) Identification of spinal circuits transmitting and gating mechanical pain. Cell 159(6):1417–1432

79. Stemkowski P et al (2016) TRPV1 nociceptor activity initiates USP5/T-type channel-mediated plasticity. Cell Rep 17(11): 2901–2912

80. Sophia V et al (2013) Genetic identification of C fibres that detect massage-like stroking of hairy skin in vivo. Nature 493(7434): 669–673

81. Huang T et al (2019) Identifying the pathways required for coping behaviours associated with sustained pain. Nature 565(7737):86–90

82. Beaudry H et al (2017) Distinct behavioral responses evoked by selective optogenetic stimulation of the major TRPV1+ and MrgD + subsets of C-fibers. Pain 158(12): 2329–2339

83. Bennett GJ, Xie YK (1988) A peripheral mononeuropathy in rat that produces disorders of pain sensation like those seen in man. Pain 33(1):87–107

84. De Vry J et al (2004) Pharmacological characterization of the chronic constriction injury model of neuropathic pain. Eur J Pharmacol 491(2–3):137–148

85. Hogan Q et al (2004) Detection of neuropathic pain in a rat model of peripheral nerve injury. Anesthesiology 101(2):476–487

86. Kim SH, Chung JM (1992) An experimental model for peripheral neuropathy produced by segmental spinal nerve ligation in the rat. Pain 50(3):355–363

87. Chung JM, Kim HK, Chung K (2004) Segmental spinal nerve ligation model of neuropathic pain. Methods Mol Med 99:35–45

88. Shields SD, Eckert WA 3rd, Basbaum AI (2003) Spared nerve injury model of neuropathic pain in the mouse: a behavioral and anatomic analysis. J Pain 4(8):465–470

89. Malmberg AB, Basbaum AI (1998) Partial sciatic nerve injury in the mouse as a model of neuropathic pain: behavioral and neuroanatomical correlates. Pain 76(1–2):215–222

90. Seltzer Z, Dubner R, Shir Y (1990) A novel behavioral model of neuropathic pain disorders produced in rats by partial sciatic nerve injury. Pain 43(2):205–218

91. Lindenlaub T, Sommer C (2000) Partial sciatic nerve transection as a model of neuropathic pain: a qualitative and quantitative neuropathological study. Pain 89(1): 97–106

92. Djouhri L et al (2006) Spontaneous pain, both neuropathic and inflammatory, is related to frequency of spontaneous firing in intact C-fiber nociceptors. J Neurosci 26(4): 1281–1292

93. Prkachin KM (1992) Dissociating spontaneous and deliberate expressions of pain: signal detection analyses. Pain 51(1):57–65

94. Ma L et al (2019) Spontaneous pain disrupts ventral hippocampal CA1-Infralimbic cortex connectivity and modulates pain progression in rats with peripheral inflammation. Cell Rep 29(6):1579–1593 e6

95. Baliki MN et al (2006) Chronic pain and the emotional brain: specific brain activity associated with spontaneous fluctuations of intensity of chronic back pain. J Neurosci 26(47): 12165–12173

96. Geha PY et al (2007) Brain activity for spontaneous pain of postherpetic neuralgia and its modulation by lidocaine patch therapy. Pain 128(1–2):88–100

97. Qu C et al (2011) Lesion of the rostral anterior cingulate cortex eliminates the aversiveness of spontaneous neuropathic pain following partial or complete axotomy. Pain 152(7):1641–1648

98. Parks EL et al (2011) Brain activity for chronic knee osteoarthritis: dissociating evoked pain from spontaneous pain. Eur J Pain 15(8):843 e1–14

99. Foss JM, Apkarian AV, Chialvo DR (2006) Dynamics of pain: fractal dimension of temporal variability of spontaneous pain differentiates between pain states. J Neurophysiol 95(2):730–736

100. Haroutounian S et al (2014) Primary afferent input critical for maintaining spontaneous pain in peripheral neuropathy. Pain 155(7): 1272–1279

101. King T et al (2011) Contribution of afferent pathways to nerve injury-induced spontaneous pain and evoked hypersensitivity. Pain 152(9):1997–2005

102. Tuttle AH et al (2018) A deep neural network to assess spontaneous pain from mouse facial expressions. Mol Pain 14: 1744806918763658

103. Wiltschko AB et al (2015) Mapping sub-second structure in mouse behavior. Neuron 88(6):1121–1135

104. Sperry MM et al (2018) Grading facial expression is a sensitive means to detect grimace differences in orofacial pain in a rat model. Sci Rep 8(1):13894

105. Rossi HL et al (2020) Evoked and spontaneous pain assessment during tooth pulp injury. Sci Rep 10(1):2759

106. Leach MC et al (2012) The assessment of post-vasectomy pain in mice using behaviour and the mouse grimace scale. PLoS One 7(4): e35656

107. Matsumiya LC et al (2012) Using the mouse grimace scale to reevaluate the efficacy of postoperative analgesics in laboratory mice. J Am Assoc Lab Anim Sci 51(1):42–49

108. Sotocinal SG et al (2011) The rat grimace scale: a partially automated method for quantifying pain in the laboratory rat via facial expressions. Mol Pain 7:55

109. Miller A et al (2015) The effect of isoflurane anaesthesia and buprenorphine on the mouse grimace scale and behaviour in CBA and DBA/2 mice. Appl Anim Behav Sci 172: 58–62

110. Oliver V et al (2014) Psychometric assessment of the rat grimace scale and development of an analgesic intervention score. PLoS One 9(5): e97882

111. Akintola T et al (2017) The grimace scale reliably assesses chronic pain in a rodent model of trigeminal neuropathic pain. Neurobiol Pain 2:13–17

INDEX

Rebecca P. Seal (ed.), *Contemporary Approaches to the Study of Pain: From Molecules to Neural Networks*, Neuromethods,
vol. 178, https://doi.org/10.1007/978-1-0716-2039-7, © Springer Science+Business Media, LLC, part of Springer Nature 2022